Lecture Notes in Compu  37

T0217121

Commenced Publication in 1973
Founding and Former Series Editors:
Gerhard Goos, Juris Hartmanis, and Jan van Leeuwen

Chengde Mao   Takashi Yokomori (Eds.)

DNA Computing

12th International Meeting on DNA Computing, DNA12
Seoul, Korea, June 5-9, 2006
Revised Selected Papers

 Springer

Volume Editors

Chengde Mao
Purdue University
Department of Chemistry
560 Oval Drive, West Lafayette, IN 47907-2084, USA
E-mail: mao@purdue.edu

Takashi Yokomori
Waseda University
Faculty of Education and Integrated Arts and Sciences
Department of Mathematics
1-6-1 Nishiwaseda, Shinjuku-ku, Tokyo 169-8050, Japan
E-mail: yokomori@waseda.jp

Library of Congress Control Number: 2006938335

CR Subject Classification (1998): F.1, F.2.2, I.2.9, J.3

LNCS Sublibrary: SL 1 – Theoretical Computer Science and General Issues

ISSN   0302-9743
ISBN-10   3-540-49024-8 Springer Berlin Heidelberg New York
ISBN-13   978-3-540-49024-1 Springer Berlin Heidelberg New York

Springer is a part of Springer Science+Business Media

springer.com

© Springer-Verlag Berlin Heidelberg 2006
Printed in Germany

Typesetting: Camera-ready by author, data conversion by Scientific Publishing Services, Chennai, India
Printed on acid-free paper   SPIN: 11925903   06/3142   5 4 3 2 1 0

Preface

This volume is based on papers presented at the 12th International Meeting on DNA Computing (DNA12), which was held during June 5–9, 2006 at Seoul National University, Seoul, South Korea. DNA computing is an interdisciplinary field across computer science, mathematics, molecular biology, chemistry, physics, and nanotechnology. The central theme is to develop novel computing paradigms based on DNA. The annual meeting on DNA computing provides a major forum for scientists to present and discuss their latest results and promotes interactions between experimentalists and theoreticians.

The DNA12 Program Committee received 72 submissions and the current volume contains a selection of 34 papers from the preliminary proceedings. All selected papers were significantly revised by the authors according to the discussion during the meeting. It is our intention to cover all major areas in DNA computing, including demonstrations of biomolecular computing, theoretical models of biomolecular computing, biomolecular algorithms, in vitro and in vivo computational processes, analysis and theoretical models of laboratory techniques, biotechnological and other applications of DNA computing, DNA nanostructures, DNA nanodevices, DNA error evaluation and correction, in vitro evolution, molecular design, self-assembly systems, nucleic acid chemistry, and simulation tools. However, some papers on experimental works are not included because the authors would like to publish their works in more traditional journals.

We have organized the current volume by classifying 34 papers into 8 categories whose topical section headings (and breakdowns) are: Molecular and Membrane Computing Models (6), Complexity Analysis (3), Sequence and Tile Designs and Their Properties (5), DNA Tile Self-Assembly Models (4), Simulator and Software for DNA Computing (4), DNA Computing Algorithms and New Applications (4), Novel Experimental Approaches (3), and Experimental Solutions (5).

The editors would like to thank all participants, referees, the Program Committee, the Organization Committee, all assistants, and all sponsors for making this conference and this volume possible.

September 2006

Chengde Mao
Takashi Yokomori

Organization

Steering Committee

Lila Kari (Chair)	University of Western Ontario
Loenard Adleman	University of Southern California
(Honorary member)	
Anne Condon	University of British Columbia
Masami Hagiya	University of Tokyo
Natasha Jonoska	University of South Florida
Chengde Mao	Purdue University
Giancarlo Mauri	University of Milan, Bicocca
Satoshi Murata	Tokyo Institute of Technology
Gheorghe Paun	Romanian Academy and Sevilla University
John H. Reif	Duke University
Grzegorz Rozenberg	University of Leiden
Nadrian Seeman	New York University
Andrew Turberfield	University of Oxford
Erik Winfree	California Institute of Technology

Program Committee

Yaakov Benenson	Harvard University
Junghuei Chen	University of Delaware
Anne Condon	University of British Columbia
Robert M. Corn	University of California, Irvine
Max H. Garzon	University of Memphis
Hendrik Jan Hoogeboom	Leiden University
Natasha Jonoska	University of South Florida
Lila Kari	University of Western Ontario
Thomas H. LaBean	Duke University
Chengde Mao (Co-chair)	Purdue University
Satoshi Murata	Tokyo Institute of Technology
Gheorghe Paun	Romanian Academy and Sevilla University
Nadrian C. Seeman	New York University
Dipankar Sen	Simon Fraser University
William M. Shih	Harvard Medical School
Friedrich C. Simmel	University of Munich
Lloyd M. Smith	University of Wisconsin, Madison
Petr Sosik	Opava University
Milan N. Stojanovic	Columbia University
Erik Winfree	California Institute of Technology

Masahito Yamamoto	Hokkaido University
Hao Yan	Arizona State University
Takashi Yokomori (Co-chair)	Waseda University
Bernard Yurke	Lucent Technologies
Claudio Zandron	University of Milan, Bicocca
Byoung-Tak Zhang	Seoul National University

External Reviewers

A. Alhazov	S. Kashiwamura	J. Schaeffer
F. Bernardini	M. Hagiya	A. Suyama
D. Besozzi	A. Kelemenov	S. Sahu
R. Brijder	S. Kobayashi	F. Tanaka
L. Cienciala	K. Komiya	D. Tulpan
P. Dario	H. Ono	M. Yamamura
C. Ferretti	A. Leporati	M. Yamashita
R. Freund	U. Majumder	P. Yin
T. Fujii	A. Paton	
A. Kameda	K. Sadakane	

Sponsoring Institutions

Center for Bioinformation Technology (CBIT) of Seoul National University
CT & D, Inc.
Digital Genomics, Inc.
GenoProt, Inc.
Japan Ministry of Education, Sports and Culture and Sciences (MEXT)
Korea Information Science Society (SIG Bioinformation Tech.)
Ministry of Industry, Commerce and Energy of Korea
Ministry of Science and Technology of Korea (KOSEF/NRL Program)
Nano Systems Institute (NSI) of Seoul National University
Research Foundation of Seoul National University
Super Intelligence Technology Center (SITC) of Inha University
US Air Force Research Laboratory (AFRL/IFTC)

Table of Contents

Molecular and Membrane Computing Models

Complexity Analysis

Sequence and Tile Designs and Their Properties

DNA Tile Self-assembly Models

Simulator and Software for DNA Computing

DNA Computing Algorithms and New Applications

Novel Experimental Approaches

Experimental Solutions

Computing with Spiking Neural P Systems: Traces and Small Universal Systems

Mihai Ionescu[1], Andrei Păun[2],
Gheorghe Păun[3,4], and Mario J. Pérez-Jiménez[4]

[1] Research Group on Mathematical Linguistics
Universitat Rovira i Virgili
Pl. Imperial Tàrraco 1, 43005 Tarragona, Spain
armandmihai.ionescu@urv.net
[2] Department of Computer Science, Louisiana Tech University
Ruston, PO Box 10348, Louisiana, LA-71272 USA, and
Universidad Politécnica de Madrid – UPM, Faculdad de Informatíca
Campus de Montegancedo s/n, Boadilla del Monte
28660 Madrid, Spain
apaun@latech.edu
[3] Institute of Mathematics of the Romanian Academy
PO Box 1-764, 014700 Bucharest, Romania
george.paun@imar.ro
[4] Department of Computer Science and AI, University of Sevilla
Avda Reina Mercedes s/n, 41012 Sevilla, Spain
gpaun@us.es, marper@us.es

Abstract. Recently, the idea of spiking neurons and thus of computing by spiking was incorporated into membrane computing, and so-called spiking neural P systems (abbreviated SN P systems) were introduced. Very shortly, in these systems neurons linked by synapses communicate by exchanging identical signals (spikes), with the information encoded in the distance between consecutive spikes. Several ways of using such devices for computing were considered in a series of papers, with universality results obtained in the case of computing numbers, both in the generating and the accepting mode; generating, accepting, or processing strings or infinite sequences was also proved to be of interest.

In the present paper, after a short survey of central notions and results related to spiking neural P systems (including the case when SN P systems are used as string generators), we contribute to this area with two (types of) results: (i) we produce small universal spiking neural P systems (84 neurons are sufficient in the basic definition, but this number is decreased to 49 neurons if a slight generalization of spiking rules is adopted), and (ii) we investigate the possibility of generating a language by following the trace of a designated spike in its way through the neurons.

1 Introduction

Spiking neural P systems (in short, SN P systems) were introduced in [6], with the motivation coming from two directions: the attempt of membrane computing

C. Mao and T. Yokomori (Eds.): DNA12, LNCS 4287, pp. 1–16, 2006.

to pass from cell-like architectures to tissue-like or neural-like architectures (see [15], [12]), and the intriguing possibility of encoding information in the duration of events, or in the interval of time elapsed between events, as vividly investigated in recent research in neural computing (of "third generation") [8], [9].

This double challenge led to a class of P systems based on the following simple ideas: let us use only one object, the symbol denoting a *spike*, and one-membrane cells (called *neurons*) which can hold any number of spikes; each neuron fires in specified conditions (after collecting a specified number of spikes) and then sends one spike along its axon; this spike passes to all neurons connected by a *synapse* to the spiking neuron (hence it is replicated into as many copies as many target neurons exist); between the moment when a neuron fires and the moment when it spikes, each neuron needs a time interval, and this time interval is the essential ingredient of the system functioning (the basic information carrier – with the mentioning that also the number of spikes accumulated in each moment in the neurons provides an important information for controlling the functioning of the system); one of the neurons is considered the output one, and its spikes provide the output of the computation. The sequence of time moments when spikes are sent out of the system is called a *spike train*. The rules for spiking take into account *all* spikes present in a neuron not only part of them, but not all spikes present in a neuron are consumed in this way; after getting fired and before sending the spike to its synapses, the neuron is idle (biology calls this the refractory period) and cannot receive spikes. There are also rules used for "forgetting" some spikes, rules which just remove a specified number of spikes from a neuron.

In the spirit of spiking neurons, as the result of a computation (not necessarily a halting one) in [6] one considers the number of steps elapsed between the first two spikes of the output neuron. Even in this restrictive framework, SN P systems turned out to be Turing complete, *able to compute all Turing computable sets of natural numbers*. This holds both in the generative mode (as sketched above, a number is computed if it represents the interval between the two consecutive spikes of the output neuron) and in the accepting mode (a number is introduced in the system in the form of the interval of time between the first two spikes entering a designated neuron, and this number is accepted if the computation halts). If a bound is imposed on the number of spikes present in any neuron during a computation, then *a characterization of semilinear sets of numbers is obtained*.

These results were extended in [13] to several other ways of associating a set of numbers with an SN P system: taking into account the interval between the first k spikes of each spike train, or all spikes, taking only alternately the intervals, or all of them, considering halting computations. Then, the spike train itself (the sequences of symbols 0, 1 describing the activity of the output neuron: we write 0 if no spike exits the system in a time unit and 1 if a spike is emitted) was considered as the result of a computation; the infinite case is investigated in [14], the finite one in [2]. A series of possibilities of handling infinite sequences of bits are discussed in [14], while morphic representations of regular and of recursively enumerable languages are found in [2]. The results from [2] are briefly recalled in Section 5 below.

In this paper we directly continue these investigations, contributing in two natural directions. First, the above mentioned universality results (the possibility to compute all Turing computable sets of numbers) do not give an estimation on the number of neurons sufficient for obtaining the universality. Which is the size of the smallest universal "brain" (of the form of an SN P system)? This is both a natural and important (from computer science and, also, from neuro-science point of view) problem, reminding the extensive efforts paid for finding small universal Turing machines – see, e.g., [16] and the references therein.

Our answer is rather surprising/encouraging: *84 neurons ensure the universality* in the basic setup of SN P systems, as they were defined in [6], while this number is decreased to 49 if slightly more general spiking rules are used (rules with the possibility to produce not only one spike, *but also two or more spikes at the same time* – such rules are called *extended*). The proof is based on simulating a small universal register machine from [7]. (The full details for the proof of these results about small universal SN P systems will be provided elsewhere – see [11].)

Extended rules are also useful when generating strings: we associate a symbol b_i with a step when the system outputs i spikes and in this way we obtain a string over an arbitrary alphabet, not only on the binary one, as in the case of standard rules. Especially flexible is the case when we associate the empty string with a step when no spike is sent out of the system we associate (that is, b_0 is interpreted as λ). Results from [3], concerning the power of extended SN P systems as language generators, are also recalled in Section 5.

Then, another natural issue is to bring to the SN P systems area a notion introduced for symport/antiport P systems in [5]: mark a spike and follow its path through the system, recording the labels of the visited neurons until either the marking disappears or the computation halts. Because of the very restrictive way of generating strings in this way, there are simple languages which cannot be computed, but, on the other hand, there are rather complex languages which can be obtained in this framework.

Due to space restrictions, we do not give full formal details in definitions and proofs (we refer to the above mentioned papers for that); such details are or will be available in separate papers to be circulated/announced through [19].

2   Formal Language Theory Prerequisites

We assume the reader to be familiar with basic language and automata theory, e.g., from [17] and [18], so that we introduce here only some notations and notions used later in the paper.

For an alphabet V, V^* denotes the set of all finite strings of symbols from V; the empty string is denoted by λ, and the set of all nonempty strings over V is denoted by V^+. When $V = \{a\}$ is a singleton, then we write simply a^* and a^+ instead of $\{a\}^*, \{a\}^+$. If $x = a_1 a_2 \ldots a_n$, $a_i \in V$, $1 \leq i \leq n$, then the mirror image of x is $mi(x) = a_n \ldots a_2 a_1$.

A morphism $h : V_1^* \longrightarrow V_1^*$ such that $h(a) \in \{a, \lambda\}$ for each $a \in V_1$ is called a projection, and a morphism $h : V_1^* \longrightarrow V_2^*$ such that $h(a) \in V_2 \cup \{\lambda\}$ for each $a \in V_1$ is called a weak coding.

If $L_1, L_2 \subseteq V^*$ are two languages, the left and right quotients of L_1 with respect to L_2 are defined by $L_2 \backslash L_1 = \{w \in V^* \mid xw \in L_1 \text{ for some } x \in L_2\}$, and respectively $L_1 / L_2 = \{w \in V^* \mid wx \in L_1 \text{ for some } x \in L_2\}$. When the language L_2 is a singleton, these operations are called left and right derivatives, and denoted by $\partial_x^l(L) = \{x\} \backslash L$ and $\partial_x^r(L) = L / \{x\}$, respectively.

A Chomsky grammar is given in the form $G = (N, T, S, P)$, where N is the nonterminal alphabet, T is the terminal alphabet, $S \in N$ is the axiom, and P is the finite set of rules. For regular grammars, the rules are of the form $A \to aB, A \to a$, for some $A, B \in N, a \in T$.

We denote by FIN, REG, CF, CS, RE the families of finite, regular, context-free, context-sensitive, and recursively enumerable languages; by MAT we denote the family of languages generated by matrix grammars without appearance checking. The family of Turing computable sets of numbers is denoted by NRE (these sets are length sets of RE languages, hence the notation).

Let $V = \{b_1, b_2, \ldots, b_m\}$, for some $m \geq 1$. For a string $x \in V^*$, let us denote by $val_m(x)$ the value in base $m + 1$ of x (we use base $m + 1$ in order to consider the symbols $b_1, \ldots, b_m$ as digits $1, 2, \ldots, m$, thus avoiding the digit 0 in the left hand of the string). We extend this notation in the natural way to sets of strings.

All universality results of the paper are based on the notion of a register machine. Such a device – in the non-deterministic version – is a construct $M = (m, H, l_0, l_h, I)$, where m is the number of registers, H is the set of instruction labels, l_0 is the start label (labeling an ADD instruction), l_h is the halt label (assigned to instruction HALT), and I is the set of instructions; each label from H labels only one instruction from I, thus precisely identifying it. The instructions are of the following forms:

- $l_i : (\text{ADD}(r), l_j, l_k)$ (add 1 to register r and then go to one of the instructions with labels l_j, l_k non-deterministically chosen),
- $l_i : (\text{SUB}(r), l_j, l_k)$ (if register r is non-empty, then subtract 1 from it and go to the instruction with label l_j, otherwise go to the instruction with label l_k),
- $l_h : \text{HALT}$ (the halt instruction).

A register machine M generates a set $N(M)$ of numbers in the following way: we start with all registers empty (i.e., storing the number zero), we apply the instruction with label l_0 and we continue to apply instructions as indicated by the labels (and made possible by the contents of registers); if we reach the halt instruction, then the number n present in register 1 at that time is said to be generated by M. (Without loss of generality we may assume that in the halting configuration all other registers are empty; also, we may assume that register 1 is never subject of SUB instructions, but only of ADD instructions.) It is known (see, e.g., [10]) that register machines generate all sets of numbers which are Turing computable.

A register machine can also be used as a number accepting device: we introduce a number n in some register r_0, we start working with the instruction with label l_0, and if the machine eventually halts, then n is accepted (we may also assume that all registers are empty in the halting configuration). Again, accepting register machines characterize NRE.

Furthermore, register machines can compute all Turing computable functions: we introduce the numbers $n_1, \ldots, n_k$ in some specified registers $r_1, \ldots, r_k$, we start with the instruction with label l_0, and when we stop (with the instruction with label l_h) the value of the function is placed in another specified register, r_t, with all registers different from r_t being empty. Without loss of generality we may assume that $r_1, \ldots, r_k$ are the first k registers of M, and then the result of the computation is denoted by $M(n_1, \ldots, n_k)$.

In both the accepting and the computing case, the register machines can be *deterministic*, i.e., with the ADD instructions of the form $l_i : (\text{ADD}(r), l_j)$ (add 1 to register r and then go to the instruction with label l_j).

In the following sections, when comparing the power of two language generating/accepting devices the empty string λ is ignored.

3   Spiking Neural P Systems

We give here the basic definition we work with, introducing SN P systems in the form considered in the small universal SN P systems, hence computing functions (which, actually, covers both the generative and accepting cases).

A computing *spiking neural membrane system* (abbreviated SN P system), of degree $m \geq 1$, is a construct of the form

$$\Pi = (O, \sigma_1, \ldots, \sigma_m, syn, in, out),$$

where:

1. $O = \{a\}$ is the singleton alphabet (a is called *spike*);
2. $\sigma_1, \ldots, \sigma_m$ are *neurons*, of the form

$$\sigma_i = (n_i, R_i), 1 \leq i \leq m,$$

where:
   a) $n_i \geq 0$ is the *initial number of spikes* contained in σ_i;
   b) R_i is a finite set of *rules* of the following two forms:
      (1) $E/a^c \rightarrow a; d$, where E is a regular expression[1] over a, $c \geq 1$, and $d \geq 0$;
      (2) $a^s \rightarrow \lambda$, for $s \geq 1$, with the restriction that for each rule $E/a^c \rightarrow a; d$ of type (1) from R_i, we have $a^s \notin L(E)$;
3. $syn \subseteq \{1, 2, \ldots, m\} \times \{1, 2, \ldots, m\}$ with $(i, i) \notin syn$ for $1 \leq i \leq m$ (*synapses* between neurons);
4. $in, out \in \{1, 2, \ldots, m\}$ indicate the *input* and the *output* neurons of Π.

---

[1] The regular language defined by E is denoted by $L(E)$.

The rules of type (1) are *firing* (we also say *spiking*) *rules*, and they are applied as follows. If the neuron σ_i contains k spikes, and $a^k \in L(E), k \geq c$, then the rule $E/a^c \to a; d \in R_i$ can be applied. This means consuming (removing) c spikes (thus only $k - c$ remain in σ_i), the neuron is fired, and it produces a spike after d time units (as usual in membrane computing, a global clock is assumed, marking the time for the whole system, hence the functioning of the system is synchronized). If $d = 0$, then the spike is emitted immediately, if $d = 1$, then the spike is emitted in the next step, etc. If the rule is used in step t and $d \geq 1$, then in steps $t, t + 1, t + 2, \ldots, t + d - 1$ the neuron is *closed* (this corresponds to the refractory period from neurobiology), so that it cannot receive new spikes (if a neuron has a synapse to a closed neuron and tries to send a spike along it, then that particular spike is lost). In the step $t + d$, the neuron spikes and becomes again open, so that it can receive spikes (which can be used starting with the step $t + d + 1$).

The rules of type (2) are *forgetting* rules; they are applied as follows: if the neuron σ_i contains exactly s spikes, then the rule $a^s \to \lambda$ from R_i can be used, meaning that all s spikes are removed from σ_i.

If a rule $E/a^c \to a; d$ of type (1) has $E = a^c$, then we will write it in the following simplified form: $a^c \to a; d$. If all spiking rules are of this form, then the system is said to be *finite* (it can handle only a bounded number of spikes in each of its neurons).

In each time unit, if a neuron σ_i can use one of its rules, then a rule from R_i *must* be used. Since two firing rules, $E_1/a^{c_1} \to a; d_1$ and $E_2/a^{c_2} \to a; d_2$, can have $L(E_1) \cap L(E_2) \neq \emptyset$, it is possible that two or more rules can be applied in a neuron, and in that case, only one of them is chosen non-deterministically. Note however that, by definition, if a firing rule is applicable, then no forgetting rule is applicable, and vice versa.

Thus, the rules are used in the sequential manner in each neuron, but neurons function in parallel with each other (the system is synchronized).

The initial configuration of the system is described by the numbers $n_1, n_2, \ldots, n_m$, of spikes present in each neuron, with all neurons being open. During the computation, a configuration is described by both the number of spikes present in each neuron and by the state of the neuron, more precisely, by the number of steps to count down until it becomes open (this number is zero if the neuron is already open).

A computation in a system as above starts in the initial configuration. In order to compute a function $f : \mathbf{N}^k \longrightarrow \mathbf{N}$, we introduce k natural numbers $n_1, \ldots, n_k$ in the system by "reading" from the environment a binary sequence $z = 0^b 10^{n_1-1} 10^{n_2-1} 1 \ldots 10^{n_k-1} 10^f$, for some $b, f \geq 0$; this means that the input neuron of Π receives a spike in each step corresponding to a digit 1 from the string z. Note that we input exactly $k + 1$ spikes. The result of the computation is also encoded in the distance between two spikes: we impose to the system to output exactly two spikes and halt (sometimes after the second spike), hence producing a train spike of the form $0^{b'} 10^{r-1} 10^{f'}$, for some $b', f' \geq 0$ and with $r = f(n_1, \ldots, n_k)$.

If we use an SN P system in the generative mode, then no input neuron is considered, hence no input is taken from the environment; we start from the initial configuration and the distance between the first two spikes of the output neuron (or other numbers, see the discussion in the Introduction) is the result of the computation. Dually, we can ignore the output neuron, we input a number in the system as the distance between two spikes entering the input neuron, and if the computation halts, then the number is accepted.

We do not give here examples, because in the next section we show the four basic modules of our small universal SN P system.

4   Two Small Universal SN P Systems

In both the generating and the accepting case, SN P systems are universal, they compute the Turing computable sets of numbers. The proofs from [6], [13] are based on simulating register machines, which are known to be equivalent to Turing machines when computing (generating or accepting) sets of numbers, [10]. In [7], the register machines are used for computing functions, with the universality defined as follows. Let $(\varphi_0, \varphi_1, \ldots)$ be a fixed admissible enumeration of the set of unary partial recursive functions. A register machine M_u is said to be universal if there is a recursive function g such that for all natural numbers x, y we have $\varphi_x(y) = M_u(g(x), y)$. In [7], the input is introduced in registers 1 and 2, and the result is obtained in register 0 of the machine.

$$
\begin{aligned}
&l_0 : (\text{SUB}(1), l_1, l_2), &&l_1 : (\text{ADD}(7), l_0), \\
&l_2 : (\text{ADD}(6), l_3), &&l_3 : (\text{SUB}(5), l_2, l_4), \\
&l_4 : (\text{SUB}(6), l_5, l_3), &&l_5 : (\text{ADD}(5), l_6), \\
&l_6 : (\text{SUB}(7), l_7, l_8), &&l_7 : (\text{ADD}(1), l_4), \\
&l_8 : (\text{SUB}(6), l_9, l_0), &&l_9 : (\text{ADD}(6), l_{10}), \\
&l_{10} : (\text{SUB}(4), l_0, l_{11}), &&l_{11} : (\text{SUB}(5), l_{12}, l_{13}), \\
&l_{12} : (\text{SUB}(5), l_{14}, l_{15}), &&l_{13} : (\text{SUB}(2), l_{18}, l_{19}), \\
&l_{14} : (\text{SUB}(5), l_{16}, l_{17}), &&l_{15} : (\text{SUB}(3), l_{18}, l_{20}), \\
&l_{16} : (\text{ADD}(4), l_{11}), &&l_{17} : (\text{ADD}(2), l_{21}), \\
&l_{18} : (\text{SUB}(4), l_0, l_h), &&l_{19} : (\text{SUB}(0), l_0, l_{18}), \\
&l_{20} : (\text{ADD}(0), l_0), &&l_{21} : (\text{ADD}(3), l_{18}), \\
&l_h : \text{HALT}.
\end{aligned}
$$

Fig. 1. The universal register machine from [7]

The constructions from [6] do not provide a bound on the number of neurons, but such a bound can be found if we start from a specific universal register machine. We will use here the one with 8 registers and 23 instructions from [7] – for the reader convenience, this machine is recalled in Figure 1, in the notation and the setup introduced in the previous section.

Theorem 1. *There is a universal SN P system with 84 neurons.*

Proof. (Outline) We follow the way used in [6] to simulate a register machine by an SN P system. This is done as follows: neurons are associated with each register (r) and with each label (l_i) of the machine; if a register contains a number n, then the associated neuron will contain $2n$ spikes; modules as in Figures 2 and 3 are associated with the ADD and the SUB instructions (each of these modules contains two neurons – with primed labels – which do not correspond to registers and labels of the simulated machine).

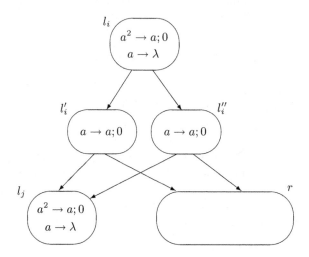

Fig. 2. Module ADD (simulating $l_i : (\mathtt{ADD}(r), l_j)$)

The work of the system is triggered by introducing two spikes in the neuron σ_{l_0} (associated with the starting instruction of the register machine). In general, the simulation of an ADD or SUB instruction starts by introducing two spikes in the neuron with the instruction label. We do not describe here in detail the (pretty transparent) way the modules from Figures 2 and 3 work – the reader can consult [6] in this respect.

Starting with neurons σ_1 and σ_2 already loaded with $2g(x)$ and $2y$ spikes, respectively, and introducing two spikes in neuron σ_{l_0}, we can compute in our system in the same way as M_u; if the computation halts, then neuron σ_0 will contain $2\varphi_x(y)$ spikes. What remains to do is to construct input and output modules, for reading a sequence of bits and introducing the right number of spikes in the neurons corresponding to registers 1 and 2, and, in the end of the computation, to output the contents of register 0. Modules of these types are given in Figures 4, 5, having seven and two additional neurons, respectively.

After this direct construction, we get a system with 91 neurons (9 for the registers of the starting register machine – one further register is necessary for technical reasons, 25 for its labels, 24×2 for the ADD and SUB instructions, 7 in the input module, and 2 in the output module). However, some "code optimization" is possible, based on certain properties of the register machine from

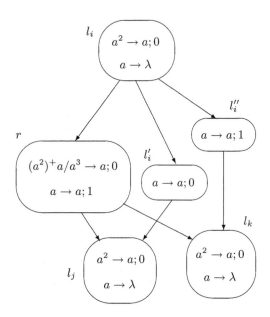

Fig. 3. Module SUB (simulating $l_i : (\text{SUB}(r), l_j, l_k)$)

[7] (for instance, consecutive ADD instructions can be simulated by a specific module, smaller than two separate ADD modules); we skip the technical details and we only mention that the final SN P system will contain 84 neurons.

This is a small number (a small "brain", compared to the human one; it would be nice to know where in the evolution scale there are animals with about 84 neurons in their brain), but we do not know whether it is optimal or not. Anyway, we believe that in the previous setup, we cannot significantly decrease the number of neurons from a universal SN P system.

However, we can do better starting from the following observation. In many modules mentioned above we need pairs of intermediate neurons for duplicating the spike to be transmitted further (this is the case for neurons $\sigma_{l'_i}, \sigma_{l''_i}$ in Figure 2), and this suggests to consider a slight extension of the rules of SN P systems: to allow spiking rules of the form $E/a^c \rightarrow a^p; d$, where all components are as usual, and $p \geq 1$. The meaning is that c spikes are consumed and p spikes are produced. To be "realistic", we impose the restriction $c \geq p$ (the number of produced spikes is not larger than the number of consumed spikes).

Theorem 2. *There is a universal SN P system with 49 neurons, using rules of the form $E/a^c \rightarrow a^p; 0$, with $p \geq 1$.*

(Note that the delay is zero in the rules of the extended form used in the theorem.) As above, we do not know whether this result is optimal, but we again believe that it cannot be significantly improved (without, maybe, changing the definition of SN P systems in an essential way).

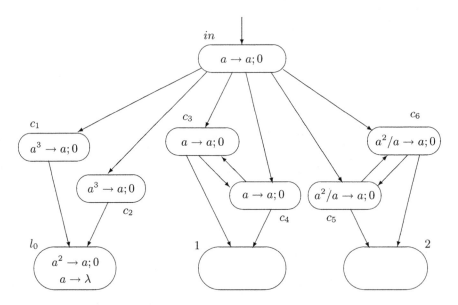

Fig. 4. Module INPUT

5 SN P Systems as String Generators

Following [2] we can also consider as the result of a computation the spike train itself, thus associating a language with an SN P system. Specifically, like in [2], we can consider the language $L_{bin}(\Pi)$ of all binary strings associated with halting computations in Π: the digit 1 is associated with a step when one or more spikes exit the output neuron, and 0 is associated with a step when no spike is emitted by the output neuron. We denote $B = \{0, 1\}$.

Because (in the case of extended systems) several spikes can exit at the same time, we can also work on an arbitrary alphabet: let us associate the symbol b_i with a step when the output neuron emits i spikes. We have two cases: interpreting b_0 (hence a step when no spike is emitted) as a symbol or as the empty string. In the first case we denote the generated language by $L_{res}(\Pi)$ (with "res" coming from "restricted"), in the latter one we write $L_\lambda(\Pi)$.

The respective families are denoted by $L_\alpha SN^e P_m(rule_k, cons_p, prod_q)$, where $\alpha \in \{bin, res, \lambda\}$ and parameters m, k, p, q are as above. We omit the superscript e and the parameter $prod_q$ when working with standard rules (in this case we always have $q = 1$).

We recall from [2] the following results:

Theorem 3. (i) *There are finite languages (for instance, $\{0^k, 10^j\}$, for any $k \geq 1$, $j \geq 0$) which cannot be generated by any SN P system with restricted rules, but for any $L \in FIN$, $L \subseteq B^+$, we have $L\{1\} \in L_{bin}SNP_1(rule_*, cons_*)$, and if $L = \{x_1, x_2, \ldots, x_n\}$, then we also have $\{0^{i+3}x_i \mid 1 \leq i \leq n\} \in L_{bin}SNP_*(rule_*, cons_*)$.*

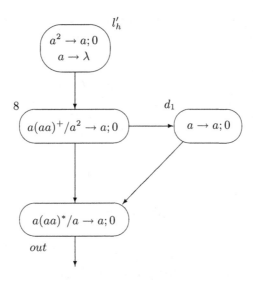

Fig. 5. Module OUTPUT

(ii) *The family of languages generated by finite non-extended SN P systems is strictly included in the family of regular languages over the binary alphabet, but for any regular language $L \subseteq V^*$ there is a finite SN P system Π and a morphism $h : V^* \longrightarrow B^*$ such that $L = h^{-1}(L(\Pi))$.*

(iii) *$L_{bin}SNP_*(rule_*, cons_*) \subset REC$, but for every alphabet $V = \{b_1, b_2, \ldots, b_s\}$ there are a morphism $h_1 : (V \cup \{b, c\})^* \longrightarrow B^*$ and a projection $h_2 : (V \cup \{b, c\})^* \longrightarrow V^*$ such that for each language $L \subseteq V^*$, $L \in RE$, there is an SN P system Π such that $L = h_2(h_1^{-1}(L(\Pi)))$.*

These results show that the language generating power of non-extended SN P systems is rather eccentric; on the one hand, finite languages (like $\{0, 1\}$) cannot be generated, on the other hand, we can represent any RE language as the direct morphic image of an inverse morphic image of a language generated in this way. This eccentricity is due mainly to the restricted way of generating strings, with one symbol added in each computation step, and this again naturally suggests the idea of extended rules, with the possibility of having λ as output in steps when no spike exits the system. As we will see immediately, this possibility considerably enlarges the generated families of languages.

The next results were obtained in [3], as counterparts of the results from Theorem 3; as expected, the extended rules are useful, the obtained families of languages are larger, and finite, regular, and recursively enumerable can be directly obtained, without additional symbols and squeezing mechanisms.

We consider at the same time both the restricted case (with b_0 associated with a step when no spike is sent out) and the non-restricted one (with b_0 interpreted as λ); V is the alphabet $\{b_1, \ldots, b_s\}$:

Theorem 4. (i) $FIN = L_\alpha SN^e P_1(rule_*, cons_*, prod_*), \alpha \in \{res, \lambda\}$, and this result is sharp, because $L_{res}SN^e P_2(rule_2, cons_3, prod_3)$ contains infinite languages.

(ii) *If* $L \in REG$, *then* $\{b_0\}L \in L_{res}SN^eP_4(rule_*, cons_*, prod_*)$ *and* $L\{b_0\} \in$ $L_{res}SN^eP_3(rule_*, cons_*, prod_*)$, *but there are minimal linear languages which are not in the family* $L_{res}SN^eP_*(rule_*, cons_*, prod_*)$.

(iii) $L_\lambda SN^eP_2(rule_*, cons_*, prod_*) \subseteq REG \subset L_\lambda SN^eP_3(rule_*, cons_*, prod_*)$; *the second inclusion is proper, because* $L_\lambda SN^eP_3(rule_4, cons_4, prod_2)$ *contains non-regular languages; actually, the family* $L_\lambda SN^eP_3(rule_3, cons_6, prod_4)$ *contains non-semilinear languages.*

(iv) $RE = L_\lambda SN^eP_*(rule_*, cons_*, prod_*)$.

It is an open problem to find characterizations (even only representations) of other families of languages in the Chomsky hierarchy.

6   Following the Traces of Spikes

We have seen above that SN P systems can be used also for generating or accepting languages, and even infinite sequences [14], by just taking the spike trains as generated strings/sequences. Here we consider yet another idea for defining a language, taking into account the traces of a distinguished spike through the system. This is a direct counterpart of trace languages from [5], and also has some similarity with the idea of "computing by observing", as recently considered in [1].

Specifically, in the initial configuration of the system we "mark" one spike from a specified neuron – the intuition is that this spike has a "flag" – and we follow the path of this flag during the computation, *recording the labels of the neurons where the flag is present in the end of each step*. Actually, for neuron σ_i we consider the symbol b_i in the trace string.

The previous definition contains many delicate points which need clarifications – and we use a simple example to do this.

Assume that in neuron σ_i we have three spikes, one of them marked; we write aaa' to represent them. Assume also that we have a spiking rule $aaa/aa \to a; 0$. When applied, this rule consumes two spikes, one remains in the neuron and one spike is produced and sent along the synapses going out of neuron σ_i. Two cases already appear: the marked spike is consumed or not. If not consumed, then it remains in the neuron. If consumed, then the flag passes to the produced spike. Now, if there are two or more synapses going out of neuron σ_i, then again we can have a branching: only one spike is marked, hence only on one of the synapses (i, j), non-deterministically chosen, we will transmit a marked spike. If σ_j is an open neuron, then the marked spike ends in this neuron. If σ_j is a closed neuron, then the marked spike is lost, and the same happens if the marked spike exits in the environment. Anyway, if the marked spike is consumed, at the end of this step it is no longer present in neuron i; it is in neuron σ_j if $(i, j) \in syn$ and neuron σ_j is open, or it is removed from the system in other cases.

Therefore, if in the initial configuration of the system neuron σ_i contains the marked spike, then the trace can start either with b_i (if the marked spike is not consumed) or with b_j (if the marked spike was consumed and passed to neuron σ_j); if the marked spike is consumed and lost, then we generate the empty string, which is ignored in our considerations. Similarly in later steps.

If the rule used is of the form $aaa/aa \rightarrow a; d$, for some $d \geq 1$, and the marked spike is consumed, then the newly marked spike remains in neuron σ_i for d steps, hence the trace starts/continues with b_i^d. Similarly, if no rule is used in neuron σ_i for k steps, then the trace records k copies of b_i.

If a forgetting rule is used in the neuron where the marked spike is placed, then the trace string stops (and no symbol is recorded for this step).

Therefore, when considering the possible branchings of the computation, we have to take into account the non-determinism not only in using the spiking rules, but also in consuming the marked spike and in sending it along one of the possible synapses.

The previous discussion has, hopefully, made clear what we mean by *recording the labels of the neurons where the flag is present in the end of each step*, and why choosing the end of a step and not the beginning: in the latter case, all traces would start with the same symbol, corresponding to the input neuron, which is a strong – and artificial – restriction.

Anyway, we take into account only halting computations: irrespective whether or not a marked spike is still present in the system, the computation should halt (note that it is possible that the marked spike is removed and the computation still continues for a while – but this time without adding further symbols to the trace string).

For an SN P system Π we denote by $T(\Pi)$ the language of all strings describing the traces of the marked spike in all halting computations of Π. Then, we denote by $TSNP_m(rule_k, cons_p, forg_q)$ the family of languages $T(\Pi)$, generated by systems Π with at most m neurons, each neuron having at most k rules, each of the spiking rules consuming at most p spikes, and each forgetting rule removing at most q spikes. As usual, a parameter m, k, p, q is replaced with $*$ if it is not bounded.

We pass now to investigating the relationship with the families of languages from Chomsky hierarchy, starting with a counterexample result (whose simple proof is omitted).

Lemma 1. *There are singleton languages which are not in $TSNP_*(rule_*, cons_*, forg_*)$.*

Theorem 5. *The family of trace languages generated by SN P systems by means of computations with a bounded number of spikes present in their neurons is strictly included in the family of regular languages.*

The inclusion follows from the fact that the transition diagram associated with the computations of an SN P system which use a bounded number of spikes is finite and can be interpreted as the transition diagram of a finite automaton. The fact that the inclusion is proper is a consequence of Lemma 1.

As expected, also non-regular languages can be generated – as well as much more complex languages.

Theorem 6. *Every unary language $L \in RE$ can be written in the form $L = h(L') = (b_1^* \setminus L') \cap b_2^*$, where $L' \in TSNP_*(rule_*, cons_*, forg_*)$ and h is a projection.*

14   M. Ionescu et al.

Proof. (Sketch) This result is a consequence of the fact that SN P systems can simulate register machines. Specifically, starting from a register machine M, we construct an SN P system Π which halts its computation with $2n$ spikes in a specified neuron σ_{out} if and only if n can be generated by the register machine M; in the halting moment, a neuron σ_{l_h} of Π associated with the label of the halting instruction of M gets two spikes and fires. The neuron σ_{out} contains no rule used in the simulation of M (the corresponding register is only incremented, but never decremented – see the details of the construction from [6], as well as Figures 2 and 3).

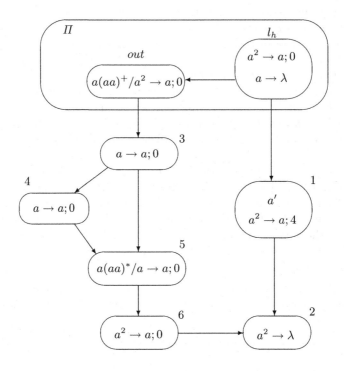

Fig. 6. The SN P system from the proof of Theorem 6

Now, consider a language $L \subseteq b_2^*, L \in RE$. There is a register machine M such that $n \in N(M)$ if and only if $b_2^n \in L$. Starting from such a machine M, we construct the system Π as in [6], having the properties described above. We append to the system Π six more neurons, as indicated in Figure 6. There is a marked spike in neuron σ_1, and it will stay here during all the simulation of M. In the moment when neuron σ_{l_h} of Π spikes, its spike goes both to neuron σ_{out} and to neuron σ_1.

Neurons $\sigma_3, \sigma_4, \sigma_5, \sigma_6$ send a spike to neuron σ_2 only when neuron σ_{out} has finished its work (this happens after n steps of using the rule $a(aa)^+/a^2 \rightarrow a; 0$, for $2n$ being the contents of neuron σ_{out} in the moment when neuron σ_{l_h} spikes).

The marked spike leaves neuron σ_1 four steps after using the rule $a^2 \to a; 4$, hence five steps after the spiking of neuron σ_{l_h}. This means that the marked spike waits in neuron σ_2 exactly n steps. When the spike of neuron σ_6 reaches neuron σ_2, the two spikes present here, the marked one included, are forgotten.

Thus, the traces of the marked spike are of the form $b_1^r b_2^n$, for some $r \geq 1$ and $n \in N(M)$. By means of the left derivative with the regular language b_1^* we can remove prefixes of the form b_1^k and by means of the intersection with b_2^* we ensure that the maximal prefix of this form is removed. Similarly, the projection $h : \{b_1, b_2\}^* \longrightarrow \{b_1, b_2\}^*$ defined by $h(b_1) = \lambda$, $h(b_2) = b_2$, removes all occurrences of b_1. Consequently, $L = (b_1^* \backslash T(\Pi)) \cap b_2^* = h(T(\Pi))$.

Corollary 1. *The family $TSNP_*(reg_*, cons_*, forg_*)$ is incomparable with each family of languages FL which contains the singleton languages, is closed under left derivative with regular languages and intersection with regular languages, and does not contain all unary recursively enumerable languages.*

Families FL as above are FIN, REG, CF, CS, MAT etc.

7   Final Remarks

After a brief informal survey of main results related to SN P systems as number generating or accepting devices, we have produced small universal SN P systems (with 84 and 49 neurons, depending on the type of spiking rules used), and we have introduced and preliminarily investigated the possibility of using SN P systems as language generators by following the trace of a marked spike across the neurons. Many topics remain open for further research, and other suggestions from biology are worth considering.

Acknowledgements

Thanks are due to Takashi Yokomori for many discussions about spiking neural P systems, as well as for many comments about a previous version of this paper.

The authors gratefully acknowledge the following (partial) support of their research. M. Ionescu: Programa Nacional para la Formación del Profesorado Universitario from the Spanish Ministry of Education. A. Păun: LA BoR RSC grant LEQSF (2004-07)-RD-A-23, and NSF Grants IMR-0414903 and CCF-0523572. Gh. Păun and M.J. Pérez-Jiménez: Project TIN2005-09345-C03-01 of Ministerio de Educación y Ciencia of Spain, cofinanced by FEDER funds, and Project of Excellence TIC 581 of Junta de Andalucia.

References

1. M. Cavaliere, P. Leupold: Evolution and observation – A new way to look at membrane systems. In *Membrane Computing. Intern. Workshop 'WMC 2003, Tarragona, Spain, July 2003. Revised Papers* (C. Martin-Vide, G. Mauri, Gh. Păun, G. Rozenberg, A. Salomaa, eds.), LNCS 2933, Springer, Berlin, 2004, 70–87.

2. H. Chen, R. Freund, M. Ionescu, Gh. Păun, M.J. Pérez-Jiménez: On string languages generated by spiking neural P systems. In *Proc. of Fourth. Brainstorming Week on Membrane Computing*, Sevilla, 2006, vol. I, 169–193 (also available at [19]).

3. H. Chen, T.-O. Ishdorj, Gh. Păun, M.J. Pérez-Jiménez: Spiking neural P systems with extended rules. In *Proc. of Fourth. Brainstorming Week on Membrane Computing*, Sevilla, 2006, vol. I, 241–266 (also available at [19]).

4. O.H. Ibarra, A. Păun, Gh. Păun, A. Rodríguez-Patón, P. Sosik, S. Woodworth: Normal forms for spiking neural P systems. In *Fourth Brainstorming Week on Membrane Computing*, Febr. 2006, Fenix Editora, Sevilla, 2006, vol. II, 105–136 (also available at [19]).

5. M. Ionescu, C. Martin-Vide, A. Păun, Gh. Păun: Membrane systems with symport/antiport: (unexpected) universality results. In *Proc. 8th International Meeting of DNA Based Computing* (M. Hagiya, A. Ohuchi, eds.), Japan, 2002, 151–160.

6. M. Ionescu, Gh. Păun, T. Yokomori: Spiking neural P systems. *Fundamenta Informaticae*, 71, 2-3 (2006), 279–308.

7. I. Korec: Small universal register machines. *Theoretical Computer Science*, 168 (1996), 267–301.

8. W. Maass: Computing with spikes. *Special Issue on Foundations of Information Processing of TELEMATIK*, 8, 1 (2002), 32–36.

9. W. Maass, C. Bishop, eds.: *Pulsed Neural Networks*, MIT Press, Cambridge, 1999.

10. M. Minsky: *Computation – Finite and Infinite Machines*. Prentice Hall, Englewood Cliffs, NJ, 1967.

11. A. Păun, Gh. Păun: Small universal spiking neural P systems. *BioSystems*, to appear.

12. Gh. Păun: *Membrane Computing – An Introduction*. Springer-Verlag, Berlin, 2002.

13. Gh. Păun, M.J. Pérez-Jiménez, G. Rozenberg: Spike trains in spiking neural P systems. *Intern. J. Found. Computer Sci.*, to appear (also available at [19]).

14. Gh. Păun, M.J. Pérez-Jiménez, G. Rozenberg: Infinite spike trains in spiking neural P systems. Submitted, 2006.

15. Gh. Păun, Y. Sakakibara, T. Yokomori: P systems on graphs of restricted forms. *Publicationes Mathematicae Debrecen*, 60 (2002), 635–660.

16. Y. Rogozhin: Small universal Turing machines. *Theoretical Computer Science*, 168 (1996), 215–240.

17. G. Rozenberg, A. Salomaa, eds.: *Handbook of Formal Languages*, 3 volumes. Springer-Verlag, Berlin, 1997.

18. A. Salomaa: *Formal Languages*. Academic Press, New York, 1973.

19. The P Systems Web Page: http://psystems.disco.unimib.it.

Minimal Parallelism for Polarizationless P Systems

Tseren-Onolt Ishdorj

Research Group on Natural Computing
Department of Computer Science and AI, University of Sevilla
Avda Reina Mercedes s/n, 41012 Sevilla, Spain
tserren@yahoo.com
Computational Biomodelling Laboratory
Turku Centre for Computer Science
Åbo Akademi University, Turku 20520, Finland
tishdorj@abo.fi

Abstract. Minimal parallelism was recently introduced [3] as a way the rules of a P system are used: from each set of applicable rules associated to the same membrane, at least one must be applied. In this paper, we consider the minimal parallelism for P systems with active membranes *without polarizations*, using additional features, such as separation operations, changing membrane labels, catalytic or cooperative rules, etc. With several combinations of such features we obtain computational completeness. In cases where membrane division (of elementary or non-elementary membranes) is allowed, we show how SAT can be solved in polynomial time.

1 Introduction

Membrane systems (referred also as P systems) are a class of distributed parallel computing devices of a biochemical type, which can be seen as a general computing architecture where various types of objects can be processed in parallel by various operations. A key structural notion is that of a *membrane* by which a system is divided into compartments where chemical reactions can take place. These reactions transform multisets of objects present in the compartments into new objects, possibly transferring objects to neighboring compartments, including the environment.

For a comprehensive introduction to membrane computing we refer to [11].

As membrane systems are inspired by living cell behavior, a continuous research topic in the area was looking for as bio-realistic computing models as possible. In this framework, the application of developmental rules under different constraints is an interesting problem to study. An idea concerning the rule application that has recently attracted the attention is the *minimal parallelism*, introduced and investigated in [3]. Minimal parallelism relaxes the condition of using the rules in a maximally parallel way. More precisely, the rules are used in the *non-deterministic minimally parallel* manner: in each step, from each set of

C. Mao and T. Yokomori (Eds.): DNA12, LNCS 4287, pp. 17–32, 2006.

rules R_i (associated with a membrane i of a P system) we use at least one rule (without specifying how many) provided that this is possible. The rules to be used, as well as the objects to which they are applied, are non-deterministically chosen.

In the original paper [3], certain open problems have been formulated, such as: removing the polarizations of membranes and/or decreasing the number of membranes in the universality proofs; whether or not membrane division for only elementary membranes suffices when solving computationally hard problems in polynomial time; looking for uniform constructions, or for deterministic systems.

We address here some of these problems, considering the minimal parallelism within the framework of P systems with active membranes, without using polarizations. In this framework, we have obtained the Turing completeness in accepting as well as in generative cases (Section 4) by simulating Minsky register machines, [6], and matrix grammars, [4]. Moreover, computational complexity issues (solving **NP**-complete problems) are also considered for polarizationless P systems in Section 5, where both uniform and semi-uniform solutions to **SAT** are provided.

2   Preliminaries

We assume the reader to be familiar with the basic elements of formal languages, Turing computability [5], computational complexity, [9], and membrane computing, [11]. We introduce here only some of the necessary notions and notation.

An *alphabet* is a finite set of symbols (letters), and a word (string) over an alphabet Σ is a finite sequence of letters from Σ. We denote the empty word by λ. A *multiset* over an alphabet Σ is a mapping from Σ to $\mathbf{N}$, the set of natural numbers; we represent a multiset by a string from Σ^*, where the number of occurrences of a symbol $a \in \Sigma$ in a string w represents the multiplicity of a in the multiset represented by w (hence all strings obtained by permuting symbols in the string w represent the same multiset). The family of Turing computable sets of natural numbers is denoted by NRE (with RE coming from "recursively enumerable"; RE denotes the family of recursively enumerable languages).

For the Turing computability proofs in the next sections, we use the characterization of NRE by means of *matrix grammars* (in a precise normal form) and *register machines*.

A matrix grammar in the *binary normal form* is a construct $G = (N, T, S, M, F)$, where $N = N_1 \cup N_2 \cup \{S, \#\}$, with these three sets mutually disjoint, and the matrices in M are in one of the following forms:

1. $(S \to XA)$, with $X \in N_1, A \in N_2$;
2. $(X \to Y, A \to x)$, with $X, Y \in N_1, A \in N_2, x \in (N_2 \cup T)^*, |x| \le 2$;
3. $(X \to Y, A \to \#)$, with $X, Y \in N_1, A \in N_2$;
4. $(X \to \lambda, A \to x)$, with $X \in N_1, A \in N_2$, and $x \in T^*, |x| \le 2$.

Moreover, there is only one matrix of type 1 (that is why one uses to write it in the form $(S \to X_{init} A_{init})$, in order to fix the symbols X, A present in it), and F consists exactly of all rules $A \to \#$ appearing in matrices of type 3; $\#$ is a trap-symbol, because once introduced, it is never removed. A matrix of type 4 is used only once, in the last step of a derivation.

For $w, z \in (N \cup T)^*$ we write $w \Longrightarrow z$ if there is a matrix in $m \in M$ such that starting from w and applying once each rule of m in the order specified by m, one can obtain z; a rule can be skipped if it is in F and it is not applicable.

The language generated by G is defined by $L(G) = \{w \in T^* \mid S \Longrightarrow^* w\}$. The family of languages of this form is denoted by MAT_{ac}. It is known that $MAT_{ac} = RE$.

In turn, an *n-register machine* is a construct $M = (n, B, l_0, l_h, I)$, where n is the number of registers, B is a set of labels, and I is a set of labeled instructions of the form $l_i : (op(r), l_j, l_k)$, where $op(r)$ is an operation on register r of M, and l_i, l_j, l_k are labels from the set B; l_0 is the label of the initial instruction, and l_h is the label of the halting instruction. The machine is capable of the following instructions:

1. $l_i : (\text{ADD}(r), l_j, l_k)$: Add one to the content of register r and proceed, in a non-deterministic way, to instruction with label l_j or to instruction with label l_k; in the deterministic variant we demand $l_j = l_k$ and then the instruction is written in the form $l_i : (\text{ADD}(r), l_j)$.
2. $l_i : (\text{SUB}(r), l_j, l_k)$: If register r is not empty, then subtract one from its contents and go to instruction with label l_j, otherwise proceed to instruction with label l_k.
3. $l_h : halt$: This instruction stops the machine and can only be assigned to the final label l_h.

A deterministic n-register machine can analyze an input $m \in \mathbf{N}$, introduced in register 1, which is accepted if and only if the machine finally stops by the halt instruction with all its registers being empty. If the machine does not halt, then the analysis was not successful. We denote by $N(M)$ the set of numbers accepted by the register machine M. If Q is a Turing computable set, then there exists a deterministic register machine M with at most three registers, such that $N(M) = Q$, [6].

It is not enough that a problem can be solved algorithmically, it is necessary that the solution comes in a reasonable time and using reasonable computing resources. Defining what "reasonable time and resources" means and classifying problems from these points of view are the main tasks of computational complexity. For details we refer to [9]. In Section 5 we show that the **NP**-complete problem SAT can be solved in polynomial time by P systems from various classes, that is why we introduce here this problem.

In SAT we are given Boolean variables $x_1, x_2, \ldots, x_n$ and a Boolean formula β involving such variables; the formula is given in a particular format called *conjunctive normal form*, that we will explain in a moment. The question is whether there is a way to assign Boolean $(true/false)$ values to the variables so that the formula is satisfied.

Boolean formulas are constructed starting from variables and applying the operators $\vee$ (that stands for OR), $\wedge$ (that stands for AND), and $\neg$ (that stands for NOT).

A *clause* is formed by taking one or more variables and connecting them with OR; for example, $(x_2 \vee \neg x_4 \vee x_5)$ is a clause. A *formula in conjunctive normal form* is the AND of a given set of clauses. For example, $(x_3 \vee \neg x_4) \wedge (x_1) \wedge (\neg x_3 \vee x_2)$ is a formula in conjunctive normal form. Note that the above formula is satisfiable, and, for example, it is satisfied by setting all the variables to *true* (there are also other possible assignments of values to the variables that would satisfy the formula). On the other hand, the formula $(x_1) \wedge (\neg x_1 \vee x_2) \wedge (\neg x_2)$ is not satisfiable, as it can easily be observed.

3   P Systems with Active Membranes

We briefly recall now the notion of P systems with active membranes.

Informally speaking, in P systems with active membranes, the following types of rules are used: (a) multiset rewriting rules (including non-cooperative and cooperative evolution rules), (b) rules for introducing objects into membranes, (c) rules for sending objects out of membranes, (d) rules for dissolving membranes, (e) rules for dividing elementary membranes, and (f) rules for dividing non-elementary membranes. In these rules, a single object is involved. Occasionally, also (g) membrane merging rules, (h) membrane separation rules, (i) membrane release rules, (k) replicative-distribution rules for sibling membranes, and (l) replicative-distribution rules for nested membranes were used, e.g., in [1,7,8]. Their common feature is that they involve multisets of objects.

Formally, a *polarizationless P system with active membranes* is a construct

$$\Pi = (O, C, H, \mu, w_1, w_2, \ldots, w_m, R_1, R_2, \ldots, R_m, h_0),$$

where: $m \geq 1$ is the initial degree of the system; O is the alphabet of *objects*; $C \subseteq O$ is the set of *catalysts* (when $C = \emptyset$, we omit writing it); H is a finite set of *labels* for membranes; μ is a *membrane structure*, consisting of m membranes, labeled (not necessarily in a one-to-one manner) with elements of H; $w_1, \ldots, w_m$ are strings over O, describing the *multisets of objects* placed in the m regions of μ; h_0 is the output membrane of Π; $R_i, 1 \leq i \leq m$, are finite sets of *developmental rules* of the forms described below.

In these rules, objects a, b, d are from the alphabet O, c is a catalyst from C, h is a label from H, and u, v are multisets of objects over the alphabet O. In membrane separation operation (sep), $Q \subseteq O$ and $K \subset Q$; we denote the subtraction operation $Q - K$ by $\neg K$.

It is important to note that we do not use polarizations for membranes. (As considered in [10,12], the membranes can have one of the negative, positive, neutral, "electrical charges", represented by $-$, $+$, and 0, respectively, but we do not use this feature here.) In the literature, one uses to add the subscript 0 to indicate that the rules do not use polarizations, writing, for instance, sep_0, mer_0, etc., but here we omit this subscript.

Action & Identification	Type of Rule
separate (sep)	$[\, Q\,]_h \rightarrow [\, K\,]_h[\, \neg K\,]_h$
merge (mer)	$[\ ]_h[\ ]_h \rightarrow [\ ]_h$
move in (in)	$a[\ ]_h \rightarrow [\, b\,]_h$
move out (out)	$[\, b\,]_h \rightarrow [\ ]_h a$
promoter (pro)	$[\, a \rightarrow v|b\,]_h$
catalytic (cat)	$[\, ca \rightarrow cv\,]_h$
cooperative (coo)	$[\, v \rightarrow u\,]_h$
non-cooperative ($ncoo$)	$[\, a \rightarrow v\,]_h$
divide for elementary ($ediv$)	$[\, a\,]_h \rightarrow [\, b\,]_h[\, d\,]_h$
divide for non-elementary ($ndiv$)	$[\, a\,]_h \rightarrow [\, b\,]_h[\, d\,]_h$

The actions and the notations of rules are as follows: sep – separation rules for elementary membranes; the membrane h, containing objects from Q, is separated into two membranes with the same labels; the objects from K are placed in the first membrane, those from $Q - K$ are placed in the other membrane; we request that both K and $Q - K = \neg K$ are not empty and also that both membranes $[\, K\,]_h$, $[\, \neg K\,]_h$ are non-empty (the rule is not applied otherwise). mer – merging rules for elementary membranes; the two membranes are merged into a single membrane; the objects of the former membranes are put together in the new membrane. in – communication rules; an object is introduced in the membrane during this process. out – communication rules; an object is sent out of the membrane during this process. pro – evolution rule with promoter; the rule is applied only if the promoter object b is present in the region. cat – catalytic rules; $ca \rightarrow cv$, where $c \in C, a \in O - C, v \in (O - C)^*$; the catalyst c is never modified, it only assists the evolution of other objects. coo – evolution rules of radius greater than one; a particular case is that of catalytic rules. $ncoo$ – object evolution rules, associated with membranes and depending on the label, but not directly involving the membranes, in the sense that the membranes are neither taking part in the application of these rules nor are they modified by them. $ediv$ – 2-division rules for elementary membranes; in reaction with an object, the membrane is divided into two membranes with the same label; the object specified in the rule is replaced in the two new membranes by possibly new objects and the remaining objects are duplicated. $ndiv$ – 2-division rules for non-elementary membranes; in reaction with an object, the membrane is divided into two membranes with the same label; the object specified in the rule is replaced in the two new membranes by possibly new objects; the remaining objects *and membranes* contained in this membrane are duplicated, and then are part of the contents of both new copies of the membrane [2].

Each copy of an object and each copy of a membrane can be used by only one rule (a promoter can promote several rules at the same time, and a membrane can appear in several evolution rules at the same time).

In each step, the use of rules is done in the bottom-up manner (first the inner objects and inner membranes evolve, and the result is duplicated if any surrounding membrane is divided or separated).

When the rules of a given type (α) are able to change the label(s) of the involved membranes, we denote that type of rules by (α'). For example, the primed versions of the merging and the separation rules are of the following forms:

(mer'): $[\ ]_{h_1}[\ ]_{h_2} \rightarrow [\ ]_{h_3}$, for $h_1, h_2, h_3 \in H$.
(sep'): $[\ Q\ ]_{h_1} \rightarrow [\ K\ ]_{h_2}[\ \neg K\ ]_{h_3}$, for $h_1, h_2, h_3 \in H, K \subset Q$.

Here we use the rules in the *minimally parallel* manner [3]. All rules of any type involving a membrane h form a set R_h. Moreover, if a membrane h appears several times in a given configuration of the system, then for each occurrence i of the membrane we consider a different set R_{h_i}. Then, in each step, from each set $R_{h_i}, h \in H$, from which at least a rule *can* be used, at least one rule *must* be used.

As usual for P systems with active membranes, each membrane and each object can be involved in only one rule, and the choice of rules to use and of objects and membranes to evolve is done in a non-deterministic way.

A halting computation gives a result, in the form of the number of objects present in membrane h_0 in the end of the computation; for a computation to be successful, exactly one membrane with label h_0 should be present in the halting configuration. The set of numbers generated in this way by a system Π is denoted by $N_{gen}(\Pi)$ and the family of such sets, generated by systems having initially at most n_1 membranes and using during the computation configurations with at most n_2 membranes is denoted by $N_{mp}^{gen}OP_{n_1,n_2}(\textit{types-of-rules})$, with the subscript *"mp"* indicating the "minimal parallelism" used in computations, and *types-of-rules* indicating the allowed types of rules. A P system can be also used in the accepting mode: we introduce a number in the system in the form of a multiset a^n, for some $a \in O$, in region h_0 and start computing; if the system halts, then the number n is accepted. The set of all numbers accepted in this way by Π is denoted by $N_{acc}(\Pi)$, and the family of such sets is denoted by $N_{mp}^{acc}OP_{n_1,n_2}(\textit{types-of-rules})$, with the obvious meaning of the used parameters. When the number of membranes does not increase during the computation we use only the subscript n_1, denoting the maximal number of membranes initially present in the used systems. When using systems with at most r catalysts, we write cat_r for the respective type of rules.

In what follows we give several accepting and generative universality results, as well as efficiency results for polarizationless P systems working in the *minimally parallel mode*.

4   Computational Completeness Results

The Accepting Case. When we remove polarizations from active membranes, additional features are in general necessary in order to reach the universality. There are several such features, for instance, cooperative rewriting rules, changing membrane labels, using promoter/inhibitor objects, priorities among rules, etc. Some of these tools will be also used in what follows.

In the proofs we will simulate register machines. In all constructions, with each register r of a register machine we associate a membrane with label r, and the number stored in register r is represented by the number of copies of object a present in membrane r.

Theorem 1. $N_{mp}^{acc}OP_n(cat_1, pro, in, out) = NRE, n \geq 7$.

Proof. Let us consider a deterministic register machine $M = (3, B, l_0, l_h, I)$ accepting an arbitrary set $N(M) \in NRE$. We construct the P system

$$\Pi = (O, C, H, \mu, (w_h)_{h \in H}, (R_h)_{h \in H}, 1) \text{ with}$$
$$O = \{a, c\} \cup \{l, \bar{l}, l', l'', l''', l^{iv}, l^v, l^{vi}, l^{vii} \mid l \in B\},$$
$$C = \{c\},$$
$$H = \{0, 1, 1', 2, 2', 3, 3'\},$$
$$\mu = [[[\ ]_{1'}]_1 [[\ ]_{2'}]_2 [[\ ]_{3'}]_3]_0,$$
$$w_0 = l_0, w_1 = w_2 = w_3 = c, w_1' = w_2' = w_3' = \lambda,$$

and with the following rules in $R_h, h \in H$. (Remember that the number to be analyzed is introduced in region 1 in the form a^n.)

An instruction $l_1 : (\text{ADD}(r), l_2)$ is simulated by means of the following rules:

step	R_0	R_r
1.	—	$l_1[\ ]_r \to [l_1]_r$
2.	—	$[l_1 \to l_2'a]_r$
3.	—	$[l_2']_r \to [\ ]_r l_2'$
4.	$[l_2' \to l_2]_0$	—

The label-object l_1 enters the correct membrane r, produces one further copy of a and the label l_2, primed, inside membrane r, then the label l_2' exits to the skin region, loses its prime, and the process can be iterated.

For the simulation of a SUB instruction $l_1 : (\text{SUB}(r), l_2, l_3)$ we use the next rules:

step	R_0	R_r	$R_{r'}$
1.	$[l_1 \to l_1'l_1'']_0$	—	—
2.	$[l_1' \to l_1''']_0$	$l_1''[\ ]_r \to [l_1'']_r$	—
3.	$[l_1''' \to l_1^{iv}]_0$	$[ca \to ca' \mid_{l_1''}]_r$	$l_1''[\ ]_{r'} \to [l_1'']_{r'}$
4.	—	$l_1^{iv}[\ ]_r \to [l_1^{iv}]_r$	$[l_1'' \to l_1^v]_{r'}$
5.	—	$[l_1^{iv} \to \bar{l}_2 \mid a']_r$	$[l_1^v \to l_1^{vi}]_{r'}$
6.	—	$[a' \to \lambda \mid_{\bar{l}_2}]_r$	$[l_1^{vi}]_{r'} \to l_1^{vi}[\ ]_{r'}$
7.	—	$[\bar{l}_2 \to l_2 \mid_{l_1^{vi}}]_r$ or $[l_1^{iv} \to l_3 \mid_{l_1^{vi}}]_r$	$l_1^{vi}[\ ]_{r'} \to [l_1^{vii}]_{r'}$
8.	—	$[l_2]_r \to l_2[\ ]_r$ or $[l_3]_r \to l_3[\ ]_r$	$[l_1^{vii} \to \lambda]_{r'}$

We start the computation by producing a couple of objects l_1' and l_1'' in the skin region by rule 1. Then l_1' evolves to l_1''' in the skin region, while object l_1''

enters into membrane r. In the third step, while l_1''' changes to the next primed version l_1^{iv} in the skin region, object l_1'' enters into the inner membrane r' – this happens in both cases irrespective whether object a exists or not in membrane r. If an object a was present, it evolves to a' in the presence of promoter object l_1'' by means of the catalytic evolution rule $[\,ca \rightarrow ca'|_{l_1''}\,]_r$ and the promoter leaves the membrane r. The catalyst c is used to prevent more objects a to evolve in the same step. At the fourth step, object l_1'' evolves to l_1^v in membrane r' and object l_1^{iv} enters into membrane r, respectively. Since no object remains in the skin region, this region will stay idle until the end of the computation. In the step five, object l_1^v evolves to l_1^{vi}, which will be sent out of the membrane r' in the following step; in membrane r, if a promoter object a' is present, then object l_1^{iv} produces the object $\bar{l}_2$. Otherwise, there is no rule to be applied here in this or the next step. In step 6, object l_1^{vi} arrives into membrane r, where the object $\bar{l}_2$ promotes the deletion rule $[\,a' \rightarrow \lambda|_{\bar{l}_2}\,]_r$ and removes the object a' previously produced. In membrane r, object l_1^{vi} promotes either object $\bar{l}_2$ or l_1^{iv}, to introduce the corresponding label-object l_2 or l_3, respectively. The correct label-object l_2 or l_3 is moved to skin region at the 8th step of the computation. The object l_1^{vi} enters membrane r' and is removed.

Note that in the simulation of instructions ADD and SUB of M in each computation step at most one rule from each set R_h has been used, hence the system works both in the minimally and in the maximally parallel mode. The starting configuration of the system is restored after each simulation, hence another instruction can be simulated. If the computation in M halts, hence l_h is reached, this means that the halting label-object l_h is introduced in the skin region, and also the computation in Π halts. Consequently, $N(M) = N_{acc}(\Pi)$, and this concludes the proof. □

Because of space restriction, the proofs of the next two results are omitted.

Theorem 2. $N_{mp}^{acc}OP_n(ncoo, in', out') = NRE, n \geq 4$.

In the next theorem we use the rather strong tool of cooperative evolution rules (with radius at most 2), but the membrane labels will not be changed during the computation.

Theorem 3. $N_{mp}^{acc}OP_n(coo, in, out) = NRE, n \geq 4$.

The Generative Case. We consider now P systems working in the generative mode.

The next universality result is based on the simulation of a matrix grammar. Catalytic evolution rules, membrane merging and separation, also changing membrane labels, are used in the proof.

Theorem 4. $N_{mp}^{gen}OP_{2,5}(cat_1, sep', mer') = NRE$.

Proof. Let us consider a matrix grammar $G = (N, T, S, M, F)$ with appearance checking, in the *binary normal form*, hence with $N = N_1 \cup N_2 \cup \{S, \#\}, T = \{a\}$, and with the matrices of the forms as mentioned in Section 2.

Assume that all matrices are injectively labeled with elements of a set B. We construct the P system of degree 2

$$\Pi = (O, C, H, \mu, w_0, w_s, (R_h)_{h \in H}, 0) \text{ with}$$
$$O = N_1 \cup N_2 \cup \{c, c_1, c_2, a, f, \#\},$$
$$C = \{c\},$$
$$H = \{m, m_1, m_2, m'_1, m''_1, m'_2, m''_2, m'''_2 \mid m \in B\} \cup \{0, 0', 0'', s\},$$
$$\mu = [\,[\ ]_0\,]_s,$$
$$w_0 = X Acc_1 c_2, w_s = \lambda,$$

and the sets R_h containing the rules below.

The simulation of a matrix $m : (X \to Y, A \to x)$, with $X \in N_1$, $Y \in N_1$, and $A \in N_2, x \in (N_2 \cup T)^*, |x| \le 2$, is done in four steps, using the following rules (the matrix with label m is encoded in the labels of the membranes created by the separation operation; we start from a configuration of the form $[\,[\ X w c c_1 c_2]_0\,]_s$, for some $w \in (N_2 \cup T)^*$):

step	$(X \to Y,$	$A \to x)$
1.	$[\ O]_0 \to [\ \{X\}]_{m_1} [\ \neg\{X\}]_{m_2}$	—
2.	$[\ X \to Y]_{m_1}$	$[\ O]_{m_2} \to [\ \{c, A\}]_{m'_2} [\ \neg\{c, A\}]_{m''_2}$, or
		$[\ c_1 \to \#]_{m_2}, [\ \# \to \#]_{m_2}$
3.	$[\ ]_{m_1} [\ ]_{m''_2} \to [\ ]_{m'_1}$	$[\ cA \to cx]_{m'_2}$
4.	$[\ ]_{m'_1} [\ ]_{m'_2} \to [\ ]_0$	—
	$[\ \alpha \to \#]_{m'_1}, \alpha \in O$	$[\ \# \to \#]_{m'_1}$

We start the simulation of matrix m by separating membrane 0 under the control of object X. After separation, object X takes place in a new membrane m_1, and other objects (including c, c_1, c_2) are placed in a membrane m_2. In the second step, object X evolves to Y while objects A, c makes membrane m_2 separate. At the third step, one copy of object A evolves to x in the new membrane m'_2. Thus, the rules of matrix m have been simulated. In the meantime, membranes with labels m_1 and m''_2 are merged into a new membrane labeled m'_1. At the fourth step, objects Y, x, c, c_1, c_2 returns to membrane 0, obtained by merging membranes m'_1 and m'_2. If this does not happen, i.e., the rule $[\ cA \to cx]_{m'_2}$ is used again in step 4, then the trap object $\#$ is introduced (there is at least the object Y in membrane m'_1). If the object X is present, hence membrane 0 is separated, but object A not, then in step 2 one introduces the trap object $\#$ by the rule $[\ c_1 \to \#]_{m_2}$, which must be used, because $[\ O]_{m_2} \to [\ \{c, A\}]_{m'_2} [\ \neg\{c, A\}]_{m''_2}$ cannot be used.

Thus, the matrix m is correctly simulated, and the system can pass to the simulation of another matrix.

The simulation of a matrix with appearance checking $m : (X \to Y, A \to \#)$, with $X, Y \in N_1$, and $A \in N_2$, is done in four steps using the following rules:

step	$(X \to Y,$	$A \to \#)$
1.	$[\,O\,]_0 \to [\,\{X, c_1\}\,]_{m_1}[\,\neg\{X, c_1\}\,]_{m_2}$	
2.	$[\,O\,]_{m_1} \to [\,\{X\}\,]_{m_1'}[\,\neg\{X\}\,]_{m_1''}$	$[\,O\,]_{m_2} \to [\,\{A\}\,]_{m_2'}[\,\neg\{A\}\,]_{m_2''}$
3.	$[\,X \to Y\,]_{m_1'}$	$[\,A \to \#\,]_{m_2'}$
	$[\,\,]_{m_1''}[\,\,]_{m_2} \to [\,\,]_{m_2'''}$	$[\,\# \to \#\,]_{m_2'}$
4.	$[\,\,]_{m_1'}[\,\,]_{m_2'''} \to [\,\,]_0$	

The computation starts with objects X, c_1 making membrane 0 to separate. At the second step, if membrane m_2 includes an object A, then it will separate into membranes m_2' including A and m_2'' including the auxiliary object c_2. If A existed, then the computation never stop. At the same time, objects X and c_1 take place in the membranes m_1' and m_1'', respectively. In the third step, X evolves to Y, and if membrane m_2 is still present, then membrane m_1'' merges with it, creating a new membrane m_2'''. Finally, membranes m_1' and m_2''' are merged into membrane 0 including all correct objects and the system returns to a configuration as the starting one.

The simulation of a matrix $m : (X \to \lambda, A \to x)$, with $X \in N_1, A \in N_2$, and $x \in T^*, |x| \le 2$, is done in six steps, using the following rules:

step	$(X \to f,$	$A \to x)$
1.	$[\,O\,]_0 \to [\,\{X\}\,]_{m_1}[\,\neg\{X\}\,]_{m_2}$	—
2.	$[\,X \to f\,]_{m_1}$	$[\,O\,]_{m_2} \to [\,\{c, A\}\,]_{m_2'}[\,\neg\{A\}\,]_{m_2''},$ or
		$[\,c_1 \to \#\,]_{m_2}, [\,\# \to \#\,]_{m_2}$
3.	$[\,\,]_{m_1}[\,\,]_{m_2''} \to [\,\,]_{m_1'}$	$[\,cA \to cx\,]_{m_2'}$
4.	$[\,\,]_{m_1'}[\,\,]_{m_2'} \to [\,\,]_{0'}$	—
—	$[\,\alpha \to \#\,]_{m_1'}, \alpha \in O$	$[\,\# \to \#\,]_{m_1'}$
5.	$[\,O\,]_{0'} \to [\,\{f\} \cup T\,]_0[\,\neg(\{f\} \cup T)\,]_{0''}$	—
6.	$[\,f \to \lambda\,]_0$	$[\,Z \to \#\,]_{0''}, Z \in N_2$
		$[\,\# \to \#\,]_{0''}$

We omit here the detailed explanation for the first 4 steps, because they are the same as in the simulation of matrix $m : (X \to Y, A \to x)$. At step 5, membrane $0'$ is separated into membranes with labels 0 and $0''$. The former one includes the terminal objects and the special object f. Object f evolves to λ at step 6. If there are objects Z from N_2 in membrane $0''$, then the rule $[\,Z \to \#\,]_{m_1'}$ is applied and the computation will never halt. Thus, the simulation is correctly completed.   □

5   Computational Complexity Results

We present here both uniform and semi-uniform linear time solutions of **SAT** based on polarizationless P systems working in the minimally parallel mode. Three results are given, with the following features (types of rules, label changing, type of construction):

no.	evolution	division	communication	label changing	construction
1.	cooper.	elementary	move out	no	semi-uniform
2.	non-coop.	non-element.	move in, out	yes	uniform
3.	cooper.	non-element.	move in, out	no	uniform

A rigorous framework for dealing with complexity matters in our area is that of *recognizing P systems*, which we introduce here following [12]. First, let us consider P systems *with input*, which will allow to input a multiset encoding a decision problem, in a special membrane. Such a device is a tuple (Π, V, i_0), where:

- Π is a usual P system, with the alphabet of objects O and initial multisets $w_1, \ldots, w_m$ (associated with membranes labeled by $1, \ldots, m$, respectively).
- V is an (input) alphabet strictly contained in O and such that $w_1, \ldots, w_m$ are multisets over $O - V$.
- $i_0 \in \{1, 2, \ldots, m\}$ is the label of a distinguished membrane (of input).

If w is a multiset over V, then the initial configuration of (Π, V, i_0) with input w is $(\mu, w'_1, \ldots, w'_m)$, where $w'_i = w_i$ for $i \neq i_0$, and $w'_{i_0} = w_{i_0} \cup w$.

The computations of a P system with input are defined in a natural way, the only change is that the initial configuration is obtained by adding the input multiset w over V to the initial configuration of the system Π.

Then, a *recognizing* P system is a P system with input, (Π, V, i_0), such that:

1. The alphabet O of Π contains two distinguished elements, yes, no.
2. All computations of the system halt.
3. If $\mathcal{C}$ is a computation of Π, then either the object yes or the object no (but not both) is sent out to the environment, and only in the last step of the computation.

We say that $\mathcal{C}$ is an accepting (respectively, rejecting) computation if the object yes (respectively, no) appears in the environment in the halting configuration of $\mathcal{C}$.

To understand what solving a problem in a semi-uniform/uniform way means, we consider a decision problem X. A family $\Pi_X = (\Pi_X(1), \Pi_X(2), \ldots)$ of P systems (with active membranes in our case) is called *semi-uniform* (*uniform*) if its elements are constructible in polynomial time starting from $X(n)$ (from n, respectively), where $X(n)$ denotes the instance of size n of X. We say that X can be solved in polynomial (linear) time by the family Π_X if the system $\Pi_X(n)$ will always stop in a polynomial (linear, respectively) number of steps, sending out the object yes if and only if the instance $X(n)$ has a positive answer. For more details about complexity classes for P systems see [11,12].

Note that always we have an answer, one of yes and no, but we have said nothing about the way the computations evolve, the only restriction we impose is that all of them halt (in a number of steps bounded by a known function). That is why we say that such systems are *confluent* (they may be non-deterministic, but the answer is obtained in a finite time irrespective of the possible branchings of the computations). The *deterministic* systems, where no branching is possible, are a particular case of confluent systems.

Theorem 5. *P systems constructed in a uniform manner and working in the minimally parallel mode using rules of types* $(ncoo, ndiv, in, out')$ *can solve* SAT *in linear time.*

Proof. Let us consider a propositional formula in the conjunctive normal form, $\alpha = C_1 \wedge \cdots \wedge C_m$, such that each clause $C_i, 1 \le i \le m$, is of the form $C_i = y_{i,1} \vee \cdots \vee y_{i,k_i}, k_i \ge 1$, where $y_{i,j} \in \{x_k, \neg x_k \mid 1 \le k \le n\}$.

The instance α is encoded as a set over

$$\Sigma(\langle n, m \rangle) = \{x_{i,j}, x'_{i,j} \mid 1 \le i \le m, 1 \le j \le n\}.$$

The object $x_{i,j}$ represents the variable x_j appearing in the clause C_i without negation, and object $x'_{i,j}$ represent the variable x_j appearing in the clause C_i with negation. Thus, the input multiset is

$$\begin{aligned}
w = {} & \{x_{i,j} \mid x_j \in \{y_{i,j} \mid 1 \le k \le l_i\}, 1 \le i \le m, 1 \le j \le n\} \\
& \cup \{x'_{i,j} \mid \neg x_j \in \{y_{i,j} \mid 1 \le k \le l_i\}, 1 \le i \le m, 1 \le j \le n\}.
\end{aligned}$$

For given $(n, m) \in \mathbf{N}^2$, we construct a recognizing P system $(\Pi(\langle n, m \rangle), \Sigma(\langle n, m \rangle), 0)$ with:

$$\begin{aligned}
\Pi(\langle n, m \rangle) = {} & (O(\langle n, m \rangle), H, \mu, (w_h)_{h \in H}, (R_h)_{h \in H}), \text{where} \\
O(\langle n, m \rangle) = {} & \{x_{i,j}, x'_{i,j}, t_{j,i}, f_{j,i} \mid 1 \le i \le m, 1 \le j \le n\} \\
& \cup \{b_i, c_i, a'_i, a''_i, a'''_i \mid 1 \le i \le n\} \cup \{\text{yes}, \text{no}, d\} \\
& \cup \{l'_i \mid 0 \le i \le n\} \cup \{l_i, e_i, \mid 1 \le i \le m\} \\
& \cup \{a_i \mid 0 \le i \le n + 2\} \cup \{d_i \mid 0 \le i \le 6n + m + 7\},
\end{aligned}$$

$$\begin{aligned}
\mu = {} & [\,[\,[\,[\ ]_m \cdots [\ ]_2 [\ ]_1]_0]_a [\ ]_d]_s, \\
& w_a = a_0, w_d = d_0, w_0 = l'_0, w_s = w_i = \lambda, 1 \le i \le m, \\
H = {} & \{s, a, b, d, y, n\} \cup \{i \mid 0 \le i \le m\} \cup \{i', i'' \mid 0 \le i', i'' \le n\},
\end{aligned}$$

and the following rules (we also give explanations about their use):

1. $[\, d_i \to d_{i+1} \,]_d, 0 \le i \le 6n + m + 5$.

Object d_i counts the computation steps in membrane d.

Initialization phase:

2. $x_{i,j}[\ ]_i \to [\, x_{i,j} \,]_i,$
   $x'_{i,j}[\ ]_i \to [\, x'_{i,j} \,]_i, 1 \le i \le m, 1 \le j \le n.$

We re-encode the problem α into m membranes in n steps.

3. $[\, a_i \to a_{i+1} \,]_a, 0 \le i \le n + 2$.
4. $[\, l'_i \to l'_{i+1} \,]_0, 0 \le i \le n - 1$.

Simultaneously with the use of rule 2, the evolution rules 3 and 4 are applied in membranes a and 0, respectively.

5. $[\, l'_n \rightarrow l_1 l_2 \ldots l_m \,]_0$.
6. $l_i [\ ]_j \rightarrow [\, l_i \,]_j$, $1 \leq i, j \leq m$.
7. $[\, l_i \,]_j \rightarrow [\ ]_{0'} d$, $1 \leq i, j \leq m$.
8. $a_{n+2} [\ ]_0 \rightarrow [\, a_1 \,]_0$.

At step $n+1$, object l'_n evolves to $l_1, l_2, \ldots, l_m$. Consequently, in order to change the membrane labels $j, 1 \leq j \leq m$, by $0'$, those objects enter into and are sent out of membranes in two steps, by means rules 6 and 7. Meanwhile, object a_{n+2} evolves to a_1 and enters into membrane 0, by rule 8. Thus, the initialization phase has been completed in $n + 3$ steps.

Checking phase:

9. $[\, a_i \,]_0 \rightarrow [\, b_i \,]_0 [\, c_i \,]_0$, $1 \leq i \leq n$.
10. $[\, b_i \rightarrow t_{1,i} t_{2,i} \ldots t_{m,i} a'_i \,]_0$,
    $[\, c_i \rightarrow f_{1,i} f_{2,i} \ldots f_{m,i} a'_i \,]_0$, $1 \leq i \leq n$.

We generate 2^n membranes with label 0 by using non-elementary membrane division rule 9. In each step, b_i and c_i correspond to the truth values *true* and *false*, respectively, for variable x_i. By rule 10, objects b_i and c_i evolve. Rules 9 and 10 are performed in 2 steps.

11. $t_{i,j} [\ ]_{(j-1)'} \rightarrow [\, t_{i,j} \,]_{(j-1)'}$, or
    $t_{i,j} [\ ]_{(j-1)''} \rightarrow [\, t_{i,j} \,]_{(j-1)''}$,
    $f_{i,j} [\ ]_{(j-1)'} \rightarrow [\, f_{i,j} \,]_{(j-1)'}$, or
    $f_{i,j} [\ ]_{(j-1)''} \rightarrow [\, f_{i,j} \,]_{(j-1)''}$, $1 \leq i \leq m, 1 \leq j \leq n$.
12. $[\, a'_j \rightarrow a''_j \,]_0$.

The groups of objects $t_{1,j}, t_{2,j}, \ldots, t_{m,j}$ and $f_{1,j}, f_{2,j}, \ldots, f_{m,j}$, corresponding to variable x_j, are introduced into inner membranes with labels $(j-1)'$ or $(j-1)''$ by rule 11, and object a'_j is evolved to a''_j by rule 12. These two rules are applied simultaneously.

13. $[\, t_{i,j} \,]_{(j-1)'} \rightarrow [\ ]_{j'} d$, $[\, t_{i,j} \,]_{(j-1)''} \rightarrow [\ ]_{j'} d$,
    $[\, f_{i,j} \,]_{(j-1)'} \rightarrow [\ ]_{j''} d$, $[\, f_{i,j} \,]_{(j-1)''} \rightarrow [\ ]_{j''} d$, $1 \leq i \leq m, 1 \leq j \leq n$.
14. $[\, a''_j \rightarrow a'''_j \,]_0$.

By using rule 13, objects $t_{i,j}$ and $f_{i,j}$ (corresponding to the truth value *true* and *false* for variable x_j of a clause C_i) leave membranes with labels $(j-1)'$ or $(j-1)''$ and change the labels to j' and j'', respectively. Object a'''_j is produced in membrane 0 by rule 14. Now (using rules 15) we will check which clauses are satisfied by truth values.

15. $[\, x_{i,j} \rightarrow e_i \,]_{j'}$,

$$[\, x'_{i,j} \rightarrow e_i \,]_{j''}, 1 \leq i \leq m, 1 \leq j \leq n.$$
16. $[\, a'''_j \rightarrow a_{j+1} \,]_0.$

We remind that single and double primes of the labels j' and j'' indicate *true* and *false* values, respectively. By rule 15, object $x_{i,j}$ in membrane j' and object $x'_{i,j}$ in membrane j'' evolve to e_i. More precisely, the object without negation has to evolve to object e_l, because the membrane label k' has a single prime and this indicates the *true* value, and the subscript of e_l must be the same with the first subscript of the object $x_{l,k}$. Similarly, the object with negation has to evolve to e_l because of k'' which is double-primed and indicates the value *false*.

In the membranes 0, the objects a'''_i lose their primes and increase the subscript a_{i+1}. We continue in this way the operations of membrane division and construction of truth assignments corresponding to the next variable x_{i+1} (rules 9-16). Between two divisions 5 steps are performed. For having all possible 2^n truth assignments, we need $5n$ division processes. In this way, we check the satisfiability of all clauses and this process ends at the $6n + 3$th step of the computation. At that step, objects a_{n+1} are removed from membrane 0 changing its label to 1, by means of rule 17.

17. $[\, a_{n+1} \,]_0 \rightarrow [\ ]_1 d.$

Recognizing phase:

After the $6n + 3$ steps, there are 2^n membranes with label 1 and each of them contains m membranes labeled by n' or n''.

18. $[\, e_i \,]_{n'} \rightarrow [\ ]_n e_i,$
    $[\, e_i \,]_{n''} \rightarrow [\ ]_n e_i, 1 \leq i \leq m.$

By rule 18, objects $e_i, 1 \leq i \leq m$, are introduced into membrane 1 changing the former membrane labels from n' and n'' to n. If all objects $e_1, e_2, \ldots, e_m$ are present in a membrane with label 1, that means that all clauses $C_1, C_2, \ldots, C_m$ are satisfied.

19. $[\, e_i \,]_i \rightarrow [\ ]_{i+1} e_i, 1 \leq i \leq m.$

Objects $e_i, 1 \leq i \leq n$, leave membrane i one by one increasing the label. In m steps, we can get objects e_m to appear in membrane a. One of them is non-deterministically chosen and expelled out of membrane a, which changes the label to b, by means of rule 20.

20. $[\, e_m \,]_a \rightarrow [\ ]_b e_m.$
21. $[\, e_m \rightarrow \text{yes} \,]_s.$
22. $[\, \text{yes} \,]_s \rightarrow [\ ]_y \text{yes}.$

If an object e_m had appeared in the skin membrane, then it will evolve to yes; this object will be sent to the environment, changing the skin membrane label to y by using rules 21, 22, in the $6n + m + 7$th step of the computation.

23. $[\ d_{6n+m+6}]_d \to [\ ]_d d_{6n+m+7}$.
24. $[\ d_{6n+m+7}]_s \to [\ ]_n \mathbf{no}$.

If the formula is not satisfiable, the object **no** is sent to environment in the step $6n + m + 8$ by rule 24.

Thus, the problem is solved in a uniform manner, in a linear time, with the system working in both the minimally parallel and the maximally parallel modes.

$\square$

Due to space restriction, we again give next theorems without the proofs. The proof ideas are the same as for Theorem 5, that is, we re-encode the input multiset in m membranes, and in this way we avoid the maximal parallel applications of rules in the same membrane.

Theorem 6. *P systems constructed in a uniform manner, working in the minimally parallel mode using rules of types $(coo, ndiv, in, out)$ can solve* SAT *in linear time.*

Theorem 7. *P systems working in the minimally parallel mode with rules of types $(coo, ediv, out)$ and constructed in a semi-uniform manner can solve* SAT *in linear time with respect to the number of the variables and the number of clauses.*

6   Final Remarks

The main contribution of this paper is the use of the minimal parallelism in the framework of P systems with active membranes, without using membrane polarizations. Both universality and efficiency results were proved in this framework, for various combinations of types of rules.

Besides possible improvements of the previous theorems, it also remains to investigate the possibility of obtaining similar universality and efficiency results for other classes of P systems working in the minimally parallel mode, in particular, for the case of using rules of the forms $ncoo, out, sep$.

Acknowledgment. The work of the author is supported by the Center for International Mobility (CIMO) Finland, grant TM-06-4036.

References

1. A. ALHAZOV, T.-O. ISHDORJ, Membrane Operations in P Systems with Active Membranes. In: GH. PĂUN, et al. (eds.) *Second Brainstorming Week on Membrane Computing*, Sevilla, 2-7 February, 2004, **TR** 01/2004, University of Sevilla, 37–44.
2. A. ALHAZOV, L. PAN, GH. PĂUN, Trading Polarizations for Labels in P Systems with Active Membranes, *Acta Informaticae*, 41, 2-3 (2004), 111–144.
3. G. CIOBANU, L. PAN, GH. PĂUN, M. J. PÉREZ-JIMÉNEZ, P Systems with Minimal Parallelism. Submitted 2005.

4. J. DASSOW, GH. PĂUN, *Regulated Rewriting in Formal Language Theory*. Springer-Verlag, Berlin, 1989.
5. J. E. HOPCROFT, J. D. ULLMAN, *Introduction to Automata Theory, Languages, and Computation*. Addison-Wesley, Reading, MA, 1979.
6. M. L. MINSKY, *Finite and Infinite Machines*. Prentice Hall, Englewood Cliffs, 1967.
7. L. PAN, A. ALHAZOV, T.-O. ISHDORJ, Further Remarks on P Systems with Active Membranes, Separation, Merging and Release Rules. *Soft Computing*, 8 (2004), 1–5.
8. L. PAN, T.-O. ISHDORJ, P Systems with Active Membranes and Separation Rules. *Journal of Universal Computer Science*, 10, 5 (2004), 630–649.
9. CH. P. PAPADIMITRIOU, *Computational Complexity*. Addison-Wesley, Reading, MA, 1994.
10. GH. PĂUN, P Systems with Active Membranes: Attacking NP-Complete Problems. *Journal of Automata, Languages and Combinatorics*, 6, 1 (2001), 75–90.
11. GH. PĂUN, *Membrane Computing: An Introduction*. Springer-Verlag, Berlin, 2002.
12. M. J. PÉREZ-JIMÉNEZ, A. ROMERO-JIMÉNEZ, F. SANCHO-CAPARRINI, Complexity Classes in Models of Cellular Computation with Membranes. *Natural Computing*, 2, 3 (2003), 265–285.

P Systems with Active Membranes Characterize PSPACE

Petr Sosík[1,2] and Alfonso Rodríguez-Patón[1]

[1] Universidad Politécnica de Madrid – UPM, Facultad de Informática
Campus de Montegancedo s/n, Boadilla del Monte, 28660 Madrid, Spain
arpaton@fi.upm.es
[2] Institute of Computer Science, Silesian University
74601 Opava, Czech Republic
petr.sosik@fpf.slu.cz

Abstract. A P system is a natural computing model inspired by information processes in cells and a control role of cellular membranes.
We show that uniform families of P systems with active membranes are
able to solve, in polynomial time, exactly the class of decisional problems **PSPACE**. Similar results were achieved also with other models of
bio-inspired computers, such as DNA computing. Together they suggest
that **PSPACE** naturally characterizes the computational potential of
biological information processing.

1 Introduction

Membrane systems, called also P systems, are bio-inspired computing models
trying to capture the computational aspects of cell metabolism and information
interchange. P systems are focusing on, among other things, molecular synthesis within cells, selective particle recognition by membranes, controlled transport
through protein channels, or membrane division and dissolution. These processes
are modeled in P systems by means of multiset processing in separate cell-like
regions. The aim of these models is to identify operations which give to cellular systems their information-processing strength, and to prepare their possible
implementation *in vitro* or *in silico*. The bibliography of P systems started with
seminal paper [1]. For an introduction and overview of P systems, we refer the
reader to [2], while a complete and up-to-date bibliography can be found online
at the url [3].

Here we investigate the family of *P systems with active membranes*, introduced in [4]. Until now, a sequence of research papers studied the class of
problems solvable by this family of P systems in polynomial time. We cite e.g.
[5,6,7,8,9,10,11,12], to name a few. The main ingredients of this model of membrane computer are the polarization, division and dissolution of membranes.
Rather surprisingly, by imposing restrictions on these operations one obtains
a scale of computational power corresponding to the complexity classes **P**, **NP**
and **coNP** just by tuning simple parameters of the P system. We refer the reader
to Section 3 for exact specification of these results.

C. Mao and T. Yokomori (Eds.): DNA12, LNCS 4287, pp. 33–46, 2006.
© Springer-Verlag Berlin Heidelberg 2006

In this paper, an upper bound on the power of unrestricted P systems with active membranes is given. We show that the class of problems they compute in polynomial time is equal to **PSPACE**. The same property is possessed by the class of standard parallel computational models as alternating Turing machine, SIMDAG (also known as SIMD PRAM) and others [13].

This result is related to properties of other natural computing models. For example, in the field of DNA computing a similar result was first given in [14]. Later, in [15] another, more robust DNA computing model capturing **PSPACE** in polynomial time was presented. Also, when focusing on the operation of genetic crossing-over, one obtains an equally powerful model called *genetic Turing machine* [16]. Altogether, these studies suggest that **PSPACE** is indeed the limiting class of power of the natural computing machinery.

2 Definitions

In this section we give a brief description of P systems with active membranes due to [4] or [2], where more details can also be found. A *membrane structure* is represented by a Venn diagram (or a rooted tree) and is identified by a string of correctly matching parentheses, with a unique external pair of parentheses corresponding to the external membrane, called *the skin*. A membrane without any other membrane inside is said to be *elementary*. The following example from [4] illustrates the situation: the membrane structure in Figure 1 is identified by the string

$$\mu = [_1[_2[_5\ ]_5[_6\ ]_6]_2[_3\ ]_3[_4[_7[_8\ ]_8]_7]_4]_1.$$

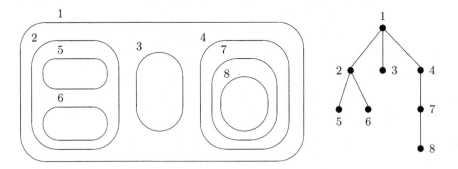

Fig. 1. A membrane structure and its associated tree

In what follows, we occasionally refer to membranes as nodes in the associated tree. The membranes can be further marked with $+$, $-$ or 0, and this is interpreted as an "electrical charge". We will write $[_i\ ]_i^+, [_i\ ]_i^-, [_i\ ]_i^0$ in the three cases, respectively.

The membranes delimit *regions*, precisely identified by the membranes. In these regions we place *objects* which are represented by symbols of an alphabet.

Several copies of the same object can be present in a region, so we work with *multisets* of objects. A multiset over an alphabet V can be represented by a string $x \in V^*$ (by V^* we denote the free monoid generated by V with respect to the concatenation and the identity λ). For a string $x \in V^*$ we denote by $m(x)$ its associated multiset over V.

A *P system with active membranes* is a construct

$$\Pi = (V, H, \mu, w_1, \ldots, w_m, R),$$

where:

(i) $m \geq 1$;

(ii) V is an alphabet;

(iii) H is a finite set of *labels* for membranes;

(iv) μ is a *membrane structure*, consisting of m membranes, labelled (not necessarily in a one-to-one manner) with elements of H; all membranes in μ are supposed to be neutral;

(v) $w_1, \ldots, w_m$ are strings over V, describing the *multisets of objects* placed in the regions of μ;

(vi) R is a finite set of *developmental rules*, of the following forms:

(a) $[_h a \to v]_h^\alpha$, for $h \in H, \alpha \in \{+, -, 0\}, a \in V, v \in V^*$
(object evolution rules);

(b) $a[_h \ ]_h^{\alpha_1} \to [_h b]_h^{\alpha_2}$, for $h \in H, \alpha_1, \alpha_2 \in \{+, -, 0\}, a, b \in V$
(communication rules);

(c) $[_h a \ ]_h^{\alpha_1} \to [_h \ ]_h^{\alpha_2} b$, for $h \in H, \alpha_1, \alpha_2 \in \{+, -, 0\}, a, b \in V$
(communication rules);

(d) $[_h a \ ]_h^\alpha \to b$, for $h \in H, \alpha \in \{+, -, 0\}, a, b \in V$
(dissolving rules);

(e) $[_h a \ ]_h^{\alpha_1} \to [_h b \ ]_h^{\alpha_2} [_h c \ ]_h^{\alpha_3}$, for $h \in H, \alpha_1, \alpha_2, \alpha_3 \in \{+, -, 0\}, a, b, c \in V$
(division rules for elementary membranes);

(f) $[_{h_0} [_{h_1} \ ]_{h_1}^+ \cdots [_{h_k} \ ]_{h_k}^+ [_{h_{k+1}} \ ]_{h_{k+1}}^- \cdots [_{h_n} \ ]_{h_n}^-]_{h_0}^{\alpha_2}$
$\to [_{h_0} [_{h_1} \ ]_{h_1}^{\alpha_3} \cdots [_{h_k} \ ]_{h_k}^{\alpha_3}]_{h_0}^{\alpha_5} [_{h_0} [_{h_{k+1}} \ ]_{h_{k+1}}^{\alpha_4} \cdots [_{h_n} \ ]_{h_n}^{\alpha_4}]_{h_0}^{\alpha_6}$,
for $n > k \geq 1, h_i \in H, 0 \leq i \leq n$, and $\alpha_2, \ldots, \alpha_6 \in \{+, -, 0\}$;
(division of non-elementary membranes);

At each computational step, first the rules are assigned in a maximally parallel way to objects and membranes. At one step, an object a can be subject to only one rule of type (a)–(e) and a membrane h can be subject to only one rule of type (b)–(f). In the case of type (f) rules this means that none of the membranes labelled $h_0, \ldots, h_n$ listed in the rule can be simultaneously subject to another rule of type (b)–(f). The membrane to which a (c), (d) or (e)-type rule is assigned may contain also other objects than a which may simultaneously evolve by rules of type (a).

Then all the rules are applied in parallel. In the case of rules (d) the results of evolution of all objects are released to the parent membrane of h. In the case of rules (e) the results of evolution of all objects except a contained in the membrane h are copied to both resulting copies of h. In the case of rules (f) the membrane h_0 can contain also other membranes than $h_1, \ldots, h_n$, which must have polarization 0. The results of evolution of all these membranes and all objects contained in the membrane h_0 are copied to both resulting copies of h_0. We refer the reader to [4] for a more detailed description.

The membrane structure of Π at a given moment, together with all multisets of objects contained in its regions, form the *configuration* of the system. The $(m + 1)$-tuple $(\mu, w_1, \ldots, w_m)$ is the *initial configuration*. We can pass from one configuration to another by using the rules from R according to the principles given above. The computation stops when there is no rule which can be applied to objects and membranes in the last configuration. The result of the computation is the collection of objects expelled from the skin membrane during the whole computation. In such a way P systems with active membranes can generate any recursively enumerable set [4].

An accepting variant of P systems also exists which solves decision problems. A distinguished region contains, at the beginning of the computation, an input – a description of an instance of a problem. The result of the computation (a solution to the instance) is "yes" if a distinguished object *yes* is expelled during the computation, otherwise the result is "no."

A P system is called *deterministic* if for each input a unique sequence of configurations exists which it can pass through. Each problem decidable by Turing machines is also decidable by deterministic P systems with active membranes [5]. A P system is called *confluent* if it always halts and, starting from the same initial configuration, it always returns the same result "yes" or "no". Therefore, given a fixed initial configuration, a confluent P system can non-deterministically choose from various sequences of configurations, but all of them must lead to the same result.

3   Complexity Classes of P Systems

Consider a decision problem X, i.e. a set of instances $\{x_1, x_2, \ldots\}$ such that to each x_i there is uniquely assigned an answer "yes" or "no". A typical situation in many biocomputing models is that each "machine" (e.g. a system of DNA molecules and reactions, or a P system) can solve only a limited number of instances. However, the problem X can still be solvable by a given class of biocomputing systems in such a sense that each instance is solved by some member of the class. Hence it is more natural to consider *families* of P systems for solving computational problems [10].

We denote by $|x_i|$ the size of an instance x_i of a problem X. In a usual representation x_i, $i = 1, 2, \ldots$, are words over a fixed finite alphabet and $|x_i|$ is the length of x_i.

Definition 1. *Let $\mathcal{D}$ be a class of P systems and let $f : \mathbb{N} \longrightarrow \mathbb{N}$ be a total function. The* class of problems solved by uniform families of P systems of type $\mathcal{D}$ in time f, *denoted by* $\mathbf{MC}_{\mathcal{D}}(f)$, *contains all problems X such that:*

1. *there exists a* uniform *family of P systems $\Pi_X = (\Pi_X(1); \Pi_X(2); \ldots)$ of type $\mathcal{D}$: each $\Pi_X(n)$ can be constructed by a deterministic Turing machine with input n in time polynomial to n.*

2. *Each $\Pi_X(n)$ is sound: there exists a distinguished object* yes *such that $\Pi_X(n)$ starting with a (properly encoded) input $x \in X$ of size n expels out object* yes *if and only if the answer to x is "yes".*
3. *Each $\Pi_X(n)$ is confluent: all computations of $\Pi_X(n)$ with the same input x of size n give the same result: "yes" or "no".*
4. *Π_X is f-efficient: $\Pi_X(n)$ always halts in at most $f(n)$ steps.*

Alternatively we can consider *semi-uniform families* of P systems $\Pi_X = (\Pi_X(x_1); \Pi_X(x_2); \ldots)$ whose members $\Pi_X(x_n)$ can be constructed by a deterministic Turing machine with input x_n in a polynomial time w.r.t. $|x_n|$. In this case, for each instance of X we have a special P system which therefore does not need an input. The resulting class of problems is denoted by $\mathbf{MC}_\mathcal{D}^S(f)$. Obviously, $\mathbf{MC}_\mathcal{D}(f) \subseteq \mathbf{MC}_\mathcal{D}^S(f)$ for a given class $\mathcal{D}$ and a constructible function f.

Particularly, we denote by

$$\mathbf{PMC}_\mathcal{D} = \bigcup_{k \in \mathbb{N}} \mathbf{MC}_\mathcal{D}(\mathcal{O}(n^k)), \qquad \mathbf{PMC}_\mathcal{D}^S = \bigcup_{k \in \mathbb{N}} \mathbf{MC}_\mathcal{D}^S(\mathcal{O}(n^k)),$$

the classes of problems solvable by uniform (semi-uniform, respectively) families of P systems in polynomial time.

We also denote by $\mathcal{AM}$ ($\mathcal{EAM}$, $\mathcal{NAM}$) the classes of P systems with active membranes (with elementary membrane division only and without membrane division, respectively). The following relations are known [6,8,10,11,12]:

$$\mathbf{P} = \mathbf{PMC}_{\mathcal{NAM}}, \tag{1}$$

$$\mathbf{P} = \mathbf{PMC}_{\mathcal{AM}, n\delta, n\pi}, \tag{2}$$

$$\mathbf{NP} \cup \mathbf{coNP} \subseteq \mathbf{PMC}_{\mathcal{EAM}}, \tag{3}$$

$$\mathbf{PSPACE} \subseteq \mathbf{PMC}_{\mathcal{AM}} \subseteq \mathbf{PMC}_{\mathcal{AM}}^S, \tag{4}$$

where in (2) we denote by $\mathcal{AM}, n\delta, n\pi$ P systems with active membranes, without membrane dissolution and polarization.

4   The Characterization of PSPACE

In this section we show that the inclusions reverse to (4) hold as well. To demonstrate that a result of any $f(n)$ time-bounded computation of an accepting P system with active membranes can be found in a space polynomial to $f(n)$, we employ the technique of reverse-time simulation. Instead of simulating a computation of a P system from its initial configuration onwards (which would require an exponential space for storing configurations), we create a recursive function which returns the state of any membrane h after a given number of steps. The recursive calls evaluate contents of the membranes interacting with h in a reverse time order (towards the initial configuration). In such a manner we do not need to store a state of any membrane, but instead we calculate it recursively whenever it is needed. The simulation algorithm is described in a high-level language; in principle, however, it could be performed by a Turing machine, too.

Notice also that the simulated P system is confluent, but the simulation is done in a deterministic way. As the rules in membranes are simulated always in the same order, we simulate only one possible sequence of configurations of the P system. This corresponds to introducing a weak priority between rules: (i) bottom-up priority between rules associated to different membranes, (ii) priority between rules in the same membrane, given by the order in which they are listed, including the priority between types (a), (b), (c), (d), (e), (f), in this order. The confluency condition ensures that such a simulation leads always to a correct result.

Theorem 1. $\mathbf{PMC}^{S}_{\mathcal{AM}} \subseteq \mathbf{PSPACE}$.

Proof. Consider a confluent P system $\Pi = (V, H, \mu, w_1, \ldots, w_m, R)$ with active membranes. For any membrane h of Π, we define its *state* $S = (M, p)$, where M is the multiset characterizing the contents of membrane h and p is its polarization. We use the notation $S.M$ and $S.p$ to refer to these two components of state.

A crucial element of our construction is the function $\mathtt{State}(h, n)$ which computes the state (M, p) of any membrane h of Π after n steps of computation. Its construction is based on the observation that $\mathtt{State}(h, n)$ depends solely on $\mathtt{State}(h, n-1)$, on the state of its parent membrane and on the states of all its embedded membranes (at all the lower levels of the membrane structure tree) after $n-1$ steps.

Simplified simulation without non-elementary membrane division

We can assume that the original labeling of membranes of Π in μ is one-to-one and the labels identify the membranes uniquely. However, during the computation of Π the membranes may be divided, keeping their original labels. To identify membranes uniquely, we add an index to each membrane label. In the initial configuration, each index is an empty string. If a membrane is not divided in a computational step, the digit 1 is added to its index. If it is divided using a rule of type (e), the resulting membranes have added the digits 1 or 2, respectively, to their indices. Hence, after n steps of computation the index of each membrane is an n-tuple of digits from $\{1, 2\}$.

We construct the function $\mathtt{State}(h_{i_1 i_2 \ldots i_n}, n)$ which computes the state of a membrane $h_{i_1 i_2 \ldots i_n}$ after n computational steps of Π. (The second argument of the function is redundant but it adds clarity to the further description.) If the membrane h has been dissolved, the returned value is *dissolved*. If it does not exist after n steps, the returned value is *nil*. Observe that our indexing system allows us to denote all membranes which could exist when all the elementary membranes divide at every step. In a particular computation of Π this need not hold, and hence some indices may denote non-existing membranes.

Function $\mathtt{State}$

Parameters (in all the functions passed by reference): $h_{i_1 i_2 \ldots i_n}$, n

Local variables: $S, S', X, X', \text{Elementary}$

1. If $n = 0$ then return the state of membrane h in the initial configuration and exit.

2. $S \leftarrow \text{State}(h_{i_1 \ldots i_{n-1}}, n - 1)$

3. If $S = nil$ then return nil and exit.
   /* If membrane $h_{i_1 \ldots i_{n-1}}$ did not exist after $(n - 1)$ steps, then after n steps its successor $h_{i_1 i_2 \ldots i_n}$ cannot exist as well. */

4. If $S = dissolved$ then:
   if $i_n = 2$ then return nil, else return $dissolved$, and exit.
   /* If membrane $h_{i_1 \ldots i_{n-1}}$ has been dissolved during the first $(n - 1)$ steps of Π, then it cannot divide during step n. */

5. $S'.M \leftarrow \emptyset$, $S'.p \leftarrow S.p$, $Elementary \leftarrow false$
   /* S' will contain the final state of membrane h after the n-th step of Π */

6. /* We calculate the contribution to the state of $h_{i_1 \ldots i_n}$ from membranes embedded in $h_{i_1 \ldots i_{n-1}}$ due to possible application of rules (b), (c), (d) in these membranes. */
   $\text{Contribution\_from\_children}(h_{i_1 \ldots i_{n-1}}, S, S', Elementary)$

7. $X.M \leftarrow \emptyset$, $X'.M \leftarrow \emptyset$, $X.p \leftarrow 0$, $X'.p \leftarrow 0$

8. /* We calculate the state of the parent membrane of h. */
   If h is not the skin membrane, then $X \leftarrow \text{State}(\text{Parent}(h_{i_1 \ldots i_{n-1}}), n - 1)$

9. /* Now we simulate evolution of membrane $h_{i_1 \ldots i_{n-1}}$ at step n. */
   (a) $\text{Try\_rules\_a}(h, S, S', X)$

   (b) $\text{Try\_rules\_b}(h, S, S', X)$; if any rule was applied, go to step 10.

   (c) $\text{Try\_rules\_c}(h, S, S', X)$; if any rule was applied, go to step 10.

   (d) $\text{Try\_rules\_d}(h, S, S', X)$; if any rule was applied, go to step 10.

   (e) If $Elementary$ then:
       if $i_n = 1$, then $\text{Try\_rules\_e}(h, S, S', X)$, else $\text{Try\_rules\_e}(h, S, X, S')$

10. If $i_n = 2$ and a rule of type (e) was not applied, then $S' \leftarrow nil$
    /* If $i_n = 2$, then membrane $h_{i_1 i_2 \ldots i_n}$ could only be created by an application of an (e)-type rule at the n-th step. */

11. If $S' \neq nil$ and $S' \neq dissolved$ then $S'.M \leftarrow S'.M \cup S.M$

12. Return S' and exit.

Procedures Try_rules_a – Try_rules_e are implemented as follows:

Parameters:
h – label of the membrane processed
S – original state of the membrane
S'– final state of the membrane
T – state of another membrane eventually acting at the operation

(a) For each rule $[_h a \to v]_h^\alpha$ in R such that $S.p = \alpha$:
   – remove all the occurrences of a from $S.M$
   – add to $S'.M$ the same number of occurrences of $m(v)$.

(b) For each rule $a[_h \ ]_h^{\alpha_1} \to [_h b]_h^{\alpha_2}$ in R : if $S.p = \alpha_1$ and $a \in T.M$ then:
   $T.M \leftarrow T.M \setminus \{a\}, \quad S'.M \leftarrow S'.M \cup \{b\}, \quad S'.p \leftarrow \alpha_2$, and skip all other applicable rules.

(c) For each rule $[_h a \ ]_h^{\alpha_1} \to [_h \ ]_h^{\alpha_2} b$ in R : if $S.p = \alpha_1$ and $a \in S.M$ then:
   $S.M \leftarrow S.M \setminus \{a\}, \quad T.M \leftarrow T.M \cup \{b\}, \quad S'.p \leftarrow \alpha_2$, and skip all other applicable rules.

(d) For each rule $[_h a \ ]_h^\alpha \to b$ in R : if $S.p = \alpha$ and $a \in S.M$ then:
   $S.M \leftarrow S.M \setminus \{a\}, \quad S'.M \leftarrow S'.M \cup \{b\}, \quad T.M \leftarrow T.M \cup S.M \cup S'.M,$
   $S' \leftarrow dissolved$, and skip all other applicable rules.

(e) For each rule $[_h a \ ]_h^{\alpha_1} \to [_h b]_h^{\alpha_2} [_h c \ ]_h^{\alpha_3}$ in R : if $S.p = \alpha_1$ and $a \in S.M$ then:
   $S.M \leftarrow S.M \setminus \{a\},$
   $S'.M \leftarrow S'.M \cup \{b\}, \quad S'.p \leftarrow \alpha_2,$
   $T.M \leftarrow T.M \cup \{c\}, \quad T.p \leftarrow \alpha_3,$
   and skip all other applicable rules.

Observe that, with the aid of the function $\texttt{State}$, we can uniquely determine the parent and the children (in terms of the membrane structure tree) of a given membrane $h_{i_1 i_2 \ldots i_n}$, without actually storing the membrane structure of Π after the n-th step.

Function $\texttt{Parent}$

Parameters:
$h_{i_1 i_2 \ldots i_n}$ – a membrane whose parent is searched for

1. Let g be the parent membrane of h in the initial membrane structure μ.

2. If $\texttt{State}(g_{1 \ldots 1}, n) = dissolved$ then return $\texttt{Parent}(g_{1 \ldots 1}, n)$, else return $g_{1 \ldots 1}$.

Procedure $\texttt{Contribution\_from\_children}$ calculates the interaction of a membrane with its children membranes in a given step, by:
– sending objects into children membranes by rules of type (b),
– receiving objects from children membranes by rules of type (c),
– receiving contents of children membranes by rules of type (d).
Furthermore, it tests whether the membrane is elementary.

Parameters:
$h_{i_1 \ldots i_n}$ – a membrane to which its children contribute
S – set of objects in h which have not been yet subject to any rule
S' – set of objects which are produced in h as a result of rule applications
$Elementary$ – a logical variable to be set to true if $h_{i_1 \ldots i_n}$ is an elementary membrane

Local variables: X, X', g, *Elementary'*

Initialize *Elementary* $\leftarrow$ *true*. For each children membrane g of h in the initial membrane structure μ, and for each index $j_1 \ldots j_n$ such that $j_i \in \{1, 2\}$, $1 \leq i \leq n$, repeat:

1. $X \leftarrow$ State$(g_{j_1 \ldots j_n}, n)$, $X' \leftarrow (\emptyset, 0)$

2. If $X = nil$ then skip the remaining steps.

3. If $X = dissolved$ then:
   Contribution_from_children$(g_{j_1 \ldots j_n}, S, S', Elementary')$,
   Elementary $\leftarrow$ *Elementary* and *Elementary'*,
   skip the remaining steps.
   /* *If membrane $g_{j_1 \ldots j_n}$ is dissolved, then its children are actually children of $h_{i_1 \ldots i_n}$.* */

4. *Elementary* $\leftarrow$ *false*

5. Contribution_from_children$(g_{j_1 \ldots j_n}, X, X', Elementary')$
   /* *We need this for the case that $g_{j_1 \ldots j_n}$ is dissolved at the n-th step and releases its contents to $h_{i_1 \ldots i_n}$.* */

6. Try_rules_a(g, X, X', X')

7. Try_rules_b(g, X, X', S); if any rule was applied, skip remaining steps.

8. Try_rules_c(g, X, X', S'); if any rule was applied, skip remaining steps.

9. Try_rules_d(g, X, X', S')
   /* *If $g_{j_1 \ldots j_n}$ is dissolved, then its recursively calculated contents (including the contents of eventual lower-level dissolved membranes) is added to S'.* */

Observe that the recursive function State was defined correctly because all its recursive calls during the computation of State$(h_{i_1 i_2 \ldots i_n}, n)$ were of the form State$(g_{j_1 j_2 \ldots j_{n-1}}, n - 1)$, i.e. referring to a state of (another) membrane at the previous step. The same holds for the recursive calls of State in procedures Parent and Contribution_from_children.

Adding the non-elementary membrane division

When the division of non-elementary membranes is allowed, a division may simultaneously take place at various levels of the membrane structure tree. Therefore, the existence of a membrane $h_{i_1 \ldots i_n}$ does *not* depend solely on the existence of its parent membrane $g_{i_1 \ldots i_{(n-1)}}$ as before. Instead, one has to take into the account the potential divisions of all the upper-level membranes containing h. As a consequence, the above described procedures State, Parent, and Contribution_from_children have to be extended substantially. Due to page restriction we give here only a brief description of these extensions.

- A more sophisticated indexing of membranes is used, taking into the account not only a particular step n of the system but also a depth k of the membrane.

The compound index has the form $i_{11} \ldots i_{1n}, i_{21} \ldots i_{2n}, \ldots, i_{k1} \ldots i_{kn}$ and contains implicitly the values of n and k. However, these two parameters determine the computational complexity of the functions State, Parent and Contribution_from_children. Hence, in the following description we use simplified notation of these functions with the parameters n and k only.

- The function State tests at the beginning the existence of the membrane specified by its parameter. If in the previous step any of the parents of this membrane was divided, then some children (together with all their objects and embedded membranes) of that parent were moved into only one of the resulting two copies and some might be copied to both. Hence this existence test results in mutual recursive calls of procedures Parent and State on decreasing levels of the membrane structure tree up to the skin membrane.
- The remaining procedures are in principle similar as in the previous section, with the exception that one must deal with possible application of (f)-type rules. These rules set polarization of membranes in two adjacent levels and change the position of all their membrane subtrees.
- The resulting structure of the recursive calls is the following:
  - State(n, k) calls State$(n-1, k)$, Contribution_from_children$(n-1, k)$, Parent$(n - 1, k)$ and Parent(n, k),
  - Parent(n, k) calls State$(n, k - 1)$, and Parent$(n, k - 1)$,
  - Contribution_from_children(n, k) calls Contribution_from_children $(n, k + 1)$ and State$(n, k + 1)$.

Space complexity of the simulation

A result of any computation of a confluent P system Π with active membranes can be calculated with the aid of function State. Let h_0 be the skin membrane of Π. One can subsequently calculate State(h_0, n), until the object yes is expelled using the rule of type (c), or until the computation halts. Halting can be tested by computing State(h, n) for all the membranes h which could potentially exist after n steps, $n = 1, 2, 3, \ldots$, until no rule can be applied in any of them.

We determine the space complexity of the function State(h, n) which can simulate n steps of a P system $\Pi = (V, H, \mu, w_1, \ldots, w_m, R)$ of size $n^{\mathcal{O}(1)}$. Let

d be the depth of the initial membrane structure tree μ,
$p = \max\{|v|; (a \rightarrow v) \in R\}$,
$q = \mathrm{card}(V)$,
o_n denote the number of objects within the system after n steps. Hence,
$o_0 = |w_1| + \ldots + |w_m|$.

By assumption, the values d, p, q and o_0 are bounded from above by $n^{\mathcal{O}(1)}$. In the rest of the proof we treat them as constants as they are fixed for a given Π. For o_n one can write that, due to possible membrane division and object multiplication,

$$o_n \leq o_0 m (p \, 2^d)^n.$$

The value of o_n is also a sufficient upper bound for the number of objects in a single membrane after n steps. Then the number of bits sufficient to store the contents of an arbitrary membrane is

$$s_n \leq q\lceil \log o_n \rceil \leq q\lceil \log(o_0 m) \rceil + nq(\lceil \log p \rceil + d) = c_0 + c_1 n \tag{5}$$

for positive constants c_0 and c_1 of size $n^{\mathcal{O}(1)}$.

Denote by $S(n,k)$ ($P(n,k)$, $C(n,k)$) the space complexity of the functions State (Parent, Contribution_from_children, respectively) with a parameter $h_{\mathcal{I}}$, where the $\mathcal{I}$ is a compound index referring to n-th computational step of a membrane h located at a level k of the membrane structure tree. These procedures store in local variables contents of the manipulated membranes requiring s_n bits. Another kn bits are needed to store identification (i.e., compound indices) of membranes, plus there are also some other variables of a constant size. The structure of these procedures correspond to the following recurrences:

$$S(0,k) = s_0, \quad 0 \leq k \leq d \tag{6}$$
$$S(n,0) = C(n-1,1) + 4s_n + c, \quad n \geq 1 \tag{7}$$
$$S(n,k) = \max\{P(n,k), P(n-1,k), C(n-1,k), S(n-1,k)\}$$
$$+ 4s_n + kn + c, \quad n \geq 1, \ 1 \leq k \leq d \tag{8}$$

$$C(n,d) = 0 \tag{9}$$
$$C(n,k) \leq \max\{C(n,k+1) + 3s_n + c, S(n,k+1)\} + n(k+1),$$
$$0 \leq k < d, n \geq 0 \tag{10}$$

$$P(n,0) = S(n,0) \tag{11}$$
$$P(n,k) \leq \max\{P(n,k-1), S(n,k-1)\} + n(k-1) + c,$$
$$1 \leq k \leq d, n \geq 0. \tag{12}$$

By expanding (10) to a series for $k, k+1, \ldots, d$ we obtain

$$C(n,k) \leq \max\{S(n,i) \mid k < i \leq d\} + \mathcal{O}(d^2 n + ds_n) \tag{13}$$

for $0 \leq k \leq d$, $n \geq 0$. Let us define

$$S(n) = \max\{S(n,k) \mid 0 \leq k \leq d\}. \tag{14}$$

Observe that one can omit $P(n-1,k)$ in (8) as obviously $P(n-1,k) \leq P(n,k)$. By (13) and (14) we can rewrite (8) in the form

$$S(n,k) \leq \max\{P(n,k), S(n-1) + \mathcal{O}(d^2 n + ds_n)\} + \mathcal{O}(s_n + kn),$$
$$n \geq 1, \ 1 \leq k \leq d \tag{15}$$

By substituting $P(n,i)$, $i = k, k-1, \ldots$ with (12), and $S(n,i)$, $i = k-1, k-2, \ldots$ with (15), we can expand (15) as follows:

$$S(n,k) \leq \max\{P(n,k-1), S(n,k-1), S(n-1) + \mathcal{O}(d^2 n + ds_n)\} + 2\mathcal{O}(s_n + kn)$$
$$\leq \max\{P(n,k-2), S(n,k-2), S(n-1) + \mathcal{O}(d^2 n + ds_n)\} + 3\mathcal{O}(s_n + kn)$$

$$\vdots$$

$$\leq \max\{P(n,0), S(n,0), S(n-1) + \mathcal{O}(d^2n + ds_n)\} + (k+1)\mathcal{O}(s_n + kn)$$
$$\leq \max\{C(n-1,1) + 4s_n + c, S(n-1) + \mathcal{O}(d^2n + ds_n)\} +$$
$$+(k+1)\mathcal{O}(s_n + kn)$$
$$\leq S(n-1) + \mathcal{O}(d^2n + ds_n).$$

The next-to-last step was done by substituting $P(n,0)$ and $S(n,0)$ with (11) and (7), respectively. In the last step we substituted $C(n-1,1)$ with (13) and (14). Therefore, the recurrence (6)–(8) can be rewritten with the aid of (14) as follows:

$$S(0) = s_0$$
$$S(n) \leq S(n-1) + \mathcal{O}(d^2n + ds_n)$$

A solution to this recurrence is $S(n) = \mathcal{O}(d^2n^2 + nds_n)$, and by (5) one gets $S(n) = n^{\mathcal{O}(1)}$. Therefore, a $n^{\mathcal{O}(1)}$ time-bounded computation of Π can be indeed simulated in space $n^{\mathcal{O}(1)}$ which concludes the proof. □

Corollary 1. PMC$_{\mathcal{AM}}$ = PMC$_{\mathcal{AM}}^S$ = PSPACE.

5   Discussion

The results presented in this paper establish a theoretical upper bound on the power of confluent P systems with active membranes. We note that the characterization of power of non-confluent P systems with active membranes remains open. The presented proof cannot be simply adapted to this case by using a non-deterministic Turing machine for simulation. Observe that in our recursive algorithm the same configuration of the P system can be re-calculated many times during one simulation run. If the simulation was non-deterministic, we could obtain different results for the same configuration which would make the simulation non-consistent.

Other variants of P systems with active membranes exist. For example, in [7] the non-elementary membrane division is controlled by rules of the form $[_h a\,]_h^{\alpha_1} \to [_h b\,]_h^{\alpha_2}[_h c\,]_h^{\alpha_3}$. Eventual simulation of this variant of P systems with active membranes would be almost the same as the first part of our proof describing the simplified simulation. Therefore, we claim that Theorem 1 also holds for this variant of membrane systems. Another idea in [7] is the use of *minimal parallelism* (in contrast to maximal parallelism employed here – all objects that can evolve also should evolve). As long as the P system remained confluent, the presented result would remain valid in the case of minimal paralelism.

Finally, one could consider further variants or extension originally designed for other models of P systems. The presented simulation algorithm could be easily adapted to many of them. For example, consider a P system with input computing optimization problems. Such a P system would output a sequence of objects (corresponding to a word or a number). One could also introduce cooperative rules (involving more than one object) or priorities among rules. The number of possible polarizations may be increased from three $(0, +, -)$ to a larger set, other features like promoters, inhibitors, impermeable membranes etc. may be used.

Acknowledgements

Research was supported by the Czech Science Foundation, grants 201/06/0567, 201/04/0528, and by the Programa Ramón y Cajal, Ministerio de Ciencia y Tecnología, Spain. We thank G. Păun and anonymous referees for valuable comments helping to improve the paper.

References

1. Gh. Păun, Computing with Membranes, J. Comput. System Sci., 61 (2000), 108–143.
2. Gh. Păun, Membrane Computing: an Introduction, Springer-Verlag, Berlin, 2002.
3. The P Systems Web Page at http://psystems.disco.unimib.it
4. Gh. Păun, P systems with active membranes: attacking NP complete problems, J. Automata, Languages and Combinatorics, 6, 1 (2001), 75–90.
5. A. Alhazov, R. Freund, A. Riscos-Núñez, One and two polarizations, membrane creation and objects complexity in P systems, in: G. Ciobanu, Gh. Păun (Eds.), First Int. Workshop on Theory and Application of P Systems (TAPS), Timişoara, Romania, 2005, pp. 9–18.
6. A. Alhazov, C. Martin-Vide, L. Pan, Solving a PSPACE-complete problem by P systems with restricted active membranes, Fundamenta Informaticae, 58, 2 (2003), 67–77.
7. G. Ciobanu, L. Pan, G. Păun, M.J. Pérez-Jiménez, P Systems with Minimal Parallelism, submitted.
8. M.A. Gutiérrez-Naranjo, M.J. Pérez-Jiménez, A. Riscos-Núñez, F.J. Romero-Campero, On the power of dissolution in P systems with active membranes, in: R. Freund et al. (Eds.), Membrane Computing, 6th International Workshop, WMC 2005, Berlin, Springer-Verlag, LNCS 3850 (2006), pp. 224–240.
9. M.A. Gutiérrez-Naranjo, M.J. Pérez-Jiménez, A. Riscos-Núñez, F.J. Romero-Campero, Characterizing standard tractability by cell-like membrane systems, in: K.G. Subramanian (Ed.), Formal Models, Languages and Applications, World Scientific, 2006, in press.
10. M.J. Pérez-Jiménez, A.R. Jiménez, F. Sancho-Caparrini, Complexity classes in models of cellular computing with membranes, Natural Computing, 2 (2003), 265–285 .
11. P. Sosík, The computational power of cell division: beating down parallel computers? Natural Computing, 2–3 (2003), 287–298.
12. C. Zandron, C. Ferretti, G. Mauri, Solving NP-complete problems using P systems with active membranes, in: I. Antoniou, C.S. Calude, M.J. Dinneen (Eds.), Unconventional Models of Computation, Springer-Verlag, London, 2000, pp. 289–301.
13. J.L. Balcazar, J. Diaz, J. Gabarro, Structural Complexity II, Springer-Verlag, Berlin, 1991.

14. D. Beaver, A universal molecular computer, in: R.J. Lipton, E.B. Baum (Eds.), DNA Based Computers, vol. 27 of DIMACS: Series in Discrete Mathematics and Theoretical Computer Science, American Mathematical Society, 1995, pp. 29–36.

15. E. Dantsin, A. Wolpert, A robust DNA computation model that captures PSPACE. Int. J. Foundations Comp. Sci., 14(5) (2003), 933–951.

16. P. Pudlák, Complexity theory and genetics: The computational power of crossing-over. Information and Computation, 171 (2001), 201–223.

All NP-Problems Can Be Solved in Polynomial Time by Accepting Networks of Splicing Processors of Constant Size

Florin Manea[1], Carlos Martín-Vide[2], and Victor Mitrana[1,2]

[1] Faculty of Mathematics and Computer Science, University of Bucharest
Str. Academiei 14, 70109, Bucharest, Romania
`flmanea@funinf.cs.unibuc.ro`
[2] Research Group in Mathematical Linguistics, Rovira i Virgili University
Pça. Imperial Tarraco 1, 43005, Tarragona, Spain
{`carlos.martin, vmi`}`@urv.cat`

Abstract. In this paper, we present two new results regarding ANSPs. The first one states that every recursively enumerable language can be accepted by an ANSP of size 7 out of which 6 do not depend on the given language. Then we propose a method for constructing, given an **NP**-language, an ANSP of size 7 accepting that language in polynomial time. Unlike the previous case, all nodes of this ANSP depend on the given language. Since each ANSP may be viewed as a problem solver as shown in [6], the later result may be interpreted as a method for solving every **NP**-problem in polynomial time by an ANSP of size 7.

1 Introduction

The origin of networks of evolutionary processors (NEP for short) is a basic architecture for parallel and distributed symbolic processing, related to the Connection Machine [5] as well as the Logic Flow paradigm [2], which consists of several processors, each of them being placed in a node of a virtual complete graph, which are able to handle data associated with the respective node. Each node processor acts on the local data in accordance with some predefined rules, and then local data becomes a mobile agent which can navigate in the network following a given protocol. Only such data can be communicated which can pass a filtering process. This filtering process may require to satisfy some conditions imposed by the sending processor, by the receiving processor or by both of them. All the nodes send simultaneously their data and the receiving nodes handle also simultaneously all the arriving messages, according to some strategies, see, e.g., [3, 5].

In a series of papers (see [7] for a survey) we considered that each node may be viewed as a cell having genetic information encoded in DNA sequences which may evolve by local evolutionary events, that is point mutations. Each node is specialized just for one of these evolutionary operations. Furthermore, the data in each node is organized in the form of multisets of words (each word appears in an arbitrarily large number of copies), and all the copies are processed in

C. Mao and T. Yokomori (Eds.): DNA12, LNCS 4287, pp. 47–57, 2006.
© Springer-Verlag Berlin Heidelberg 2006

parallel such that all the possible events that can take place do actually take place. Obviously, the computational process just described is not exactly an evolutionary process in the Darwinian sense. But the rewriting operations we have considered might be interpreted as mutations and the filtering process might be viewed as a selection process. Recombination is missing but it was asserted that evolutionary and functional relationships between genes can be captured by taking only local mutations into consideration [10].

In [6] we replaced the point mutations associated with each node by the missing operation mentioned above, that of splicing. This new processor was called splicing processor. This computing model, called accepting network of splicing processors (shortly ANSP), is similar to some extent to the test tube distributed systems based on splicing introduced in [1] and further explored in [9]. However, there are several differences: first, the model proposed in [1] is a language generating mechanism while ours is an accepting one; second, we use a single splicing step, while every splicing step in [1] is actually an infinite process consisting of iterated splicing steps; third, each splicing step in our model is reflexive; fourth, the filters of our model are based on random context conditions while those considered in [1] are based on membership conditions; fifth, at every splicing step a set of auxiliary words, always the same and proper to every node, is available for splicing. We want to stress from the very beginning that the splicing processor we discuss here is a mathematical object only and the biological hints presented above are intended to explain in an informal way how some biological phenomena are *sources of inspiration* for our mathematical computing model.

In [6] we presented a characterization of the complexity class **NP** based on ANSP and discussed how these networks can be considered as problem solvers. Along these lines we presented a uniform linear time solution to SAT. In this paper, we present two new results regarding ANSPs:

1. Every recursively enumerable language can be accepted by an ANSP of size 7. It has 6 fixed nodes which do not depend on the given language and one node only, the input one, which depends on the given language.

2. A method for constructing, given an **NP**-language, an ANSP of size 7 accepting that language in polynomial time. Unlike the previous case, all nodes of this ANSP depend on the given language. Since each ANSP may be viewed as a problem solver as shown in [6], the later result may be interpreted as a method for solving every **NP**-problem in polynomial time by an ANSP of size 7. The fact that all **NP**-problems can be solved in polynomial time by ANSPs actually follows from the linear solution to SAT presented in [6], but the novelty here is given by the constant size of the networks solving these problems.

2   Basic Definitions

We start by summarizing the notions used throughout the paper. An *alphabet* is a finite and nonempty set of symbols. The cardinality of a finite set A is written $card(A)$. Any finite sequence of symbols from an alphabet V is called *word* over V.

The set of all words over V is denoted by V^* and the empty word is denoted by ε. The length of a word x is denoted by $|x|$ while $alph(x)$ denotes the minimal alphabet W such that $x \in W^*$.

A *splicing rule* over the alphabet V is a quadruple written in the form $\sigma = [(x, y); (u, v)]$, where x, y, u, v are words over V. Given a splicing rule σ over V as above and a pair of words (w, z) over the same alphabet V we define the action of σ on (w, z) by:

$$\sigma(w, z) = \begin{cases} \{t \mid w = \alpha x y \beta, z = \gamma u v \delta \text{ for some words } \alpha, \beta, \gamma, \delta \text{ over } V \\ \qquad \text{and } t = \alpha x v \delta \text{ or } t = \gamma u y \beta\} \\ \{w\} \cup \{z\}, \text{ if the set above is empty.} \end{cases}$$

This action on pair of words can be naturally extended to a language L by $\sigma(L) = \bigcup\limits_{w, z \in L} \sigma(w, z)$. Furthermore, if M is a finite set of splicing rules over V, then we set $M(L) = \bigcup\limits_{\sigma \in M} \sigma(L)$.

For two disjoint subsets P and F of an alphabet V and a word x over V, we define the predicates

$$\varphi^s(x; P, F) \equiv P \subseteq alph(x) \qquad \wedge \quad F \cap alph(x) = \emptyset$$
$$\varphi^w(x; P, F) \equiv alph(x) \cap P \neq \emptyset \wedge \ F \cap alph(x) = \emptyset.$$

The construction of these predicates is based on *random-context conditions* defined by the two sets P (*permitting contexts/symbols*) and F (*forbidding contexts/symbols*). Informally, the former condition requires (s stands for strong) that all permitting symbols are present and no forbidding symbol is present in x, while the latter (w stands for weak) is a weaker variant such that at least one permitting symbol appears in x but still no forbidding symbol is present in x.

For every language $L \subseteq V^*$ and $\beta \in \{s, w\}$, we define:

$$\varphi^\beta(L, P, F) = \{x \in L \mid \varphi^\beta(x; P, F)\}.$$

A *splicing processor* over V is a 6-tuple (S, A, PI, FI, PO, FO), where:

– S is a finite set of splicing rules over V.

– A is a finite set of *auxiliary words* over V. These auxiliary words are to be used by this splicing processor for splicing.

– $PI, FI \subseteq V$ are the *input* permitting/forbidding contexts of the processor, while $PO, FO \subseteq V$ are the *output* permitting/forbidding contexts of the processor (with $PI \cap FI = \emptyset$ and $PO \cap FO = \emptyset$).

We denote the set of splicing processors over V by SP_V.

An *accepting network of splicing processors* (ANSP for short) is a 9-tuple $\Gamma = (V, U, \langle, \rangle, G, \mathcal{N}, \alpha, x_I, x_O)$, where:

• V and U are the input and network alphabet, respectively, $V \subseteq U$, and, also, $\langle, \rangle \in U \setminus V$ are two special symbols.

- $G = (X_G, E_G)$ is an undirected graph without loops with the set of nodes X_G and the set of edges E_G. Each edge is given in the form of a binary set. G is called the *underlying graph* of the network.
- $\mathcal{N} : X_G \longrightarrow SP_U$ is a mapping which associates with each node $x \in X_G$ the splicing processor $\mathcal{N}(x) = (S_x, A_x, PI_x, FI_x, PO_x, FO_x)$.
- $\alpha : X_G \longrightarrow \{s, w\}$ defines the type of the *input/output filters* of a node. More precisely, for every node, $x \in X_G$, the following filters are defined:

$$\text{input filter: } \rho_x(\cdot) = \varphi^{\alpha(x)}(\cdot; PI_x, FI_x),$$
$$\text{output filter: } \tau_x(\cdot) = \varphi^{\alpha(x)}(\cdot; PO_x, FO_x).$$

That is, $\rho_x(z)$ (resp. τ_x) indicates whether or not the word z can pass the input (resp. output) filter of x. More generally, $\rho_x(L)$ (resp. $\tau_x(L)$) is the set of words of L that can pass the input (resp. output) filter of x.

- $x_I, x_O \in X_G$ are the *input* and the *output* node of Γ, respectively.

We say that $card(X_G)$ is the size of Γ. We focus here on *complete* ANSPs, i.e., ANSPs having a complete underlying graph (every two nodes are connected) denoted by K_n, where n is the number of nodes.

A *configuration* of a ANSP Γ as above is a mapping $C : X_G \longrightarrow 2^{U^*}$ which associates a set of words with every node of the graph. A configuration may be understood as the sets of words which are present in any node at a given moment. Given a word $z \in V^*$, the initial configuration of Γ on z is defined by $C_0^{(z)}(x_I) = \{\langle z \rangle\}$ and $C_0^{(z)}(x) = \emptyset$ for all $x \in X_G \setminus \{x_I\}$. Notice that the auxiliary words do not appear in any configuration.

A configuration can change either by a *splicing step* or by a *communication step*. When changing by a splicing step, each component $C(x)$ of the configuration C is changed in accordance with the set of splicing rules M_x associated with the node x and the set A_x. Formally, we say that the configuration C' is obtained in *one splicing step* from the configuration C, written as $C \Longrightarrow C'$, iff

$$C'(x) = S_x(C(x) \cup A_x) \text{ for all } x \in X_G.$$

Since each word present in a node, as well as each auxiliary word, appears in an arbitrarily large number of identical copies, all possible splicings are assumed to be done in one splicing step.

When changing by a communication step, each node processor $x \in X_G$ sends one copy of each word it has, which is able to pass the output filter of x, to all the node processors connected to x and receives all the words sent by any node processor connected with x providing that they can pass its input filter.

Formally, we say that the configuration C' is obtained in *one communication step* from configuration C, written as $C \vdash C'$, iff

$$C'(x) = (C(x) \setminus \tau_x(C(x))) \cup \bigcup_{\{x,y\} \in E_G} (\tau_y(C(y)) \cap \rho_x(C(y))) \text{ for all } x \in X_G.$$

Notice that, according to this definition, all the words that can go out of a node go out even if they cannot pass any input filter (so they are "lost").

Let Γ be an ANSP, the computation of Γ on the input word $z \in V^*$ is a sequence of configurations $C_0^{(z)}, C_1^{(z)}, C_2^{(z)}, \ldots$, where $C_0^{(z)}$ is the initial configuration of Γ on z, $C_{2i}^{(z)} \implies C_{2i+1}^{(z)}$ and $C_{2i+1}^{(z)} \vdash C_{2i+2}^{(z)}$, for all $i \geq 0$. By the previous definitions, each configuration $C_i^{(z)}$ is uniquely determined by the configuration $C_{i-1}^{(z)}$. In other words, each computation in an ANSP is deterministic. A computation *halts* (and it is said to be *finite*) if one of the following two conditions holds:

(i) There exists a configuration in which the set of words existing the output node x_O is non-empty. In this case, the computation is said to be an *accepting computation*.
(ii) There exist two consecutive identical configurations.

The *language accepted* by Γ is

$$L_a(\Gamma) = \{z \in V^* \mid \text{the computation of } \Gamma \text{ on } z \text{ is an accepting one}\}.$$

We say that an ANSP Γ decides the language $L \subseteq V^*$, and write $L(\Gamma) = L$ iff $L_a(\Gamma) = L$ and the computation of Γ on every $z \in V^*$ halts.

3   Encoding Complete ANSPs

In this section we describe a straightforward way of encoding an arbitrary complete ANSP using the fixed alphabet:

$$A = \{\$, \#, *, s, w, 0, 1, \spadesuit, \bullet\}.$$

Let $\Gamma = (V, U, \langle, \rangle, G, \mathcal{N}, \alpha, x_I, x_O)$ be an ANSP, where
  – $V = \{a_1, a_2, \ldots, a_m\}$, and $U = \{a_1, a_2, \ldots, a_p\}$, $p \geq m$,
  – the nodes of K_n are $x_1, x_2, \ldots, x_n$, with $x_1 = x_I$ and $x_2 = x_O$.
We encode every symbol a_i of U, and denote this encoding by $code(a_i)$, in the following way:

$$code(a_i) = \begin{cases} 10^i, 1 \leq i \leq m \\ \spadesuit 0^i, m+1 \leq i \leq p \end{cases}$$

We assume that $code(\langle) = \spadesuit 0^{m+1}$ and $code(\rangle) = \spadesuit 0^{m+2}$. Given z, a word over U as above, we define its encoding $code(z)$ as follows:

$$code(\varepsilon) = 1, code(b_1 b_2 \ldots b_k) = code(b_1) code(b_2) \ldots code(b_k),$$

$k \geq 1, b_i \in U, 1 \leq i \leq k$. Let $L \subseteq U^*$ be a finite language, $L = \{z_1, \ldots, z_k\}$. We encode this language by the word $code(L) = \bullet code(z_1) \bullet code(z_2) \bullet \ldots \bullet code(z_k) \bullet$. The empty language is encoded by $code(\emptyset) = \bullet$.

As a direct consequence of the above considerations, the splicing rule $r = [(x, y); (u, v)]$ is encoded as: $code(r) = *code(x) * code(y) * code(u) * code(v)*$. A set of splicing rules $R = \{r_1, \ldots, r_m\}$ is encoded:

$$code(R) = \bullet code(r_1) \bullet code(r_2) \bullet \ldots \bullet code(r_m) \bullet.$$

For each node x, $code(\mathcal{N}(x))$ is

$$\#code(S_x)\#code(A_x)\#code(PI_x)\#code(FI_x)\#code(PO_x)\#code(FO_x)\#,$$

and $code(x) = \#code(\mathcal{N}(x))\alpha(x)\#$. We now describe the way Γ is encoded. This is:

$$code(\Gamma) = \$code(K_n)\$code(x_1)\$code(x_2)\$\ldots\$code(x_n)\$,$$

where $code(K_n) = \spadesuit^n$.

Note that a similar encoding of an ANSP can be accomplish using the binary alphabet $A = \{0, 1\}$ only. However, in order to make the further exposure more readable we do not use this binary encoding, though the results we present in the sequel can be easily carried over this setting.

4 How to Decide All Recursively Enumerable Languages with ANSPs of Constant Size

In this section we prove that there exists an ANSP Γ_U, such that if the input word of Γ_U is $code(\Gamma)code(z)$, for some ANSP Γ and word z the followings hold:

- Γ_U halts on the input $code(\Gamma)code(z)$ if and only if Γ halts on the input z.
- $code(\Gamma)code(z)$ is accepted by Γ_U if and only if z is accepted by Γ.

The first step of this construction is to define a Turing machine that behaves as described in the next theorem.

Theorem 1. *There exists a Turing machine T_U, with the input alphabet A, satisfying the following conditions on any input $code(\Gamma)code(z)$, where $\Gamma = (V, U, \langle, \rangle, G, \mathcal{N}, \alpha, x_I, x_O)$ is an arbitrary ANSP and z is a word over the input alphabet of Γ:*

(i) T_U halts on the input $code(\Gamma)code(z)$ if and only if Γ halts on the input z.

(ii) $code(\Gamma)code(z)$ is accepted by T_U if and only if z is accepted by Γ.

Proof. We briefly describe the way T_U is obtained. Let T_U' be a 4-tapes Turing Machine, with the tapes labeled W, X, Y, Z. The algorithm that this machine implements is the following:

1. On tape W it is found the input word: $code(\Gamma)code(z)$. We assume that $\Gamma = (V, U, \langle, \rangle, G, \mathcal{N}, \alpha, x_I, x_O)$, the nodes of K_n being $x_1, x_2, \ldots, x_n$, with $x_1 = x_I$ and $x_2 = x_O$. We copy on tape X the encoding of the graph K_n by copying the part of $code(\Gamma)$ between the first and the second occurrence of $\$$. Each symbol $\spadesuit$ on this tape will be used to keep track of the node that is processed at a given moment. On tape Y we construct the initial configuration of Γ. Then, our strategy is the following one.

2. The encoding of the configuration of the i-th node will be memorized between the i-th and $(i+1)$-th occurrences of symbol $\$$ on tape Y.

3. Tapes Y and Z, the latter containing initially $n + 1$ symbols $\$$, will be used in the simulation of both splicing and communication steps which run alternatively one after another each being preceded by the the following acceptance

phase: If the configuration associated with x_O after a splicing or communication step is not empty, then the computation stops and our machine accepts the input word. Otherwise, if, before a splicing or communication step, the words from tapes Y and Z are identical, then the computation also stops, but the input word is not accepted.

Clearly, this strategy can be formally implemented by a Turing machine. We obtain that T'_U implements the desired behavior. From a classical result, it follows that there exist a 1-tape Turing Machine, T_U, with the same behavior as T'_U. This concludes the proof of the theorem. □

The final step of the construction of a universal ANSP is based on the following theorem:

Theorem 2. [6] *For any recursively enumerable language L, accepted by the Turing machine $M = (Q, V, U, \delta, q_0, B, F)$, there exists an ANSP Γ such that $L(\Gamma) = L$ and $size(\Gamma) = 2card(U) + 2$.*

From the proof of Theorem 2, as presented in [6], it follows that the ANSP Γ halts on exactly the same input words as M does. Consequently, we can construct an ANSP Γ_U, that implements the same behavior as T_U, which is the universal ANSP. Therefore, we have shown:

Theorem 3. *There exists an ANSP Γ_U, with the input alphabet A, satisfying the following conditions on any input $code(\Gamma)code(z)$, where Γ is an arbitrary ANSP and z is a word over the input alphabet of Γ:*

(i) Γ_U halts on the input $code(\Gamma)code(z)$ if and only if Γ halts on the input z.

(ii) $code(\Gamma)code(z)$ is accepted by Γ_U if and only if z is accepted by Γ. Moreover, $size(\Gamma_U) = 2card(A) + 2$.

Recall that A can be reduced to a binary alphabet, hence the size of Γ_U can be reduced to 6.

We can now prove the main result of this section:

Theorem 4. *For every recursively enumerable language L there exists an ANSP Γ such that $L(\Gamma) = L$ and $size(\Gamma) = 2card(A) + 3$.*

Proof. Let $\Gamma_U = (A, W_U, \langle, \rangle, K_{2card(A)+2}, \mathcal{N}, \alpha_U, x_I^{(U)}, x_O^{(U)})$. We may assume without loss of generality that the language L is over an alphabet V, which has no common symbol with the working alphabet of the universal ANSP Γ_U, namely W_U. Since L is recursively enumerable, due to Theorem 2, there exists an ANSP Π such that $L(\Pi) = L$. The following statements are immediate: $z \in L$ iff $z \in L(\Pi)$, hence, $z \in L$ iff $code(\Pi)code(z) \in L(\Gamma_U)$. From these remarks we can proceed with the construction of the ANSP

$$\Gamma = (V, W, \langle, \rangle, K_{2card(A)+3}, \mathcal{N}, \alpha, In, x_O^{(U)}).$$

as follows. Let $V = \{a_1, a_2, \ldots, a_n\}$. We first define the working alphabet of Γ: $W = V \cup W_U \cup \{\#', \#''\}$.

The nodes of the network Γ consist of the input node In and the nodes of the universal ANSP Γ_U. Hence, we should only give the definitions for the node In:

- $S(In) = \{[(\langle, a), (\langle code(\Pi), \#')] \mid a \in V \cup \{\rangle\}\} \cup$
  $\{[(code(\Pi), a), (\#'', code(a)\#')] \mid a \in V\} \cup$
  $\{[(code(b), a), (\#'', code(a)\#')] \mid b, a \in V\} \cup$
  $\{[(d\ code(a), \#')(\#''a, b)] \mid a \in V, b \in V \cup \{\rangle\}\}, d \in W_U\},$
- $A(In) = \{\langle code(\Pi)\#'\} \cup \{\#''code(a)\#' \mid a \in V\},$
- $\begin{cases} PI(In) = \emptyset & FI(In) = W_U \\ PO(In) = A \cup \{\langle,\rangle\} & FO(In) = V \cup \{\#', \#''\} \end{cases}$
- $\alpha(In) = w.$

For the rest of the nodes x of the network, those of Γ_U, we assume that $\alpha(x) = \alpha_U(x)$. We modify only the input filters of the node $x_I^{(U)}$ of Γ_U by allowing to enter the words containing only the symbols in $A \cup \{\langle,\rangle\}$. Note that the output node of Γ is the output node of Γ_U. We mention also that, due to the form of the input filters of the nodes of Γ_U, any word leaving In can enter only $x_I^{(U)}$.

In the following we describe the way Γ works. Let us assume that $z = z_1 \ldots z_k$ is the input word. Hence, at the beginning of the computation, the only not-void configuration will be that associated with the node In: $C_0^{(z)}(In) = \{\langle z\rangle\}$.
It can be shown by induction on i that:

(i) After $2i + 1, 0 \leq i \leq k$, splicing steps all the sets associated with nodes different from In will be empty, and:
$C_{2i+1}^{(z)}(In) = \{\langle code(\Pi)code(z_1) \ldots code(z_i)z_{i+1} \ldots z_k\rangle\} \cup \{\#''z_j\#' \mid 0 \leq j \leq i\}$
$\cup$
$\{\langle\#'\} \cup \{\#''code(a)\#' \mid a \in V\} \cup \{\langle code(\Pi)\#'\}.$
No word from this set can leave the node In until $i = k$.

(ii) After $2i, 1 \leq i \leq k$, splicing steps all the sets associate with nodes different from In will be empty, and:
$C_{2i}^{(z)}(In) = \{\langle code(\Pi)code(z_1) \ldots code(z_i)\#'\} \cup \{\#''z_j\#' \mid 1 \leq j < i\} \cup \{\langle\#'\}$
$\cup$
$\{\#''code(a)\#' \mid a \in V \setminus \{x_i\}\} \cup \{\langle code(\Pi)\#'\}.$
No word from this set can leave the node In.

After $2k + 1$ splicing steps, only $\langle code(\Pi)code(z_1) \ldots code(z_k)\rangle$ will leave In in the next communication step entering $x_I^{(U)}$, while the other words in In will remain unchanged in this node until the end of computation. No other words that might leave In will ever appear in this node, since no further word can enter this node. Hence, this node will not affect the rest of computation which consists in the steps performed by the universal ANSP Γ_U on the input $\langle code(\Pi)code(z)\rangle$.

From the considerations above it follows that the word z is accepted by Γ if, and only if, $z \in L(\Pi) = L$.   □

We make several remarks regarding the construction above. Only the rules in the node In depend on the language L, and the encoding that we use for its symbols. The parameters of the other nodes do not depend in any way on the language L, on the encoding of the ANSP Π, or on the symbols of V. Finally, by the considerations on the cardinality of A in Section 3, the size of the ANSP Γ proposed in the proof of the last theorem can be decreased to 7.

5   How to Decide NP-Languages in Polynomial Time with ANSPs of Constant Size

In this section we approach the problem of finding a constant-size ANSP for every recursively enumerable language in a different manner: we try to find an architecture that is both descriptive and computational efficient for a given language, even if this means to obtain a construction in which every node depends effectively on the language that it accepts. We recall from [6] the following definitions. We consider an ANSP Γ with the input alphabet V that halts on every input. The *time complexity* of the finite computation $C_0^{(x)}$, $C_1^{(x)}$, $C_2^{(x)}$, ... $C_m^{(x)}$ of Γ on $x \in V^*$ is denoted by $Time_\Gamma(x)$ and equals m. The time complexity of Γ is the partial function from $I\!N$ to $I\!N$:

$$Time_\Gamma(n) = \max\{Time_\Gamma(x) \mid x \in V^*, |x| = n\}.$$

Theorem 5. *For every language L in* **NP***, there exists an ANSP of size 7 that decides L in polynomial time.*

Proof. Assume that we are given a language L, recognized in polynomial time by a non-deterministic Turing machine $M = (Q, V, U, \delta, q_0, B, F)$. Let $U = \{x_1, \ldots, x_k\}$; we construct the morphism $h : U \to \{0,1\}^*$ defined by $h(x_i) = 0^{i-1}10^{k-i}$. It is not hard to construct a Turing machine M', with working alphabet $\{0,1\}$, that accepts $h(L)$ in polynomial time. From theorem 2 it follows that there exists an ANSP $\Gamma_L = (\{0,1\}, \{0,1\}, \langle, \rangle, K_6, \mathcal{N}, \alpha, x_I, x_O)$ that accepts $h(L)$ in polynomial ANSP-time.

We construct $\Gamma = (V, \{0,1\}, \langle, \rangle, K_7, \alpha, In, x_O)$, such that the nodes of Γ are the nodes of Γ_L and the new input node In. The processors in the nodes of Γ_L are not modified, except for the node x_I, whose input filters are modified in order to permit the access of the words that have only the symbols: $\{\langle, \rangle, 0, 1\}$. We should only define the processor placed in the node In:

- $S(In) = \{[(\langle, a), (\#'', h(a)\#')] \mid a \in V\} \cup \{[(h(b), a), (\#'', h(a)\#')] \mid b, a \in V\} \cup \{[(c\, h(a), \#')(\#''a, b)] \mid a \in V, b \in V \cup \{\rangle\}\}, c \in \{0, 1, \langle\}\}$
- $A(In) = \{\#''h(a)\#' \mid a \in V\}$
- $\begin{cases} PI(In) = \emptyset \\ PO(In) = \{0, 1, \langle, \rangle\} \end{cases}$ $\quad\quad FI(In) = W_U$ $\quad\quad FO(In) = V \cup \{\#', \#''\}$
- $\alpha(In) = w.$

Note that, due to the form of the input filters of the nodes of Γ_U, adjusted as described above, any word leaving In can enter only x_I.

In the following we describe the way Γ works. Let us assume that $z = z_1 \ldots z_k$ is the input word. It can be proved by induction on i that:

(i) After $2i - 1, 1 \leq i \leq k$, splicing steps all the sets associate with nodes different from In will be empty, and:
$C_{2i-1}^{(z)}(In) = \{\langle h(z_1) \ldots h(z_i)\#', \#''z_i \ldots z_k\rangle\} \cup \{\#''z_j\#' \mid 1 \leq j < i\} \cup \{\#''h(a)\#' \mid a \in V \setminus \{z_i\}\}.$

No word from this set can leave the node In in any of the communication steps, consequently, all the sets associated with other nodes remain empty.

(ii) After $2i, 1 \leq i \leq k$, splicing steps all the sets associated with nodes different from In will be empty, and:
$$C_{2i}^{(z)}(In) = \{\langle h(z_1) \ldots h(z_i) z_{i+1} \ldots z_k \rangle\} \cup \{\#'' z_j \#' \mid 1 \leq j \leq i\} \cup \{\#'' h(a) \#' \mid a \in V \setminus \{z_i\}\}.$$
No word from this set can leave the node In in any of the communication steps, until $i = k$, when only $\langle h(z_1) \ldots h(z_k) \rangle$ will leave In in the next communication step and enter x_I, while the other words will remain unchanged in this node until the end of computation. No other words that can leave In will ever appear in this node, since no word can enter this node. Hence, this node will not affect the further computation. The rest of the computation consists on the steps performed by the ANSP Γ_L on the input $\langle h(z) \rangle$. From the considerations above it is straightforward that the word z is accepted by Γ if, and only if, $h(z) \in L(\Gamma_L) = h(L)$, which is equivalent to $z \in L$.

We assume now that $Time_{\Gamma_L}(n) \leq P(n)$, for all $n \in I\!N$, where P is a polynomial function. Such an assumption holds as $L \in \mathbf{NP}$, and $\mathbf{NP}$ equals the class of all languages decided by ANSPs in polynomial time [6]. According to the facts already presented, it is clear that the number of steps performed by the ANSP Γ on an input word z is bounded by $4|z| + P(|z|)$, hence it is polynomial in the length of z. This concludes the proof.   □

We discuss briefly and informally how ANSPs could be used as problem solvers. A possible correspondence between decision problems and languages can be done via an encoding function which transforms an instance of a given decision problem into a word, see, e.g., [4]. We say that a decision problem P is solved in time $O(f(n))$ by ANSPs if there exists a family $\mathcal{A}$ of ANSPs such that the following conditions are satisfied:

1. The encoding function of any instance p of P having size n can be computed by a deterministic Turing machine in time $O(f(n))$.
2. For each instance p of size n of the problem one can effectively construct, in time $O(f(n))$, an ANSP $\Gamma(p) \in \mathcal{A}$ which decides, again in time $O(f(n))$, the word encoding the given instance. This means that the word is accepted if and only if the solution to the given instance of the problem is "YES". This effective construction is called an $O(f(n))$ time solution to the considered problem.

If an ANSP $\Gamma \in \mathcal{A}$, constructed as above, decides the language of words encoding all instances of the same size n, then the construction of $\mathcal{A}$ is called a uniform solution. Intuitively, a solution is uniform if for problem size n, we can construct a unique ANSP solving all instances of size n taking the (reasonable) encoding of instance as "input". This view of ANSPs as problem solvers is consistent for, as shown in [6], each language decided by a Turing machine (deterministic or not) in time $O(f(n))$ is decided by an ANSP in time $O(f(n))$. Consequently, the last result may be interpreted as a method for solving every $\mathbf{NP}$-problem in polynomial time by an ANSP of size 7.

6   Final Remarks

We briefly discuss here a few stimulating problems, in our view, that remained unsolved in this work. First, we consider that further research might be aimed at investigating whether the constant size we proposed here is optimal. Second, the problem of minimal number of nodes of an ANSP deciding an NP-language that necessarily depend on the language appears rather difficult and quite attractive to us. It might be the case that if one allowed a bigger size than 7, then some nodes would be independent of the language, that is we may have a sort of "trade-off" between the size and the independent nodes.

References

[1] E. Csuhaj-Varjú, L. Kari, G. Păun, *Test tube distributed systems based on splicing*, Computers and AI, **15** (1996), 211–232.

[2] L. Errico, C. Jesshope, *Towards a new architecture for symbolic processing*, in *Artificial Intelligence and Information-Control Systems of Robots '94*, World Sci. Publ., Singapore, 31–40, 1994.

[3] S. E. Fahlman, G. E. Hinton, T. J. Seijnowski, *Massively parallel architectures for AI: NETL, THISTLE and Boltzmann machines*, in *Proc. AAAI National Conf. on AI*, William Kaufman, Los Altos, 109–113, 1983.

[4] M. Garey, D. Johnson, *Computers and Intractability. A Guide to the Theory of NP-completeness*, Freeman, San Francisco, CA, 1979.

[5] W. D. Hillis, *The Connection Machine*, MIT Press, Cambridge, 1985.

[6] F. Manea, C. Martín-Vide, V. Mitrana, Accepting networks of splicing processors: complexity results, *Theoretical Computer Science*, to appear.

[7] C. Martín-Vide, V. Mitrana, Networks of evolutionary processors: results and perspectives, chapter in *Molecular Computational Models: Unconventional Approaches*, Idea Group Publishing, Hershey, 78–114, 2005.

[8] G. Păun, G. Rozenberg, A. Salomaa, *DNA Computing. New Computing Paradigms*, Springer-Verlag, Berlin, 1998.

[9] G. Păun, *Distributed architectures in DNA computing based on splicing: Limiting the size of components*, in *Unconventional Models of Computation*, Springer-Verlag, Berlin, 323–335, 1998.

[10] D. Sankoff et al. *Gene order comparisons for phylogenetic inference:Evolution of the mitochondrial genome*, Proc. Natl. Acad. Sci. USA, **89** (1992), 6575–6579.

Length-Separating Test Tube Systems

Erzsébet Csuhaj-Varjú[1] and Sergey Verlan[2]

[1] Computer and Automation Research Institute,
Hungarian Academy of Sciences
H-1111 Budapest, Kende u. 13-17.
[2] Department of Algorithms and Their Applications,
Faculty of Informatics, Eötvös Loránd University,
H-1117, Budapest, Pázmány Péter sétány 1/c.
csuhaj@sztaki.hu
[3] LACL, Département Informatique, Université Paris 12,
61, av. Général de Gaulle, 94010 Créteil, France
verlan@univ-paris12.fr

Abstract. In this article we propose a formalization of protocols simulating the separation of molecules by gel electrophoresis. In our model, we introduce a new concept, namely, filtering by length – a direct formalization of the gel electrophoresis action. We also define a distributed computational model based on this action and on the splicing operation, called length-separating splicing test tube systems. We prove that these constructs, even with restricted size parameters, can simulate the Turing machines. We also discuss different natural restrictions and generalizations of the model which may be used to find efficient ways to realize DNA transformations in the laboratory.

1 Introduction

Gel electrophoresis is a technique for separation of molecules which is widely used in the laboratory. It is usually performed for analytical purposes at the final stage of the experiment. However, gel electrophoresis can be used for extracting molecules having a particular length from the solution. This property makes it attractive to use for transformation of molecules. For example, a molecule may be cut off by a restriction enzyme and after that the gel may be used to find one of the parts that was cut off. This part may be recuperated and ligated in another test tube with another molecule. Finally, the transformed molecule may be extracted from a new gel.

In a more general sense, the above property can be used for designing a computational model where gel electrophoresis is used as a filtering device which provides the possibility to perform branching in the computation. The model can also be considered as a transducer, *i.e.*, receiving an amount of DNA molecules as input, it will transform these molecules in a desired way. Thus, it would be theoretically possible to design experiments that will do necessary transformations in laboratory.

C. Mao and T. Yokomori (Eds.): DNA12, LNCS 4287, pp. 58–70, 2006.

In this article, we investigate a formalization of protocols based on operations simulating the gel electrophoresis filtering. In an informal way, our model corresponds to the following experiment. Let us suppose that there is a set of test tubes. Each of these test tubes may transform DNA molecules (cut, ligate, multiply etc). The tubes are selective and they can do their transformations only on specific molecules (for example, in a tube DNA molecules may be cut with a specific enzyme, hence only molecules having a corresponding site will be modified). Taking a tube, we may put some amount of DNA molecules into it. After the transformation, all molecules from a tube are put in a gel electrophoresis. After the separation, the gel is cut at some points corresponding to some molecular lengths. Hence, molecules will be grouped by some length intervals. After that, molecules are extracted from the gel and distributed among other test tubes depending on their molecular length interval. All tubes are organized in a network. Initial DNA molecules are put in some fixed tube, and the transformed molecules are collected in the output tube.

We remark that the above filtering is not necessarily based on gel electrophoresis. Indeed, there exist other methods, like size exclusion chromatography, that permit to separate molecules depending on their size.

We investigate a model which uses the splicing operation in the test tubes. The choice of splicing, a notion introduced by T. Head in [2], was directed by a huge number of theoretical results concerning variants of splicing systems, see [5] for an overview. The existence of similar models – splicing test tube systems [1] – helped us to concentrate on the new filtering mechanism offered by the gel electrophoresis.

Firstly, we show that these constructs, even with very restricted size parameters, are able to simulate Turing machines. This result corresponds to our expectations, due to the nature of the splicing operation. In particular, the discussion of the relation between splicing, elimination, and universality may be found in [8] and [9].

Although our construction has the power of Turing machines, this does not help in efficiently solving practical problems. For example, given a particular molecule, can we design a system that will perform a particular transformation on it? Moreover, this transformation should be efficient, *i.e.*, it shall be done in the smallest possible number of steps, involving the smallest number of high-cost operations. This problem is difficult to solve. Here we provide only some ideas that may be useful for its solution. More precisely, we discuss natural restrictions and extensions of the model which may speed-up the computation and reduce the number of operations having high costs. We mainly focus on the network structure and length filtering, the operation-related improvements remain to be further investigated.

2   Preliminaries

In this section we recall some very basic notions and notations we use throughout the paper. We assume the reader to be familiar with the basics of formal language theory. For more details, we refer to [3,6]. The set of non-empty words over an alphabet V is denoted by V^+; if the empty word, ε, is included, we use the notation V^*. Furthermore, by $|w|$ we denote the length of word w.

In the paper, we consider non-stationary deterministic Turing machines, *i.e.*, those ones where at each step of the computation the head moves either to the left or to the right. These machines are given as $M = (Q, T, a_0, q_0, F, \delta)$, where Q is the set of states, T is the tape alphabet, $a_0 \in T$ is the blank symbol, $q_0 \in Q$ is the initial state, $F \subseteq Q$ is the set of final (halting) states, and δ denotes the set of instructions. Each instruction is of the form (q_i, a_k, D, q_j, a_l) which is interpreted as follows: if the head of M being in state q_i is scanning a cell which contains a_k, then the contents of the scanned cell is replaced by a_l, the head moves to the left $(D = L)$ or to the right $(D = R)$ and the state of the machine changes to q_j.

By a *configuration* of a Turing machine we mean a string $w_1 q w_2$, where $w_1 \in T^*$, $w_2 \in T^+$ and $q \in Q$. A configuration represents the contents of non-empty cells of the working tape of the machine and all the blank symbols in between them, from left to right, (*i.e.* all other cells to the left and to the right are blank), its state, and the position of the head on the tape. The machine head is assumed to read the leftmost letter of w_2. Initially all cells on the tape are blank except finitely many cells.

It is known that non-stationary deterministic Turing machines are as powerful as the generic (non-restricted) ones.

Now we briefly recall the basic notions concerning the splicing operation and related constructs [5].

By an (abstract) *molecule* we mean a word over an alphabet.

A *splicing rule* (over V) is a 4-tuple (u_1, u_2, u_3, u_4) where $u_1, u_2, u_3, u_4 \in V^*$. It is also written as $u_1 \# u_2 \$ u_3 \# u_4$, where $\$, \# \notin V$, or in the form $\dfrac{u_1 \mid u_2}{u_3 \mid u_4}$.

Strings $u_1 u_2$ and $u_3 u_4$ are called splicing *sites*.

We say that a word x *matches* a splicing rule r if x contains an occurrence of one of the two sites of r. We also say that x and y are *complementary* with respect to a rule r if x contains one site of r and y contains the other one. In this case we also say that x or y may *enter* rule r. When x and y enter a rule $r = u_1 \# u_2 \$ u_3 \# u_4$, *i.e.*, we have $x = x_1 u_1 u_2 x_2$ and $y = y_1 u_3 u_4 y_2$, it is possible to define the application of r to the couple x, y. The result of this application is w and z where $w = x_1 u_1 u_4 y_2$ and $z = y_1 u_3 u_2 x_2$. We can also say that x and y are spliced and w and z are the result of this splicing, written as follows: $(x, y) \vdash_r (w, z)$ or

$$\frac{x_1 u_1 \mid u_2 x_2}{y_1 u_3 \mid u_4 y_2} \quad \vdash_r \quad \frac{x_1 u_1 u_4 y_2}{y_1 u_3 u_2 x_2} \, .$$

The pair $\sigma = (V, R)$, where V is an alphabet and R is a set of splicing rules, is called a *splicing scheme* or an H-scheme. For given $R' \subseteq R$, $\sigma_{R'}$ is the restriction of σ to R'.

For a splicing scheme $\sigma = (V, R)$ and for a language $L \subseteq V^*$ we define:
$$\sigma(L) = \{w, z \in V^* \mid (x, y) \vdash_r (w, z), x, y \in L, r \in R\}.$$

The iteration of the application of the splicing operation to L, according to σ, is defined as follows:

$\sigma^0(L) = L$,

$\sigma^{i+1}(L) = \sigma^i(L) \cup \sigma(\sigma^i(L))$, $i \geq 0$,

$\sigma^*(L) = \cup_{i \geq 0} \sigma^i(L)$.

It is known, see [5], that iterated splicing preserves the regularity of a language: For a regular language $L \subseteq T^*$ and a splicing scheme $\sigma = (T, R)$, it holds that $\sigma^*(L)$ is a regular language.

3   Length-Separating Splicing Test Tube Systems

In the following we define the notion of *length-separating splicing test tube systems*. This notion inherits the distributed architecture and most of the main features of the known variants of test tube systems based on splicing, but the communication among the splicing schemes (test tubes) is defined in a significantly different manner, based on filtering by length, *i.e.*, using the formalization of gel electrophoresis action.

Before turning to the model, we need some auxiliary notions. Let V be an alphabet.

1. Let $\pi_{=k} : V^* \to \{\underline{true}, \underline{false}\}$, $k \geq 1$, be a mapping (a predicate) defined by
$$\pi_{=k}(w) = \begin{cases} \underline{true} & \text{if } |w| = k, \\ \underline{false} & \text{otherwise.} \end{cases}$$
Mappings (predicates) $\pi_{\leq k}$, $\pi_{\geq k}$, $\pi_{>k}$, $\pi_{<k}$, $\pi_{\neq k}$ are defined by modifying the condition $|w| = k$ to $|w| \leq k$, $|w| \geq k$, $|w| > k$, $|w| < k$, and $|w| \neq k$, respectively.

2. For the sake of completeness, we define $\pi_{\geq 0} : V^* \to \{\underline{true}, \underline{false}\}$ with $\pi_{\geq 0}(w) = \underline{true}$ for any w in V^*.

3. Let $\pi_{\max} : V^* \times 2^{V^*} \to \{\underline{true}, \underline{false}\}$ where
$$\pi_{\max}(w, L) = \begin{cases} \underline{true} & \text{if } w \in L \text{ and } |w| \geq |u|, \text{ for any } u \in L \\ \underline{false} & \text{otherwise.} \end{cases}$$

4. Let $\pi_{\min} : V^* \times 2^{V^*} \to \{\underline{true}, \underline{false}\}$ where
$$\pi_{\min}(w, L) = \begin{cases} \underline{true} & \text{if } w \in L \text{ and } |w| = \min_{w' \in L} |w'| \\ \underline{false} & \text{otherwise.} \end{cases}$$

5. We define $\pi_{\neg\max} : V^* \times 2^{V^*} \to \{\underline{true}, \underline{false}\}$ and $\pi_{\neg\min} : V^* \times 2^{V^*} \to \{\underline{true}, \underline{false}\}$ as the negations of predicates of $\pi_{\max}$ and $\pi_{\min}$, respectively.

If no confusion arises, instead of $\pi_{\leq k}$, $\pi_{\geq k}$, $\pi_{>k}$, $\pi_{<k}$, $\pi_{=k}$, $\pi_{\neq k}$ and $\pi_{\max}$, $\pi_{\min}$, $\pi_{\neg\max}$, $\pi_{\neg\min}$, we might also use the notations $\leq k$, $\geq k$, $> k$, $< k$, $= k$, $\neq k$ and max, $\neg$max, min, $\neg$min, respectively.

A *length-separating splicing test tube system* is the following construct $\Delta = (n, V, T, G, A, R, O)$, where V is an alphabet, $T \subseteq V$ is the terminal alphabet, G is a labelled graph, called the *communication graph* of Δ, whose nodes are also called *test tubes* (tubes for short), n is the size of the graph G, and O is a subset of nodes of G, called the set of *output nodes*. $A = (A_1, \ldots, A_n)$ is the initial configuration of Δ, where A_i is a finite subset of V^*, for $1 \leq i \leq n$, called the set of *axioms* at node i, and $R = (R_1, \ldots, R_n)$, where each R_i, $1 \leq i \leq n$, is a finite set of splicing *rules* over V^*, *i.e.*, the set of rules associated to node i of G.

Each edge from node i to node j of G, for $1 \leq i, j \leq n$, denoted by (i, j), is labelled by a mapping $p_{i,j}$ from the set $\{\pi_{\leq k}, \pi_{\geq k}, \pi_{>k}, \pi_{<k}, \pi_{=k}, \pi_{\neq k} \mid k \geq 1\} \cup \{\pi_{\geq 0}, \pi_{\max}, \pi_{\neg\max}, \pi_{\min}, \pi_{\neg\min}\}$.

Furthermore, we require the labels of the edges going out from the same node to be non-contradictory, *i.e.*, no word in V^* can satisfy more than one predicate associated to different edges going out from the same node. Thus, the labels define a subpartition of the set of natural numbers.

The *computation* in Δ is a sequence of two subsequent steps, a computation step and a communication step, which are repeated iteratively and change the configuration of the system.

By a *configuration* of Δ, above, we mean an n-tuple $(L_1, \ldots, L_n)$, where $L_i \in V^*$, $1 \leq i \leq n$. (The initial configuration of Δ is $(A_1, \ldots, A_n)$).

The *computation* step consists in an iterative application of R_i, $\sigma_{R_i}^*$, at each node i of G to molecules found there.

We say that configuration $(L_1', \ldots, L_n')$ is obtained from configuration $(L_1, \ldots, L_n)$ by a computation step in Δ, denoted by $(L_1, \ldots, L_n) \Longrightarrow_{comp} (L_1', \ldots, L_n')$, if $L_i' = \sigma_{R_i}^*(L_i)$ holds for $1 \leq i \leq n$.

During the *communication* step, the actual contents of the test tubes, *i.e.*, the set of molecules found at the nodes, is re-distributed according to the communication graph G and to the labels of the edges, *i.e.* to the associated predicates.

In order to define this step more formally, we need an auxiliary notion. Let $(L_1, \ldots, L_n)$ be a configuration of Δ. We say that $w \in L_i$ can be communicated from node i to node j in Δ, if (i, j) is an edge in G and either

- $p_{i,j}(w)$ is <u>true</u> where $p_{i,j} \in \{\pi_{\geq 0}, \pi_{\leq k}, \pi_{\geq k}, \pi_{>k}, \pi_{<k}, \pi_{=k}, \pi_{\neq k} \mid k \geq 1\}$, or
- $p_{i,j}(w, L_i)$ is <u>true</u> where $p_{i,j} \in \{\pi_{\max}, \pi_{\neg\max}, \pi_{\min}, \pi_{\neg\min}\}$.

We say that configuration $(L_1', \ldots, L_n')$ is obtained from configuration $(L_1, \ldots, L_n)$ by a communication step in Δ, denoted by $(L_1, \ldots, L_n) \Longrightarrow_{comm} (L_1', \ldots, L_n')$, if L_i' consists of all words $w \in V^*$ which satisfy one of the following conditions:

- $w \in L_j$ and w can be communicated from node j to node i according to $p_{j,i}$,
- $w \in L_i$ and there is no edge (i, j), $1 \leq i, j \leq n$, in G such that w can be communicated to node j from node i according to $p_{i,j}$.

Thus, after performing the corresponding iterated splicing, the words are communicated to other nodes. Due to the non-contradictoriness of the labels, each word is sent to exactly one node. For example, if edge (i, j) is labelled $\leq k$ (respectively, $\geq k$, $< k, > k, = k, \neq k$), then a word w at node i with $|w| \leq k$ (respectively, $|w| \geq k$, $|w| < k$, $|w| > k$, $|w| = k$, $|w| \neq k$) is sent to node j. Predicate $p_{i,j}(w, L_i)$ with $p_{i,j} = \max$ (respectively, $p_{i,j} = \neg\max$) makes possible w to be communicated to node j if w is (respectively, is not) a string of maximal length at node i. Similarly, predicates min (respectively $\neg$min) are for communicating a string w to node j if w is (respectively, is not) a string of smallest length at node i.

The result of a computation is the set of strings over the terminal alphabet collected at the nodes in O after a communication step in Δ, i.e.,

$$L(\Delta) = \{w \in T^* \mid (A_1, \ldots, A_n) \Longrightarrow_{comp} (L_1^{(1)}, \ldots, L_n^{(1)}) \Longrightarrow_{comm}$$
$$(L_1^{(2)}, \ldots, L_n^{(2)}) \ldots \Longrightarrow_{comm} (L_1^{(2s)}, \ldots, L_n^{(2s)}),$$
$$\text{for some } s \geq 0 \text{ such that } w \in L_i^{(2s)}, i \in O\}.$$

We illustrate the above notions by an example.

Example 1. Let us define Δ as follows. Consider the following communication graph G and predicates associated to the edges. We highlighted a node that has only incoming edges and we denoted it by X (it corresponds to a trash can).

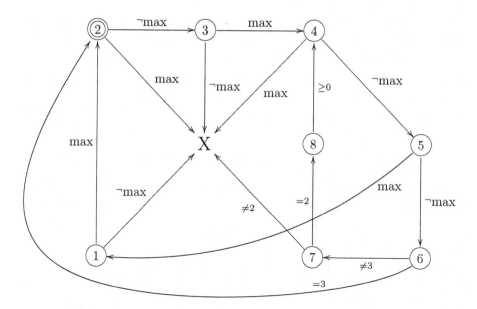

Let $T = \{a, b, c\}$, let the only output node be node 2, and let the axioms to be given as $A_1 = \{c'ab, cab'\}$, $A_6 = \{XXZ, ZYY\} \cup \{c'Z, Zb'\}$, with all other axioms being empty. Let us define the rule sets associated to the nodes as follows:

$$R_1 = \left\{ \frac{a \mid b'}{c' \mid a} \right\}$$
$$R_2 = \left\{ \frac{c \mid a}{XX \mid Z}, \frac{a \mid b}{Z \mid YY} \right\}.$$
$$R_4 = \left\{ \frac{XX \mid a}{c' \mid Z}, \frac{a \mid YY}{Z \mid b'} \right\}.$$

All other rule sets are empty, thus no operation is performed there.

Starting from $c'ab$ and cab' at node 1, string $caab$ is obtained which is sent to node 2 (satisfying communication condition max). At node 2, it is transformed to molecules $XXaab$, $caaYY$ and $XXaaYY$. Then, $XXaaYY$ is sent to node X, and the other strings to node 3. After that, these latter strings arrive at node 4, since they satisfy condition max. There we obtain molecules $c'aab$, $caab'$, $XXaab$ and $caaYY$. As before, we send $XXaab$ and $caaYY$ to the "trash" node (X), and then strings $c'aab$ and $caab'$ via node 5, arrive at node 1. At the same time, strings XXZ, ZYY, $c'Z$ and Zb' return to their original location. The above procedure can be repeated. Hence, we obtain the language $\{ca^{2^n}b \mid n > 0\}$, which is a well-known context-sensitive language.

4   Computational Completeness

In this section we demonstrate how length-separating splicing test tube systems simulate Turing machines.

We first define a coding function ϕ as follows. For any configuration w_1qw_2 of a Turing machine $M = (Q, T, a_0, q_0, F, \delta)$, we define $\phi(w_1qw_2) = X^6w_1qw_2Y^6$, where X and Y are symbols not in T.

Theorem 1. *Let $M = (Q, T, a_0, s_0, F, \delta)$ be a Turing machine and w be an input of M. Then, we can construct a length-separating splicing test tube system $\Delta = (11, V, T', G, A, R, \{10\})$ which, given the axiom $\phi(w)$ at a distinguished node, simulates the behavior of M on input w, i.e. in the following sense:*

1. *for any word w on which M halts in configuration w_1qw_2, the system Δ produces a unique output $\phi(w_1qw_2)$.*
2. *for any word w, on which M does not halt, the system Δ computes the empty language.*

Proof. Let $M = (Q, T, a_0, q_0, F, \delta)$ be a Turing machine. Firstly, we give the construction of Δ and explain the simulation.

Let $T = \{a_0, \ldots, a_{m-1}\}$ and $Q = \{q_0, \ldots, q_{n-1}\}$. Throughout the proof, let $\mathbf{a}, \mathbf{b}, \mathbf{c}$ denote arbitrary letters in T, and let us use the notations $\mathbf{d} \in T \cup \{Y\}$, $\mathbf{q} \in Q$ in the same manner. Let a_0 be the blank symbol.

Without the loss of generality, we may assume that w_1w_2 (remember, a configuration is given as w_1qw_2) does not contain the symbol a_0 and that M is not stationary.

We construct Δ simulating M as follows.

Let $V = T \cup \{X, Y, R_1, R_2, R_3, R_4, L_1, L_2, L_3, L_4, R'_1, R'_2, L'_3, L'_4, F, Z\}$ and $T' = T \cup \{X, Y\} \cup \{q_f | q_f \in F\}$.

The communication graph G is defined as follows:

The sets of axioms are given as $A_1 = \{X^6a_0q_0wY^6\} \cup \{R_1L_1, R_2L_2, R_3L_3\} \cup \cup \{R_4L_4, R_4L'_4\}$, $A_7 = \{FFR'_1a_lq_jZ, FR'_2a_lq_ja_0Z\} \cup \{Zq_j\mathbf{aa}_lL'_3F, Za_0q_j\mathbf{aa}_lL'_4\}$ and $A_8 = \{X^6Z, ZY^6\}$. the other axiom sets are empty.

We define the rules as follows (the first number is the node where the rule is located).

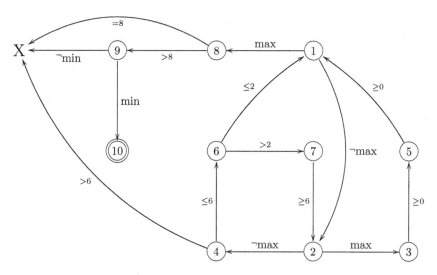

I. For any instruction $(q_i, a_k, R, q_j, a_l) \in \delta$ there is the following group of rules in Δ.

$$1.1 : \frac{\mathbf{a} \ \big| q_i a_k \mathbf{b}}{R_1 \big| \ L_1} \ ; \qquad 1.2 : \frac{\mathbf{a} \ \big| q_i a_k Y}{R_2 \big| \ L_2} \ ;$$

$$2.1 : \frac{R_1 q_i a_k \ \big| \mathbf{b}}{FFR_1' a_l q_j \big| Z} \ ; \qquad 2.2 : \frac{R_1 q_i a_k \ \big| Y}{FR_2' a_l q_j a_0 \big| Z} \ ;$$

$$2.3 : \frac{FFR_1' \big| \mathbf{aqb}}{\mathbf{c} \ \big| \ L_1} \ ; \qquad 2.4 : \frac{FR_2' \big| \mathbf{aqa_0} Y}{\mathbf{c} \ \big| \ L_2} \ ;$$

II. For any instruction $(q_i, a_k, L, q_j, a_l) \in \delta$ there is the following group of rules in Δ.

$$1.3 : \frac{\mathbf{ab} q_i a_k \mathbf{c} \big| \mathbf{d}}{R_3 \big| L_3} \ ; \qquad 1.4 : \frac{X \mathbf{a} q_i a_k \big| \mathbf{d}}{R_4 \big| L_4} \ ;$$

$$2.5 : \frac{\mathbf{b} \big| \mathbf{a} q_i a_k L_3}{Z \big| q_j \mathbf{aa}_l L_3' F} \ ; \qquad 2.6 : \frac{X \big| \mathbf{a} q_i a_k L_4}{Z \big| a_0 q_j \mathbf{aa}_l L_4'} \ ;$$

$$2.7 : \frac{\mathbf{b} q_j \mathbf{aa}_l \big| L_3' F}{R_3 \big| \mathbf{d}} \ ; \qquad 2.8 : \frac{X a_0 q_j \mathbf{aa}_l \big| L_4'}{R_4 \big| \mathbf{d}} \ ;$$

III. There are also following rules:

$$8.1 : \frac{X a_0 \big| \mathbf{a}}{X \big| Z} \ ; \qquad 8.2 : \frac{\mathbf{a} \big| a_0 Y}{Z \big| Y} \ ;$$

Any other rule sets, R_i, where $i \neq 1, 2, 8$ are empty. The part of the tape of M containing information is encoded in the following way: for any configuration

w_1qw_2 of M, a string $X^6w_1qw_2Y^6$ is found at node 1 in a configuration of Δ. Thus, the tape is enclosed by strings X^6 and Y^6 and a marker symbol q refers both to the position of the head and to the current state of the machine.

The construction above is similar to constructions from [4] and [7]. Using max and ¬max predicates it is possible to filter out strings which do not correspond to a correct simulation of the Turing machine. Hence, only strings representing (through encoding) configurations of M will be kept.

Now let us explain the simulation in details.

Suppose that current configuration of M is $w_1q_ia_kw_2$ and that M contains instruction (q_i, a_k, R, q_j, a_l), i.e., a right move shall be performed. Suppose also that $w_2 \neq \varepsilon$ (the case $w_2 = \varepsilon$ is treated analogously).

The configuration of M is represented by the string $X^6w_1q_ia_kw_2Y^6$ in tube 1. The configuration of Δ is the following (since there are no edges going out from node X, we do not show this node):

L_1	$\{X^6w_1q_ia_kw_2Y^6\} \cup \{R_1L_1, R_2L_2, R_3L_3, R_4L_4\}$
$L_2, \ldots, L_6$	$\emptyset$
L_7	$\{FFR_1'a_lq_jZ, FR_2'a_lq_ja_0Z\} \cup \{Zq_jaa_lL_3'F, Za_0q_jaa_lL_4'\}$
L_8	$\{X^6Z, ZY^6\}$
$L_9, \ldots, L_{10}$	$\emptyset$

By applying rule 1.1 string $X^6w_1q_ia_kw_2Y^6$ is split in front of q_i:

$$\frac{X^6w_1 \mid q_ia_kw_2Y^6}{R_1 \mid L_1} \quad \vdash_{1.1} \quad \frac{X^6w_1L_1}{R_1q_ia_kw_2Y^6} \ .$$

The original string, $X^6w_1q_ia_kw_2Y^6$, is sent to tube 8 during the communication step, because it is the longest. This string is used in the procedure obtaining the result. We will discuss its evolution later and we hide for the moment tubes 8, 9 and 10, because they behave independently from the rest of the system. All other strings are sent to tube 2. Strings from tube 7 also forwarded to tube 2.

The next configuration of Δ is the following:

L_1	$\emptyset$
L_2	$\{X^6w_1L_1, R_1q_ia_kw_2Y^6\} \cup \{R_1L_1, R_2L_2, R_3L_3, R_4L_4\} \cup$ $\{FFR_1'a_lq_jZ, FR_2'a_lq_ja_0Z\} \cup \{Zq_jaa_lL_3'F, Za_0q_jaa_lL_4'\}$
$L_3, \ldots, L_7$	$\emptyset$

Now, q_ia_k is replaced by q_ja_l (rule 2.1) and the two parts of the configuration are joined by rule 2.3:

$$\frac{R_1q_ia_k \mid w_2Y^6}{FFR_1'a_lq_j \mid Z} \quad \vdash_{2.1} \quad \frac{R_1q_ia_kZ}{FFR_1'a_lq_jw_2Y^6} \quad \text{and}$$

$$\frac{FFR_1'a_lq_jw_2Y^6}{X^6w_1 \mid L_1} \quad \vdash_{2.3} \quad \frac{FFR_1'L_1}{X^6w_1a_lq_jw_2Y^6} \ .$$

Now, the longest string, $X^6w_1a_lq_jw_2Y^6$, represents the next configuration of M. It is sent to tube 3. However, all other strings should not be eliminated (by sending them to tube X) because it is necessary to recuperate strings which are used for computation in tubes 1 and 2. It is easy to observe that strings that need to be in tube 1 are of size 2 and strings that need to be sent to tube 2 (via tube 7) are of size 6.

The next configuration of Δ is the following:

$L_1, \ldots, L_2$	$\emptyset$
L_3	$\{X^6 w_1 a_l q_j w_2 Y^6\}$
L_4	$\{R_1 L_1, R_2 L_2, R_3 L_3, R_4 L_4\} \cup \{FFR_1' a_l q_j Z, FR_2' a_l q_j a_0 Z\} \cup$ $\{Z q_j aa_l L_3' F, Z a_0 q_j aa_l L_4', X^6 w_1 L_1, R_1 q_i a_k w_2 Y^6\}$
$L_5, \ldots, L_7$	$\emptyset$

Now, string $X^6 w_1 a_l q_j w_2 Y^6$ moves to tube 5, while strings from tube 4 are redistributed. Strings having length smaller than 6 are sent to tube 6, while the other strings are forwarded to tube X.

The next configuration of Δ is the following:

$L_1, \ldots, L_4$	$\emptyset$
L_5	$\{X^6 w_1 a_l q_j w_2 Y^6\}$
L_6	$\{R_1 L_1, R_2 L_2, R_3 L_3, R_4 L_4\} \cup \{FFR_1' a_l q_j Z, FR_2' a_l q_j a_0 Z\}$
L_7	$\emptyset$

Finally, string $X^6 w_1 a_l q_j w_2 Y^6$ as well as strings from tube 6 having length 2 move to tube 1. The other strings from tube 6 move to tube 7.

The next configuration of Δ is the following:

L_1	$\{X^6 w_1 a_l q_j w_2 Y^6\} \cup \{R_1 L_1, R_2 L_2, R_3 L_3, R_4 L_4\}$
$L_2, \ldots, L_6$	$\emptyset$
L_7	$\{FFR_1' a_l q_j Z, R_2' a_l q_j a_0 ZZ\}$

This configuration may be obtained from the first configuration by replacing string $X^6 w_1 q_i a_k w_2 Y^6$ with $X^6 w_1 a_l q_j w_2 Y^6$. Hence, the corresponding instruction of M is simulated.

If necessary (when $w_2 = \varepsilon$), the tape can be extended to the right and rules 1.2, 2.2 and 2.4, which work similarly to rules 1.1, 2.1 and 2.3, take it into account.

The case of a left move can be treated analogously. Therefore, the system closely simulates the behavior of M and it will reach a halting configuration only when M halts.

Now, we discuss the evolution of string of form $X^6 w_1 q_i a_k w_2 Y^6$ after its arrival in tube 8. The corresponding configuration of Δ is (we show only tubes 8, 9 and 10, because they behave independently from the rest of the system):

L_8	$\{X^6 w_1 q_i a_k w_2 Y^6\} \cup \{X^6 Z, ZY^6\}$
L_9	$\emptyset$
L_{10}	$\emptyset$

Firstly, we remark that, due to the rules of our system, the prefix of w_1 and the suffix of w_2 may contain symbols a_0. By using rules 8.1 and 8.2 these symbols are removed. Hence, new strings $X^6 w_1^s q_i a_k w_2^s Y^6$, containing less symbols a_0 will be obtained. There will be also a string $X^6 w_1' q_i a_k w_2' Y^6$ where the prefix of w_1' and the suffix of w_2' do not contain any symbols a_0.

We also remark that at least one a_0 is present at the beginning of w_1 (the computation preserves the invariant that there is at least one symbol a_0 after X). Therefore, after the computation step there will certainly be a string $X^6 w_1' q_i a_k w_2' Y^6$ in tube 8 and it will be shorter than the original string. During the communication step, strings $X^6 a_0 Z$ and $Z a_0 Y^6$ are sent to tube X, while all other strings, except strings from A_8, are sent to node 9. The corresponding configuration of Δ is:

L_8	$\{X^6 Z, Z Y^6\}$
L_9	$\{X^6 w_1 q_i a_k w_2 Y^6\} \cup \{X^6 w_1^s q_i a_k w_2^s Y^6\} \cup \{X^6 w_1' q_i a_k w_2' Y^6\}$
L_{10}	$\emptyset$

Now it is clear that $X^6 w_1' q_i a_k w_2' Y^6$ is the string with the minimal length, because it does not contain surrounding symbols a_0. Therefore, it will be sent to tube 10, while all other strings will be discarded by sending them to node X. The corresponding configuration of Δ is:

L_8	$\{X^6 Z, Z Y^6\}$
L_9	$\emptyset$
L_{10}	$\{X^6 w_1' q_i a_k w_2' Y^6\}$

Here, if q_i is in F, then this string will be the result (we remind that in this case, in the remaining of the system no rules can be applied any more).

Now, in order to complete the proof it is sufficient to show that no other strings of length 2 and 6, except strings from A_1 and A_7, can be produced in the "computation" part of the system (tubes 1–7) and that no strings of length 7 can be produced in the "result" part of the system (tubes 8–10). Indeed, any splitting rule (1.x) will produce strings which are at least of length 7. Substitution rules (2.1, 2.2, 2.5 and 2.6) may generate strings of size 4, 5 and at least 11. Rules 2.3, 2.4 and 2.7 produce strings of size 3,4 and at least 16. Rule 2.8 produces $R_4 L_4'$ of length 2 which is in A_1. For the "result" part it is sufficient to observe that rules 8.1 and 8.2 produce strings of length 8 and at least 13.

Thus, the theorem is proved.

5   Discussion of the Model

The reader can observe that, in fact, the main ingredient of the presented model is the *communication* controlled by the length of the words. Thus, the idea of a length separating test tube system can also be given in a *general form*, i.e., using an arbitrary operation, γ, defined on words (or sets of words.) Moreover, we may associate different operations to different test tubes, thus obtaining a *hybrid system*. In this way, we need only to change the definition of the computing step. The study and comparison of length-separating *test tube systems based on different operations* would be of interest.

These systems can also be considered as *transducers*, i.e., having one input test tube as well as one output tube which are used only for the input and the output. In this case, several systems may be chained one after another and their action can be combined. From a practical point of view, this also means that $V = T$, *i.e.*, because in this case no final filtering of the strings (terminal strings) is done. Notice, that the distinction of the terminal alphabet is traditionally used only for selecting some particular strings among the results of the computation and it is not an inherent property of the construction.

Another important point for the definition of the model is the detection of a *halting* configuration. In most of the constructions from DNA computing the end of the computation is defined by halting. Usually, this condition means that the system cannot evolve, *i.e.*, there is no applicable rule in the system. However, from practical point of view, it is difficult to detect the fulfillment of this condition. Therefore, we propose to use a special *acknowledgement test tube*. When anything arrives in this tube, the computation is considered to be finished and the result is read in the output test tube. It is also possible to combine the acknowledgement and the output test tubes. In this case the computation stops when a first molecule arrives in that tube.

6 Practical Optimizations

In this section we discuss some optimizations of the model that can be performed for practical reasons.

A *practical restriction* would be the application of only *one elementary rule per test tube*. In the splicing case, this would mean that only *one restriction enzyme* is used. We think that this restriction may better reflect the reality.

Gel-related operations are time-consuming. It is important to minimize the number of such operations. In terms of our model, this correspond to the minimization of the number of predicates on edges. We also think that min and max predicates are easier to be implemented in laboratory than the numerical predicates, since in the first case we need not to wait until the gel operation fully completes.

Another idea is to *permit contradictory predicates*. In this case when a molecule validates several predicates, a copy of it is sent to corresponding tubes. In the simplest case, when two connections with ≥ 0 predicate are present, this corresponds to the duplication of the contents of the test tube.

Finally, it would be interesting to consider predicates like max_k which denotes the k-th maximal string. This corresponds to the extraction of the k-th molecule from the gel electrophoresis lane. From the computational point of view, these operations are more powerful than the extraction by a fixed length.

7 Conclusions

In this article we formalized the action of an operation commonly used in the laboratory: the gel electrophoresis. This operation is usually used in combination with other molecular operations that modify DNA molecules. We proposed and discussed protocols that use gel electrophoresis for branching during the transformation. We formalized these kinds of protocols as a two-steps procedure consisting of a transformation (computation) step and a communication step. During the transformation step, molecules are modified in test tubes, while during the communication step molecules are redistributed to other test tubes depending on their length. As an example of a concrete model, we chose a model based on the splicing operation. We have shown that this construction is able to simulate any Turing machine. We have also discussed possible modifications of the model in order to increase the efficiency of the length-driven communication strategy.

The formalization that we provided can be further used to investigate different properties of protocols based on gel electrophoresis. Moreover, since some ideas from previous section may be realized in practice, it can be used for engineering transducers (protocols) that will do some particular transformation on DNA molecules in laboratory. However, in this case, size exclusion chromatography becomes more interesting as it permits to separate molecules much faster.

Acknowledgements. This publication was supported by the Hungarian Foundation for Research and Technological Innovation (project no. TéT F-19/04) and the EGIDE in France (project no. Balaton 09000TC, year 2005) in the frame of the Hungarian-French Intergovernmental Scientific and Technological Cooperation.

References

1. E. Csuhaj-Varjú, L. Kari, and G. Păun. Test tube distributed systems based on splicing. *Computers and AI*, 15(2–3):211–232, 1996.
2. T. Head. Formal language theory and DNA: an analysis of the generative capacity of specific recombinant behaviors. *Bulletin of Mathematical Biology*, 49(6):737–759, 1987.
3. J. Hopcroft, R. Motwani, and J. Ullman. *Introduction to Automata Theory, Languages, and Computation*. Addison-Wesley, Reading, Mass., 2nd edition, 2001.
4. M. Margenstern and Y. Rogozhin. A universal time-varying distributed H system of degree 2. *Biosystems*, 52:73–80, 1999.
5. G. Păun, G. Rozenberg, and A. Salomaa. *DNA Computing: New Computing Paradigms*. Springer Verlag, Berlin, Heidelberg, New York, 1998.
6. G. Rozenberg and A. Salomaa. *Handbook of Formal Languages, 3 volumes*. Springer Verlag, Berlin, Heidelberg, New York, 1997.
7. S. Verlan. Communicating distributed H systems with alternating filters. In N. Jonoska, G. Paun, and G. Rozenberg, editors, *Aspects of Molecular Computing. Essays Dedicated to Tom Head on the Occasion of His 70th Birthday*, volume 2950 of *LNCS*, pages 367–384. Springer Verlag, Berlin, Heidelberg, New York, 2004.
8. S. Verlan. *Head Systems and Applications to Bio-Informatics*. PhD thesis, University of Metz, France, 2004.
9. S. Verlan. A boundary result on enhanced time-varying distributed H systems with parallel computations. *Theoretical Computer Science*, 344(2-3):226–242, 2005.

Gene Assembly Algorithms for Ciliates

Lucian Ilie*,** and Roberto Solis-Oba***

Department of Computer Science, University of Western Ontario
London, Ontario, N6A 5B7, Canada
{ilie, solis}@csd.uwo.ca

Abstract. The micronuclear genes in stichotrichous ciliates are interrupted by multiple non-coding DNA segments. The coding segments are in scrambled disorder and can also be inverted. Identical short sequences (pointers) at the ends of the coding segments undergo homologous recombination to excise the non-coding segments and splice the coding ones. We consider the intramolecular model of Prescott, Ehrenfeucht, and Rozenberg for gene assembly in stichotrichous ciliates from the algorithmic point of view. We give a quadratic time algorithm for finding a successful sequence of operations to assemble a gene. We also prove an $\Omega(n \log n)$ lower bound on the amount of work needed to assemble genes, even when any pair of identical pointers have the same orientation. For the problem of finding the minimum number of operations needed to assemble a given gene, we give a heuristic quadratic algorithm which works well in practice. The complexity of this problem remains open.

Keywords: ciliates, stichotrichs, gene assembly, algorithm, dlad-first greedy strategy.

1 Ciliates

Ciliates form a large ancient group of unicellular organisms which can be found almost in any place where there is water. Their name comes from one of their common characteristics, namely, they all possess cilia projecting from their surfaces, used to propel them through an aqueous environment. A second common feature is the presence of two kinds of nuclei in the same cell: a micronucleus and a macronucleus; the former is used in mating whereas the later produces the RNA needed for cell operations.

One group of ciliates, known as *stichotrichs*, are of special interest because of the amazing DNA manipulations which take place within their micronuclei. The micronuclei are activated only during reproduction where at some stage their genome transforms into the macronuclear genome; the process is called *gene assembly*. This is the most involved DNA processing known in living organisms and its discovery expanded our view of the versatility of DNA changes in biological evolution. What makes this process so complex is the unusual arrangement

---

* Research partially supported by NSERC.
** Corresponding author.
*** Research partially supported by NSERC grant 227829-04.

C. Mao and T. Yokomori (Eds.): DNA12, LNCS 4287, pp. 71–82, 2006.
© Springer-Verlag Berlin Heidelberg 2006

of the genes in the micronucleus. While the genes in the macronucleus are contiguous sequences of DNA, the ones in the micronucleus are broken into pieces, called *macronuclear destined segments (MDS)*, separated by non-coding blocks, called *internal eliminated segments (IES)*. In addition, the MDSs may appear in scrambled disorder and some may be inverted. An example is shown in Fig. 1(a) where rectangles represent MDSs and the lines connecting them are IESs. (The similarity with a doubly linked list is remarkable.)

(a)

(b)

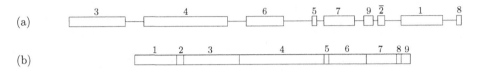

Fig. 1. (a) Structure of the micronuclear gene encoding actin protein in the stichotrich *Sterkiella nova*. Notice that the MDSs are in scrambled disorder and that MDS number 2 is inverted. (b) The same gene with IESs excised and MDSs assembled in the right order.

During the development of a micronucleus into a macronucleus the IESs are excised from micronuclear genes and the MDSs are spliced into the sequentially correct order. According to [9], there are on average four IESs per gene and thus, among the approximately 25,000 micronuclear genes, there may be roughly 100,000 IESs to be excised, therefore requiring massive processing of DNA.

Two models have been proposed for gene assembly: one intermolecular, due to Landweber and Kari [6,7], and another intramolecular, due to Prescott, Ehrenfeucht and Rozenberg [8,4].

We are concerned in this paper with the intramolecular model. The reader is referred to the book [3] for more details and further references. The operations of this model, which we describe in the next section, were proven to be able to assemble any micronuclear gene. However, very little seems to be known about the algorithmic aspect of the gene assembly problem. Questions such as how fast we can find a correct sequence of operations to assemble a given gene, or what the smallest number of operations is in any such sequence, appear naturally. For the former question, we give a quadratic time algorithm and an $\Omega(n \log n)$ lower bound that holds even for the case when the gene contains no inverted pointers (see the next section for definitions). For the latter problem, we give a heuristic quadratic algorithm which works well in practice. The complexity of both problems remains to be determined.

2   Gene Assembly in Ciliates

As mentioned above, gene assembly is the process by which the IESs are excised from a micronuclear gene and the MDSs are arranged into the sequentially correct order. An example is shown in Fig. 1 where (a) is transformed into (b) by the gene assembly process. The lengths and positions of the IESs, as well as the correct order of the MDSs, is determined by the fact that at the MDS-IES junction consecutive MDSs contain the same sequence of 3 to 20 nucleotides, which from now on we shall call *pointers*. Put differently, the same pointer is both a suffix of the ith MDS and a prefix of the $(i+1)$st MDS. Notice also that some MDSs may be inverted, that is, their sequence is read backwards; e.g., MDS number 2 in Fig. 1(a).

By a process called homologous recombination, two identical pointers may align side-by-side and an enzyme produces identical staggered cuts at the same position in both. The cut pieces may then be switched and ligated; the result being two DNA molecules that have recombined. Based on homologous recombination, Prescott, Ehrenfeucht, and Rozenberg introduced three molecular operations that account for the process of gene assembly.

The first one is *loop recombination* in which the IES between two consecutive MDSs is eliminated and the two MDSs are joined together; see Fig. 2(a): the sequence $u_1 \, MDS_i \, p \, IES \, p \, MDS_{i+1} \, u_2$ becomes $u_1 \, MDS_i \, p \, MDS_{i+1} \, u_2$. The operation is technically called ld, from *loop, direct repeat*.

The second operation is *hairpin recombination*, when one pointer appears inverted; see Fig. 2(b): the sequence $u_1 \, MDS_i \, p \, u_2 \, \overline{MDS_{i+1}} \, \overline{p} \, u_3$ becomes $u_1 \, MDS_i \, MDS_{i+1} \, \overline{u}_2 \, u_3$. The operation is technically called hi, from *hairpin, inverted repeat*.

The third and last operation introduced is *double loop recombination*, applicable when two pairs of pointers appear alternating; see Fig. 2(c): for instance, the sequence $u_1 \, MDS_i \, p \, u_2 \, q \, MDS_{j+1} \, u_3 \, p \, MDS_{i+1} \, u_4 \, MDS_j \, q \, u_5$ is transformed into $u_1 \, MDS_i \, MDS_{i+1} \, u_4 \, MDS_j \, MDS_{j+1} \, u_3 \, u_2 \, u_5$. The operation is technically called dlad, from *double loop, alternating direct repeat*.

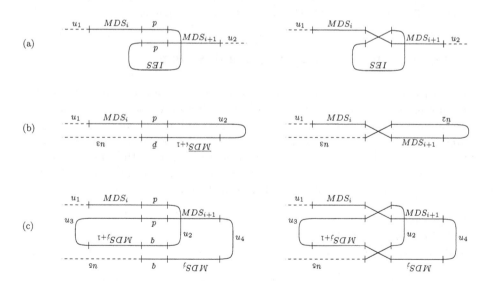

Fig. 2. The gene assembly operations: (a) ld, (b) hi, (c) dlad

3   Legal Strings

We shall use an abstract representation due to [2,1] which captures the sequence of pointers in the order they appear in a micronuclear gene. It is based on legal strings which we define next.

For a finite alphabet A, A^* denotes the set of all finite strings over A. Let n be a positive integer; n will be the number of different pointers in a micronuclear gene. For each pointer i, its inverted version is denoted $\bar{i}$ (signed i). Denote $[n] = \{1, 2, \ldots, n\}$ and $\overline{[n]} = \{\bar{1}, \bar{2}, \ldots, \bar{n}\}$. The union $[n] \cup \overline{[n]}$ contains all pointers and their inverted versions. A *signed string* is any string w over $[n] \cup \overline{[n]}$. A signed string is called *legal* if, for any $i \in [n]$, w contains exactly two letters (which need not be different) from the set $\{i, \bar{i}\}$; that means, each pointer appears twice and any occurrence may be inverted. For a string u of pointers, $u = i_1 i_2 \cdots i_k$, the inverted version of u is $\bar{u} = \bar{i}_k \bar{i}_{k-1} \cdots \bar{i}_1$.

A pointer i occuring in a legal string w is called *negative* if w contains two i's or two $\bar{i}$'s; otherwise i is called *positive* for w. A legal string w is called *negative* if all of w's pointers are negative.

It is not difficult to see that the number of legal strings with n pairs of pointers is $2^n(2n)!$. Indeed, the number of all permutations with $2n$ elements is $(2n)!$. Consider the permutations of the multiset $\{1, 1, 2, 2, \ldots, n, n\}$. We can permute the elements in $(2n)!$ ways but exchanging the places of the two i's will result in the same permutation, for any $1 \le i \le n$. As there are always 2^n possibilities of doing that, the number of permutations of the above multiset is $2^{-n}(2n)!$. Now, the legal strings are obtained precisely by adding signs in all possible ways to all permutations of the multiset. For each such permutation, there are exactly 2^{2n} ways of signing and each produces always a different legal string. Therefore, the number of legal strings is $2^{-n}2^{2n}(2n)! = 2^n(2n)!$, as claimed.

4   Operations for Reducing Legal Strings

Legal strings can model MDS arrangements; see [3]. In order to describe the operations corresponding to the above ld, hi, and dlad, we need to see that each pair of pointers is used only once. Therefore, for legal strings, we shall attempt to *reduce* rather than arrange them. This means, once a pair of pointers is used, it will be removed. The goal is to reduce a given legal string to the empty string Λ.

The operations for reducing legal strings, which we abusively denote the same as the ones already introduced, are defined as follows:

1. $\mathsf{ld}_p(u_1 p p u_2) = u_1 u_2$,
2. $\mathsf{hi}_p(u_1 p u_2 \bar{p} u_3) = u_1 \bar{u}_2 u_3$, and
3. $\mathsf{dlad}_{p,q}(u_1 p u_2 q u_3 p u_4 q u_5) = u_1 u_4 u_3 u_2 u_5$,

where $p, q \in [n] \cup \overline{[n]}$, $u_i \in ([n] \cup \overline{[n]})^*$. The correspondence with the previous operations should be obvious. A *reduction* of a legal string is a sequence of any of the above operations which reduces the legal string to the empty string.

As proved in [2,1], any legal string has a successful reduction using the above three operations. Also, any negative legal string has a successful reduction using ld and dlad operations only.

5   An Algorithm for Reducing Legal Strings

Our algorithm works in two phases: in the first phase it applies hi operations until the input string is transformed into a negative string; in the second phase the negative string is reduced to the empty string.

5.1   Reducing a Legal String to a Negative String

We assume that the string s is implemented as a doubly-linked list in which every node stores one of the pointers of the string. For each node i of this list, $i.pointer$ denotes the pointer stored in i, $i.right$ and $i.left$ are references to the right and left neighbours of i in the list, and $i.twin$ is a reference to the node in s storing the other occurrence of $i.pointer$. The algorithm is as follows.

> **Algorithm** produce_negative(s)
> $i \leftarrow$ first node of s
>   **while** $i \neq$ NULL **do**
>   **if** exactly one of $i.pointer$, $(i.twin).pointer$ is inverted **then** {
>   //Perform an hi operation
>   Invert the sublist (i, j) of s and all pointers in this sublist.
> $i \leftarrow$ first node of s
>   }
>   **else** $i \leftarrow i.right$

Every hi operation performed by the algorithm needs $O(n)$ time and at most n hi operations are performed, so the time complexity of this algorithm is $O(n^2)$.

5.2   An Algorithm for Reducing Negative Strings

Let s be a negative string of length n. Each pointer p in s appears twice. The *first occurrence* of a pointer p in the string s is the leftmost occurrence of p. The *second occurrence* of p is the rightmost occurrence of p in s.

Given a string $s = u_1 p u_2 q u_3 p u_4 q u_5$, a dlad operation $dlad_{p,q}(s)$ has *minimal overlap* if for no pointer r occurring in u_3 the dlad operation $dlad_{p,r}(s)$ is applicable. For example, consider the legal string $s = abcdbacd$. Operation $dlad_{a,d}(s)$ has minimal overlap, while $dlad_{a,c}(s)$ does not.

Given a string s for which two dlad operations, $dlad_{p,q}(s)$ and $dlad_{r,t}(s)$, are applicable, we say that $dlad_{p,q}(s)$ *appears before* $dlad_{r,t}(s)$ if the second occurrence of p in s appears before the second occurrence of r in s. For example, consider the same string s as above. Operation $dlad_{b,d}(s)$ appears before $dlad_{a,c}(s)$.

Given a string s we say that $dlad_{p,q}(s)$ is the *first* dlad operation applicable to s if no other applicable dlad operation appears before $dlad_{p,q}(s)$ and $dlad_{p,q}(s)$ has minimal overlap. For the same string as above, for example, $dlad_{b,d}(s)$ is the first dlad operation applicable to s.

Lemma 1. *Let $s = u_1pu_2qu_3pu_4qu_5$ be a negative legal string and let $\mathsf{dlad}_{p,q}(s)$ be the first* dlad *operation applicable to s.*

(a) If r is a pointer occurring in u_3, then r occurs twice in u_3.

(b) No dlad *operation* $\mathsf{dlad}_{r,t}(s)$ *is applicable to s for any pair of pointers r, t occurring in u_1, u_2 or u_3.*

Proof. (a) We prove the claim by contradiction. Let r be a pointer that occurs only once in u_3. Then, $s = u_1pu_2qu_3'ru_3''pu_4qu_5$. If r is the second occurrence of this pointer in s, then $\mathsf{dlad}_{q,r}(s)$ would be an applicable dlad operation that appears before $\mathsf{dlad}_{p,q}(s)$, contradicting the assumption that $\mathsf{dlad}_{p,q}(s)$ is the first dlad operation in s. On the other hand, if r is the first occurrence of this pointer in s, then $\mathsf{dlad}_{r,p}(s)$ is an applicable operation, so $\mathsf{dlad}_{p,q}(s)$ cannot have minimal overlap.

(b) The claim follows from the fact that $\mathsf{dlad}_{p,q}(s)$ is the first dlad operation in s. $\qquad\square$

5.3   Finding the First dlad Operation Applicable to s

The first dlad operation applicable to the string s can be found and performed using the following algorithm.

Algorithm `firstdlad`(s)
    $S \leftarrow$ empty stack
    $i \leftarrow$ first node of s
    while $i \neq$ null **do** {
        $p \leftarrow i.pointer$
        if this is the first occurrence of p **then** {
            Put p in the stack S
            $i \leftarrow i.right$
        }
        else {
            $q \leftarrow$ pointer at the top of S
            if $p = q$ **then** { // Apply $\mathsf{ld}_p(s)$ operation
                $i \leftarrow i.right$
                Remove the nodes storing p and q from s
                Remove q from S.
            }
            else { // Apply $\mathsf{dlad}_{p,q}(s)$ operation
                Let j_0 be the node storing q.
                $j \leftarrow j_0.twin$
                $i_0 \leftarrow i.twin$
                Swap the sublists (i, j) and (i_0, j_0) in s.
                Remove nodes i, i_0, j, and j_0 from s.
                return
            }
        }
    }

The algorithm scans the legal string s from left to right until it finds the first pointer p whose second occurrence is preceded by the first occurrence of another pointer q. While scanning s all ld operations encountered are performed. Therefore, the algorithm correctly finds the first dlad operation applicable to s.

We can check in $O(1)$ time whether some node i contains the first or second occurrence of a pointer p by using a binary vector storing one bit for each pointer. Initially all bits are set to zero. When a pointer p is processed by the algorithm, the corresponding bit is checked. If the bit is zero then this is the fist occurrence of p, so the bit is set to 1; this way when the second occurrence of p is found the corresponding bit in the binary vector will be 1.

The algorithm spends $O(1)$ time processing each pointer in s, so its total running time is $O(n)$. Repeated application of this algorithm would reduce any negative legal string s to the empty string. However, up to $O(n)$ iterations might be needed (like, for example, for the input string $p_1 p_2 \cdots p_n p_1 p_2 \cdots p_n$) yielding an $O(n^2)$ time algorithm.

6 Lower Bounds for Reducing Negative Strings

We consider only algorithms that need to compare pointers or positions (of pointers) to find applicable dlad and ld operations. Any such algorithm A can be modeled as a decision tree T_A. Every branch of the decision tree corresponds to an execution path of A, i.e., the set of instructions executed by A on a given input. In this tree every internal node is labelled with two indices $i : j$. This label might either mean the comparison of the pointers at positions i and j in s or the comparison of the two positions i, j in s. As A executes, the instructions performed by the algorithm define a unique path in the decision tree: when A compares pointers (or positions) i and j, the current state of A is represented by a node a of T_A labelled $i : j$; if $i \leq j$ the algorithm moves to the left child of node a, otherwise it moves to the right child of a.

Proceeding in this manner with every comparison made by A, the execution of the algorithm defines a path $\rho(r, \ell)$ from the root r to a leaf ℓ of T_A. A leaf of the decision tree models the termination of the execution of A. Every leaf ℓ is labelled with a set S_ℓ of negative strings with n pointers; any string $t \in S_\ell$ causes the algorithm to follow the path $\rho(r, \ell)$ from r to ℓ.

Some edges of T_A will be labelled with a set of ld and dlad operations that the algorithm performs when traversing these edges. For example, if A has computed the positions i, j of the two occurrences of some pointer p, then A might check if $j = i + 1$ (by testing whether $j \leq i + 1$ and $i + 1 \leq j$), for then the $\mathsf{ld}_p(s)$ operation can be performed. This part of the algorithm might be modeled with the decision tree shown in Figure 3.

In the decision tree an ld operation is recorded as $\mathsf{ld}_p(i, i+1)$ to make it explicit that the same pointer p appears at positions i and $i + 1$ of the input string s. A dlad operation is expressed as $\mathsf{dlad}_{p,q}(i, j, i_1, j_1)$, where $i < j < i_1 < j_1$, p appears at positions i and i_1 of s, and q appears at positions j and j_1.

For simplicity, let us consider only legal strings s that do not contain inverted pointers. Note that this assumption does not affect Theorem 1.

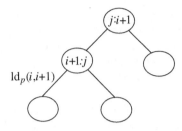

Fig. 3. Execution of A leading to an ld operation

Lemma 2. *For every leaf ℓ of T_A, consider the set S'_ℓ obtained from S_ℓ by removing all strings that contain inverted pointers. Then S'_ℓ contains at most n^n strings.*

Proof. Consider a leaf ℓ of T_A and let $u \in S'_\ell$. Consider the path $\rho(r, \ell)$ from the root r of T_A to the leaf ℓ. As the algorithm A follows this path, ld and/or dlad operations are applied to u so at the end it is reduced to the empty string.

Let us consider a reversed version A^r of the algorithm A that starts with an empty string and it follows the path $\rho(r, \ell)$, from ℓ moving up towards r. Every time that A^r follows an edge e labelled by a sequence $op_1(e), op_2(e), \ldots, op_k(e)$ of ld and dlad operations, the algorithm will perform over the current string s' a sequence $op_k^{-1}(e), \ldots, op_2^{-1}(e), op_1^{-1}(e)$ of operations, where

- $\mathsf{ld}_p^{-1}(i, i+1)$ adds a new pointer p (different from the other pointers in s') to positions i and $i + 1$ of s', and
- $\mathsf{dlad}_{p,q}^{-1}(i, j, i_1, j_1)$ adds new pointers p and q to s' so that the resulting string s'' has pointer p in positions i and i_1, q appears in positions j and j_1, and $\mathsf{dlad}_{p,q}(s'') = s'$.

Note that $\mathsf{ld}_p^{-1}(i, i+1)$ has at most n different outcomes as p could be any of the pointers $1, 2, \ldots, n$ and $\mathsf{ld}_{p,q}^{-1}(i, j, i_1, j_1)$ has at most $n(n-1)$ different outcomes. As every path $\rho(r, \ell)$ from the root of T_A to a leaf is labelled by a sequence of ld and dlad operations involving n pointers, then the application of the inverse operation $op_i^{-1}(e)$ along the path can produce at most n^n different strings. $\square$

Theorem 1. *Every algorithm which reduces negative strings requires $\Omega(n \log n)$ time.*

Proof. By the above lemma, each leaf of the decision tree is labelled by at most n^n strings without inverted pointers, and different leaves are labelled by disjoint sets of strings. Since the number of strings without inverted pointers is $(2n)!2^{-n}$, by Lemma 2 the decision tree has at least $(2n)!2^{-n}/n^n = (2n)!(2n)^{-n} > (2n)^{2n}e^{-2n}(2n)^{-n} = (2n)^n e^{-2n}$ leaves and, thus, the height of the decision tree is at least $\log_2((2n)^n e^{-2n}) = \Theta(n \log n)$. This implies that any reduction algorithm needs to perform at least $\Omega(n \log n)$ comparisons and, hence, to reduce any negative string, at least $\Omega(n \log n)$ time is required. $\square$

Corollary 1. *Every algorithm for reducing legal strings needs $\Omega(n \log n)$ time.*

7 Optimal Reductions

So far we have been concerned with fast algorithms for finding reductions of legal strings. Another basic question, seemingly more difficult, is how to find reductions which are themselves fast in reducing the legal strings, e.g., have minimum number of operations. (Clearly, some strings have many possible reductions; see below for examples.) One can also imagine weighted variants, where a certain operation has a given cost. This could be fixed for each operation or even depend on, for instance, the relative position of the pointers involved; the farther apart the pointers, the higher the cost.

The basic variant where the total number of operations needs to be minimized looks already difficult. We don't even know whether it is NP-hard or not. We discuss next a greedy approach to this problem.

7.1 Dlad-First Heuristic

It is clear that minimizing the total number of operations amounts to maximizing the number of dlad operations used. That is simply because a ld or a hi eliminates one pair of pointers whereas a dlad eliminates two. This raises the idea for the dlad-first strategy: perform a dlad whenever possible. Clearly, it may not be unique. Any of these will be called dlad-first reductions. However, there are cases when no dlad-first reduction is optimal. Here is an example. Consider the legal string $1\,3\,2\,\bar5\,\bar4\,\bar6\,\bar3\,1\,2\,6\,4\,5$. The only dlad we can perform is $\mathsf{dlad}_{1,2}$ and then we have the following reduction:

$$1\,3\,2\,\bar5\,\bar4\,\bar6\,\bar3\,1\,2\,6\,4\,5 \overset{\mathsf{dlad}_{1,2}}{\Longrightarrow} \bar5\,\bar4\,\bar6\,\bar3\,3\,6\,4\,5 \overset{\mathsf{hi}_{\bar3}}{\Longrightarrow} \bar5\,\bar4\,\bar6\,6\,4\,5 \overset{\mathsf{hi}_{\bar6}}{\Longrightarrow} \bar5\,\bar4\,4\,5 \overset{\mathsf{hi}_{\bar4}}{\Longrightarrow} \bar5\,5 \overset{\mathsf{hi}_{\bar5}}{\Longrightarrow} \Lambda.$$

Notice that the hi operations can be performed in any order producing different reductions but with the same number of operations. Also, after $\mathsf{dlad}_{1,2}$, nothing else except for hi operations can be performed, and so the smallest number of operations we can obtain with a dlad-first strategy is 5.

On the other hand, the following (optimal) reduction has four operations only, proving the claim:

$$1\,3\,2\,\bar5\,\bar4\,\bar6\,\bar3\,1\,2\,6\,4\,5 \overset{\mathsf{hi}_{\bar3}}{\Longrightarrow} 1\,6\,4\,5\,\bar2\,1\,2\,6\,4\,5 \overset{\mathsf{dlad}_{4,5}}{\Longrightarrow} 1\,6\,\bar2\,1\,2\,6 \overset{\mathsf{dlad}_{1,6}}{\Longrightarrow} 2\,\bar2 \overset{\mathsf{hi}_{\bar2}}{\Longrightarrow} \Lambda.$$

This example has six pairs of pointers. It is an interesting combinatorial exercise to prove that there is no such legal string with five pairs or less. Notice that it is not very easy to verify such properties using a computer program due to the high number of legal strings.

We start with a lemma which can be proved by a pretty straightforward (but tedious) exhaustive case analysis. (The second part of the lemma can also be derived from a result by Harju et al. [5].) The notation $\mathsf{dlad}_{\{i,j\}}$, where $i \neq j$, stands for the dlad involving both the i-pair and j-pair.

Lemma 3. *Consider a legal string w with n pairs and four pairwise different integers i, j, k, ℓ.*

(i) *If both* $\mathsf{dlad}_{\{i,j\}}$ *and* $\mathsf{dlad}_{\{i,k\}}$ *are applicable and* $w \stackrel{\mathsf{dlad}_{\{i,j\}}}{\Longrightarrow} u_1 k u_2 k u_3$, $w \stackrel{\mathsf{dlad}_{\{i,k\}}}{\Longrightarrow}$ $v_1 j v_2 j v_3$, *then* u_2 *and* v_2 *are a permutation of each other.*

(ii) *If both* $\mathsf{dlad}_{\{i,j\}}$ *and* $\mathsf{dlad}_{\{k,\ell\}}$ *are applicable and* $w \stackrel{\mathsf{dlad}_{\{i,j\}}}{\Longrightarrow} w_1$, $w \stackrel{\mathsf{dlad}_{\{k,\ell\}}}{\Longrightarrow}$ w_2, *then the reductions* $w_1 \stackrel{\mathsf{dlad}_{\{k,\ell\}}}{\Longrightarrow} \cdots$ *and* $w_2 \stackrel{\mathsf{dlad}_{\{i,j\}}}{\Longrightarrow} \cdots$ *are simultaneously possible or impossible.*

Assume that w is a legal string with 4 pairs which can be reduced by two operations, say: $w \stackrel{\mathsf{dlad}_{\{i,j\}}}{\Longrightarrow} w_1 \stackrel{\mathsf{dlad}_{\{k,\ell\}}}{\Longrightarrow} \Lambda$. If we can perform a $\mathsf{dlad}_{\{r,s\}}$ different from $\mathsf{dlad}_{\{i,j\}}$ then we can always continue with another dlad operation. Indeed, this follows from Lemma 3(i) in the case when $\{i,j\}$ and $\{r,s\}$ are not disjoint and by Lemma 3(ii) if they are. We proved

Corollary 2. *If a legal string* w *with 4 pairs can be reduced in two steps, then any* dlad *applied to* w *can be followed by another* dlad.

Proposition 1. *For all legal strings with five pairs or less, all dlad-first reductions are optimal.*

Proof. Assume there is a legal string w with five pairs or less which has a dlad-first reduction that is not optimal. Assume w has also the minimum number of pairs. Then, no ld can be performed as the resulting legal string would have the same property as w but fewer pairs.

In order to be able to make a difference in the number of operations, the optimal reduction must have at least two dlad operations. As w has at most five pairs, the optimal reduction will have exactly two dlad operations whereas the dlad-first one will not have any further dlad operations (which actually means that no dlad operations will be possible, as otherwise it would be optimal). In particular w must have five pairs.

We may assume, by possibly renaming some letters, that the two reductions, the first optimal and the second dlad-first non-optimal are:

$$w = x 1 y \bar{1} z \stackrel{\mathsf{hi}_1}{\Longrightarrow} w_1 = x \bar{y} z \stackrel{\mathsf{dlad}}{\Longrightarrow} w_2 \stackrel{\mathsf{dlad}}{\Longrightarrow} \Lambda \text{ and}$$
$$w \stackrel{\mathsf{dlad}_{2,3}}{\Longrightarrow} w_3 \Longrightarrow \Lambda.$$

We have that all $2, 3, 4, 5$ are negative in w_1, so that we can perform the two remaining dlad operations. On the other hand, 2 and 3 are negative already in w since $\mathsf{dlad}_{2,3}$ is applicable.

We notice that y cannot be empty as in that case we would find a shorter w with the same properties. Therefore, hi_1 switches some signs. If y does not contain any of $2, \bar{2}, 3, \bar{3}$ then hi_1 and $\mathsf{dlad}_{2,3}$ are independent and so the dlad-first reduction is optimal as well, a contradiction. If this is not the case, then assume y contains 2 (or $\bar{2}$). (All the other cases are analogous.) Then it has to contain the other 2 (or $\bar{2}$) as well since otherwise 2 would not be negative in w_1. Since $\mathsf{dlad}_{2,3}$ can be applied in w, it means there is a 3 or $\bar{3}$ between the 2's and therefore inside y. But, for the same reason, y must also contain the other 3 (or $\bar{3}$).

Assume $y = y_1 2 y_2 3 y_3 2 y_4 3 y_5$. We have then $w = x 1 y_1 2 y_2 3 y_3 2 y_4 3 y_5 \bar{1} z$ and the following reductions:

$$w \overset{\text{hi}_1}{\Longrightarrow} w_1 = x \bar{y}_5 \bar{3} \bar{y}_4 \bar{2} \bar{y}_3 \bar{3} \bar{y}_2 \bar{2} \bar{y}_1 z \overset{\text{dlad}_{3,2}}{\Longrightarrow} w_2 = x \bar{y}_5 \bar{y}_2 \bar{y}_3 \bar{y}_4 \bar{y}_1 z \text{ and}$$
$$w \overset{\text{dlad}_{2,3}}{\Longrightarrow} w_3 = x 1 y_1 y_4 y_3 y_2 y_5 \bar{1} z \overset{\text{hi}_1}{\Longrightarrow} w_2.$$

Since w_3 can be reduced necessarily in no less than three operations, no dlad is applicable to w_2. On the other hand, there is a reduction of w_1 by two dlad operations only. This contradicts Corollary 2. $\qquad\square$

7.2 Algorithms and Experiments for Dlad-First

Using the algorithms in Section 5 we can construct a quadratic time algorithm which finds a dlad-first reduction. Finding a best dlad-first reduction seems much harder.

Also, note that a sub-quadratic algorithm for performing all hi operations would not immediately imply a sub-quadratic algorithm for finding a dlad-first reduction. That is because we may have to alternate many times between dlad operations and hi operations. Here is a simple example. In the legal string $1213\bar{2}\bar{3}$ $4546\bar{5}\bar{6}$ $7879\bar{8}\bar{9}$ ... each group is independent from the others and a hi needs to be done first. If we use first hi_3, then we must have $\text{dlad}_{1,2}$ and so they can alternate until the end. (Notice that this particular example can also be reduced using hi operations only.)

Experimentally, the dlad-first strategy behaves quite well. Our tests on random legal strings with 10 pairs show that most strings (over 98%) possess dlad-first optimal reductions and for the remaining ones dlad-first reductions are usually longer by only one operation and very rarely by 2.

Also, all known micronuclear genes do have a dlad-first optimal reduction.

8 Conclusions and Further Research

We have investigated a number of algorithmic issues concerning gene assembly in ciliates. We gave a quadratic time algorithm for reducing strings and discussed a heuristic algorithm for approximating the optimal reductions. A number of interesting problems remain to be investigated. First, is it possible to reduce a string in sub-quadratic time? Second, can we compute a dlad-first reduction in linear time? Third, is the problem of computing optimal reductions NP-hard?

References

1. A. Ehrenfeucht, T. Harju, I. Petre, D.M. Prescott, and G. Rozenberg, Formal systems for gene assembly in ciliates, *Theoret. Comput. Sci.* **292** (2003) 199 – 219.
2. A. Ehrenfeucht, I. Petre, D.M. Prescott, and G. Rozenberg, String and graph reduction systems for gene assembly in ciliates, *Math. Structures Comput. Sci.* **12** (2001) 113 – 134.

3. A. Ehrenfeucht, T. Harju, I. Petre, D.M. Prescott, and G. Rozenberg, *Computation in Living Cells*, Springer, Berlin, Heidelberg, New-York, 2004.
4. A. Ehrenfeucht, D.M. Prescott, and G. Rozenberg, Computational aspects of gene (un)scrambling in ciliates, in: L.F. Landweber and E. Winfree (eds.), *Evolution as Computation*, Springer, Berlin, Heidelberg, 2001, 216 - 256.
5. T. Harju, C. Li, I. Petre, and G. Rozenberg, Parallelism in Gene Assembly, in: *Proceedings of the Tenth International Meeting on DNA Computing*, LNCS 3384, 2005, 138–148.
6. L.F. Landweber and L. Kari, The evolution of cellular computing: Nature's solution to a computational problem, in: *Proceedings of the 4th DIMACS Meeting on DNA Based Computers*, Philadelphia, PA, 1998, 3 - 15.
7. L.F. Landweber and L. Kari, Universal molecular computation in ciliates, in: L.F. Landweber and E. Winfree (eds.), *Evolution as Computation*, Springer, Berlin Heidelberg, 2002, 257 – 274.
8. D.M. Prescott, A. Ehrenfeucht, and G. Rozenberg, Molecular operations for DNA processing in hypotrichous ciliates, *Europ. J. Protistology* **37** (2001) 241 - 260.
9. D.M. Prescott and G. Rozenberg, Encrypted genes and their reassembly in ciliates, in: M. Amos (ed.), *Cellular Computing*, Oxford University Press, Oxford, 2003, 171–201.

Spectrum of a Pot for DNA Complexes

Nataša Jonoska, Gregory L. McColm, and Ana Staninska

Department of Mathematics
University of South Florida
{jonoska, mccolm, staninsk}@math.usf.edu

Abstract. Given a set of flexible branched junction DNA molecules (building blocks) with sticky ends we consider the question of determining the proper stoichiometry such that all sticky ends could end up connected. The idea is to determine the proper proportion (spectrum) of each type of molecules present, which in general is not uniform. We classify the pot in three classes: weakly satisfiable, satisfiable and strongly satisfiable according to possible components that assemble in complete complexes. This classification is characterized through the spectrum of the pot, which can be computed in PTIME using the standard Gauss-Jordan elimination method.

1 Introduction

DNA molecules through weak chemical bonds (hydrogen bonds) self assemble into larger and relatively stable nanostructures. The model described in this paper is motivated by DNA self-assembly based on the Watson-Crick complementarity pairing. Although the naturally occurring DNA molecule has a double helix structure, it can be configured in many other forms, e.g.: hairpin, branched 3 and 4 junction molecules, stick cube, truncated octahedron, etc. (see [3, 4, 11, 17, 18]). These newly formed molecules have been proposed for computational purposes [6] as well as for scaffolding for other structures [3, 4, 17, 18].

Several models for DNA self-assembly have appeared, mostly using rigid square tiles [1, 2, 9, 13, 14, 15]. In this paper, we consider another model that uses flexible tiles (each tile composed of a single branched junction molecule). This model was initially proposed in [5] and elaborated in [7, 8]. Flexible junction molecules have been used in experiments to obtain regular graph structures, such as the cube [3] and truncated octahedron [17, 18] and non-regular graph structures [6].

This model is based on DNA branched junction molecules with flexible arms extending to free sticky ends. By imposing restrictions on the number of types of tiles, one can get DNA computability classes that correspond to extant complexity classes. A "polynomial" restriction produces precisely the NPTIME queries; no restriction at all produces the classes of all computable queries [8].

In this model, a problem is encoded in branched junction molecules (tiles) and a solution is obtained if and only if a complete complex (complex without sticky ends) of appropriate size can assemble. In an experiment one can expect

C. Mao and T. Yokomori (Eds.): DNA12, LNCS 4287, pp. 83–94, 2006.
© Springer-Verlag Berlin Heidelberg 2006

to obtain many kinds of complexes, and not all of them may represent the designated structures. It can be observed experimentally that a portion of the DNA material in the pot ends up in incomplete complexes. Also the appearance of topoisomers have been reported [12, 16].

In [7] we considered questions for the probability of obtaining a complex of certain size, mainly with the use of methods from random graph theory. In this paper we consider the question how to reduce the amount of the useless material that is obtained at the end of an experiment. We propose a method and a program that determines the correct stoichiometry of the molecules in order to avoid the incomplete complexes present at the end of an experiment in ideal conditions. This means, we identify the right proportions of each molecule such that at the end of the experiment only desired structures are obtained.

A pot is defined to be a collection of tiles. For every pot we define the "spectrum" of the pot to be the set of the vectors of "right proportions" for the molecules. The spectrum of a pot is a convex subset of $\mathbb{Q}^m$ ($\mathbb{Q}$ is the set of rational numbers and m is the number of distinct tiles) hence if it is not empty, it is either a singleton or infinite. We use the Gauss-Jordan elimination algorithm for matrices to prove that the spectrum of a given pot is PTIME computable. This process also identifies the tiles that are "useless" in the sense, that will not appear in a complete complex.

The description of the model is presented in Section 2. It contains the main definitions of complexes and structures that are built up by junction molecules. In Section 3 we classify the pots in three categories depending whether every junction (tile) can be included into a complete complex or whether every sticky end can be used for assembly of a complete complex, or whether the pot gives rise to a complete complex. The definition of the spectrum of a given pot and its algebraic and geometric properties are given in Section 4.

Although the method developed is mainly applied to DNA self assembly, the idea behind it is very general and it is applicable to any other self-assembly process.

2   Model Description

Let H be a finite set called *sticky end types* and let $\theta : H \to H$ be such that $\theta(\theta(\mathbf{h})) = \mathbf{h}$ for all $\mathbf{h} \in H$. We call $\theta(\mathbf{h}) \in H$ the *complementary sticky end type* to $\mathbf{h}$ such that sticky ends of type $\mathbf{h}$ and type $\theta(\mathbf{h})$ bond. For each $\mathbf{h} \in H$ we assume that $\theta(\mathbf{h}) \neq \mathbf{h} = \theta(\theta(\mathbf{h}))$. Thus H can be partitioned into two sets H^+ and H^- such that if $\mathbf{h}$ is an element of H^+ then $\theta(\mathbf{h})$ is an element of H^-.

We simplify the notation by writing $\hat{\mathbf{h}}$ for $\theta(\mathbf{h})$ and we fix H for the rest of the paper. We use notation $[n] = \{1, 2, 3 \ldots n\}$ and $\mathbb{N} = \{0, 1, 2, \ldots\}$ in what follows.

Definition 1. *A tile type over (H, θ) is a function* $\mathbf{t} : H \to \mathbb{N}$. *A tile of type* $\mathbf{t}$ *has* $\mathbf{t}(\mathbf{h})$ *sticky ends of type* $\mathbf{h}$.

Informally, a tile type represents a type of branched junction molecule. Schematically it can be presented as a star-like graph (see Fig.1) with sticky end labels at the one degree vertices. A _tile_ is a copy of a certain tile type, and if a tile t is of type $\mathbf{t}$ then for each $\mathbf{h} \in H$, $t(\mathbf{h}) = \mathbf{t}(\mathbf{h})$, meaning that the tile t has exactly $\mathbf{t}(\mathbf{h})$ sticky ends of type $\mathbf{h}$. In a pot with DNA molecules there are many copies of a given type of junction molecule, and hence we can assume potentially an infinite supply of tiles for each type of tile.

Definition 2. _A_ pot type _over_ (H, θ) _is a set_ $\mathbf{P}$ _of tile types over_ (H, θ) _such that for any_ $\mathbf{h} \in \overline{H}$ _and_ $\mathbf{t} \in \mathbf{P}$, _if_ $\mathbf{t}(\mathbf{h}) > 0$ _then there exists_ $\mathbf{t}' \in \mathbf{P}$ _such that_ $\mathbf{t}'(\hat{\mathbf{h}}) > 0$. _To ease the notation we write_ $\mathbf{P}(H, \theta)$ _for a pot type over_ (H, θ).

Thus no pot type admits tiles with unattachable sticky ends. A _pot_ P is a collection of tiles from types in $\mathbf{P}$. We will presume that we are working with a pot P of type $\mathbf{P}$, where P contains sufficiently many distinct tiles of each tile type to permit the construction described in this abstract.

Definition 3. _A_ complex _over a pot type_ $\mathbf{P}$ _is a pair_ $C = \langle T, J \rangle$ _where_ T _is a set of tiles with tile types in_ $\mathbf{P}$ _and_ J _is a set of unordered pairs_ $c = \{(t, \mathbf{h}), (t', \mathbf{h}')\}$ _satisfying the following properties:_

- _for each_ $c = \{(t, \mathbf{h}), (t', \mathbf{h}')\} \in J$, $t, t' \in T$, $t(\mathbf{h}), t'(\mathbf{h}') > 0$, _and_ $\mathbf{h}' = \hat{\mathbf{h}}$ (c _indicates the connection between two complementary sticky ends) and_
- _the cardinality_ $|\{c : (t, \mathbf{h}) \in c\}| \leq t(\mathbf{h})$ _(this prevents the tile from making more connections than it has sticky ends)._

Definition 4. _The_ type _of a complex_ $C = \langle T, J \rangle$ _is the function_ $type(C) : H \rightarrow \mathbb{N}$ _defined by_

$$type(C)(\mathbf{h}) = \sum_{t \in T} t(\mathbf{h}) - |\{c : (t, \mathbf{h}) \in c\}|.$$

Informally, a complex type records the number and the types of the sticky ends that are free.

Note: A tile is also a complex $t = \langle \{t\}, \emptyset \rangle$ and a tile type is also a complex type.

Definition 5. _A complex_ C _is called_ complete _if it has no free sticky ends, i.e., for all sticky ends_ $\mathbf{h}$, $type(C)(\mathbf{h}) = 0$.

For a pot type $\mathbf{P}$ _we denote by_ $\mathcal{C}(\mathbf{P})$ _the set of all complete complexes that can be obtained by tiles of tile types in_ $\mathbf{P}$.

3 Pot Type Classification

A complex $C = \langle T, J \rangle$ is _embedded_ in a complex $C' = \langle T', J' \rangle$ if $T \subseteq T'$ and $J \subseteq J'$.

For pot types we consider three levels of "satisfiability".

Definition 6.   _A pot type_ $\mathbf{P}$ _is_ weakly satisfiable _if it admits a complete complex, i.e,_ $\mathcal{C}(\mathbf{P}) \neq \emptyset$.

A pot type $\mathbf{P}$ _is_ satisfiable _if, for each_ $\mathbf{h} \in H$, _there is a complete complex_ $C \in \mathcal{C}(\mathbf{P})$ _of the pot containing at least one sticky end of type_ $\mathbf{h}$.

A pot type **P** *is strongly satisfiable if every complex that can be generated by* **P** *can be embedded into a complete complex of* **P**.

Strong satisfiability is the notion of most immediate interest in this paper and in general for describing self-assembly processes.

Lemma 7. *A pot type* **P** *is strongly satisfiable iff every tile of a type in* **P** *can be embedded into a complete complex in* $\mathcal{C}(\mathbf{P})$.

Proof. (Sketch)

One implication of the lemma is trivial; if **P** is strongly satisfiable, since every tile is a complex, it can be embedded into a complete complex.

The converse is obtained by mathematical induction on the number of tiles in a complex. If the statement holds for k-tile complexes, then as a $(k+1)$-tile complex C can be "separated" into a k-tile complex C' and a tile t. These complexes C' and t can join with complexes $\hat{C}'$ and $C_{\hat{t}}$ such that $\hat{C}'$ and $C_{\hat{t}}$ have complementary sticky ends to C' and t respectively, to form complete complexes. Then C could join an amalgam of $\hat{C}'$ and $C_{\hat{t}}$ and form a complete complex. Therefore every complex can be embedded into a complete complex.   □

It is straightforward to see that all strongly satisfiable pot types are also satisfiable and that all satisfiable pot types are weakly satisfiable. But the converse is not necessary true. Fig.1a shows a pot type that is satisfiable, but not strongly satisfiable, since a tile of tile type t_3 can never be embedded into a complete complex. The pot type of Fig.1b is an example of a pot type that is weakly satisfiable, but not satisfiable since the sticky end type c can never be a part of any complete complex.

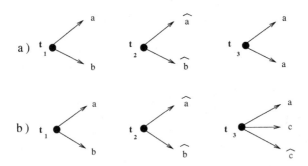

Fig. 1. a) Satisfiable pot type that is not strongly satisfiable (t_3 cannot be a part of a complete complex) b) weakly satisfiable pot type that is not satisfiable (the sticky end c cannot be a part of a complete complex)

Note that the number of sticky end types doesn't depend on the number of tile types. The pot types in all three examples in Fig 2 are strongly satisfiable; in the first example the number of tile types and sticky end types are equal; in the second one there are less tile types then free sticky end types; and in the third example there are more tile types then sticky end types.

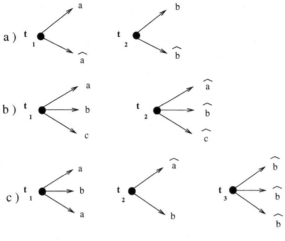

Fig. 2. a) $|\mathbf{P}| = |H^+|$ b) $|\mathbf{P}| < |H^+|$ c) $|\mathbf{P}| > |H^+|$

From reports on DNA assemblies we know that when one runs an experiment, the desired complexes are not the only thing which show in the pot, for there may be a lot of incomplete complexes. They increase the error rate and certainly increase the cost of the experiment.

If the stoichiometry in the test tube is bad, i.e., an improper ratio of each of the molecules is used, then any conditions there will be incomplete complexes. We propose a method which theoretically (ignoring all dynamic considerations such as those in [10]) eliminates the presence of incomplete complexes assuming that assembly occurs ideally in a well mixed diluted pot.

The main motive of this research is to study the optimization of the assembly of complete complexes and to show how to predesign a pot type in which at the end of an experiment only complete complexes would be expected.

To ease notation, *for the rest of the paper we will work with a pot type* $\mathbf{P} = \{\mathbf{t}_1, \mathbf{t}_2 \dots, \mathbf{t}_m\}$ *consisting of m tile types, and n sticky end types* $H^+ = \{\mathbf{h}_1, \mathbf{h}_2, \dots, \mathbf{h}_n\}$.

To each complex type C we associate a vector $z_C = (z_1, z_2, \dots, z_n)$ from $\mathbb{Z}^n$ such that $z_i : H^+ \to \mathbb{Z}$

$$z_i = \text{type}(C)(\mathbf{h}_i) - \text{type}(C)(\hat{\mathbf{h}}_i).$$

We assume that the pot is diluted and the thermodynamic conditions are such that all sticky ends that can connect would be able to. In this sense z_C gives information about the remaining free sticky end types on the complex C.

4   Spectrum of a Pot

4.1   Definitions

To solve a problem with DNA molecules, first the molecules that encode the problem need to be synthesized, and then combined into one pot. Often, best stoichiometry is assumed by uniform distribution of the predesigned molecules. If instead of uniform distribution, the proportion of every molecule in the pot is carefully selected, the incomplete complexes may be eliminated from the pot. The collection of vectors that represent the selected proportions of tile types such that only complete complexes are assembled is called the *spectrum of the pot*. Formally,

Definition 8. *The* spectrum *of* **P** *is the set* S *of all vectors* $\mathbf{r} = (r_\mathbf{t} : \mathbf{t} \in \mathbf{P})$ *such that:*

1) *For each* $\mathbf{t}$, $r_\mathbf{t} \geq 0$ *and*

$$\sum_{\mathbf{t} \in \mathbf{P}} r_\mathbf{t} = 1, \tag{1}$$

2) *for each* $\mathbf{h}$,

$$\sum_{\mathbf{t} \in \mathbf{P}} r_\mathbf{t} \mathbf{t}(\mathbf{h}) = \sum_{\mathbf{t} \in \mathbf{P}} r_\mathbf{t} \mathbf{t}(\hat{\mathbf{h}}), \tag{2}$$

i.e, for each $\mathbf{h} \in H$ *there are as many sticky ends of type* $\hat{\mathbf{h}}$ *as there are of type* $\mathbf{h}$.

Using the vector $z_\mathbf{t}(\mathbf{h}) = \mathbf{t}(\mathbf{h}) - \mathbf{t}(\hat{\mathbf{h}})$ associated to tile t when it is considered as a complex, the second part of the definition can be rewritten in the following form

2*)

$$\sum_{\mathbf{t} \in \mathbf{P}} r_\mathbf{t} z_\mathbf{t}(\mathbf{h}) = 0, \quad \text{for } z_\mathbf{t}(\hat{\mathbf{h}}) = -z_\mathbf{t}(\mathbf{h}). \tag{3}$$

If a pot has a mixture of tiles whose proportions correspond to a vector in the spectrum, then theoretically, "ignoring" dynamics, only complete complexes need be expected. If the used stoichiometry is different then the proportions of any of the vectors of the spectrum, there are no conditions in the tube that could avoid incomplete complexes. A vector in the spectrum can also be considered as a vector of the probabilities for each tile to be part of a complete complex.

Example 9. The spectrum of the pot types given in Fig.1a,b each containing three tile types, is the solution of the following systems of equations for $r_1, r_2, r_3 \geq 0$.

(a) $r_1 + r_2 + r_3 = 1$	(b) $r_1 + r_2 + r_3 = 1$
$r_1 - r_2 + 2r_3 = 0$	$r_1 - r_2 = 0$
$r_1 - r_2 = 0$	$r_1 - r_2 = 0$
	$2r_3 = 0$.

Both systems have the same solution, i.e., the spectrum of both pot types is $S = \{(\frac{1}{2}, \frac{1}{2}, 0)\}$. These two examples show that it cannot be distinguished between the spectra of a satisfiable pot and a weakly satisfiable pot. Note that a pot type admits a complete complex iff its spectrum is nonempty. So in the above example the spectrum points out that the use of equal number of molecules of the first two types, and no use of any molecules from the third type for no sticky ends to remain free.

There are finite number of tiles in a given pot type, so the proportion of each tile is a rational number, i.e. $\mathcal{S}(\mathbf{P}) \subseteq \mathbb{Q}^m$, $\mathbb{Q}$ being the set of rational numbers.

4.2   Geometric Representation of the Spectrum

The first part of Definition 8 shows that the spectrum of a pot with m tile types is a subset of the set $\mathcal{H}_m$ defined as the intersection of the subspace of $\mathbb{Q}^m_+ = \{(r_1, r_2, r_3 \ldots r_m) : r_i \in \mathbb{Q},\ r_i \geq 0 \text{ for } i \in [m]\}$ and the hyperplane $r_1 + r_2 + \ldots + r_m = 1$. This hyperplane is a simplex, and the spectrum of any given pot is dense in a simplex and includes the vertices of that simplex. [1]

From part (2) of Definition 8 it follows that the spectrum is an intersection of n hyperplanes (for each $\mathbf{h}$, $\sum_{\mathbf{t} \in \mathbf{P}} r_\mathbf{t} z_\mathbf{t}(\mathbf{h}) = 0$) and $\mathcal{H}_m$.

Proposition 10. *The spectrum $\mathcal{S}(\mathbf{P})$ of a pot type $\mathbf{P}$ with $|\mathbf{P}| = m$ and corresponding set of sticky ends H with $|H| = n$ is an intersection of n hyperplanes and the set $\mathcal{H}_m$. Moreover the spectrum is a simplex, since $\mathcal{H}_m$ is a simplex, and its vertices are rational points.*

Proposition 11. a) *A pot type is weakly satisfiable iff it admits a nonempty spectrum.*
b) *The closure of the spectrum $\mathcal{S}(\mathbf{P})$ of a pot type $\mathbf{P}$ is a convex set of vectors: if $\mathbf{u}, \mathbf{v} \in \mathcal{S}(\mathbf{P})$, and if $z \in [0,1]$, then $z\mathbf{u} + (1 - z)\mathbf{v} \in \mathcal{S}(\mathbf{P})$.*

Proof

a) If a weakly satisfiable pot type $\mathbf{P}$ admits a complete complex, then at least one of the tiles in the pot is a part of that complex. The proportion of that tile type is nonzero and hence $\mathcal{S}(\mathbf{P}) \neq \emptyset$.

Conversely, if $\mathcal{S}(\mathbf{P}) \neq \emptyset$, there exists a nonzero vector $r = (r_\mathbf{t} : \mathbf{t} \in \mathbf{P})$ of rational numbers in $\mathcal{S}(\mathbf{P})$. Multiply r with a positive integer to get a vector $(q_\mathbf{t} : \mathbf{t} \in \mathbf{P})$ of positive integer. Take $q_\mathbf{t}$ tiles of type $\mathbf{t}$ for each $\mathbf{t}$; and connect the complementary sticky ends. The result is a collection of complete complexes.
b) Follows immediately from the fact that the spectrum is a simplex.   □

Proposition 12. *The spectrum $\mathcal{S}(\mathbf{P})$ of a given pot $\mathbf{P}$ is either empty, a singleton or an infinite set.*

---

[1] Let $\{a_1, a_2, \ldots, a_n\}$ be a geometrically independent set in $\mathbb{R}^N$. An n-simplex spanned by $a_1, \ldots, a_n$ is the set of all points $x \in \mathbb{R}^N$ such that $x = \sum_{i=0}^{n} t_i a_i$ where $\sum_{i=0}^{n} t_i = 1$. An intersection of simplices is a simplex . We will further require that $t_i \geq 0$ for each i. In this abstract, the vectors a_i are all rational, and hence the simplices are convex polytopes (n-dimensional polygonal solids) with rational vertices.

Proof. The spectrum of a non-weakly satisfiable pot is empty (Proposition 11). Since the spectrum is a convex set, if it contains two points then it contains every other rational point between those two. Hence the spectrum is infinite. □

Proposition 13. *If the spectrum consists of only one point, $S(\mathbf{P}) = \{(r_{t_1}, r_{t_2}, \dots, r_{t_m})\}$, and if $r_{t_k} > 0$ for some $k \in [m]$, then every complete complex in $\mathcal{C}(\mathbf{P})$ contains a tile of type $\mathbf{t}_k$. Moreover, if $r_{t_k} > 0$ for all $k \in [m]$, $\mathbf{P}$ is strongly satisfiable.*

Proof. First suppose that $r_{t_k} > 0$, but towards contradiction, suppose that there is a complete complex C that does not contain t_k. If C contains q_l tiles of type $\mathbf{t}_l$ for each l, and q tiles altogether, then $(\frac{q_1}{q}, \frac{q_2}{q}, \dots, \frac{q_m}{q}) \in S(\mathbf{P})$, contradicting that $|S(\mathbf{P})| = 1$ and $r_{t_k} \neq o = \frac{q_k}{q}$. □

Now let's consider examples for the spectra of strongly satisfiable pot types.

Example 14. Consider the pot types depicted in Fig.2. Their spectra can be computed similarly as in Example 9. The spectrum of the pot type in Fig.2a is $S = \{(u, 1 - u) | u \in \mathbb{Q}\}$, the spectrum of the pot type in Fig.2b is $S = \{(\frac{1}{2}, \frac{1}{2})\}$, while the one for Fig.2c is $S = \{(\frac{1}{4}, \frac{1}{2}, \frac{1}{4})\}$. In the pot type of Fig.2a every proportion of the molecules of each type ends up with complete complexes only, while for the other two pot types it is not the case.

In two dimensional space (corresponding to a pot type with exactly two tile types) since all of the equations for the hyperplanes are equations of a line, the spectrum is a part of the line segment $(r_1 + r_2 = 1, 0 \le r_1 \le 1, 0 \le r_2 \le 1)$ connecting the points $(0, 1)$ and $(1, 0)$. So the spectrum is either a point of that line segment, or it is the entire line segment, or it is the empty set.

In three dimensional space, (corresponding to a pot type with exactly three tiles) the spectrum is the intersection of the planes $\sum_{i=1}^{3} r_i \mathbf{z}_{\mathbf{t}_i}(\mathbf{h}) = 0$, the plane $r_1 + r_2 + r_3 = 1$ and the subspace $\{(r_1, r_2, r_3)| \ r_i \ge 0 \text{ for } i \in \{1, 2, 3\}\}$.

It means the spectrum of a pot with three tiles types can be either a point in the triangle $\{(r_1, r_2, r_3)| \ r_1 + r_2 + r_3 = 1\} \cap \{(r_1, r_2, r_3)| \ r_i \ge 0 \text{ for } i \in \{1, 2, 3\}\}$ (which is the triangle that connects points $(1, 0, 0)$, $(0, 1, 0)$ and $(0, 0, 1)$) (See 4), a line segment within this triangle, or the entire interior of the triangle with the triangle itself. The following example illustrates all three cases.

Example 15. All three examples have a three tile type pot type $\mathbf{P} = \{\mathbf{t}_1, \mathbf{t}_2, \mathbf{t}_3\}$, and the set of sticky end types $H = \{\mathbf{a}, \mathbf{b}, \hat{\mathbf{a}}, \hat{\mathbf{b}}\}$.

For two dimensional space the spectrum is part of the line segment connecting the points $(1, 0)$ and $(0, 1)$, for three dimensional space it is part of the triangle connecting the points $(1, 0, 0)$, $(0, 1, 0)$ and $(0, 0, 1)$, for four dimensional it is part of the tetrahedron connecting the points $(1, 0, 0, 0)$, $(0, 1, 0, 0)$, $(0, 0, 1, 0)$ and $(0, 0, 0, 1)$, etc.

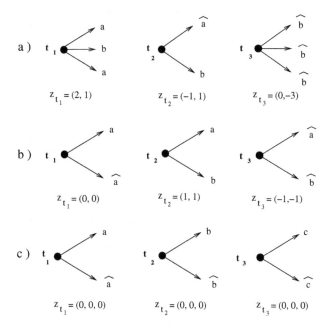

Fig. 3. a) Strongly satisfiable pot type with spectrum $\{(\frac{1}{4}, \frac{1}{2}, \frac{1}{4})\}$, b) Strongly satisfiable pot type with spectrum $\{(1 - 2u, u, u) : 0 \le u \le \frac{1}{2}\}$ c) Strongly satisfiable pot type with spectrum $\{(1 - u - v, u, v) : 0 \le u \le 1, 0 \le v \le 1, u + v \le 1\}\}$

Note the spectrum contains vectors with rational entries, its closure is bounded and therefore compact subset of $\mathbb{R}^m$.

4.3   Algebraic Representation of the Spectrum

The spectrum is the intersection of n hyperplanes (for each $\mathbf{h}$, $\displaystyle\sum_{t \in P} r_t z_t(\mathbf{h}) = 0$) and $\mathcal{H}_m$. Hence it is the solution of n homogeneous and 1 inhomogeneous equations with m variables over $\mathbb{Q}^+$.

$$
\begin{aligned}
r_1 + \quad\quad r_2 + \ldots + \quad\quad\quad r_m &= 1 \\
z_{t_1}(\mathbf{h}_1)r_1 + z_{t_2}(\mathbf{h}_1)r_2 + \ldots + z_{t_m}(\mathbf{h}_1)r_m &= 0 \\
z_{t_1}(\mathbf{h}_2)r_1 + z_{t_2}(\mathbf{h}_2)r_2 + \ldots + z_{t_m}(\mathbf{h}_2)r_m &= 0 \\
\vdots \quad\quad\quad\quad \vdots \quad\quad\quad\quad\quad\quad \vdots \quad &\vdots \\
z_{t_1}(\mathbf{h}_n)r_1 + z_{t_2}(\mathbf{h}_n)r_2 + \ldots + z_{t_m}(\mathbf{h}_n)r_m &= 0.
\end{aligned}
\tag{4}
$$

An efficient way to do that is by the Gauss-Jordan elimination, which transforms the augmented matrix of system (4) into the row-echelon form. The computational complexity of solving this system with the aid of Gauss-Jordan elimination is $O(m^2 n)$.

From Example 9 and Example 14 it can be seen that for satisfiable and weakly satisfiable pots (and not strongly satisfiable pots) vectors of the spectrum may

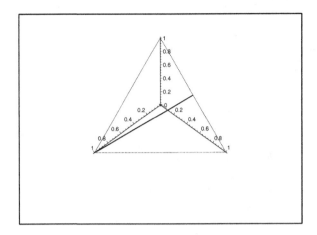

Fig. 4. The closure of the spectrum of the pot type given in Example 15b) is the line segment; the closure of the spectrum of the pot type given in Example 15c) is the triangle bounded by the dotted lines along with its interior

have zero coordinates. If the spectrum of a strongly satisfiable pot is a singleton, then all of its coordinates are positive numbers (Proposition 17).

Definition 16. *Let A be a set of n dimensional vectors. The* support *of A is the set $supp(A) = \{i \in [n] : \text{there exists a vector } \mathbf{u} = (u_1, u_2, \ldots, u_n) \in A \text{ such that } u_i \neq 0\}$. In other words, if $i \notin supp(A)$, then the i th coordinate of every point in A is 0.*

Proposition 17. *Suppose $\mathcal{S}(\mathbf{P})$ is the spectrum of a given pot type $\mathbf{P}$ with $|\mathbf{P}| = m$*

a) $supp(\mathcal{S}(\mathbf{P})) = [m]$ *iff $\mathbf{P}$ is strongly satisfiable.*
b) $\emptyset \neq supp(\mathcal{S}(\mathbf{P})) \subsetneq [m]$ *iff (P is weakly satisfiable that is not strongly satisfiable).*

Proof. a) $supp(\mathcal{S}(\mathbf{P})) = [m]$ iff every tile has a positive probability of being on a complete complex, i.e., every tile can be embedded into a complete complex, which means that the pot type is strongly satisfiable.
b) $supp(\mathcal{S}(\mathbf{P})) \subsetneq [m]$ iff there is a coordinate that is zero in every vector of the spectrum, i.e. at least one tile type cannot be embedded into a complete complex, so the pot is not strongly satisfiable, but it is weakly satisfiable since it has a nonempty spectrum.

□

Proposition 18. *Weak satisfiability, satisfiability and strong satisfiability of pot types are in PTIME.*

Proof. Let $\mathbf{P}$ be a pot type with m tile types and n sticky end types. In order to obtain the spectrum for the given pot we need to solve system (4) of $n + 1$

equations with m variables. If there is a solution with all positive coordinates, then the spectrum is nonempty and from Proposition 17 it follows that the pot is strongly satisfiable. If there is a solution to the system (4) and $supp(\mathcal{S}(\mathbf{P})) \subsetneq [m]$, then the spectrum is nonempty and from Propositions 11 and 17 follows that the pot is weakly satisfiable but not strongly satisfiable. Therefore the weak satisfiability and strong satisfiability are in PTIME.

Now suppose that the pot is weakly satisfiable but not strongly satisfiable, i.e. $supp(\mathcal{S}(\mathbf{P})) \subsetneq [m]$.

In order to determine whether a pot is satisfiable or only weakly satisfiable, for every $\mathbf{h} \in H$ we consider the following m dimensional vectors: $l_{\mathbf{h}} = (l_{\mathbf{t}}(\mathbf{h}) : \mathbf{t} \in \mathbf{P})$ such that

$$l_{\mathbf{t}}(\mathbf{h}) = \begin{cases} 1 & \text{if } \mathbf{t}(\mathbf{h}) \geq 1 \text{ or } \mathbf{t}(\hat{\mathbf{h}}) \geq 1 \\ 0 & \text{otherwise} \end{cases}$$

If there exists a sticky end $\mathbf{h} \in H$ such that $l_{\mathbf{h}}(i) = 0$ for all $i \in supp(\mathcal{S}(\mathbf{P}))$ are 0's, then $\mathbf{h}$ could not be embedded into any complete complex, hence the pot is weakly satisfiable, but not satisfiable (otherwise the pot is satisfiable). Hence, the computational complexity for classifying pot types according to their type of satisfiability is $O(m^2 n) + O(mn) = O(m^2 n)$, i.e it is in PTIME.   □

Acknowledgment. Authors thank Stephen W. Suen, David A. Rabson, and the referees for providing valuable suggestions. The work is supported in part by NSF grants CCF #0432009 and CCF #0523928.

References

[1] L.M. Adleman, Q. Cheng, A. Goel, M-D. Huang, D. Kempe, P. Moisset de Espanes, P.W.K. Rothemund. *Combinatorial optimization problems in self-assembly*, STOC'02 Proceedings, Montreal Quebec, Canada, 2002.

[2] L.M. Adleman, J. Kari, L. Kari, D. Reishus. *On the decidability of self-assembly of infinite ribons*, Proceedings of FOCS 2002, IEEE Symposium on Foundations of Computer Science, Washington (2002) 530-537.

[3] J.H. Chen, N.C. Seeman. *Synthesis from DNA of a molecule with the connectivity of a cube*, Nature **350** (1991).

[4] R. P. Goodman, I. A. T. Schaap, C. F. Tardin, C. M. Erben, R. M. Berry, C. F. Schmidt, A. J. Turberfield. *Rapid Chiral Assembly of Rigid DNA Building Blocks for Molecular Nanofabrication*, Science **310** (2005).

[5] N. Jonoska, S. Karl, M. Saito. *Three dimensional DNA structures in computing*, BioSystems **52** (1999) 143-153.

[6] N. Jonoska, P. Sa-Ardyen, N.C. Seeman. *Computation by self-assembly of DNA graphs*, Genetic Programming and Evolvable Machines **4** (2003) 123-137.

[7] N. Jonoska, G. McColm, A. Staninska. *Expectation and variance of self-assembled graph structures*, A. Carbone, N. Pierce (eds) Springer LNCS 3892 (2006)144-157.

[8] N. Jonoska, G.L. McColm. *A computational model for self-assembling flexible titles*, C. S. Calude et al. (eds) Springer LNCS 3699 (2005).

[9] M-Y. Kao, V. Ramachandran. *DNA self-assembly for constructing 3D boxes*. Algorithms and Computations, ISAC 2001 Preceedings, Springer LNCS **2223** (2001) 429-440.

[10] S. A. Kurtz, S. R. Mahaney, J. S. Royer, J. Simon. *Active transport in biological computing.* L. Landweber and E. Baum (eds) DIMACS Vol 44 (1997) 171-181.

[11] C. Mao, W. Sun, N.C. Seeman. *Designed two-dimensional DNA holliday junction arrays visualized by atomic force microscopy,* Journal of American Chemical Society 121(23) (1999) 5437-5443.

[12] J. Qi, X. Li, X. Yang, N.C. Seeman. *Ligation of triangles built from bulged 3-arm DNA branched junctions,* Journal of American Chemical Society 120 (1996) 6121-6130.

[13] J. H. Reif, S. Sahu, P. Yin. *A self-assembly model of time-dependent glue strength,* A. Carbone, N. Pierce (eds) Springer LNCS 3892 (2006).

[14] P.W.K. Rothemund, P. Papadakis, E. Winfree. *Algorithmic self-assembly of DNA Sierpinski triangles,* PLoD Biology 2 (12) e424 (2004).

[15] P.W.K. Rothemund, E. Winfree. *The program-size complexity of self-assembled squares,* Proceedings of 33rd ACM meeting STOC 2001, Portland, Oregon, May 21-23 (2001) 459-468.

[16] P. Sa-Ardyen, N. Jonoska, N.C. Seeman. *Self-assembly of graphs represented by DNA helix axis topology,* Journal of American Chemical Society 126(21) (2004) 6648-6657.

[17] W.M. Shih, J.D. Quispe, G.F. Joyce., *A 1.7-kilobase single stranded DNA folds into a nanoscale octahedron,* Nature **427** (2004) 618-621.

[18] Y. Zhang, N.C. Seeman. *The construction of a DNA truncated octahedron,* Journal of American Chemical Society 116(5) (1994) 1661-1669.

On the Complexity of Graph Self-assembly in Accretive Systems

Stanislav Angelov, Sanjeev Khanna, and Mirkó Visontai

Department of Computer and Information Science
School of Engineering and Applied Sciences
University of Pennsylvania, Philadelphia, PA 19104, USA
{angelov, sanjeev, mirko}@cis.upenn.edu

Abstract. We study the complexity of the Accretive Graph Assembly Problem (AGAP). An instance of AGAP consists of an edge-weighted graph G, a seed vertex in G, and a temperature τ. The goal is to determine if there is a sequence of vertex additions which constructs G starting from the seed. The edge weights model the forces of attraction and repulsion, and determine which vertices can be added to a partially assembled graph at the given temperature.

Our first result is that AGAP is NP-complete even on degree 3 planar graphs when edges have only two different types of weights. This resolves the complexity of AGAP in the sense that the problem is polytime solvable when either the degree is bounded by 2 or the number of distinct edge weights is one, and is NP-complete otherwise. Our second result is a dichotomy theorem that completely characterizes the complexity of AGAP on degree 3 bounded graphs with two distinct weights: w_p, w_n. We give a simple system of linear constraints on w_p, w_n, and τ that determines whether the problem is NP-complete or is polytime solvable. In the process of establishing this dichotomy, we give the first polytime algorithm to solve a non-trivial class of AGAP. Finally, we consider the optimization version of AGAP where the goal is to realize a largest-possible subgraph of the given input graph. We show that even on constructible graphs of degree at most 3, it is NP-hard to realize a $(1/n^{1-\epsilon})$-fraction of the input graph for any $\epsilon > 0$; here n denotes the number of vertices in G.

1 Introduction

Self-assembly is a process in which small objects interact autonomously with each other to form intricate complexes. The self-assembly approach is particularly appealing for constructing molecular scale objects with nano-scale features [1]. Examples of its application and practical modeling can be found in [2,3,4,5,6,7,8,9].

Based on the Wang Tiling Models [10], Rothemund and Winfree [11] proposed the *Tile Assembly Model* to formalize and facilitate theoretical study of the self-assembly process. In this model, DNA tiles are abstracted as oriented squares, where each side has a glue type and (non-negative) strength. An assembly starts from a designated *seed* tile and can be augmented by a tile if the sides of the tile match the glue types of its already assembled neighbors and the total glue

C. Mao and T. Yokomori (Eds.): DNA12, LNCS 4287, pp. 95–110, 2006.
© Springer-Verlag Berlin Heidelberg 2006

strength is no less than a threshold parameter τ, referred to as the *temperature* of the assembly.

Recently, Reif, Sahu, and Yin [1] proposed a generalization of the Tile Assembly Model, i.e., assembly on two-dimensional grids, to one on general graphs, named the *Accretive Graph Self-Assembly Model*. The accretive graph assembly is a *sequential* process where a weighted graph is assembled one vertex at a time starting from a seed vertex. The weight of each positive (resp. negative) edge specifies the magnitude of attraction (resp. repulsion) between the incident vertices. Again, a vertex is added to the assembly if the net attraction minus repulsion of the built neighbors is at least τ. *Accretive* here suggests the monotone property of the process, i.e., an added vertex cannot be removed later, in contrast to the *Self-Destructive Graph Assembly Model* [1], or the *Kinetic Tile Assembly Model* where tiles can fall off [12,13].

The Accretive Graph Self-Assembly Model addresses some of the deficiencies of the Tile Assembly Model. Namely, it models repulsion and allows the assembly of general graph structures. A central problem in this model is the *Accretive Graph Assembly Problem* (AGAP): Given a weighted graph on n vertices and a seed vertex, the problem asks for a sequence of vertex additions respecting τ that builds the graph. Among other results, it was shown in [1] that AGAP is NP-complete for graphs with maximum degree 4 and for planar graphs (PAGAP) with maximum degree 5. The authors in [1] posed several natural open problems related to AGAP which we address in this paper. The first question was to determine the precise degree bound for which AGAP and PAGAP change in complexity from polytime solvable to NP-complete, and the second one was to determine the difficulty of the optimization versions of these problems.

Our Results and Techniques. The complexity of a graph assembly system can be measured by the degree of the underlying graph G and the number of possible weights an edge can take. Similarly to the number of different tiles in the Tile Assembly Model, here we can bound the number of different vertex *types*, where a type is determined by the weights of the edges incident to a vertex. A natural question is under what conditions we can solve AGAP in polytime and what is the smallest complexity for which we can show the problem to be NP-complete. Our main results settle open problems posed in [1] and are as follows:

- We show that PAGAP (and hence AGAP) is NP-complete even if the maximum degree of the input graph is 3 and edges can take only two different weights. This result is tight in the sense that AGAP is polytime solvable if either the maximum degree is bounded by two or all edges have identical weights.
- We prove a dichotomy theorem that completely characterizes the complexity of AGAP on degree 3 bounded graphs with two distinct weights: w_p, w_n. We give a simple system of linear constraints on w_p, w_n, and τ that determines whether the problem is NP-complete or polytime solvable. In the process of establishing this dichotomy, we give a polytime algorithm to solve a nontrivial class of AGAP instances.

- We show that MAX AGAP, the optimization version of AGAP, is hard to approximate within a factor of $O(n^{1-\epsilon})$ for any $\epsilon > 0$ even if the degree of the input graph is 3; here n denotes the number of vertices of the underlying graph G. When the graph edges are restricted to only two weights, we show the same hardness of approximation for degree 5 graphs via a novel reduction from the directed Hamiltonian path problem on cubic (degree 3) graphs. The results hold even if the seed vertex is not part of the input.

Our technique for showing NP-hardness extends the reduction from P3SAT shown in [1] with a modular design using gadgets. Note that these gadgets might be easy to understand but are hard to find. We also show that some of the NP-hardness results can be obtained independently by reduction from the Hamiltonian path problem [14]. Our polytime algorithm for AGAP arises by a reduction to a problem called the Rainbow Spanning Tree Problem which is known to be solvable by using a result from matroid theory [15]. The hardness of approximation relies on the NP-hardness of the underlying problem combined with additional gadget constructions.

Related Work. Much of the theoretical work on self-assembly to date has focused on analyzing the complexity of the original Tile Assembly Model. Adleman et al. [16] showed that determining the minimum number of distinct tiles required to produce a given shape is NP-complete in general, and polytime solvable for trees and squares. The authors also gave an $O(\log n)$-approximation algorithm for determining the relative concentration of tile types that achieved optimal assembly time in *partial order systems*. In the case of $n \times n$ squares, an optimal assembly requiring $\Theta(n)$ time and $\Theta(\frac{\log n}{\log \log n})$ tile types was described in [11,17] based on simulation of binary counters. Extensions to the Tile Assembly Model include consideration of flexible glue-strengths and temperature programming [18,19,20], fault tolerance and self-correction [12,13,21,22,23,24,25,26], patterning (of components) and self-replication [4,6,27,28,29,30].

Among the first works to study the self-assembly process on general graphs are [31,32,33,34,35,36]. It was shown that 3SAT and 3-Vertex-Colorability can be solved by self-assembly of DNA graphs using a constant number of laboratory steps [31,32]. A generalization of the Tile Assembly Model, where *flexible* tiles may connect to more than 4 tiles in a not necessarily planar arrangement, was investigated in [33]. Graph grammars were used to model self-assembly on planar graphs [34,35]. Experiments on construction of non-regular graphs were presented in [36].

Organization. We begin by formally describing the AGAP problem and providing the necessary notations and definitions. Building on the ideas in [1], in Section 3 we show that AGAP is NP-complete on degree 4 planar graphs. In Section 4 we introduce new types of constructions showing that AGAP is NP-complete even on degree 3 planar graphs with two distinct edge weights. In Section 5 we show our hardness of approximation results for AGAP. We summarize our results and discuss some open problems in Section 6.

2   Preliminaries

2.1   Model and Problem Statements

We adopt the *Accretive Graph Self-Assembly Model* introduced in [1]. A graph assembly system is a quadruple $\langle G, v_s, w, \tau \rangle$, where $G = (V, E)$ is undirected weighted graph, $v_s \in V$ is a seed vertex, w is a weight function: $w : E \to \mathbb{Z}$, and $\tau \in \mathbb{N}$ is the temperature of the assembly. Here the weight of an edge represents the strength of attraction between adjacent vertices if positive, and their repulsion if negative. An analogue of the weight function in the Tile Assembly Model [11] is the glue function (cf. glue strength is non-negative).

The self-assembly process in the Accretive Graph Self-Assembly Model proceeds as follows. The graph G serves as a template of construction and initially only the seed vertex v_s of G is built. For a vertex v, let $\Gamma(v)$ represent the set of neighbors of v in G that are already built. A new vertex v of G can be attached to the construction if and only if $\sum_{u \in \Gamma(v)} w(u, v) \geq \tau$, i.e., the sum of the weights from v to its already built neighbors is at least equal to the temperature of assembly. The assembly is *sequential*, i.e., vertices are built one at a time, and *accretive*, i.e., once a vertex is built it cannot be detached from the construction. We will use $u \prec v$ to denote that vertex v is built after vertex u.

We consider the following problems:

Definition 1 (Accretive Graph Assembly Problem (AGAP)). *Given an accretive graph assembly system* $\langle G, v_s, w, \tau \rangle$, *determine if G is* sequentially constructible *(in short,* constructible*) starting from the seed vertex v_s, and provide a feasible order of construction,* $\pi : v_s = v_{\pi_1} \prec v_{\pi_2} \prec \ldots \prec v_{\pi_n}$, *if one exists.*

Definition 2 (Planar Accretive Graph Assembly Problem (PAGAP)). *The* AGAP *problem restricted to planar graphs.*

We also consider the following restrictions of AGAP (PAGAP). The k-WEIGHT AGAP (k-WEIGHT PAGAP) is a special instance of AGAP (PAGAP) such that, there are at most k different edge weights in G. When the degree of G is restricted to d, we refer to the problem as d-DEGREE AGAP (d-DEGREE PAGAP).

Since AGAP is NP-complete in general [1], it is natural to consider polytime approximation algorithms that seek to build a largest subset of vertices of the input graph.

Definition 3 (Maximum AGAP (MAX AGAP)). *Given an instance of AGAP on a graph G, find a largest subgraph of G which is constructible starting from the seed vertex, and provide an order of construction.*

An α-*approximation* algorithm for MAX AGAP is a polytime algorithm that on any given input instance G computes a constructible subgraph of G, say H, such that $|H| \geq |H^*|/\alpha$, where H^* is a largest constructible subgraph of G.

2.2   Background Results and Definitions

The two propositions below characterize some simple cases where AGAP is polytime solvable.

Proposition 1. [1] AGAP *with only positive edge weights can be solved in* $O(|V| + |E|)$ *time.*

Proposition 2. 2-DEGREE AGAP *can be solved in* $O(|V| + |E|)$ *time.*

We, therefore, focus on graphs with maximum degree at least 3 and with at least one edge with negative weight. In order G to be constructible there must be an edge of weight at least $\tau \geq 0$, otherwise the first vertex other than the seed cannot be built. Hence, we consider graphs with at least two different weights.

We will only show NP-hardness in the NP-completeness proofs for PAGAP since (P)AGAP is easily shown to be in NP: given an ordering of the vertices, it can be verified in polynomial time if it is feasible.

Planar 3SAT. In our results we will mostly use a reduction from planar 3SAT (P3SAT), similar to [1]. Lichtenstein [37] proved that P3SAT, i.e., 3SAT with the restriction that the *identifying graph* is planar, remains NP-complete. The identifying graph of a 3SAT formula ϕ is a graph $G = (V, E)$ where vertices correspond to literals and clauses; and there is an edge between a literal vertex and a clause vertex if and only if the literal participates in the clause. Also, there is an edge between every literal and its complement. Middleton [38] showed that deciding the satisfiability of a P3SAT formula with an identifying graph (see Fig. 1) obeying the following restrictions is still NP-complete:

(1) There is a cyclic path, called the *loop*, that can be drawn in the plane such that it passes between all pairs of complementary literals, but does not intersect any other edges of G.
(2) The boolean formula contains only clauses in which the literals are either all positive or all negative.
(3) The graph G can be arranged so that interior (resp. exterior) clauses have only positive (resp. negative) literals.
(4) Let $C(\ell)$ denote the set of clauses in which a literal ℓ participates, then $|C(\ell)| \leq 2$ for all ℓ in ϕ.

The dashed circle in Fig. 1 corresponds to the loop described above, which we assume to be *directed* and denote by L. The loop provides a natural (cyclic) ordering of the variables, e.g., $L = x, y, z, w, x$. For variables u and v we will use $uv \in L$ to denote that v follows u in L, e.g., $xy \in L$, but $xz \notin L$.

3   4-DEGREE PAGAP Is NP-Complete

The construction in this section is similar to that of [1] but it reduces the degree of the resulting graph from 5 to 4. We start our reduction from a P3SAT formula ϕ and its identifying graph. For every variable x and its negation $\bar{x}$, we replace the edge $(x, \bar{x})$ in the graph (Fig. 1) with the gadget depicted in Fig. 2. For x and y, $xy \in L$, we connect the corresponding gadgets with edge (t_x, s_y) with weight $\tau = 2$. This gadget ensures two vital properties. First, along the loop L we can build all vertices corresponding to literals which are set to TRUE, and

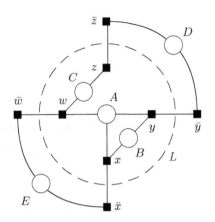

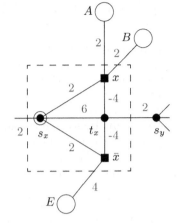

Fig. 1. The identifying graph for the formula $A \wedge B \wedge C \wedge D \wedge E = (x \vee y \vee w) \wedge (x \vee y) \wedge (w \vee z) \wedge (\bar{y} \vee \bar{z}) \wedge (\bar{w} \vee \bar{x})$

Fig. 2. Gadget for degree 4 planar graphs and $\tau = 2$. Here, s_x is the seed vertex.

therefore all vertices corresponding to clauses. Secondly, we can complete all the remaining vertices afterwards. On the other hand, if a vertex corresponding to a literal is built then we need to build t_x before we build the complementary literal. Now, the edges on the complementary literal are such, that we can only build it if all of its adjacent clauses are already built. This corresponds to the fact, that if we set x to TRUE, the formula is only satisfiable if the clauses in $C(\bar{x})$ can be satisfied independently of $\bar{x}$. We now describe the construction in detail.

3.1   The Gadget Construction

Figure 2 shows the gadget replacing $(x, \bar{x})$ in the identifying graph (for $\tau = 2$), where x and $\bar{x}$ participate in clauses A and B, and E, respectively. Formally, we construct the gadget for a variable x as follows. We use additional nodes s_x and t_x and introduce new edges with the following weights: $w(s_x, t_x) = 3\tau$, $w(x, s_x) = w(\bar{x}, s_x) = \tau$, $w(x, t_x) = w(\bar{x}, t_x) = -2\tau$ and $w(x, c) = 2\tau/|C(x)|$ for all clauses $c \in C(x)$, and $w(\bar{x}, c) = 2\tau/|C(\bar{x})|$ for all clauses $c \in C(\bar{x})$. (Recall that $|C(\ell)| \in \{1, 2\}$ for any literal ℓ.) Also, for $xy \in L$, we add an edge (t_x, s_y) with weight τ to connect to the gadget replacing $(y, \bar{y})$.

The following theorem shows that ϕ is satisfiable if and only if there is an ordering to assemble G.

Theorem 1. 4-DEGREE PAGAP *is* NP-*complete.*

Proof. For the first part, assume there is a satisfying assignment of the underlying formula. Consider the obtained graph G. We show that there is an ordering of vertices in G in which every vertex can be built. Starting from the seed vertex we construct the literals x or $\bar{x}$ depending on whose value is set to TRUE in σ following the loop L. After x or $\bar{x}$ is built, we construct t_x and we proceed to the next variable y in the loop by building s_y. Since σ is a satisfying assignment, each vertex corresponding to a clause is adjacent to a vertex which is built (each clause has a variable which is set to TRUE) and the edge weights connecting them are $\geq \tau$. At the final step we can build the literals ℓ which are set to FALSE in σ. We have two cases based on the cardinality of $C(\ell)$. If $|C(\ell)| = 1$, the contribution of ℓ's neighbors is $-2\tau + \tau + 2\tau = \tau$ and similarly if $|C(\ell)| = 2$ the contribution is $-2\tau + \tau + \tau + \tau = \tau$.

For the second part, consider an ordering in which we complete all the vertices of the graph G. Look at the following assignment: set x to TRUE if x is built before $\bar{x}$ and set x to FALSE otherwise. We claim this is a satisfying assignment to ϕ. We prove by contradiction. Assume there is a clause $A = x \vee y \vee z$ which is not satisfied, hence $x = y = z = $ FALSE. Thus, $\bar{x} \prec x$, $\bar{y} \prec y$ and $\bar{z} \prec z$. W.l.o.g., $x \prec y \prec z$ in this ordering. The clause A is adjacent only to x, y, z and thus $x \prec A$. But, due to the construction x can only be built after $C(x)$ is built (since it must be the case that $\bar{x} \prec t_x \prec x$), implying $A \prec x$, which is clearly a contradiction. $\qquad\square$

3.2 An Alternative Approach

We can show a stronger version of the above theorem, via a reduction from the directed Hamiltonian path problem in cubic graphs [14]. This new approach only uses 3 distinct weights, as opposed to the 4 weights in the preceding construction. In addition to that, every constructible instance has a *stable* order of construction, i.e., at every step of the construction, each built vertex has a net attraction at least τ. This is in contrast to the previous reduction using gadgets shown in Fig. 2, where no stable order of construction exists since the sum of the weights of edges incident on t_x is less than τ.

Theorem 2. 4-DEGREE 2-WEIGHT PAGAP *is* NP-*complete*.

4 The Complexity of 3-DEGREE 2-WEIGHT AGAP (PAGAP)

In this section, we prove a dichotomy theorem that completely characterizes the complexity of AGAP on degree 3 bounded graphs with two distinct weights w_p and w_n. We assume w.l.o.g. that $w_p \geq \tau$ and $w_n < 0$ since the case when both weights are positive is trivially solvable by Proposition 1, and the case when $\max\{w_p, w_n\} < \tau$ has no solution. We give a simple system of linear constraints on w_p, w_n and τ that determines whether the problem is NP-complete or solvable in polynomial time (see Table 1).

Theorem 3. 3-DEGREE 2-WEIGHT (P)AGAP *with weights* w_p *and* w_n *is* NP-*complete if and only if* $w_p + 2w_n < \tau$ *and* $w_p + w_n \geq \tau$; *otherwise it is solvable in polynomial time*.

We first improve on the NP-completeness result for 4-DEGREE PAGAP by showing that 3-DEGREE 2-WEIGHT PAGAP is NP-complete. Our construction is related

Table 1. Complexity of 3-DEGREE 2-WEIGHT AGAP with weights $w_p \geq \tau$ and $w_n < 0$

$w_p + 2w_n \geq \tau$	$w_p + w_n \geq \tau$	Results
TRUE	TRUE	Polytime solvable (Lemma 2)
FALSE	TRUE	NP-complete (planar graphs, Lemma 1)
FALSE	FALSE	Polytime solvable (Lemma 3 and Lemma 4)

to the degree 4 case but due to the imposed restrictions, it requires careful composition of more sophisticated gadgets. We use two gadgets, the *direction* and *choice* gadgets, depicted in Figs. 3(a) and 3(b), which are put together as shown in Fig. 3(c). The resulting gadget satisfies properties similar to the properties of the gadget in the degree 4 case (Fig. 2). We then give polynomial time algorithms for the remaining cases of 3-DEGREE 2-WEIGHT AGAP.

4.1   3-DEGREE 2-WEIGHT PAGAP is NP-Complete

To show NP-hardness of 3-DEGREE 2-WEIGHT PAGAP we follow closely the reduction of Section 3. Because of the restriction on the number of distinct edge weights we use different gadgets as building blocks. Their careful composition, however, preserves the desired properties of the above analysis. For ease of presentation we fix edge weights to be $w_p = 3$ and $w_n = -1$ at temperature $\tau = 2$. We note that the construction works in general for any $\langle w_n, w_p, \tau \rangle$ satisfying $w_p + w_n \geq \tau$ and $w_p + 2w_n < \tau$. In other words, building a single neighbor connected with negative edge to a vertex does not by itself make the vertex not constructible (infeasible), but building two such neighbors makes it infeasible. We now describe the gadgets in detail.

Direction gadget. The properties of the direction gadget shown in Fig. 3(a) are as follows:

- If s_d is built, we can complete the gadget: $s_d \prec a \prec d \prec b \prec c \prec a' \prec d' \prec b' \prec c' \prec \{t_d, t_d'\}$
- If t_d and t_d' are built, we can complete the gadget: $\{t_d, t_d'\} \prec a' \prec d' \prec c' \prec b' \prec a \prec d \prec c \prec b \prec s_d$
- If only t_d or t_d' are built, but not both, we cannot build s_d via the gadget unless we make d' or d infeasible. Observe that if, say, t_d is built the only way to reach s_d is via $t_d \prec c' \prec b' \prec a \prec s_d$ but this will make d' infeasible (d' will have two built neighbors contributing -1 each).

Intuitively, we will use the direction gadget to connect a literal ℓ to the clauses $C(\ell)$ for $|C(\ell)| = 2$. The gadget ensures that if ℓ is built then we can build $C(\ell)$, and if both clauses are built then we can build ℓ.

Choice gadget. The properties of the choice gadget shown in Fig. 3(b) are as follows:

- If s_c is built, we can build either t_c or t_c' but not both via the gadget without making i infeasible.

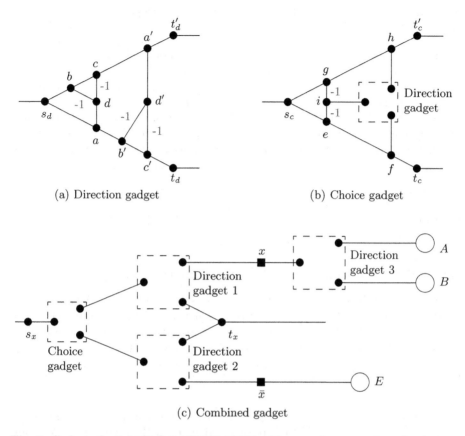

(a) Direction gadget

(b) Choice gadget

(c) Combined gadget

Fig. 3. Gadgets for 3-DEGREE 2-WEIGHT PAGAP with $w_p = 3$, $w_n = -1$, and temperature $\tau = 2$. Edges without annotation have weight 3.

- If only t_c (resp. t'_c) is built, we cannot build t'_c (resp. t_c) via the gadget without making i infeasible.
- If t_c, t'_c and only one of e and g are built, we can complete the gadget, i.e., if we used the gadget to make a choice to build t_c (or t'_c) from s_c we can complete it once t'_c (or t_c) is built independently.

In the analysis in Section 3, it is argued that given a satisfying assignment of the boolean formula used in the reduction, we can (virtually) walk the loop L and build x or $\bar{x}$ (but not both) depending on which literal is set to TRUE in the assignment. The choice gadget is used to obtain this property.

Putting the gadgets together. We compose the direction and choice gadgets to obtain a gadget (Fig. 3(c)) equivalent to the one showed in Fig. 2, decreasing the maximum degree of the resulting graph to 3. Again, we use the gadget to replace the $(x, \bar{x})$ edges in the identifying graph. We connect the gadgets corresponding to x and y, $xy \in L$, with an edge (t_x, s_y) with weight 3. The following properties hold:

- Starting from s_x we can build x or $\bar{x}$ but not both. This property ensures that if there is a satisfying assignment, we can complete G: suppose we build x, then we can build all the clauses $C(x)$ in which x participates, build t_x and continue to s_y of the next variable y in L.
- If x (resp. $\bar{x}$) is built via the gadget, the only way $\bar{x}$ (resp. x) can be built is by building first the clauses in which it participates. This ensures that if the graph is built, the corresponding ϕ formula is satisfiable. Again, we use the following satisfying assignment: x is TRUE if and only if $x \prec \bar{x}$.

Using analogous analysis to the degree 4 case, we obtain the following lemma.

Lemma 1. 3-DEGREE 2-WEIGHT PAGAP *such that* $w_p + 2w_n < \tau$ *and* $w_p + w_n \geq \tau$ *is NP-complete.*

4.2 Polynomial Time Algorithms for 3-DEGREE 2-WEIGHT AGAP

We now give polynomial time algorithms to solve 3-DEGREE 2-WEIGHT AGAP when the weights w_p and w_n are such that either $w_p + 2w_n \geq \tau$ (see Lemma 2) or $w_p + w_n < \tau$. The latter case is subdivided into two sub-cases depending on the relation between $2w_p + w_n$ and τ (see Lemmas 3 and 4).

Lemma 2. 3-DEGREE 2-WEIGHT AGAP *such that* $w_p + 2w_n \geq \tau$ *can be solved in* $O(|V| + |E|)$ *time.*

Proof. Note, $w_p + 2w_n \geq \tau$ implies negative edges cannot make a vertex infeasible as long as it is reachable through positive edges from the seed vertex. We can therefore use Proposition 1 to solve the problem on the graph induced by the positive edges (negative edges neither help nor obstruct the construction). □

Lemma 3. 3-DEGREE 2-WEIGHT AGAP *such that* $2w_p + w_n < \tau$ *is solvable in* $O(|V| + |E|)$ *time.*

Proof. The condition $2w_p + w_n < \tau$ implies the graph cannot be built if there is a negative edge. For contradiction, assume there is a negative edge (u, v) and the graph can be built. W.l.o.g. assume that in a feasible ordering u is built before v. Then, by the choice of weights, v cannot be built; a contradiction. □

We prove the remaining case (Lemma 4) in two steps. We first show a feasibility-preserving transformation that removes any negative edge (u, v) such that either u or v have a single positive edge incident on it. The resulting graph is such that every vertex has more positive than negative edges. We then show that the problem can be viewed as a special case of the Rainbow Spanning Tree Problem (see Definition 4) where at most two edges have the same color.

Lemma 4. 3-DEGREE 2-WEIGHT AGAP *such that* $w_p + w_n < \tau$ *and* $2w_p + w_n \geq \tau$ *can be solved in polynomial time.*

(a) Removing edges with negative weight and 3 positive edge neighbors.

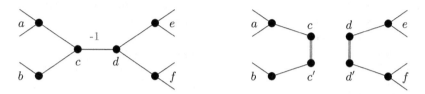

(b) Vertices c and d are replaced by edges (c, c') and (d, d') of the same unique color.

Fig. 4. Graph transformations removing negative weight edges

Claim. An instance of 3-DEGREE 2-WEIGHT AGAP satisfying the conditions of Lemma 4 can be reduced to an instance where each vertex has more positive edges than negative edges incident on it.

Proof. W.l.o.g. assume that each vertex is adjacent to at least one positive edge, otherwise this vertex cannot be built. Similarly to Lemma 3, since $w_p + w_n < \tau$, we can argue that for a negative edge, at least one of its endpoints should be adjacent to two positive edges. Now consider a negative edge (c, d) where d has only one positive edge (d might have another negative edge). It follows that d must be built before c in order for the graph to be constructible, and c is built after its both neighbors are built. We can therefore remove such negative edges, one by one, making a copy of c, c', connecting each to only one (different) of c's neighbors as shown in Fig. 4(a). In the new graph, we can assume *w.l.o.g.* that $d \prec \{c, c'\}$ since c and c' are not used to build other vertices. Now, it is not hard to see the if and only if correspondence between the two instances.   □

We next consider instances where each vertex has more positive edges than negative edges and reduce the problem to the following combinatorial problem.

Definition 4 (Rainbow Spanning Tree Problem). *Given a graph G with colors on its edges, is there a spanning tree of G that has no two edges with the same color?*

The Rainbow Spanning Tree Problem can be solved in polynomial time since it can be formulated as the problem of finding the maximum independent set of the intersection of two matroids [15].

Claim. Given 3-DEGREE 2-WEIGHT AGAP on graph G where each vertex has more positive than negative edges, we can compute a graph H such that G is constructible if and only if H has a rainbow spanning tree.

Proof. We obtain H from G by performing graph transformations to remove all negative edges. For each removed edge we split its endpoints and introduce two positive edges with the same unique color (see Fig. 4(b)). All other edges are assigned unique colors. Ignoring colors, since $w_p \geq \tau$, to build a vertex it is enough to have an already built neighbor, i.e., to build all vertices we need a spanning tree. The colors enforce that if we use an edge of a given color to build a vertex, we cannot build another vertex using an edge of the same color, i.e., we need a rainbow spanning tree.

Formally, consider a negative edge (c, d) and the transformation described in Fig. 4(b). Vertices c and d have degree 3 by the claim proposition. (Note that c and d may share some of their neighbors, e.g., $a = e$.) Assume in a feasible ordering of G, $d \prec c$. From the choice of weights, it follows that $\{a, b\} \prec c$. Therefore, if G is constructible, we can construct a spanning tree in H which includes (a, c) and (b, c') but not (c, c'). Since the color of edge (c, c') appears at most twice, the obtained tree is a rainbow spanning tree of H. Conversely, consider a rainbow spanning tree of H. If say edge (d, d') is included, we can build d before c in G. It must be the case that a and b are connected with a path of distinct colors that does not include c or c'. Therefore we can defer the building of c after a and b are built in G. The claim follows. □

This concludes the proof of Lemma 4. Combining Lemmas 1, 2, 3, and 4, we obtain Theorem 3.

5   Hardness of Approximation of MAX AGAP

We now focus on MAX AGAP where the goal is to find the maximum number of vertices that can be sequentially built of a given graph assembly system starting from the specified seed vertex. Since AGAP is NP-complete, it is natural to study polytime approximation algorithms for AGAP. We show that even in very restricted settings, any non-trivial approximation of AGAP is hard. In particular, we show that AGAP is hard to approximate within a factor of $n^{1-\epsilon}$ for any $\epsilon > 0$. This hardness of approximation holds for degree 3 graphs with three distinct edge weights as well as for degree 5 graphs with two distinct edge weights.

An approach to show hardness of approximation for degree 3 graphs is to use the NP-hardness result of 3-DEGREE 2-WEIGHT PAGAP. We would like to boost the hardness of the instance by attaching a large graph H to it and argue that a significant fraction of H can be constructed if and only if the original instance can be fully constructed. This, however, will not work since we can attach H (almost) only to the vertices corresponding to variables (see Fig. 3(c)) without increasing the overall degree of the graph. Moreover, we can easily construct all variables making vertex i of each choice gadget (see Fig. 3(b)) infeasible, and construct all of H regardless of whether we are given a hard instance or not to begin with. In fact, there is a 2-approximation algorithm for this case since we can account for the vertices that are made infeasible by their built neighbors.

Lemma 5. *There is a 2-approximation algorithm to* 3–DEGREE 2–WEIGHT MAX AGAP *when* $w_p + w_n \geq \tau$ *and* $w_p + 2w_n < \tau$.

To show hardness of approximation we allow 3 distinct weights instead of 2: τ, $\tau - 1$ and -1 for any $\tau \geq 2$. We can now define gadgets, shown in Figs. 5(a) and 5(b), equivalent to the direction and choice gadgets (Figs. 3(a) and 3(b)) such that choice and direction are enforced directly, and not by making some of the vertices infeasible. We can show in this setting that the underlying formula in the P3SAT instance is satisfiable if and only if we can build all vertices corresponding to literals in the respective instance of PAGAP. Furthermore, we can build these vertices if and only if the corresponding PAGAP can be built.

We now proceed to establish our hardness result. Fix a parameter $\epsilon > 0$. We consider the following construction which is a composition of two graphs G and H. Consider a 3–DEGREE 3–WEIGHT PAGAP instance with the graph $G = (V, E)$ obtained from a P3SAT formula with n variables. We have $|V| = O(n)$ since each variable participates in constant number of clauses and its corresponding gadget is of constant size. The graph $H = (V', E')$ consists of $n^{2/\epsilon}$ chained copies of the *cooperation* gadget shown in Fig. 5(c). In H, the vertex t_i of the jth copy of the cooperation gadget is connected to the vertex s_i of the $(j + 1)$th copy with an edge of weight τ. To compose G and H we connect the ith variable (resp. its negation) to s_{2i-1} (resp. s_{2i}) of the first copy of the cooperation gadget as shown in Fig. 5(d). Note that the resulting graph has degree 3 since literals have degree 2 (see Fig. 3(c)). However, the resulting graph is no longer planar.

The cooperation gadget has the property that if all s_i vertices are built, we can build all t_i vertices. However, if only m of the s_i's are built, $m < 2n$, we can build at most $m - 1$ of the t_i's. Therefore, if G is constructible we can build all of H, otherwise we can build only $O(n + n^2) = O(n^2)$ vertices of G and H.

The total number of vertices in $V \cup V'$ is $N = \Theta(n^{1+2/\epsilon})$ since each of the $n^{2/\epsilon}$ copies of the cooperation gadget has $O(n)$ vertices. It follows that an $N^{1-\epsilon}$-approximation polytime algorithm for the optimization version of 3–DEGREE 3–WEIGHT AGAP can be used to decide whether or not the instance is constructible, and therefore decide P3SAT. Hence we obtain the following theorem.

Theorem 4. 3–DEGREE 3–WEIGHT MAX AGAP *is* NP-*hard to approximate within a factor of* $O(n^{1-\epsilon})$ *for any* $\epsilon > 0$, *where* n *denotes the number of vertices in* G.

If we restrict the number of weights to two, we can show a similar result by allowing the maximum degree to be 5. We replace each edge (x, y) with weight 1 (Figs. 5(a) and 5(c)) by a "triangle", adding a vertex z and using weights $w(x, z) = w(z, y) = 2$ and $w(x, y) = -1$. Note that the degree of the construction is increased by at most 2, hence we have the following corollary.

Corollary 1. 5–DEGREE 2–WEIGHT MAX AGAP *is* NP-*hard to approximate within a factor of* $O(n^{1-\epsilon})$ *for any* $\epsilon > 0$, *where* n *denotes the number of vertices in* G.

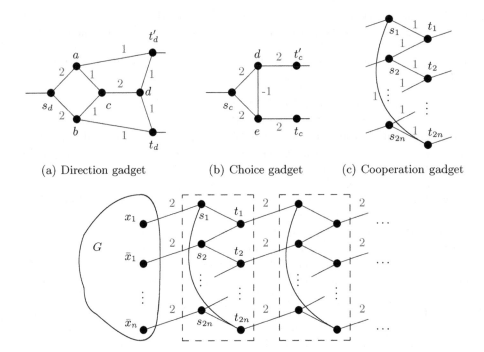

(a) Direction gadget   (b) Choice gadget   (c) Cooperation gadget

(d) Cooperation gadgets (in the dashed boxes) and their composition
with the graph G

Fig. 5. Gadgets for degree 3 planar graphs with three possible edge weights for $\tau = 2$

We note that the NP-hardness and hardness of approximation results hold even when the algorithm is allowed to choose any seed vertex of its choice. We also note that the results of this section hold even if we require that each instance has a stable order of construction, as in Section 3.2.

6   Conclusion

In this paper we resolved the complexity of AGAP and PAGAP by showing that 3-DEGREE PAGAP and hence 3-DEGREE AGAP is NP-complete even with two edge weights. We proved a dichotomy theorem completely describing the complexity of 3-DEGREE 2-WEIGHT AGAP, providing a simple system of linear constraints on the weights and the temperature of assembly to determine whether the problem is NP-complete or polytime solvable. The solution for the polytime case provides the first non-trivial algorithm for AGAP. Finally, we proved that both 3-DEGREE 3-WEIGHT AGAP and 5-DEGREE 2-WEIGHT AGAP are hard to approximate within a factor of $O(n^{1-\epsilon})$. These negative results motivate the question whether there exist better approximation algorithms for the lower degree cases as well as for planar graphs. Approximating PAGAP is especially interesting since it captures the essential geometry of 2D physical systems.

Acknowledgments. This work was supported in part by an NSF Career Award CCR-0093117. The authors would like to thank Péter Biró for helpful discussions. We also thank the anonymous reviewers for their useful suggestions and comments.

References

1. Reif, J.H., Sahu, S., Yin, P.: Complexity of graph self-assembly in accretive systems and self-destructible systems. In: DNA Computing. (2005) 101–112
2. Winfree, E., Liu, F., Wenzler, L.A., Seeman, N.C.: Design and self-assembly of two-dimensional DNA crystals. Nature **394** (1998) 539–544
3. Rothemund, P.: Using lateral capillary forces to compute by self-assembly. Proc. Nat. Acad. Sci. U.S.A. **97** (2000) 984–989
4. LaBean, T.H., Yan, H., Kopatsch, J., Liu, F., Winfree, E., Reif, J.H., Seeman, N.C.: Construction, analysis, ligation, and self-assembly of DNA triple crossover complexes. J. Amer. Chem. Soc. **122** (2000) 1848–1860
5. Yan, H., LaBean, T.H., Feng, L., Reif, J.H.: Directed nucleation assembly of DNA tile complexes for barcode-patterned lattices. Proc. Nat. Acad. Sci. U.S.A. **100** (2003) 8103–8108
6. Rothemund, P.W.K., Papadakis, N., Winfree, E.: Algorithmic self-assembly of DNA Sierpinski triangles. PLoS Biology **2** (2004) 2041–2053
7. Chelyapov, N., Brun, Y., Gopalkrishnan, M., Reishus, D., Shaw, B., Adleman, L.M.: DNA triangles and self-assembled hexagonal tilings. J. Amer. Chem. Soc. **126** (2004) 13924–13925
8. He, Y., Chen, Y., Liu, H., Ribbe, A.E., Mao, C.: Self-assembly of hexagonal DNA two-dimensional (2D) arrays. J. Amer. Chem. Soc. **127** (2005) 12202–12203
9. Malo, J., Mitchell, J.C., Vénien-Bryan, C., Harris, J.R., Wille, H., Sherratt, D.J., Turberfield, A.J.: Engineering a 2D protein-DNA crystal. Angewandte Chemie International Edition **44** (2005) 3057–3061
10. Wang, H.: Proving theorems by pattern recognition II. Bell Systems Technical Journal **40** (1961) 1–41
11. Rothemund, P.W.K., Winfree, E.: The program-size complexity of self-assembled squares (extended abstract). In: STOC. (2000) 459–468
12. Winfree, E., Bekbolatov, R.: Proofreading tile sets: Error correction for algorithmic self-assembly. In: DNA Based Computers. (2003) 126–144
13. Chen, H.L., Goel, A.: Error free self-assembly using error prone tiles. In: DNA Computing. (2004) 62–75
14. Plesník, J.: The NP-completeness of the Hamiltonian cycle problem in planar digraphs with degree bound two. Inform. Process. Lett. **8** (1979) 199–201
15. Broersma, H., Li, X.: Spanning trees with many or few colors in edge-colored graphs. Discuss. Math. Graph Theory **17** (1997) 259–269
16. Adleman, L.M., Cheng, Q., Goel, A., Huang, M.D.A., Kempe, D., de Espanés, P.M., Rothemund, P.W.K.: Combinatorial optimization problems in self-assembly. In: STOC. (2002) 23–32
17. Adleman, L.M., Cheng, Q., Goel, A., Huang, M.D.A.: Running time and program size for self-assembled squares. In: STOC. (2001) 740–748
18. Aggarwal, G., Goldwasser, M., Kao, M.Y., Schweller, R.T.: Complexities for generalized models of self-assembly. In: SODA. (2004) 880–889

19. Sahu, S., Yin, P., Reif, J.H.: A self-assembly model of DNA tiles with time dependent glue strength. In: DNA Computing. (2005) 113–124
20. Kao, M.Y., Schweller, R.: Reducing tile complexity for self-assembly through temperature programming. In: SODA. (2006) 571–580
21. Chen, H.L., Cheng, Q., Goel, A., Huang, M.D.A., de Espanés, P.M.: Invadable self-assembly: combining robustness with efficiency. In: SODA. (2004) 890–899
22. Fujibayashi, K., Murata, S.: A method of error suppression for self-assembling DNA tiles. In: DNA Computing. (2004) 113–127
23. Reif, J.H., Sahu, S., Yin, P.: Compact error-resilient computational DNA tiling assemblies. In: DNA Computing. (2004) 293–307
24. Schulman, R., Winfree, E.: Programmable control of nucleation for algorithmic self-assembly. In: DNA Computing. (2004) 319–328
25. Soloveichik, D., Winfree, E.: Complexity of self-assembled shapes. In: DNA Computing. (2004) 344–354
26. Soloveichik, D., Winfree, E.: Complexity of compact proofreading for self-assembled patterns. In: DNA Computing. (2005) 125–135
27. Lagoudakis, M.G., LaBean, T.H.: 2D DNA self-assembly for satisfiability. In: DNA Based Computers. (1999) 139–152
28. Cook, M., Rothemund, P.W.K., Winfree, E.: Self-assembled circuit patterns. In: DNA Based Computers. (2003) 91–107
29. Schulman, R., Lee, S., Papadakis, N., Winfree, E.: One dimensional boundaries for DNA tile self-assembly. In: DNA Based Computers. (2003) 108–126
30. Barish, R.D., Rothemund, P.W.K., Winfree, E.: Two computational primitives for algorithmic self-assembly: Copying and counting. Nano Letters **5** (2005) 2586–2592
31. Jonoska, N., Karl, S.A., Saito, M.: Three dimensional DNA structures in computing. BioSystems **52** (1999) 143–153
32. Jonoska, N., Sa-Ardyen, P., Seeman, N.C.: Computation by self-assembly of DNA graphs. Genetic Programming and Evolvable Machines **4** (2003) 123–137
33. Jonoska, N., McColm, G.L.: A computational model for self-assembling flexible tiles. In: UC. (2005) 142–156
34. Klavins, E., Ghrist, R., Lipsky, D.: A grammatical approach to self-organizing robotic systems. IEEE Trans. Automat. Control **51** (2006) 949–962
35. Klavins, E.: Directed self-assembly using graph grammars. In: FNANO. (2004)
36. Sa-Ardyen, P., Jonoska, N., Seeman, N.C.: Self-assembling DNA graphs. Natural Computing **2** (2003) 427–438
37. Lichtenstein, D.: Planar formulae and their uses. SIAM J. Comput. **11** (1982) 329–343
38. Middleton, A.A.: Computational complexity of determining the barriers to interface motion in random systems. Phys. Rev. E **59** (1999) 2571–2577

Viral Genome Compression

Lucian Ilie[1,*,**], Liviu Tinta[1],
Cristian Popescu[1], and Kathleen A. Hill[2]

[1] Department of Computer Science, University of Western Ontario
London, Ontario, N6A 5B7, Canada
ilie@csd.uwo.ca
[2] Department of Biology, University of Western Ontario
London, Ontario, N6A 5B7, Canada

Abstract. Viruses compress their genome to reduce space. One of the main techniques is overlapping genes. We model this process by the shortest common superstring problem, that is, we look for the shortest genome which still contains all genes. We give an algorithm for computing optimal solutions which is slow in the number of strings but fast (linear) in their total length. This algorithm is used for a number of viruses with relatively few genes. When the number of genes is larger, we compute approximate solutions using the greedy algorithm which gives an upper bound for the optimal solution. We give also a lower bound for the shortest common superstring problem. The results obtained are then compared with what happens in nature. Remarkably, the compression obtained by viruses is quite high and also very close to the one achieved by modern computers.

Keywords: viruses, viral genomes, genome compression, overlapping genes, shortest common superstring problem, exact algorithms, approximate solutions, lower bounds.

1   Introduction

According to [5], all virus genomes experience pressure to minimize their size. For example, those with prokaryotic hosts must be able to replicate quickly to keep up with their host cells. In the case of viruses with eukaryotic hosts, the pressure on the genome size comes from the small size of the virus, that is, from the amount of nucleic acid that can be incorporated.

One way to reduce the size of their genome is by overlapping genes. Some viruses show tremendous compression of genetic information when compared with the low density of information in the genomes of eukaryotic cells. As claimed in [5], overlapping genes are common and "the maximum genetic capacity is compressed into the minimum genome size." This property looks very interesting from mathematical point of view and we found it surprising that it was not much investigated. Daley and McQuillan [9] introduces and investigates a number

---

* Corresponding author.
** Research partially supported by NSERC.

of formal language theory operations motivated by the biological phenomenon. Krakauer [12] discusses genomic compression in general as achieved through reduced redundancy, overlapping genes, or translational coupling.

In this paper, we investigate this property by naturally modelling it as the shortest common superstring problem (SCS). The genes are seen as strings and we look for the shortest superstring that contains them all. A variation is also considered due to the retrograde overlaps which may be present in some viruses.

The SCS problem is known to be NP-hard. We give an algorithm to compute optimal solutions which works well when the number of strings is not too high. The algorithm is conceptually very simple and also very fast with respect to the total length of all strings. We used this algorithm for those viral genomes whose number of genes is not very high.

When the number of strings increases, we are no longer able to find optimal solutions and use a greedy algorithm for an approximation. This gives an upper bound for the length of a shortest superstring and, for a better estimate, we provide also a lower bound.

Finally, our results are compared with those obtained by viruses. The amount of compression using gene overlapping achieved by the viruses is remarkable; in all examples considered, it is the same or very close to the one obtained by modern computers. The biological significance of these results is to be investigated. Aside from the compression achieved in nature, any solution (or lower bound) for the corresponding SCS problem provides a limitation on the size of a viral genome which contains a given set of genes. Again, the biological relevance of such results remains to be clarified.

2   Basic Definitions

Let Σ be an alphabet, that is, a finite non-empty set. Such an alphabet can be the set of four nucleotides $\{A, T, C, G\}$. We denote by Σ^* the set of all finite strings over Σ. The empty word is denoted ε. Given a string $w \in \Sigma^*$, $w = a_1 a_2 \cdots a_n$, $a_i \in \Sigma$, the length of w is $|w| = n$; the length of ε is 0. We also denote $w[i] = a_i$ and $w[i..j] = a_i a_{i+1} \cdots a_j$, for all $1 \leq i \leq j \leq n$. The *reversal* of w is $a_n a_{n-1} \cdots a_1$.

If $w = xyz$, for some $w, x, y, z \in \Sigma^*$, then x, y, and z are a *prefix*, *factor* (or *substring*), and *suffix* of w, resp. The prefix (suffix) of length n of w is denoted $\mathsf{pref}_n(w)$ ($\mathsf{suff}_n(w)$).

For further notions and results on string combinatorics and algorithms we refer to [14] and [7].

3   The Shortest Common Superstring Problem

The formal definition of the shortest common superstring problem (SCS) is: given k strings $w_1, w_2, \ldots, w_k$, find a shortest string w which contains all w_is as factors; such a w is usually called a shortest common superstring. Any superstring will be called a solution, whereas a shortest one is an optimal solution.

Example 1. Consider the strings $w_1 = $ baac, $w_2 = $ aacc, and $w_3 = $ acaa. A shortest superstring has length 8; it is baacaacc.

The SCS problem has many applications. Data compression is one of the fields where the SCS problem is very useful because data may be stored very efficiently as a superstring; see [10], [15]. This superstring contains all the information in a compressed form. Computational biology is another field where SCS can be applied; see [13].

The SCS problem was proved to be NP-hard in [10] and then MAX SNP-hard in [3]. Therefore, it is unlikely to have polynomial time exact algorithms and research focussed mainly on approximation algorithms [17,8,11,1,2,4]. The best approximation algorithm to date is due to Sweedyk [16] and can reach an approximation ratio of $2\frac{1}{2}$.

Still, in practice the very simple greedy algorithm is used with very good results. Blum et al. [3] proved that greedy is a 4-approximation algorithm. The still open conjecture is that the approximation factor is 2, which would be optimal as there are examples for which greedy produces no better approximations.

4 Viral Genome Compression

As already mentioned in the introduction, viruses can overlap their genes. There are several types of overlaps. First we need to recall the DNA complementarity: the two strands of DNA are complementary and have opposite direction. The complementarity is such that whenever an A occurs on one strand, a T must appear on the other; we say that A and T are complementary. Similarly, C and G are complementary. We denote the complement of a nucleotide N by $\overline{N}$. That is, we have $\overline{A} = T$, $\overline{C} = G$, and vice versa. Also, $\overline{\overline{A}} = A$. Complementarity is needed to understand retrograde overlapping.

For a string $w = a_1 a_2 \cdots a_{|w|}$, we construct the *complemented reversal* of w, $\overline{w} = \overline{a}_{|w|} \overline{a}_{|w|-1} \cdots \overline{a}_1$. When w appears in one strand, $\overline{w}$ occurs opposite it in the other strand.

Example 2. Let $x = $ ACCGTGTAC and $y = $ GTGTACCGTAC be two hypothetical genes. The complemented reversal of x is $\overline{x} = $ GTACACGGT. The possible overlaps between x and y are shown in Figs. 1 and 2. In Fig. 1 we have overlaps on the same strand, that is, direct overlaps; one is called suffix overlap and the other prefix overlap but such a difference is irrelevant for us.

In Fig. 2 we have retrograde overlaps (as can appear, for instance, in double stranded DNA viral genomes). As seen in the figure, each x in the upper strand

Fig. 1. Direct overlaps (same strand)

Fig. 2. Retrograde overlaps (opposite strands)

correspond to an $\overline{x}$ in the lower strand. Again, one is called head-on overlap, the other end-on overlap, without relevance for our purpose.

5   Computing Overlaps

In order to give some algorithms for optimal or approximate solutions for the SCS problem, we need to compute overlaps between strings. Also, we need to eliminate those strings which are factors of others.

An *overlap* between two given strings u and v is any suffix of u that is also a prefix of v. We shall need only the longest overlaps but our algorithm computes them all in the same optimal time. The set OVERLAPS(u, v) contains the lengths of all suffixes of u that are prefixes of v. We denote by overlap(u, v) the length of the longest overlap. Here is an example.

Example 3. For the strings $u = \text{abaababa}$ and $v = \text{ababab}$ we have

$$\text{OVERLAPS}(u, v) = \{1, 3, 5\}, \quad \text{overlap}(u, v) = 5,$$
$$\text{OVERLAPS}(v, u) = \emptyset, \quad \text{overlap}(v, u) = 0.$$

To compute overlaps, we shall use a classical notion in pattern matching: a *border* of a string w is any string which is both a prefix and a suffix of w; *the border* of w, denoted border(w), is the longest non-trivial border of w, that is, different from w itself. Notice that all borders of w are: border(w), border$^2(w) = $ border(border(w)), border$^3(w), \ldots, \varepsilon$.

Denote $|w| = n$ and consider the array BORDER$_w[0..n]$, where, for all $1 \leq i \leq n$, BORDER$_w[i] = |\text{border}(w[1..i])|$; BORDER$_w[0] = -1$ for technical purposes.

Example 4. For the string $w = \text{abaababaaba}$ we have

$$\text{BORDER}_w = [-1, 0, 0, 1, 1, 2, 3, 2, 3, 4, 5, 6]$$

and all borders of w are

$$\text{BORDER}_w(|w|) = 6, \quad \text{border}(w) = \text{abaaba},$$
$$\text{BORDER}_w^2(|w|) = 3, \quad \text{border}^2(w) = \text{aba},$$
$$\text{BORDER}_w^3(|w|) = 1, \quad \text{border}^3(w) = \text{a}.$$

The array BORDER$_w$ can be computed in time linear in $|w|$ by a classical algorithm. The idea is to compute the elements from first to last. Then, when computing BORDER$_w[i]$, all previous elements are known. The border of $w[1..i]$ is either an extension of a border of $w[1..i-1]$ or empty if this is not possible.

We use borders to solve our problem. Assume we are given two strings u and v. Consider a new letters $\#$ (which does not appear in u or v) and construct the string $w = v\#u$. It is clear that any border of w gives an overlap of u and v and vice versa. Therefore, using borders, we obtain an algorithm for computing overlaps which is linear in terms of $|u| + |v|$. Notice, however, that if one of the strings is much longer than the other, then we do not need the whole long string but just a short piece of it. An algorithm which works in linear time in the size of the shorter string would simply consider the string $\mathsf{pref}_s(v)\#\,\mathsf{suff}_s(u)$, where $s = \min(|u| - 1, |v| - 1)$.

We can also do it all at once. For the SCS problem, we always exclude from calculations the strings which are included as factors in others. This is pattern searching and there are many linear time algorithms for it. We can also use the borders as above to give a simple algorithm to both identify factors and compute overlaps. We consider $w = v\#u$. Assuming $|v| \le |u|$, v is a factor of u if and only if there is i such that $\mathrm{BORDER}_w(i) = |v|$.

OVERLAPS-AND-FACTORS(u, v)

1. $w \leftarrow v\#u$
2. $n \leftarrow |w|$
3. $\mathrm{BORDER}_w[0] \leftarrow -1$
4. $b \leftarrow -1$
5.   **for** i **from** 1 **to** n **do**
6.   **while** $b \ge 0$ **and** $w[b+1] \ne w[i]$ **do**
7. $b \leftarrow \mathrm{BORDER}_w[b]$
8. $b \leftarrow b + 1$
9. $\mathrm{BORDER}_w[i] \leftarrow b$
10.   **if** $\mathrm{BORDER}_w[i] = |v|$ **and** $|v| \le |u|$ **then**
11.   **return** $\mathsf{overlap}(u, v) = -1$ [v is a factor of u]
12.   **return** $\mathsf{overlap}(u, v) = \mathrm{BORDER}_w[|w|]$

This algorithm is linear in $|u| + |v|$; this is optimal since it is the minimum required for searching.

Lemma 1. *The algorithm* OVERLAPS-AND-FACTORS(u, v) *returns* -1 *iff* v *is a factor of* u *and otherwise computes the longest overlap of* u *and* v. *It runs in time* $\mathcal{O}(|u| + |v|)$.

6   Optimal Solutions of SCS

We may assume that none of the strings w_i appears as factor of another one. (We check this in the algorithm.) Therefore, for any solution w of SCS, there is a permutation σ on k elements such that w contains each w_i as a factor starting at position p_i and

$$p_{\sigma(1)} < p_{\sigma(2)} < \cdots < p_{\sigma(k)}.$$

Example 5. For the strings in Example 1, the optimal solution is given by the permutation $(1, 3, 2)$.

Therefore, our brute-force algorithm to compute an optimal solution of SCS will try all such permutations σ; the set of all permutations on k elements is the symmetric group $\mathcal{S}_k$. For each permutation, we need the maximum overlap between $w_{\sigma(i)}$ and $w_{\sigma(i+1)}$. No other overlaps are needed. Assuming that $w_{\sigma(i)}$ and $w_{\sigma(i+1)}$ overlap each other on a length less than their maximal overlap. Then we can simply overlap them more to obtain a shorter superstring.

We shall need one more definition. For two strings u and v which are not factors of each other, we denote by $\mathsf{merge}(u, v)$ the string obtained by overlapping them as much as possible, that is, $\mathsf{merge}(u, v) = u\,\mathsf{suff}_{|v|-\mathsf{overlap}(u,v)}(v) = \mathsf{pref}_{|u|-\mathsf{overlap}(u,v)}(u)v$.

Example 6. For the strings $u = $ abaababa, $v = $ abababb we have $\mathsf{merge}(u, v) = $ abaabababb.

Here is the algorithm.

SCS-OPTIMAL$(w_1, w_2, \ldots, w_k)$

```
1.   for i from 1 to k do
2.        for j from 1 to k do
3.             if i ≠ j then
4.                  overlap(w_i, w_j) ← OVERLAPS-AND-FACTORS(w_i, w_j)
5.                  if overlap(w_i, w_j) = −1 then eliminate w_i
6.   scs ← ∑_{i=0}^{k} |w_i|        [ we use the same k but it may be smaller ]
7.   for all σ ∈ S_k do
8.        w ← w_{σ(1)}
9.        for i from 2 to k do
10.            w ← merge(w, w_{σ(i)})
11.       if scs > |w| then
12.            scs ← |w|
13.  return scs
```

Proposition 1. *The algorithm* SCS-OPTIMAL$(w_1, w_2, \ldots, w_k)$ *computes an optimal solution for SCS and runs in time* $\mathcal{O}(k!\ell)$, *where* $\ell = \sum_{i=1}^{k} |w_i|$.

Proof. The correctness follows from the fact that we try all permutations. As explained above, after eliminating strings which appear as factors of others, it it enough to consider only longest overlaps.

The time complexity for the preprocessing steps 1-5 is $\mathcal{O}(k^2\ell)$, because of Lemma 1. In the main processing part, steps 7-12, we repeat $k!$ times something linear in ℓ. This is the dominant order. $\qquad\square$

7   Approximate Solutions of SCS

As the SCS problem is NP-hard, in practice approximation algorithms are often used to find a superstring which may not be shortest but hopefully close to optimal. The most common such algorithm for SCS is the greedy algorithm, which we describe below. It uses the natural idea of considering the longer overlaps first. It may not produce an optimal solution but it cannot be too far away. Here is an example when the greedy algorithm does not give an optimal solution.

Example 7. Consider again the strings in Example 1, $w_1 = $ baac, $w_2 = $ aacc, and $w_3 = $ acaa. The overlaps are shown below:

overlap(w_i, w_j)	w_1	w_2	w_3
w_1		3	2
w_2	0		0
w_3	0	2	

The greedy algorithm chooses first the longest overlap, that is, overlap(w_1, w_2), and obtains the string baaccacaa of length 9, since merge(w_1, w_2) and w_3 have no overlap. But there is a shorter one, given by the permutation $(1, 3, 2)$, of length 8, that is baacaacc.

It is conjectured that the greedy solution is always at most twice longer than optimal; see [16] and the references therein for approximation algorithms for the SCS problem. In practice, the greedy algorithm works pretty well, as we shall see also in our experiments.

SCS-GREEDY$(w_1, w_2, \ldots, w_k)$

1.   compute overlaps and eliminate factors as before
2. $greedy\_scs \leftarrow \sum_{i=0}^{k} |w_i|$
3.   **for all** (i, j) with overlap$(w_i, w_j) = \max_{(s,t)}$ overlap(w_s, w_t) **do**
4.   eliminate w_i and w_j from the list
5.   add $w = $ merge(w_i, w_j) to the list
6.   denote the new list $w'_1, \ldots, w'_{k-1}$
7.   the overlaps of w are given by w_i for prefix and by w_j for suffix
8. $\ell \leftarrow $ SCS-GREEDY$(w'_1, w'_2, \ldots, w'_{k-1})$
9.   **if** $greedy\_scs > \ell$ **then**
10. $greedy\_scs \leftarrow \ell$
11. **return** $greedy\_scs$

The greedy algorithm gives an upper bound for the shortest length of a common superstring.

8   Lower Bounds

We give in this section a lower bound for the length of the shortest superstring. It is computed using also a greedy approach but without checking if it is possible to actually find a superstring which uses the considered overlaps. (When this is possible, we have an optimal solution of SCS.)

Any superstring w is defined by a permutation σ on k elements which gives $k - 1$ overlaps. Also, the length of the superstring is the total length of all strings minus the total length of overlaps, that is,

$$|w| = \sum_{i=1}^{k} |w_i| - \sum_{i=1}^{k-1} \text{overlap}(w_{\sigma(i)}, w_{\sigma(i+1)}).$$

For our estimate, we consider the matrix of overlaps, $(\mathsf{overlap}(w_i, w_j))_{1 \leq i \neq j \leq k}$. A permutation σ as above gives $k-1$ overlaps such that no two are in the same row or column. We relax this condition by considering only rows or only columns. Choosing $k-1$ longest overlaps such that no two are on the same row gives a lower bound. Similarly for columns.

The algorithm below computes the first one. The second is computed analogously. We assume the matrix of overlaps has already been computed.

LOWER-BOUND-ROW$(w_1, w_2, \ldots, w_k)$

1. sort all elements of the matrix $(\mathsf{overlap}(w_i, w_j))_{1 \leq i \neq j \leq k}$ decreasingly
2.   to obtain $\mathsf{overlap}(w_{i_1}, w_{j_1}), \ldots, \mathsf{overlap}(w_{i_{n^2-n}}, w_{j_{n^2-n}})$
3. $lower\_bound\_row \leftarrow 0$
4. $rows\_used \leftarrow 0$
5. $t \leftarrow 1$
6. **while** $rows\_used < k - 1$ **do**
7.   **if** row i_t not used **then**
8. $lower\_bound\_row \leftarrow lower\_bound\_row + |w_{i_t}| - \mathsf{overlap}(w_{i_t}, w_{j_t})$
9.   mark row i_t as used
10. $rows\_used \leftarrow rows\_used + 1$
11. $t \leftarrow t + 1$
12. $lower\_bound\_row \leftarrow lower\_bound\_row + |w_{j_{t-1}}|$
13. **return** $lower\_bound\_row$

Proposition 2. *The above algorithm computes a lower bound for the length of the shortest superstring in time $\mathcal{O}(k^2 \log k)$.*

Proof. The time required by the algorithm is $\mathcal{O}(k^2 \log k)$ because of sorting. The **while** cycle takes only $\mathcal{O}(k^2)$ time as it traverses the list of $k^2 - k$ elements at most once and spends constant time for each element.

For correctness, it is enough to prove that the sum of the overlaps chosen by the algorithm is larger than the sum of overlaps corresponding to an optimal solution. In both cases, we have $k - 1$ overlaps involved, no two in the same row. Assume that an optimal solution chooses all rows except for the ith whereas our algorithm for the lower bound misses only the jth row. In all rows chosen by both, the overlap included for the lower bound is at least as large. If $i = j$, this proves that we obtain indeed a lower bound. If $i \neq j$, then the overlap chosen for the lower bound from row i is larger than the one for the optimal solution in row j as the former appear first in the sorted list from step 2.   □

As already mentioned, another lower bound is obtained similarly, by choosing $k - 1$ elements from different columns in the overlap matrix; denote this lower bound by $lower\_bound\_col$. We have then the following lower bound:

$$lower\_bound\_scs = \max(lower\_bound\_row, lower\_bound\_col).$$

The next result, which summarizes the above discussed bounds, is clear.

Proposition 3. *We always have*

$$lower\_bound\_scs \leq scs \leq greedy\_scs.$$

Example 8. For the strings in Example 1, we have:

$lower\_bound\_row = 7,$ because of $\mathsf{overlap}(w_1, w_2)$ and $\mathsf{overlap}(w_3, w_2),$
$lower\_bound\_col = 7,$ because of $\mathsf{overlap}(w_1, w_2)$ and $\mathsf{overlap}(w_1, w_3),$
$lower\_bound\_scs = 7,$
$scs = 8,$
$greedy\_scs = 9.$

The lower bound cannot be achieved however, as it involves the beginning of w_2 (or the end of w_1) twice. Also, it happened that the lower bounds corresponding to rows and columns are the same; this is not true in general.

9   Retrograde Overlaps

The possibility of retrograde overlaps (see Fig. 2) further complicates the search for solutions, optimal or approximate. Each string may appear in a superstring as it is or as its complemented reversal.

Therefore, we need first to compute more overlaps. The following equalities help computing only half of all possible ones:

(i) $\mathsf{merge}(x, y) = \overline{\mathsf{merge}(\overline{y}, \overline{x})},$
(ii) $\mathsf{merge}(x, \overline{y}) = \mathsf{merge}(y, \overline{x}).$

For the exact algorithm, we need to consider, for each string w_i, whether w_i or $\overline{w}_i$ appears at position $p_{\sigma(i)}$, which makes the algorithm even slower in the number of strings.

The greedy algorithm works rather similarly. Only the overlaps for the merged strings need to be set a bit differently. For instance, if the overlap between w_i and $\overline{w}_j$ is chosen, then the string $\mathsf{merge}(w_i, \overline{w}_j)$ is added and its overlaps are taken from those given by prefixes of w_i and w_j.

The lower bound is computed similarly. When choosing a certain overlap, the proper rows or columns need to be discarded for further consideration. For instance, in case of $lower\_bound\_row$, if the overlap between w_i and $\overline{w}_j$ is chosen, then all overlaps involving the suffix of of w_i must be discarded, that is, all pairs (w_i, w_s), $(w_i, \overline{w}_s)$, $(w_s, \overline{w}_i)$ and $(\overline{w}_s, \overline{w}_i)$.

10   Viral Compression Versus Computer Compression

We show in this section our computations for a number of viral genomes which were obtained from "The National Center for Biotechnology Information," (web site www.ncbi.nlm.nih.gov). We start with a set of strings which are the genes and try to find a short superstring. Then we compare our result with the one

Table 1. Viral genome compression - optimal solutions

Family	Name	Total length	Viral	SCS
Paramyxoviridae	Human respiratory syncytial virus	13641	13609	13602
Rhabdoviridae	Bovine ephemeral fever virus	15029	14662	14650
Rhabdoviridae	Northern cereal mosaic virus	11922	11922	11917
Togaviridae	Sleeping disease virus	11745	11745	11738
Coronaviridae	SARS coronavirus	29974	29046	29040
Retroviridae	HIV-1 isolate 01IN565.11 from India	14125	8647	8646
Retroviridae	HIV-2 isolate ALI from Guinea-Bissau	14466	8809	8809

Table 2. Viral genome compression - approximate solutions

Family	Name	Total length	Viral	Greedy	Lower bound
Baculoviridae	Choristoneura fumiferana MNPV	119168	118319	117414	117228
Poxviridae	Vaccinia Virus strain Ankara	152029	150885	150588	150329
Herpesviridae	Bovine Herpesvirus 1	124819	119378	119276	119137
Adenoviridae	Human adenovirus type 5	36576	34342	34328	34322
Adenoviridae	Hemorrhagic enteritis virus	25158	23433	23414	23402
Iridoviridae	Frog virus 3	85593	84443	84248	84174

Fig. 3. Human respiratory syncytial virus

Fig. 4. Bovine ephemeral fever virus

Fig. 5. Northern cereal mosaic virus

Fig. 6. Sleeping disease virus

Fig. 7. SARS coronavirus

Fig. 8. HIV-1 isolate 01IN565.11 from India

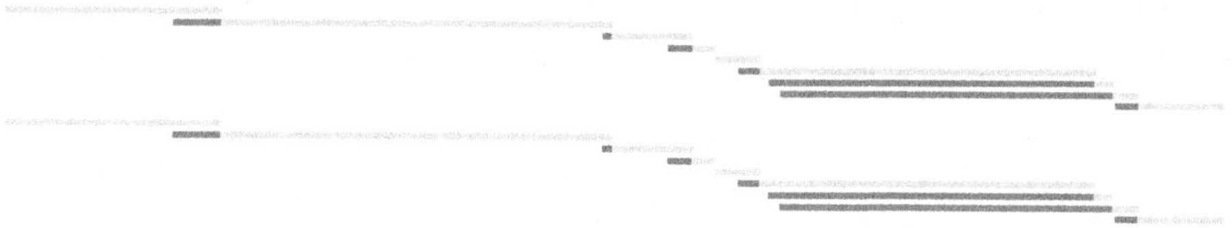

Fig. 9. HIV-2 isolate ALI from Guinea-Bissau

Fig. 10. Human adenovirus type 5

Fig. 11. Choristoneura fumiferana MNPV (left) and Vaccinia Virus strain Ankara (right)

achieved by the viruses. Notice that the time complexity of our exact algorithm grows very fast with the number of genes, but is linear in the total length.

We managed to obtain exact solutions in Table 1 for a number of single stranded RNA viral genomes with relatively few genes. The columns give, in order, the family, the name of the virus, the total length of all genes, the compression achieved by the virus (total length of coding regions), and the shortest common superstring. All lengths are given in number of nucleotides.

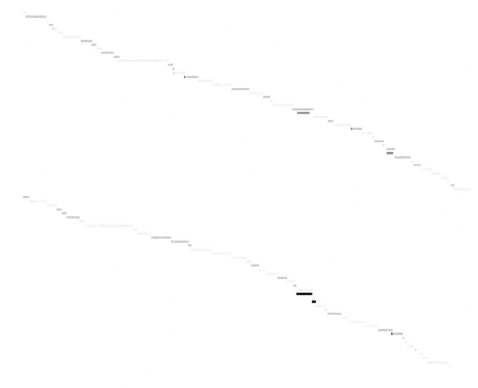

Fig. 12. Bovine Herpesvirus 1

Fig. 13. Hemorrhagic enteritis virus

For genomes with more genes, we had to use the approximation algorithms. The results for a number of double stranded DNA viral genomes are shown in Table 2. The columns have similar meaning, except that the one for the shortest common superstring is replaced by two: greedy and lower bound. All lengths are given in number of base pairs.

Fig. 14. Frog virus 3

The compression achieved by the viruses is, on average, 7.98%, that is, the (average) ratio between the reduction in size (total length of all genes minus viral coding) and the initial size (total length of genes). For the viruses in the first table, the ratio is higher, 11.95%, whereas for the second table it is 3.36%. The average compression ratio is remarkably high if we keep in mind that DNA molecules (seen as strings) are very difficult to compress in general. Commercial file-compression programs achieve usually no compression at all and the best especially designed algorithms, see [6], can achieve something like 13.73%

(that is the average for DNACompress from [6], the best such algorithm to date).

Also, the compression achieved by viruses is very close to what we can do (using overlapping only) by computers. The above averages, for all viruses considered, single stranded RNA, and double stranded DNA viruses are 8.11% (only 0.13% better than viruses), 11.99%, and 3.59%, resp. For the second table we used the greedy compression; it should also be noticed that our lower bound behaves pretty well.

To give a better idea of the overlaps, Figs. 3–14 at the end show all genomes considered above as they appear in nature with the non-coding regions removed (top) and then as computed by our programs (bottom). The overlaps and different strands are shown in different color. (The figures are most useful in the electronic version of the paper.)

References

1. C. Armen and C. Stein, Improved length bounds for the shortest superstring problem, *Proc. 5th Internat. Workshop on Algorithms and Data Structures*, Lecture Notes in Comput. Sci. **955**, Springer-Verlag, Berlin, 1995, 494 – 505.
2. C. Armen and C. Stein, A $2\frac{2}{3}$ approximation algorithm for the shortest superstring problem, *Proc. Combinatorial Pattern Matching*, Lecture Notes in Comput. Sci. **1075**, Springer-Verlag, Berlin, 1996, 87 – 101.
3. A. Blum, T. Jiang, M. Li, J. Tromp, and M. Yannakakis, Linear approximation of shortest superstrings, *J. Assoc. Comput. Mach.* **41**(4) (1994) 630 – 647.
4. D. Breslauer, T. Jiang, and Z. Jiang, Rotations of periodic strings and short superstrings, *J. Algorithms* **24** (1997) 340 – 353.
5. A.J. Cann, *Principles of Molecular Virology*, 3rd ed. Elsevier Academic Press, London, San Diego, 2001.
6. X. Chen, M. Li, B. Ma, and J. Tromp, DNACompress: fast and effective DNA sequence compression, *Bioinformatics* **18** 2002 1696 – 1698.
7. M. Crochemore and W. Rytter, *Jewels of Stringology*, World Sci. Pub., 2003.
8. A. Czumaj, L. Gasieniec, M. Piotrow, and W. Rytter, Parallel and sequential approximations of shortest superstrings, *Proc. First Scandinavian Workshop on Algorithm Theory*, Lecture Notes in Comput. Sci. **824**, Springer-Verlag, Berlin, 1994, 95 – 106.
9. M. Daley and I. McQuillan, Viral gene compression: complexity and verification, *Proc. of CIAA'04*, Lecture Notes in Comput. Sci. **3317**, Springer, Berlin, 2005, 102–112.
10. J. Gallant, D. Maier, and J. Storer, On finding minimal length superstrings, *Journal of Comput. and Syst. Sci.* **20**(1) (1980) 50 – 58.
11. R. Kosaraju, J. Park, and C. Stein, Long tours and short superstrings, *Proc. 35th Annual IEEE Symposium on Foundations of Computer Science*, IEEE Computer Society Press, Piscataway, NJ, 1994, 166 – 177.
12. D.C. Krakauer, Evolutionary principles of genomic compression, *Comments on Theor. Biol.* **7** (2002) 215 – 236.
13. A. Lesk, *Introduction to Bioinformatics*, Oxford University Press, Oxford, 2002.

14. M. Lothaire, *Algebraic Combinatorics on Words*, Cambridge Univ. Press, 2002.

15. J. Storer, *Data Compression: Methods and Theory*, Computer Science Press, 1988.

16. Z. Sweedyk, A $2\frac{1}{2}$-approximation algorithms for shortest superstring, *SIAM J. Comput.* **29**(3) (1999) 954 – 986.

17. S. Teng and F. Yao, Approximating shortest superstrings, *Proc. 34th Annual IEEE Symposium on Foundations of Computer Science*, IEEE Computer Society Press, Piscataway, NJ, 1993, 158 – 165.

DNA Codes and Their Properties

Lila Kari and Kalpana Mahalingam

University of Western Ontario,
Department of Computer Science,
London, ON N6A5B7
{lila, kalpana}@csd.uwo.ca

Abstract. One of the main research topics in DNA computing is associated with the design of information encoding single or double stranded
DNA strands that are "suitable" for computation. Double stranded or
partially double stranded DNA occurs as a result of binding between
complementary DNA single strands (A is complementary to T and C is
complementary to G). This paper continues the study of the algebraic
properties of DNA word sets that ensure that certain undesirable bonds
do not occur. We formalize and investigate such properties of sets of sequences, e.g., where no complement of a sequence is a prefix or suffix of
another sequence or no complement of a concatenation of n sequences
is a subword of the concatenation of $n + 1$ sequences. The sets of code
words that satisfy the above properties are called θ-prefix, θ-suffix and θ-
intercode respectively, where θ is the formalization of the Watson-Crick
complementarity. Lastly we develop certain methods of constructing such
sets of DNA words with good properties and compute their informational
entropy.

1   Introduction

Several attempts have been made to address the problem of encoding information
on DNA and many authors have proposed various solutions. A common approach
has been to use the Hamming distance [2,7,8,9,25]. Experimental separation of
strands with "good" sequences that avoid intermolecular cross hybridization was
reported in [5,6]. In [12], Kari et.al. introduced a theoretical approach to the
problem of designing code words. Theoretical properties of languages that avoid
certain undesirable hybridizations were discussed in [14,16,18]. Based on these
ideas and code-theoretic properties, a computer program for generating code
words is being developed [13,20]. Another algorithm, based on backtracking, for
generating such code words is also developed by Li [22]. In [21] the author used
the notion of partial words with holes for the design of DNA strands.

In this paper we continue the study of the algebraic properties of DNA languages suitable for computation. More precisely, every biomolecular protocol involving DNA or RNA generates molecules whose sequences of nucleotides form
a language over the four letter alphabet $\Delta = \{A, G, C, T\}$. The Watson-Crick
(W/C) complementarity of the nucleotides defines a natural involution mapping

C. Mao and T. Yokomori (Eds.): DNA12, LNCS 4287, pp. 127–142, 2006.
© Springer-Verlag Berlin Heidelberg 2006

θ, $A \mapsto T$ and $G \mapsto C$ which is an anti-morphism of Δ^*. Undesirable Watson-Crick bonds (undesirable hybridizations) can be avoided if the language satisfies certain coding properties. In this paper we concentrate on θ-prefix, θ-suffix and θ-intercode (i.e.) languages where no Watson-Crick complement of a word is a prefix or suffix of another word, respectively no Watson-Crick complement of a composition of n words is a subword of a composition of $n+1$ words (See Fig 1 for the types of hybridizations that are avoided if a word set satisfies these properties). We start the paper with definitions of coding properties that avoid intermolecular cross hybridizations. The notions of θ-prefix and θ-suffix languages have been defined in [16] under the names of θ-p-compliant and θ-s-compliant respectively. Here we also consider two additional coding properties namely θ-bifix code and θ-intercode. We make several observations about the closure properties of such languages. In particular, we concentrate on properties of languages that are preserved by union and concatenation.

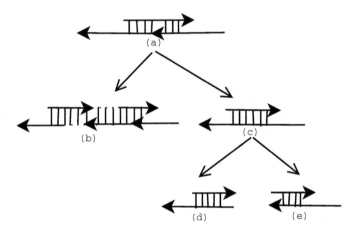

Fig. 1. Various types of intermolecular hybridization that we want to avoid: (a) a code word is the reverse complement of a subword of a concatenation of two other code words: θ-comma-free codes avoid such hybridizations, (b) the catenation of m codewords is the reverse complement of a subword of a concatenation of composition of $m+1$ code words: θ-intercodes (a new notion introduced in this paper) avoid such hybridizations (c) a code word is a reverse complement of a subword of another code word: θ-infix codes avoid such hybridizations (d) a code word is the reverse complement of a suffix of another code word: θ-suffix codes avoid such hybridizations, (e) a code word is the reverse complement of a prefix of another code word: θ-prefix codes avoid such hybiridzations. (The $3'$ end is indicated by an arrow.)

Also, we show that if a set of DNA strands has "good" coding properties that are preserved under concatenation, then the same properties will be preserved under arbitrary ligation of the strands. Section 3 investigates closure properties of various types of involution codes. Algebraic properties of θ-intercodes are discussed in Section 4. We introduce and discuss the properties of sets whose n element subsets are θ-intercodes and θ-comma-free codes in Section 5. Section 6 describes several methods to generate involution codes and also calculate their informational entropy. Since it turns out that the entropy of these generated involution codes is greater than $\log 2$, by the coding theorem ([1,23]) it follows that the constructed code words can be used to encode binary strings. We end with a few concluding remarks.

2   Definitions and Properties

An alphabet Σ is a finite non-empty set of symbols. We will denote by Δ the special case when the alphabet is $\{A, G, C, T\}$ representing the DNA nucleotides. A word u over Σ is a finite sequence of symbols in Σ. We denote by Σ^* the set of all words over Σ, including the empty word 1 and by Σ^+, the set of all non-empty words over Σ. We note that with the concatenation operation on words, Σ^* is the free monoid and Σ^+ is the free semigroup generated by Σ. The length of a word $u = a_1 \cdots a_n$ is n and is denoted by $|u|$. For words representing DNA sequences, we will use the following convention. A word u over Δ denotes a DNA strand in its $5' \to 3'$ orientation. The Watson-Crick complement of the word u, also in orientation $5' \to 3'$ is denoted by $\overleftarrow{u}$. For example if $u = AGGC$ then $\overleftarrow{u} = GCCT$.

Throughout the rest of the paper, we concentrate on finite sets $X \subseteq \Sigma^+$ that are *codes* i.e. every word in X^+ can be written uniquely as a product of words in X. For the background on codes we refer the reader to [4,26]. We will need the following definitions:

$$\mathrm{PPref}(X) = \{u \mid \exists v \in \Sigma^+, uv \in X\}$$
$$\mathrm{PSuff}(X) = \{u \mid \exists v \in \Sigma^+, vu \in X\}$$
$$\mathrm{PSub}(X) = \{u \mid \exists v_1, v_2 \in \Sigma^*, v_1 v_2 \neq 1, v_1 u v_2 \in X\}$$

We define the set of prefixes, suffixes and subwords of a set of words as Pref , Suff and Sub. Similarly, we have $\mathrm{Suff}_k(w) = \mathrm{Suff}(w) \cap \Sigma^k$, $\mathrm{Pref}_k(w) = \mathrm{Pref}(w) \cap \Sigma^k$, $\mathrm{Sub}_k(w) = \mathrm{Sub}(w) \cap \Sigma^k$. We follow the definitions initiated in [12] and used in [13]. An involution $\theta : \Sigma \to \Sigma$ of a set Σ is a mapping such that θ^2 equals the identity mapping, $\theta(\theta(x)) = x, \forall x \in \Sigma$. The mapping $\nu : \Delta \to \Delta$ defined by $\nu(A) = T, \nu(T) = A, \nu(C) = G, \nu(G) = C$ is an involution on Δ and can be extended to a morphic involution of Δ^*. Since the Watson-Crick complementarity appears in a reverse orientation, we consider another involution $\rho : \Delta^* \to \Delta^*$ defined inductively, $\rho(s) = s$ for $s \in \Delta$ and $\rho(us) = \rho(s)\rho(u) = s\rho(u)$ for all $s \in \Delta$ and $u \in \Delta^*$. This involution is antimorphism such that $\rho(uv) = \rho(v)\rho(u)$. The Watson-Crick complementarity then is the antimorphic involution obtained by the composition $\nu\rho = \rho\nu$. Hence for a DNA strand u we have that $\rho\nu(u) = \nu\rho(u) = \overleftarrow{u}$. The involution ρ reverses the order of the letters in a word and as such is used in the rest of the paper.

The following Definition 1 [14,16] introduces notions meant to formalize a variety of language properties, each of whom guarantees the absence of a certain unwanted hybridization. The notion of θ-infix and θ-comma-free code were introduced in [12] and was called θ-compliant and θ-free respectively. The definition of θ-intercode and θ-outfix code are new notions introduced here.

Definition 1. *Let* $\theta : \Sigma^* \to \Sigma^*$ *be a morphic or antimorphic involution and* $X \subseteq \Sigma^+$.

1. *The set* X *is called* θ-infix-code *if* $\Sigma^* \theta(X) \Sigma^+ \cap X = \emptyset$ *and* $\Sigma^+ \theta(X) \Sigma^* \cap X = \emptyset$.
2. *The set* X *is called* θ-comma-free-code *if* $X^2 \cap \Sigma^+ \theta(X) \Sigma^+ = \emptyset$.
3. *The set* X *is called* θ-strict-code *if* $X \cap \theta(X) = \emptyset$.
4. *The set* X *is called* θ-prefix-code *if* $X \cap \theta(X) \Sigma^+ = \emptyset$.
5. *The set* X *is called* θ-suffix-code *if* $X \cap \Sigma^+ \theta(X) = \emptyset$.
6. *The set* X *is called* θ-bifix-code *if* X *is both* θ-prefix *and* θ-suffix.
7. *The set* X *is called a* θ-intercode *if* $X^{m+1} \cap \Sigma^+ \theta(X^m) \Sigma^+ = \emptyset$, $m \geq 1$. *The integer* m *is called the index of* X.
8. *The set* X *is called* θ-outfix-code *if for* $u, \theta(u_1) x \theta(u_2) \in X$ *with* $\theta(u) = \theta(u_1) \theta(u_2)$ *implies* $x = 1$.

Note that θ-infix languages avoid undesirable hybridization of the type depicted in Fig 1c, θ-comma-free languages avoid undesirable hybridization of the type depicted in Fig 1a, θ-intercodes avoid undesirable hybridization of the type depicted in Fig 1b, θ-suffix languages avoid undesirable hybridization of the type depicted in Fig 1d, θ-prefix languages avoid undesirable hybridization of the type depicted in Fig 1e, θ-outfix languages avoid undesirable hybridization of the type depicted in Fig 2. Note that a θ-intercode of index one is θ-comma-free.

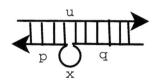

Fig. 2. Another type of intermolecular hybridization that we want to avoid: the reverse complement of a code word is a concatenation of a prefix and a suffix of another code word. A θ-outfix code (a new notion defined in this paper) avoids such hybridizations.

Also note that X is θ-intercode of index m if and only if $\theta(X)$ is θ-intercode of index m. We have defined several properties that are desirable for DNA languages to have. The properties 1 to 4 in Definition 1 have been extensively studied in [12,14,16]. Here we complete this study by proving the relationship between several properties. The following proposition shows the connection between θ-infix and θ-comma-free languages. We use $\theta : \Sigma^* \mapsto \Sigma^*$ to be either morphic or antimorphic involution throughout this paper unless specified. In the following, we list some of the properties and relations between θ-infix and θ-comma-free code.

Proposition 1. *Let* $\theta : \Sigma^* \to \Sigma^*$ *be a morphic or antimorphic involution and* $X \subseteq \Sigma^*$. *Then the following are equivalent.*

1. X is a θ-comma-free code.
2. X is θ-infix and $\theta(X) \cap PSuff(X)PPref(X) = \emptyset$.
3. X is θ-infix and $X^2 \cap PPref(X)\theta(X)PSuff(X) = \emptyset$.
4. X is θ-infix and $X^n \cap (\Sigma^+\theta(X)\Sigma^+X^{n-2}) = \emptyset$.
5. X is θ-infix and $X^n \cap (X^{n-2}\Sigma^+\theta(X)\Sigma^+) = \emptyset$.

Proposition 2. *Let $X \subseteq \Sigma^+$ be a θ-infix code. Then $X^3 \cap \Sigma^+\theta(X^2)\Sigma^+ = \emptyset$ if and only if $\theta(X^2) \cap PSuff(X)XPPref(X) = \emptyset$.*

Corollary 1. *If $X \subseteq \Sigma^+$ is θ-comma-free then $X^2 \cap PSuff(X)\theta(X)PPref(X) = \emptyset$ and $\theta(X^2) \cap PSuff(X)XPPref(X) = \emptyset$.*

3   Closure Properties of Involution Codes

In this section we discuss several properties of θ-prefix, θ-suffix, θ-bifix, θ-outfix codes and θ-strict codes. Besides being generalizations of outfix codes, the motivation behind introducing the notion of θ-outfix codes comes from the fact that a set of DNA words that is a θ-outfix code avoids any undesirable hybridization of the type in Fig 2. Ensuring that no such unwanted hybridization occurs is obviously desirable from an experimental view point. It is interesting to note that certain properties that are not satisfied by θ-prefix and θ-suffix codes are satisfied by θ-bifix codes. In particular we discuss the conditions under which such languages are closed under arbitrary concatenation. From a practical point of view, these results give conditions under which, given a small finite set of "good" codewords, we can construct arbitrarily large sets of good code words by concatenation.

Lemma 1. *Let $X \subseteq \Sigma^+$.*

1. *If X is θ-infix then X is both θ-prefix and θ-suffix and hence a θ-bifix code.*
2. *For a morphic involution θ, X is θ-prefix(suffix) if and only if $\theta(X)$ is θ-prefix(suffix).*
3. *For an antimorphic involution θ, X is θ-prefix(suffix) if and only if $\theta(X)$ is θ-suffix(prefix).*
4. *X is θ-bifix if and only if $\theta(X)$ is θ-bifix.*

In the next proposition we show that the family of θ-prefix(suffix) codes is closed under concatenation when θ is a morphic involution and hence for a θ-prefix(suffix) code X, any arbitrary power of X is also a θ-prefix(suffix) code when θ is a morphic involution.

Proposition 3. *For a morphic involution θ, the family of θ-prefix (θ-suffix) codes is closed under concatenation.*

Proof. Let X_1 and X_2 be θ-prefix. Suppose X_1X_2 is not θ-prefix then there exists $x_1x_2, y_1y_2 \in X_1X_2$ such that $x_1x_2 = \theta(y_1y_2)b = \theta(y_1)\theta(y_2)b$ for some $b \in \Sigma^+$. Note that neither x_1 is a prefix of $\theta(y_1)$ nor $\theta(y_1)$ is a prefix of x_1 since both X_1 and $\theta(X_1)$ (Please refer 1) are θ-prefix. Hence $x_1 = \theta(y_1)$ which implies that $\theta(y_2)$ is a prefix of x_2 which is a contradiction to our assumption that X_2 is θ-prefix. Similar proof works for θ-suffix.   □

Corollary 2. *Let θ be a morphic involution.*

1. *If X is θ-prefix then X^n is θ-prefix for all $n \geq 1$.*
2. *If X is θ-suffix then X^n is θ-suffix for all $n \geq 1$.*

Note that the above proposition does not hold when θ is an antimorphic involution. For example let $X_1 = \{aa, baa\}$ and $X_2 = \{bb, bbb\}$ over the alphabet set $\Sigma = \{a, b\}$ and let θ be antimorphism such that $a \mapsto b$ and $b \mapsto a$. Note that both X_1 and X_2 are θ-prefix but $X_1 X_2$ is not θ-prefix since for $aabb \in \theta(X_1 X_2)$, $aabbb \in X_1 X_2$. Hence when θ is an antimorphic involution we need an additional restriction on the sets X_1 and X_2 which is shown in the next proposition.

Proposition 4. *For an antimorphic involution θ, if X_1 and X_2 are such that $X_1 \cup X_2$ is θ-bifix, then $X_1 X_2$ and $X_2 X_1$ are θ-bifix.*

Proof. Immediate.

Corollary 3. *Let θ be morphic or antimorphic involution on Σ^*. If X is a θ-bifix code then X^n is a θ-bifix code for all $n \geq 1$.*

In the next proposition we provide with the necessary condition under which for a set X, the Kleene star of the set X is θ-prefix(suffix).

Proposition 5. *If X is such that X is θ-strict-infix code then X^+ is both θ-prefix and θ-suffix.*

Proof. To show that X^+ is θ-prefix (i.e.) to show that $X^+ \cap \theta(X^+) \Sigma^+ = \emptyset$. Suppose X^+ is not θ-prefix code then there exists $x_1 x_2 ... x_n = \theta(y_1 ... y_m) b$ for $x_i, y_j \in X, i = 1, .., n, j = 1, ..., m$ and $b \in \Sigma^+$. For a morphic θ, $x_1 x_2 ... x_n = \theta(y_1) ... \theta(y_m) b$ implies either x_1 is a subword of $\theta(y_1)$ or $x_1 = \theta(y_1)$ or $\theta(y_1)$ is a subword of x_1. All cases contradict our assumption that X is strictly θ-infix. Similarly we can prove that X^+ is θ-suffix code.

The next two propositions gives us conditions under which when a composition of some arbitrary languages satisfy good encoding properties, the right and the left context of such languages also satisfy the same good encoding properties.

Proposition 6. *Let $X \subseteq \Sigma^+$ be such that X is not a θ-strict code.*

1. *If X^m is θ-prefix for $m \geq 1$, then X is θ-prefix.*
2. *If X^m is θ-suffix for $m \geq 1$, then X is θ-suffix.*
3. *If X^m is θ-bifix for $m \geq 1$, then X is θ-bifix.*

Proposition 7. *Let X_i, $i = 1, 2, ..., m$ be non empty languages over Σ such that $X_i \cap \theta(X_i) \neq \emptyset$, $i = 1, 2, ..., m$. Let θ be a morphic involution. Then the following are true.*

1. *If $X_1 X_2 ... X_m$ is θ-prefix, then $X_2 ... X_m$, $X_3 ... X_m, ...,$ $X_{m-1} X_m$, X_m are θ-prefix codes.*

2. *If $X_1X_2...X_m$ is θ-suffix, then $X_1...X_{m-1}$, $X_1...X_{m-2},...,$ X_1X_2, X_1 are θ-suffix codes.*

Proof. We prove (i) for the case $m = 2$ and the result follows by induction. Assume X_1X_2 is θ-prefix. If X_2 is not θ-prefix code then there exists $x_2, y_2 \in X_2$ such that $x_2 = \theta(y_2)b$. Since $X_1 \cap \theta(X_1) \neq \emptyset$, there exists $x_1, y_1 \in X$ such that $x_1 = \theta(y_1)$ and hence $x_1x_2 = \theta(y_1)\theta(y_2)b = \theta(y_1y_2)b$ which is a contradiction to our assumption that X_1X_2 is θ-prefix. Similar proof works for (ii).   □

In the following propositions we investigate certain properties of θ-outfix codes. We recall the following definition of insertion into a set from [15].
    For $X \subseteq \Sigma^+$, let

$$Y = \theta(X) \leftarrow \Sigma^+ = \bigcup_{u \in \theta(X), v \in \Sigma^+} (u \leftarrow v)$$

where, $u \leftarrow v = \{u_1vu_2 : u = u_1u_2, u_1, u_2 \in \Sigma^*\}$.
    The next lemma is a direct consequence of the definition of θ-outfix codes.

Lemma 2. *For $X \subseteq \Sigma^+$ let Y be the set obtained above. Then X is a θ-outfix code iff $Y \cap X = \emptyset$.*

Corollary 4. *For a regular X, it is decidable whether X is a θ-outfix code or not.*

Proof. For a regular X, $\theta(X)$ is regular and it has been show in [15] that for regular set $\theta(X)$ and Σ^+, $Y = \theta(X) \leftarrow \Sigma^+$ is also regular. It is decidable whether a regular set is empty or not. Hence it is decidable whether X is a θ-outfix code or not.

It is easy to see that every θ-outfix code is θ-prefix and θ-suffix and hence a θ-bifix code. Also note that X is θ-outfix code if and only if $\theta(X)$ is a θ-outfix code. In the following propositions we investigate the closure properties of θ-outfix codes. In most cases we omit the proof.

Proposition 8. *For a morphic involution θ, the family of θ-outfix codes is closed under concatenation.*

Note that the above proposition does not hold when θ is an anitmorphic involution. For example let $X_1 = \{aa, baa\}$ and $X_2 = \{bb, bbb\}$ over the alphabet set $\Sigma = \{a, b\}$ and let θ be antimorphism such that $a \mapsto b$ and $b \mapsto a$. Note that both X_1 and X_2 are θ-outfix but X_1X_2 is not θ-outfix since for $aabb \in \theta(X_1X_2)$, $aa(b)bb \in X_1X_2$. But the cases become simpler if we just work with one set X. In the next proposition, we show that a θ-outfix code is closed under arbitrary concatenation with itself for both morphic and antimorphic involution.

Proposition 9. *X is a θ-outfix code iff X^+ is a θ-outfix code.*

Proposition 10. *For a morphic involution θ, let $X_1, X_2 \subseteq \Sigma^+$ be such that $X_i \cap \theta(X_i) \neq \emptyset$ for $i = 1, 2$. If X_1X_2 is θ-outfix code then both X_1 and X_2 are θ-outfix codes.*

Proof. Suppose that X_1 is not θ-outfix, then $xy, \theta(x)u\theta(y) \in X_1$ for some $u \in \Sigma^+, x, y \in \Sigma^*$. Consider $z \in X_2 \cap \theta(X_2)$. Then $xyz \in X_1X_2$ and $\theta(x)u\theta(y)z \in X_1X_2$, a contradiction. Hence X_1 must be a θ-outfix code. Similarly, we can show that X_2 is a θ-outfix code.   □

In the next proposition we investigate certain properties of θ-strict codes and their relation with other sets of codes.

Lemma 3.   1. *If $X_1, X_2 \subseteq \Sigma^+$ are θ-strict, then $X_1 \cup X_2$ is not necessarily θ-strict.*
2. *Let X_1, X_2 be θ-strict. Then $X_1 \cap \theta(X_2) = \emptyset$ and $X_2 \cap \theta(X_1) = \emptyset$ if and only if $X_1 \cup X_2$ is θ-strict.*
3. *If X_1 and X_2 are θ-strict, then $X_1 \cap X_2$ is θ-strict.*
4. *Let X_1 and X_2 be θ-strict. When θ is morphism, if one of X_1 or X_2 is θ-prefix , then X_1X_2 is θ-strict. When θ is antimorphism, if $X_1 \cup X_2$ is θ-strict-bifix, then X_1X_2 is θ-strict.*
5. *If X is θ-strict-bifix, then X^+ is θ-strict.*
6. *X is θ-strict if and only if $\theta(X)$ is θ-strict.*

4   Involution Intercodes

We now generalize the concept of θ-comma-free codes to θ-intercodes and study the properties of such codes. Note that if θ is the identity function, a θ-intercode becomes the well known notion of intercode, widely studied in the literature [26]. Besides being generalizations of intercodes, the motivation behind introducing the notion of θ-intercodes comes from the fact that a set of DNA words that is a θ-intercode avoids any undesirable hybridization of the type in Fig 1b. Ensuring that no such unwanted hybridization occurs is obviously desirable from an experimental view point.

Proposition 11. *Let X be a regular language. Then for a given $m \geq 1$, it is decidable whether or not X is a θ-intercode of index m.*

Proof. X is a θ-intercode of index m if and only if $X^{m+1} \cap \Sigma^+\theta(X^m)\Sigma^+ = \emptyset$. Since the family of regular languages is closed under catenation and intersection, X^{m+1} and $\Sigma^+\theta(X^m)\Sigma^+$, $\theta(X^m)$ and $X^{m+1} \cap \Sigma^+\theta(X^m)\Sigma^+$ are regular. It is decidable whether a regular language is empty or not.

Proposition 12. *Let $|\Sigma| \geq 2$. Then for any $m \geq 1$, every θ-intercode of index m is a θ-intercode of index $m + 1$.*

Proof. Given X is a θ-intercode of index m hence $X^{m+1} \cap \Sigma^+\theta(X^m)\Sigma^+ = \emptyset$. Suppose $X^{m+2} \cap \Sigma^+\theta(X^{m+1})\Sigma^+ \neq \emptyset$ then there exists $x_1, x_2, ...x_{m+2}, y_1, y_2, ...y_{m+1} \in X$ such that $x_1x_2...x_{m+2} = a\theta(y_1...y_{m+1})b$ for some $a, b \in \Sigma^+$. If $|a| \geq |x_1|$ then $x_2...x_{m+2} \in \Sigma^+\theta(y_2)...\theta(y_{m+1})\Sigma^+$. If $|b| \geq |x_{m+2}|$ then $x_1...x_{m+1} \in \Sigma^+\theta(y_1)...\theta(y_m)\Sigma^+$. If $|a| < |x_1|$, $|b| < |x_{m+2}|$ then $a_1x_2...x_{m+1}b_1 = \theta(y_1)...\theta(y_{m+1})$ which implies $y_1...y_{m+1} = a_2\theta(x_2...x_{m+1})b_2$ which is a contradiction.

Both cases contradict our assumption that X is a θ-intercode of index m.   □

Proposition 13. *For any involution θ, every θ-intercode X such that $X \cap \theta(X) \neq \emptyset$ is a θ-bifix code.*

Proof. X is a θ-intercode of index m, then by definition $X^{m+1} \cap \Sigma^+ \theta(X^m)\Sigma^+ = \emptyset$. Since $X \subseteq \Sigma^+$ and $X^{m+1} \cap \theta(X^{m+1})\Sigma^+ \subseteq X^{m+1} \cap \Sigma^+\theta(X^m)\Sigma^+ = \emptyset$ we have X^{m+1} is a θ-prefix code which implies X is a θ-prefix code by proposition 6. Similarly we can show that X is a θ-suffix code. □

The converse of the above proposition is not true. For example let $X = \{aab, aba\}$ over the alphabet set $\Sigma = \{a, b\}$. Let θ be a morphic involution with $a \mapsto b$ and $b \mapsto a$. Note that X is both θ-prefix and θ-suffix but $aa\theta(aba)a = aababa \in X^2$. Hence X is not θ-intercode of index one. Also it is shown in [12] that every θ-comma-free code is θ-infix. But this is not the case for θ-intercodes of index $m \geq 2$. One example is as follows: Let $X = \{b^2ab^3ab^2, a^3\}$ over the alphabet set $\Sigma = \{a, b\}$ and let θ be an antimorphic involution such that $a \mapsto b$ and $b \mapsto a$. The language X is θ-intercode of index 2 but not a θ-infix code.

Proposition 14. *If X is θ-comma-free code then X is a θ-intercode of index m for all $m \geq 1$.*

Proof. Suppose $X^{m+1} \cap \Sigma^+\theta(X^m)\Sigma^+ \neq \emptyset$, then there exists $x_1, x_2, ...x_{m+1}$, $y_1, y_2, ..., y_m \in X$ and $a, b \in \Sigma^+$ such that $x_1x_2...x_{m+1} = a\theta(y_1y_2...y_m)b$. Then either $\theta(y_i)$ is a subword of x_j which contradicts X being θ-infix and hence θ-comma-free or $\theta(y_i)$ is a subword of x_jx_{j+1} which contradicts X being θ-comma-free. □

Note that the converse of the above proposition is not true. For example let $X = \{cbaa, baad, babb\}$ over the alphabet set $\Sigma = \{a, b, c, d\}$. Let θ be an antimorphic involution with $a \mapsto b$ and $c \mapsto d$. It is easy to check that $X^3 \cap \Sigma^+\theta(X^2)\Sigma^+ = \emptyset$ but X is not θ-comma-free since $cb\theta(babb)ad = cbaabaad \in X^2$.

For any word $u = a_1a_2...a_n \in \Sigma^*$ with $a_i \in \Sigma$ define the reverse of u as $\hat{u} = a_na_{n-1}...a_2a_1$. For $X \subseteq \Sigma^+$, define $\hat{X} = \{\hat{u} : u \in X\}$. The following characterization of θ-intercodes of index m is an immediate result from the definition of θ-intercodes.

Proposition 15. *Let $X \subseteq \Sigma^+$. The following are equivalent.*

1. X is a θ-intercode of index m.
2. $\hat{X}$ is a θ-intercode of index m.
3. For any $u \in X^m$, $x, y \in \Sigma^*$, $x\theta(u)y \in X^{m+1}$ implies $x = 1$ or $y = 1$.

Proposition 16. *If X is a θ-intercode of index m then $X^k \cap \Sigma^+\theta(X^m)\Sigma^+ = \emptyset$ for all $k \leq m + 1$.*

Proof. Suppose $X^k \cap \Sigma^+\theta(X^m)\Sigma^+ \neq \emptyset$ for some $k < m + 1$, then $\emptyset \neq X^{m+1} \cap X^{m+1-k}\Sigma^+\theta(X^m)\Sigma^+ \subseteq X^{m+1} \cap \Sigma^+\theta(X^m)\Sigma^+$ which is a contradiction to our assumption X is a θ-intercode of index m.

Proposition 17. *If $X \subseteq \Sigma^+$ is a θ-intercode of index m, $m \geq 1$ and X is strictly θ-infix, then $X^n \cap \Sigma^+\theta(X^m)\Sigma^+ = \emptyset$ and $X^m \cap \Sigma\theta(X^n)\Sigma^+ = \emptyset$ for all $n \geq m$.*

Proof. We prove by induction on n. Suppose for $n = m$, $x \in X^m \cap \Sigma^+\theta(X^m)\Sigma^+$ then $x = x_1...x_m = a\theta(y_1)...\theta(y_m)b$. Then atleast one of the $\theta(y_i)$ is a subword of x_j which is a contradiction to our assumption that X is strictly θ-infix. When $n = m+1$, $X^{m+1} \cap \Sigma^+\theta(X^m)\Sigma^+ = \emptyset$ since X is a θ-intercode. Now assume that $X^n \cap \Sigma^+\theta(X^m)\Sigma^+ = \emptyset$ for all $m \leq n \leq k$. Suppose $X^{k+1} \cap \Sigma^+\theta(X^m)\Sigma^+ \neq \emptyset$, then there exists $x_1, x_2, ..., x_{k+1}, y_1, y_2..., y_m \in X$ such that $x_1x_2...x_{k+1} = a\theta(y_1)\theta(y_2)...\theta(y_m)b$. If $|a| = |x_1|$ then $x_2...x_k = \theta(y_1)...\theta(y_m)b$ which implies either x_2 is a subword of $\theta(y_1)$ or $\theta(y_1)$ is a subword of x_2 which contradicts our assumption that X is strictly θ-infix. If $|a| > |x_1|$ then $x_2...x_k = a_1\theta(y_1)...\theta(y_m)b$ which implies that $X^{k-1} \cap \Sigma^+\theta(X^m)\Sigma^+ \neq \emptyset$ which is a contradiction to our induction hypothesis. The cases when $|b| = |x_{k+1}|$ or $|b| > |x_{k+1}|$ are similar to the case when $|a| = |x_1|$ or $|a| > |x_1|$ respectively. If $|a| < |x_1|$ and $|b| < |x_k|$ then atleast one of $\theta(y_i)$ is a subword of x_j which is a contradiction to our assumption that X is strictly θ-infix. $\square$

Proposition 18. *If X is a θ-intercode of index m and X is strictly θ-infix, then X^n is a θ-intercode of index m, for all $n \geq 1$.*

Proof. We need to show that $(X^n)^{m+1} \cap \Sigma^+(\theta(X^n))^m\Sigma^+ = \emptyset$ for all $n > 1$. Then $(X^n)^{m+1} \cap \Sigma^+(\theta(X^n))^m\Sigma^+ = (X^{nm+n}) \cap \Sigma^+\theta(X^{nm})\Sigma^+$. Note that by proposition 12 X is a θ-intercode of index nm and hence $X^{nm+1} \cap \Sigma^+\theta(X^{nm})\Sigma^+ = \emptyset$. Then by proposition 17 $(X^{nm+n}) \cap \Sigma^+\theta(X^{nm})\Sigma^+ = \emptyset$. Hence X^n is a θ-intercode of index m. $\square$

Proposition 19. *If X is a θ-intercode of index m and X is strictly θ-infix then X^+ is a θ-intercode of index m.*

Proof. Suppose X^+ is not θ-intercode of index m, then there exists $x, y \in X^+$ such that $x = x_1...x_{m+1}$ and $y = y_1..y_m$ for all $x_i, y_j \in X^+$ and $x = a\theta(y)b$ for some $a, b \in \Sigma^+$. The case when $x_i, y_j \in X$ for all i, j then $x \in X^{m+1} \cap \Sigma^+\theta(X^m)\Sigma^+$ which is a contradiction. If $x \in X^p$ and $y \in X^q$ for some $p, q \geq m$ then $x \in X^p \cap \Sigma^+\theta(X^q)\Sigma^+$ which is a contradiction by Proposition 12 and 17. Hence X^+ is a θ-intercode of index m. $\square$

Proposition 20. *If $X \cup Y$ is a θ-intercode of index m then XY is a θ-intercode of index m.*

Proof. Suppose $(XY)^{m+1} \cap \Sigma^+\theta((XY)^m)\Sigma^+ \neq \emptyset$ then let $r \in (XY)^{m+1}$ such that $r = x_1y_1x_2y_2...x_{m+1}y_{m+1} = a\theta(p_1q_1...p_mq_m)b$ for $x_1, ..., x_{m+1}, p_1, ..., p_m \in X$ and $y_1, ..., y_{m+1}, q_1, ..., q_m \in Y$ and $a, b \in \Sigma^+$. But $r \in (X \cup Y)^{2(m+1)} \cap \Sigma^+\theta((X \cup Y)^{2m})\Sigma^+$ which is a contradiction by Proposition 12 and 17. $\square$

5 n-θ-Comma-Free Codes and n-θ-Intercodes

If the alphabet Σ consists of more than one letter, the partial order $\leq_c$ defined on Σ^* by $u \leq_c v$ if and only if $v = xu = ux$ for some $x \in \Sigma^*$ plays an interesting role. That is if $u \leq_c v$, then $u = f^i$ for some primitive word f (f is primitive if $f = a^i$, $a \in \Sigma^+$ for some i implies $i = 1$) and $v = f^{i+j}$ for some $j \geq 0$. Thus if $u, v \in X \subseteq \Sigma^+$ and X is an independent set with respect to $\leq_c$, then $uv \neq vu$, which is equivalent to the fact that the two element set $\{u, v\}$ is a code. Hence a $\leq_c$-independent set is called a 2-code. This notion can be generalized as follows: An n-code is a set X with the property that every n element subset of the set X is a code ([26]). The notion of n-codes, n-comma free codes and hence n-intercodes were defined and studied in [26]. Here we extend these concepts to involution comma-free and involution intercodes as follows. This section investigates these notions and algebraic properties of these codes. An n-θ-intercode of index m is a language $X \subseteq \Sigma^+$ such that every subset of X with at most n elements is a θ-intercode of index m. An n-θ-comma-free code is an n-θ-intercode of index one.

Proposition 21. *The class of 2-θ-comma-free codes is not closed under union, catenation, complement, catenation closure.*

Proof. The proof will be done by constructing some examples. We consider two languages $\{ab\}$ and $\{ba\}$ that are 2-θ-comma-free for an antimorphic θ mapping $a \mapsto b$. It is clear that the union $\{ab, ba\} = \{ab, ba\}$, the product $\{ab\}^2$ and the catenation closure $\{(ab)^+\}$ are not 2-θ-comma-free codes. Also, the class of 2-θ-comma-free codes are closed under intersection but not under union or complement.

Proposition 22. *If X is a 2-θ-comma-free code then X is θ-infix.*

Proof. Suppose X is not θ-infix then there exists $x, y \in X$ such that $x = a\theta(y)b$ which implies $\{x, y\}$ is not θ-infix and hence not θ-comma-free which is a contradiction. Hence X is θ-infix.   □

Proposition 23. *Let $X \subseteq \Sigma^+$ be such that $X \cap \theta(X) = \emptyset$ and $\theta(\mathrm{PSuff}(X)) \cap \mathrm{PPref}(X) = \emptyset$.*
Then the following are equivalent.

1. *X is a 2-θ-comma-free code.*
2. *X is θ-infix and for $u, v \in \Sigma^+$, if $uv \in \theta(X)$ then $X \cap v\Sigma^*u = \emptyset$.*

Proof. Note that from Proposition 22 X is θ-infix. Let $u, v \in \Sigma^+$ such that $uv \in \theta(X)$. If $X' = \{\theta(uv), vxu\} \subseteq X$ for some $x \in \Sigma^*$, then $(vxu)^2 \in X'^2 \cap \Sigma^+ X' \Sigma^+$. This implies that X is not a 2-θ-comma-free code.

For the converse, let X be θ-infix. Suppose there exists $x, y \in X$ such that $\{x, y\}$ is not θ-comma-free then the either $xy = a\theta(x)b$ or $x^2 = a\theta(x)b$ or $y^2 = a\theta(x)b$ or $yx = a\theta(x)b$ for some $a, b \in \Sigma^+$. Since X is θ-infix, $\theta(x)$ is not a proper subword of x or y. Also note that $\theta(x) \neq x_2x_1$ for $x_1x_2 = x$ or $\theta(x) \neq y_2y_1$ for

$y_1 y_2 = y$ since for all $uv \in \theta(X)$, $v\Sigma^* u \cap X = \emptyset$. Suppose $a\theta(x)b = xy$, then $\theta(x) = x_2 y_1$ for $x = x_1 x_2$ and $y = y_1 y_2$. When θ is morphism, $x = \theta(x_2)\theta(y_1)$ and when θ is antimorphism, $x = \theta(y_1)\theta(x_2)$ both cases contradict our assumption that $\theta(\text{PSuff}(X)) \cap \text{PPref}(X) = \emptyset$. Hence X is a 2-θ-comma-free code.

Proposition 24. X is a 3-θ-comma-free code if and only if X is a θ-comma-free code.

Proposition 25. If X is a k-θ-comma-free code then X is a i-θ-comma-free code for all $i \leq k$.

Proof. Immediate. Note that X being k-θ-comma-free code does not imply X is i-θ-comma-free code for $i \geq k+1$ and $k \leq 2$. For example let $X = \{cbaa, baad, babb\}$ over the alphabet set $\Sigma = \{a, b, c, d\}$. Let θ be an antimorphic involution such that $a \mapsto b$ and $c \mapsto d$. It is easy to check that X is not θ-comma-free but X is 2-θ-comma-free.

Proposition 26. X is a θ-intercode of index m if and only if X is a $(2m+1)$-θ-intercode of index m.

Proof. Let X be $(2m+1)$-θ-intercode of index m. Suppose X is not θ-intercode of index m then there exists $x_1, x_2, ..., x_{m+1}, y_1, ..y_m \in X$ such that $x_1 x_2...x_{m+1} = a\theta(y_1...y_m)b$ for some $a, b \in \Sigma^+$ which implies that X is not $(2m+1)$-θ-intercode of index m. The converse of the proof is immediate. $\square$

Note that every θ-intercode of index m is an n-θ-intercode of index m for all $n \geq 1$. But for $n \leq 2m$ an n-θ-intercode of index m is not neccesarily a θ-intercode of index m. For example.....

Proposition 27. If X is a 2-θ-comma-free code , then Xy and yX are 2-θ-comma-free code for all $y \in X$.

Proof. Suppose Xy is not 2-θ-comma-free then there exists $\{x_1 y, x_2 y\} \subseteq Xy$ such that atleast one of the following happens:
$x_1 y x_2 y = a\theta(x_1 y)b$ or $x_1 y x_2 y = a\theta(x_2 y)b$ or $x_2 y x_1 y = a\theta(x_1 y)b$ or $x_2 y x_1 y = a\theta(x_2 y)b$ or $x_1 y x_1 y = a\theta(x_1 y)b$ or $x_1 y x_1 y = a\theta(x_2 y)b$ or $x_2 y x_2 y = a\theta(x_1 y)b$ or $x_2 y x_2 y = a\theta(x_2 y)b$.
Note that none of the $\theta(x_1)$ or $\theta(x_2)$ or $\theta(y)$ is a subword of x_1, x_2 or y since X is θ-infix. Also none of the $\theta(x_i)$ or $\theta(y)$ is a subword of $x_i y$ or $y x_i$ since X is a 2-θ-comma-free code. Suppose in $x_1 y x_2 y = a\theta(x_1 y)b$ if $\theta(x_1)$ is a subword of $y x_2$ then either $\theta(y)$ is a subword of x_1 or $x_2 y$ which is a contradiction to the given assumption. Similarly we can show that yX is a 2-θ-comma-free code. $\square$

Note that X being 2-θ-comma-free code does not imply X^n is 2-θ-comma-free code. For example let $X = \{ebb, dae, aac, bcb\}$. Let θ be a morphic involution such that $a \mapsto b$, $c \mapsto d$ and $e \mapsto e$. It is easy to check that X is 2-θ-comma-free code but X^2 is not since $ebbdae, aacbeb \in X^2$ with $ebbdaeaacbeb = e\theta(aacbeb)acbeb$.

6   Methods for Constructing Involution Codes

With the constructions in this section we show several ways to generate involution codes with "good" properties. Many authors have realized that in the design of DNA strands it is helpful to consider three out of the four bases. This was the case with several successful experiments [3,8,24]. It turns out that this, or a variation of this technique, can be generalized in such a way that codes with some of the desired properties can be easily constructed. Methods to construct θ-infix, θ-comma-free, θ-k-code and θ-subword-k-codes were provided in [14]. In this section, we concentrate on providing methods to generate θ-prefix, θ-suffix, θ-bifix, θ-outfix and θ-intercodes X such that X^+ has the same property. Some of these methods (Proposition 28) are in some sense generalizations of the idea of considering only three out of four bases. For each code X, the entropy of X^+ is computed. The entropy measures the information capacity of the codes, i.e., the efficiency of these codes when used to represent information.

Suppose that we have a source alphabet with p symbols each occurring with probability $s_1, s_2, ...s_p$. If $s_1 = 1$, then there is no information since we know what the message must be. If all the probabilities are different then for a symbol with low probability we get more information than for a symbol with high probability. Hence information is somewhat inversely related to the probability of occurrence. Entropy is the average information over the whole alphabet of symbols.

The standard definition of entropy of a code $X \subseteq \Sigma^+$ uses a probability distribution over the symbols of the alphabet of X (see [4]). However, for a p-symbol alphabet, the maximal entropy is obtained when each symbol appears with the same probability $\frac{1}{p}$. In this case the entropy essentially counts the average number of words of a given length as subwords of the code words [19]. From the Coding Theorem ([1]), it follows that $\{0,1\}^+$ can be encoded by X^+ with $\Sigma \mapsto \{0,1\}$ if the entropy of X^+ is at least $\log 2$ (see Theorem 5.2.5 in [23]). The codes for θ-comma-free, strictly θ-comma-free, and θ-k-codes designed in this section have entropy larger than $\log 2$ when the alphabet has $p = 4$ symbols. Hence, such DNA codes can be used for encoding bit-strings.

We start with the entropy definition as defined in [23].

Definition 2. *Let X be a code. The entropy of X^+ is defined by*

$$\hbar(X) = lim_{n \to \infty} \frac{1}{n} \log |\text{Sub}_n(X^+)|.$$

If G is a deterministic automaton or an automaton with a delay (see [23]) that recognizes X^+ and A_G is the adjacency matrix of G, then by Perron-Frobenius theory A_G has a maximal positive eigen value $\bar{\mu}$ and the entropy of X^+ is $\log \bar{\mu}$ (see Chapter 4 of [23]). We use this fact in the following computations of the entropies of the designed codes. In [12], Proposition 16, authors designed a set of DNA code words that is strictly θ-comma-free. The following propositions shows that, in a similar way, we can construct codes with additional "good" properties.

In what follows we assume that Σ is a finite alphabet with $|\Sigma| \geq 3$ and $\theta : \Sigma \to \Sigma$ is an involution which is not identity. We denote by p the number of symbols in Σ.

Proposition 28. *Let $a \in \Sigma$ be such that $\theta(a) \neq a$. Let $X = \bigcup_{i=1}^{\infty} a^n(\Sigma \setminus \theta(a))^i a^n$ for a fixed integer $n \geq 1$. Then X and X^+ are both θ-prefix and θ-suffix. The entropy of X^+ is such that $\log(p-1) < \hbar(X^+) < \log(p)$.*

Proof. By Proposition 5 it is enough to show that X is strict θ-infix. Let $x, y \in X$ such that $u\theta(x)b = v$ for some $u, v \in \Sigma^*$. Then $u\theta(a^n w_1 a^n)v = a^n w_2 a^n$ for some $w_1, w_2 \in (\Sigma \setminus \{\theta(a)\})^i$ which implies $ub^n w_3 b^n v = a^n w_2 a^n$ where $\theta(a) = b$ which is not possible since $a \neq b$ and $b \neq \mathrm{Sub}(w_2)$. Therefore X is θ-strict-infix code and hence X^+ is both θ-prefix and θ-suffix.

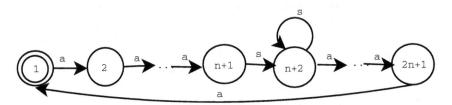

Fig. 3. Finite state automaton $\mathcal{A}$ that recognizes X^+ where $S = \Sigma \setminus \theta(a)$

Let $\mathcal{A} = (\mathcal{V}, \mathcal{E}, \lambda)$ be the automaton that recognizes X^+ where $V = \{1, ..., 2n+1\}$ is the set of vertices, $E \subseteq V \times \Sigma \times V$ and $\lambda : E \to \Sigma$ (with $(i, s, j) \mapsto s$) is the labeling function defined in the following way:

$$\lambda(i, s, j) = \begin{cases} a & \text{for } 1 \leq i \leq n, n+2 \leq i \leq 2n, \ j = i+1, \\ & \text{and } i = 2n+1, j = 1, \\ s & \text{for } i = n+1, n+2, \ j = n+2, \ s \in \Sigma \setminus \{\theta(a)\} \end{cases}$$

Then the adjacency matrix for $\mathcal{A}$ is a $(2n+1) \times (2n+1)$ matrix with ijth entry equal to the number of edges from vertex i to vertex j. Then the characteristic polynomial can be computed to be $\det(A - \mu I) = (-\mu)^{2n}(p-1-\mu) + (-1)^{2n}(p-1)$. The eigen values are solutions of the equation $\mu^{2n}(p-1) - \mu^{2n+1} + p - 1 = 0$ which gives $p - 1 = \mu - \frac{\mu}{\mu^{2n}+1}$. Hence $0 < \frac{\mu}{\mu^{2n}+1} < 1$, i.e., $p - 1 < \mu < p$.   □

In the case of the DNA alphabet, $p = 4$ and for $n = 1$ the above characteristic equation becomes $\mu^3 - 3\mu^2 - 3 = 0$. The largest real value of μ is approximately 3.27902 which means that the entropy of X^+ is greater than $\log 2$.

Proposition 29. *Let $a, b \in \Sigma$ be such that for all $\theta(a) \neq \theta(b) \neq a \neq b$. Let $X = \bigcup_{i=1}^{\infty} a^n \Sigma^i b^n$ for a fixed integer $n \geq 1$. Then X and X^+ are θ-bifix and θ-outfix. The entropy of X^+ is such that $\log(p-1) < \hbar(X^+) < \log(p)$.*

In the case of the DNA alphabet, $p = 4$ and for $n = 1$ the above characteristic equation becomes $\mu^3 - 4\mu^2 - 4 = 0$. The largest real value of μ is approximately 4.22417 which means that the entropy of X^+ is greater than $\log 2$.

Proposition 30. *Choose distinct $a, b, c \in \Sigma$ such that $\theta(a) \neq b, c$, $\theta(a) \neq a$. Let $X = \bigcup_{i=1}^{\infty} a^n(\Sigma^{n-1}c)^i b^n$ for some $n \geq 2$. Then X and so X^+ are strictly θ-intercodes of index m for all $m \geq 1$. The entropy of X^+ is such that $\log(p^{\frac{n-1}{n}}) < \hbar(X^+) < \log((p^{n-1} + 1)^{\frac{1}{n}})$.*

For the DNA alphabet, $p = 4$, and for $n = 2$, the above characteristic equation becomes $\mu^6 - 4\mu^4 - 4 = 0$. Solving for μ, the largest real value of μ is 2.05528. Hence the entropy of X^+ is greater than $\log 2$.

7 Concluding Remarks

In this paper we investigated theoretical properties of languages that avoided certain type of undesirable Watson-Crick bindings; θ-outfix codes, θ-intercodes, n-θ-intercodes and n-θ-comma-free codes. All these new concepts generalize classical notions of outfix codes, intercodes, n-intercodes and n-comma-free codes respectively. In addition, DNA word sets that are θ-outfix codes or θ-intercodes are of interest in the design of DNA computing experiments since such sets avoid unwanted hybridization Fig 1 and Fig 2. This paper We also developed certain methods to construct such sets of DNA code words with good properties and have calculated their informational entropy.

Acknowledgment. This work has been supported by NSERC and Canada Research Chair Grant for Lila Kari.

References

1. R.L. Adler, D. Coppersmith and M. Hassner, *Algorithms for sliding block codes -an application of symbolic dynamics to information theory*, IEEE Trans. Inform. Theory 29 (1983): 5-22.
2. E.B. Baum, *DNA Sequences useful for computation* unpublished article (1996).
3. R.S.Braich, N.Chelyapov, C.Johnson, P.W.K.Rothemund, L.Adleman, *Solution of a 20-variable 3-SAT problem on a DNA computer Science*, Science 19, Vol 296(5567) (2002) 499-502.
4. J. Berstel, D. Perrin, *Theory of Codes,* Academis Press, Inc. Orlando Florida, 1985.
5. R.Deaton, J.Chen, H.Bi, M.Garzon, H.Rubin, D.F.Wood, *A PCR based protocol for In vitro selection of non-crosshybridizing oligonucleotides, DNA Computing: Proceedings of the 8th International Meeting on DNA Based Computers* (M.Hagiya, A.Ohuchi editors), Springer LNCS 2568 (2003) 196-204.
6. R.Deaton, J.Chen, M.Garzon, J.Kim, D.Wood, H.Bi, D.Carpenter, Y.Wang, *Characterization of Non-Crosshybridizing DNA Oligonucleotides Manufactured in Vitro, DNA computing: Preliminary Proceedings of the 10th International Meeting on DNA Based Computers* (C.Ferretti, G.Mauri, C.Zandron editors) June7-10, (2004) 132-141.
7. R. Deaton et. al, *A DNA based implementation of an evolutionary search for good encodings for DNA computation,* Proc. IEEE Conference on Evolutionary Computation ICEC-97, (1997) 267-271.
8. D. Faulhammer, A. R. Cukras, R. J. Lipton, L. F.Landweber, *Molecular Computation: RNA solutions to chess problems,* Proceedings of the National Academy of Sciences, USA, 97 4 (2000) 1385-1389.
9. M. Garzon, R. Deaton, D. Reanult, *Virtual test tubes: a new methodology for computing,* Proc. 7th. Int. Symposium on String Processing and Information retrieval, A Coruña, Spain. IEEE Computing Society Press (2000) 116-121.

10. T. Head, *Formal language theory and DNA: an analysis of the generative capacity of specific recombinant behaviors*, Bull. Math. Biology 49 (1987) 737-759.

11. T. Head, Gh. Paun, D. Pixton, *Language theory and molecular genetics*, in Handbook of formal languages, Vol.II (G. Rozenberg, A. Salomaa editors) Springer Verlag (1997) 295-358.

12. S. Hussini, L. Kari, S. Konstantinidis, *Coding properties of DNA languages*, DNA Computing: Proceedings of the 7th International Meeting on DNA Based Computers (N. Jonoska, N.C. Seeman editors), Springer LNCS 2340 (2002) 57-69.

13. N. Jonoska, D. Kephart, K. Mahalingam, *Generating DNA code words* Congressus Numernatium **156** (2002): 99-110.

14. N.Jonoska , K.Mahalingam and J.Chen, *Involution Codes: With Application to DNA Coded Languages*, Natural Computing, Vol 4-2(2005), 141-162.

15. L.Kari, *On insertion and deletion on formal languages*, Doctoral Dissertation in Mathematics, University of Turku, Finland.

16. L. Kari, S. Konstantinidis, E. Losseva and G. Wozniak, *Sticky-free and overhang-free DNA languages*, Acta Informatica 40 (2003): 119-157.

17. L.Kari, S.Konstantinidis and P.Sosik, *Bond-free Languages: Formalizations, Maximality and Construction Methods*, Preliminary Proceedings of DNA 10 June 7-10 (2004):16-25.

18. L.Kari, S.Konstantinidis, P.Sosik, *Preventing Undesirable Bonds between DNA Codewords* Preliminary Proceedings of DNA 10 June 7-10 (2004) 375-384.

19. M.S. Keane, *Ergodic theory an subshifts of finite type* , Ergodic theory, symbolic dynamics and hyperbolic spaces (ed. T.Edford, et.al.) Oxford Univ. Press, Oxford (1991): 35-70.

20. D.Kephart and J.Lefevre, *Codegen: The generation and testing of DNA code words*, Proceedings of IEEE Congress on Evolutionary Computation, June (2004): 1865-1873.

21. P.Leupold, *Partial Words for DNA Coding* Preliminary Proceedings of DNA 10 June 7-10 (2004) 26-35.

22. Z. Li, *Construct DNA code words using backtrack algorithm*, preprint.

23. D. Lind and B. Marcus, *An introduction to Symbolic Dynamics and Coding*, Cambridge University Press, Inc. Cambridge United Kingdom (1999).

24. Q. Liu et al., *DNA computing on surfaces*, Nature 403 (2000) 175-179.

25. A. Marathe, A.E. Condon, R.M. Corn, *On combinatorial word design* Preproceedings of the 5th International Meeting on DNA Based Computers, Boston (1999) 75-88.

26. H.J.Shyr, *Free Monoids and Languages*, Hon Min Book Company 2001.

In Search of Optimal Codes for DNA Computing

Max Garzon, Vinhthuy Phan, Sujoy Roy, and Andrew Neel

Computer Science, The University of Memphis TN 38152-3240, U.S.A.
{mgarzon, vphan, sujoyroy, aneel}@memphis.edu

Abstract. Encoding and processing information in DNA-, RNA- and other biomolecule-based devices is an important requirement for DNA-based computing with potentially important applications. Recent experimental and theoretical advances have produced and tested new methods to obtain large code sets of oligonucleotides to support virtually any kind of application. We report results of a *tour de force* to conduct an exhaustive search to produce code sets that are arguably of sizes comparable to that of maximal sets while guaranteeing high quality, as measured by the minimum Gibbs energy between any pair of code words and other criteria. The method is constructive and directly produces the actual composition of the sets, unlike their counterpart *in vitro*. The sequences allow a quantitative characterization of their composition. We also present a new technique to generate code sets with desirable more stringent constraints on their possible interaction under a variety of conditions, as measured by Gibbs energies of duplex formation. The results predict close agreement with known results *in vitro* for 20−mers. Consequences of these results are bounds on the capacity of DNA for information storage and processing in wet tubes for a given oligo length, as well as many other applications where specific and complex self-directed assembly of large number of components may be required.

Keywords: Word design, encoding data on DNA, Gibbs energy, fault-tolerant DNA computing, maximal code sets, complex DNA self-assembly, shuffle codes.

1 Introduction

The problem of data and information encoding on DNA bears an increasing interest for both biological and non-biological applications of biomolecular computing (BMC). Virtually every application of DNA computing maps data to appropriate sequences to achieve intended reactions, reaction products, and yields. DNA molecules usually process information by intramolecular and (more often) intermolecular reactions, usually hybridization in DNA-based computing. Most of prior work in this area has been centered on the so-called *word design problem*, or even the encoding problem (Garzon et al., 1997) [14,15]. Encoding sets will eventually prove necessary to realize the successful applications of BMC to expected and competitive scales in a number of areas such as directed self-assembly of complex structures (Seeman et al, 2005; Reif et al., 2003; Winfree

C. Mao and T. Yokomori (Eds.): DNA12, LNCS 4287, pp. 143–156, 2006.
© Springer-Verlag Berlin Heidelberg 2006

et al., 2001), DNA memories (Garzon et al., 2005)[16], and other applications. In (Phan and Garzon, 2004) [21], we introduced an algorithm that produces codesets of nearly optimal size under the h-distance model [14]. Whether these codesets will perform accordingly *in vitro* requires experimental confirmation.

The purpose of this paper is to present two new families of oligonucleotide sets of high quality for DNA-based computing. The first family is the result of exhaustive searches to produce code sets of n-mers, for $n \leq 20$, that are very likely to be maximal sets while guaranteeing high quality as measured by the minimum Gibbs energy between any pair of code words. This is a challenging problem because of the enormous size of the search space of all n-mers (exponential 4^n) and the practically infinite size of possible candidate subsets (superexponential 2^{4^n}) under virtually any condition on the quality of the sets, even for small oligos typically used on DNA chips ($n \leq 20$). In Section 2, we first describe the tools and techniques used to cope with the combinatorial explosion, as well as analyses of the size of the sets for their near optimality and a characterization of their noncrosshybridzation quality. The methods are constructive in two important senses. First, the sets are actually produced by simulation of proven experimental techniques *in vitro* (Deaton et al., 2004) [9], but, unlike the latter method, their composition is actually obtained as well. This fact affords important advantages because of the enormous cost of synthesizing (possibly hundreds of) thousands of different strands with a protocol that produces them directly *in vitro*. Second, we show, for example, an analysis of their composition that provides interesting clues as to the nature of Gibbs energy landscapes at play in BMC applications. The code sets obtained in Section 2 exhibit minimal noncrosshybridization for actual applications. Further selection may be necessary for specific uses of a code set.

In Section 3, we present a second family of codes that satisfy desirable and more stringent constraints on their interaction under a variety of conditions by refining the previously introduced shuffle operation of (Phan and Garzon, 2004) [21]. The thermodynamic stringency maintained is that the free energies are not too low between pairs of strands in the set, pairs of strands and the complements of other strands in the set, and pairs of complements of strands in the set. The codesets obtained are generally competitive with, and at time superior to, those obtained from other methods, while being produced by a much more efficient method.

Finally, in Section 4 we conclude with some remarks about the availability of the codes, some possible applications, their implications for DNA computing, as well as the difficulties in extending these methods to produce longer codeword sets.

2   Nearly Maximal DNA Codes

At first blush, finding good code sets appears to be a straightforward extension of the analogous problem of finding error-correcting codes of maximal rate in information theory (Roman, 1995) [22]. Two differences make the problem much more difficult. First, the metric (distance function) involved in determining the quality of the code (*i.e.*, minimum distance between any two codewords), namely the Gibbs energy of duplex formation (SantaLucia and Hicks, 2004) [24], is *not* the Hamming distance, indeed not ever a metric remotely comparable to the Hamming distance. In fact, even finding a very large set of very short oligonucleotide (say 20−mers) is computationally difficult [19]. Recent efforts based on the thermodynamic estimate of free energy has only been able to produce sets of no larger than 400 strands of 12- to 20-mers (Shortreed et al., 2005; Tulpan et al., 2005) [25,26]. Second, the computation of the Gibbs energy is in itself a challenge (probably **NP**-hard in itself) which requires an approximation of its own. Although more metric-like approximations of the Hamming distance have been found (Garzon et al., 1997)[14], the strong contrast in the topology of the corresponding spaces make it quite different. Nonetheless, the Hamming distance offers an inspiring analogy to tackle the search, as discussed next.

2.1   Gibbs Energy Model

The Gibbs energy value can be thus regarded as a measure of the hybridization likelihood between any pair of strands. Of the various methods that have been proposed, we use in this section a computation based on a model that shortcuts the common dynamical programming algorithm in the nearest neighbor model by adding a penalty function for buldges (Deaton et al., 2002)[8]. This model has been proven to work well in the solution of the encoding problem *in vitro* (Chen et al., 2004; 2005)[6,4] . More details on this point can be can be found in (Garzon and Deaton, 2004)[13].

Envisioning the full space of $n-$mers as a (pseudo-)metric space, the problem is analogous to the well known sphere-packing problem in ordinary euclidean spaces as follows. The stringency conditions in the test tube (as determined by factors such as temperature, salinity, and kinetic effects) can be abstracted as an appropriate parameter τ representing the threshold under which hybridization will occur. We set this value at the standard of $\tau = -6$ Kcal/mole. Inclusion of any $n-$mer x as a codeword automatically excludes all other $n-$mers within a Gibbs energy of duplex formation with x below τ, herein named the "ball of radius τ centered at x." We are thus searching for sets C of noncrosshybridizing $n-$mers of the largest possible size with the requirement that the minimum distance between every pair of $n-$mers is at least τ.

2.2   Exhaustive Search Methods: The PCR Selection Protocol

Polymerase Chain Reaction (PCR) is a well established technique to amplify the number of copies of DNA strands by thermal cycles using a polymerase enzyme (Mullins, 1984). At low temperatures, DNA strands are primed for copying by attaching primers (short ssDNA) to their 5′-end and 3′-ends. By supplying additional *Taq* polymerase and nucleotide bases, the ssDNA will extend forming fully Double Stranded DNA (dsDNA). At high temperatures, dsDNA melts or splits into ssDNA. Thus, by repeating this basic round, new copies of original DNA are produced with amazing volume and accuracy in only a few cycles.

PCR Selection (PCRS) (Bi et al., 2003) [3] is an *in vitro* protocol that was designed to make use of PCR to obtain good code sets from a seed set by adding a filtering phase. In the experimental confirmation, the target strands in the seed set (the full population of 20−mers) were flanked by carefully chosen 20−mer primers on both sides obtained using the model in (Deaton et al., 2002) [8]. In the extension phase, the template strands that have not found sticking (pseudo)complements are extended in a PCR cycle; in the filtering phase, those not amplified are eventually filtered out of the set. After repeating this process for a few rounds (four sufficed in [3]), a code set of actual DNA strands has been experimentally confirmed to be obtained with high noncrosshybridizing quality (Chen et al., 2005) [4]. Because the seed set is the largest possible and strands are only filtered out only when they have been caught crosshybridizing with other strands, the results after an appropriate number of rounds is expected to be a maximal noncrosshybridizing set, not unlike the perfect codes well known in information theory with the Hamming distance (Roman, 1995) [22].

It is impressive how the PCR Selection protocol can thus perform an exhaustive search that cuts through the combinatorial explosion to produce good code sets in feasible time (a matter of days). It is worth pointing out that the protocol does obtain the code set *in vitro*, thus eliminating the cost of synthesizing a large number of strands in code sets designed by design or conventional means. On the other hand, the actual composition of the sequences and the quality of the full set cannot be subject to the kind of analysis and so be given the guarantees that designs and evolutionary methods *in silico* do provide. Although experimental characterization of their quality has been given (Chen et al., 2005; Chen et al., 2004) [4,6], even if one is willing to bear the cost, in terms of money, of sequencing a sample of the code set *in vitro*, it is forbiddingly infeasible in terms of time to do so for the whole set of estimated $10,000 - 50,000$ species present in the wet set.

2.3 Maximal Code Sets

In this section we present the results of an analogous exhaustive search *in silico* by running PCR selection in simulation. Monetary costs of this approach are only those associated with developing the required software and purchasing the required hardware to run it. If the simulation is feasible to perform *in silico*, the full sequences of the code sets would actually be obtained and analyses could be performed of their nature and composition. The results in terms of knowledge of the actual composition of the set and their structure would be clearly superior, if it could be done.

PCR Selection in simulation is not without its challenges. For example, the set of 20−mers contains about 1 trillion species and would hold a full terabyte of data if just one byte is associated to just one DNA species. Simulation of PCRS on a terabyte of data is no trivial task for any computer. Fortunately, we have developed a simulation tool, *EdnaCo*, in the last few years for understanding biochemical reactions involving DNA. *EdnaCo* is a virtual test tube (VTT) capable of reliably predicting wet test tube outcomes with high fidelity

(Garzon et al, 2004) [18]. Although running on a cluster of PCs, it is still a challenge for *EdnaCo* to handle a simulation of PCR Selection on 20−mers in less than six months. We thus gave PCRS a hand by adding a pre-processing step, i.e. doing PCRS in two major stages, (*a*) *filtering* and (*b*) *simulation* on the filtered set. Several filters were applied to the full seed set of *n*−mers in order to remove *a priori* those obvious candidate DNA species likely to be removed by PCR Selection. The filters remove from the full set (left bars in Fig. 1, top right) those strands which are palindromes, contain hairpins (a folding of DNA onto itself), contain 4−mer runs of a single base, or are too close in *h*-distance to other strands already passed through the filter. Filtering the DNA sets can require from several minutes (for shorter strands) to several weeks (for larger sets.) The net gain is a filtered set (middle bars) that is at least one order of magnitude smaller in size than the full set (left bars), as shown in Fig. 1 (top right).

The general fidelity of the software simulations has been documented in (Garzon et al., 2004; Garzon et al., 2003) [18,11]. As a further validation of our simulation PCRS++, we ran PCR selection on a smaller subset of about 5,400 40−mers in order to determine the number of PCR rounds necessary to filter out all crosshybridizing pairs. The protocol converged to a stable product after four (4) rounds, which is in good agreement with the number of rounds considered to be necessary with *in vitro* runs of the protocol. Fig. 1 shows that PCRS++typically converges in practice in under 20 rounds to a stable set of 20−mers on filtered seed sets of about $5K$ species, which was heuristically determined to be the optimal value for parallelization of the job. (PCRS++a randomized algorithm and it may take in principle a very long time to converge to an ideal product subset of the seed set.) Further validation of the results will be possible when a comparison can be made with the analogous characterizations of the product code sets to the available characterization of the products *in vitro* for 20−mers (Chen et al., 2005; Chen et al., 2004)[4].

Fig. 1 also shows the results of filtering the full set of DNA *n*−mers of various lengths. The size of the remaining set of strands after filtering still grows exponentially with a power law $4^{0.72n+1.66}$. This suggests that at some point, filtering will yield DNA sets that still contain too many DNA species. After projecting the curve forward, their intercept is estimated to be beyond DNA sets of length 60. This evidence does not suggest a limitation of our method since the Gibbs energy approximation of (Deaton et al., 2002)[8] breaks down beyond 60−mers. Rather, it suggests that filtering as described above may not be ideal for discovering a maximal set of strands of a specific length. Even so, a subset of strands obtained will far exceed the size of the best set know today, and by analogous arguments to the advantages of PCR selection *in vitro*, produce sets that are of comparable order of magnitude to an optimal (perfect) code.

Fig. 2 shows the noncrosshybridizing quality of the sets obtained by the simulation over a period of months, as measured by the pairwise Gibbs energy of strand pairs in the input filtered sets and across word pairs in the set resulting from the PCRS products. The length of the simulation for one round of PCRS

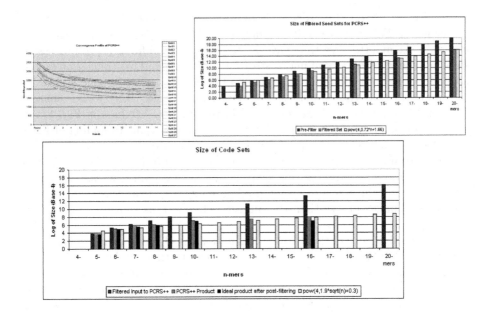

Fig. 1. Top Left: The simulations converge to a noncrosshybridizing subset of a seed set of $5K$ species in about fifteen (15) rounds of PCRS++on a single processor. Four (4) rounds could be achieved on a larger number of processors to match the experimental protocol *in vitro* (Chen et al., 2005) [4]. Top Right: Pre-filtering of the space of $n-$mers reduces the full set by at least one order of magnitude. The filtering process removes from the full input set (left bars) the DNA species that contain hairpins or short runs of the same base, and those species that are palindromes or are too close to other strands already accepted by the filter. The size of the resulting filtered seed sets (middle bars) can be estimated by the power law $4^{0.72n+1.66}$ (right bars). Bottom center: Size of the PCR Selection products obtained from the filtered input seed sets (left bars). The size of PCRS++products (middle left bars) come between 1% and 27% of the ideal PCRS subset contained in the filtered seed set if all collisions were eliminated (middle right bars). (The full runs of PCRS++are not finished in time for 16−mers and 20−mers; shown are estimates based on the partial results available but guaranteed to be within 5% error of the true values.) The size of ideal PCRS++products (middle right bars) can be estimated by the sub-powerlaw $4^{1.9\sqrt{n}+0.3}$ (light bars on the right).

has been calculated to guarantee enough time for every strand to visit every place in the test tube simulator *EdnaCo* (Garzon et al., 2004) [18]. The minimum energies are the most stringent criterion, and a value of −6 Kcal/mole is generally required to avoid crosshybridization between any pair of strands. Most sets obtained satisfy this condition on the average in under fifteen (15) rounds, and in the few that do not, an additional round of exhaustive filtering (not included in the results reported here for PCRS++) was found to satisfy this minimum criterion easily. It must be noted that the full runs of PCRS++were not finished in time for 16−mer and 20−mer; shown are estimates based on the partial results available but guaranteed to be within 5% error of the true values.

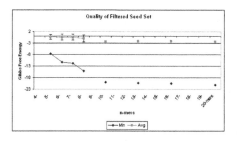 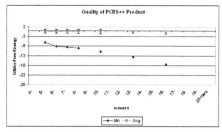

Fig. 2. High quality of PCR Selection protocol product set, as measured by two different criteria: Gibbs energies across word pairs in the input filtered set (left) and across word pairs in the PCRS product set (right). Despite their much larger size, they are competitive in quality with similar codes obtained by other methods (such as the template method (Arita et al, 2002) [1] or methods in Section 3 below.) The ideal PCRS++products have minimum quality above −6 Kcal/mole when measured at 20°C.

With the code sets in hand, we can proceed to do a number of analysis that would be impossible to perform if the results were obtained *in vitro*. For example, Fig. 3 shows the quality of the sets in terms of GC-content (occurrence of 1−mers, top left), frequency of 2−mer blocks (top right); frequency of 3−mer blocks (bottom left); and 4−mer blocks (bottom right). On this metric again, these codes are competitive with similar codes obtained by other methods (such as the template method (Arita et al, 2003) [1] or methods in Section 3 below. For example, the GC content, although not exceeding 60%, remains above 40% . There are biases in the distribution of the blocks. Some blocks seem to occur with comparably the same highest frequency if the oligos are long enough (for 13−mers, for example, GA, AG, AT, GG among 2−mers; GTA, GAG, GGT, GGA among 3−mers; and GAGG, AGGT,ATGA,AGGG among 4−mers.) These statistics lend empirical evidence for the effectiveness of the shuffle code method to build codesets (more in Section 3.).

3   Construction of More Stringent Codesets

In many applications, the requirement that for all pairs of strands u and v, their Gibbs energies are high enough to prohibit cross-hybridization is sufficient. In others, additional requirements to prohibit cross-hybridization among strands in the codesets and strands *not in the codesets* may be necessary to ensure not only sensitivity but also specificity of intended hybridizations. Namely, in addition to the requirement that Gibbs energies among all pairs in the codeset are above a certain threshold, it is desirable to require (a) the Gibbs energies between a strand in the codeset and the complements of other strands in the codeset and (b) the Gibbs energies among complements of strands in the codeset are also above a certain threshold. In other words, the goodness of strands in a codeset depends not only on interaction among strands in the set, but also on interaction

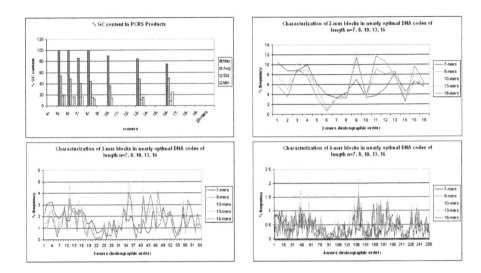

Fig. 3. A characterization of PCRS++product sets given by frequency occurrence of short blocks of size $m = 1, 2, 3, 4$: GC/AT content of words in the set(top left); frequency of 2−mer blocks (top right); frequency of 3−mer blocks (bottom left); and 4−mer blocks (bottom right).

of strands *not* in the set. In each iteration of the present PCR protocol, only interaction of strands in the set are considered. For example, complements of strands already excluded from the current set should not possibly be considered for selection into the current set. Under the h-distance model that approximate Gibbs energy by assuming stiff strands (e.g. bulges cannot form), the requirement that the h-distance $h(u, v) > \tau$ for all pairs u and v implies this additional requirement, because $h(u, v) = h(u', v')/leh(u, v')$. Under the nearest neighbor model with Gibbs energy, however, these two quantities are not the same because free energy of complementary stack pairs are not symmetric (SantaLucia and Hicks, 2004) [24].

We introduce a new method of constructing codesets that satisfy the additional criteria that require (a) the Gibbs energies among strands in the set and complementaries of other strands in the set are high enough and (b) the Gibbs energies among complementaries of strands in the set are also high enough. We obtained generally larger codesets (actually, very much larger in some cases) than those of similar quality obtained by other researchers (Tulpan et al., 2005) [26]; see Fig. 4. This method relies on the Shuffle algorithm (Phan and Garzon, 2004) [21], that produces codesets of provably near-optimal size under the h-distance model, where an *(n, τ)-codeset* S, is a set of n-mers that satisfies the condition $\forall u, v \in S, h(u, v) \geq \tau$; for applications that requires the strands to have similar melting temperatures, we extend this notion to an *(n, τ, w)-codeset*, that is an (n, τ)-codeset with a GC-content of w. The larger the value of τ, the farther apart the strands in S are, which means hybridization among strands in S is

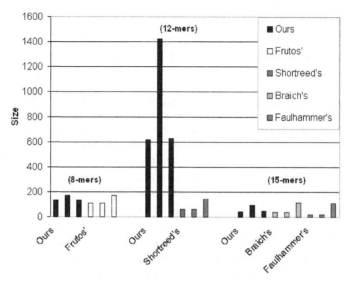

Fig. 4. The size of our codesets (mer8s, 12−mer and 15−mermers) compares favorably to that of extant codes of similar quality in the literature. A more detailed comparison is shown in Table 1.

less likely. To impose the additional criteria, we post-process the large codesets by optimizing approximately certain instances of the **NP**-hard *minimum vertex cover* problem.

3.1   Shuffle Codes

The shuffle algorithm was introduced in (Phan and Garzon, 2004) [21], as a particular case of the tensor product operation of (Garzon et al., 2004)[12], to construct provably near-optimal codesets under the h-distance model (Garzon et al., 1997)[14]. This algorithm is based on the so-called *shuffle* function, Sh, which is defined on l DNA strands $u_1, u_2, \cdots, u_l$, each of length n, as follows:

$$Sh(u_1, \cdots, u_l) = u_{11}u_{21} \cdots u_{l1}u_{12}u_{22} \cdots x_{l2} \cdots u_{1n}u_{2n} \cdots u_{ln}$$

where $u_i = u_{i1}u_{i2} \cdots u_{in}$. In other words, $Sh(u_1, \cdots, u_l)$ is a string of length ln obtained by concatenating n blocks, each of length l each obtained by concatenated together the first (second, third, $\cdots$, nth) characters of $u_1, \cdots, u_l$, respectively. Briefly, to construct a very large codeset, we started with an $(n, 1)$-codeset (or an $(n, 1, w)$-codeset with w being the GC-content). Then, the base codeset is shuffled l times to create a l (nl, l)-codeset or an (nl, l, wl)-codeset respectively. The reason this construction works as claimed is based on the property of the shuffle function, Sh, being *amplifiable* with respect to the h-distance function (Phan, 2006)[20]. A function $f : D^n \times D^n \times \cdots D^n (l \text{ times}) \rightarrow D^{nl}$, where D^m is a set of strands of length n, is *amplifiable* with respect to a distance function d if

$$d(f(u_1, \cdots, u_l), f(v_1, \cdots, v_l)) \geq \min_{1 \leq i,j \leq l} \{l \cdot d(u_i, v_j)\}$$

It is shown, in particular, that using any amplifiable function, one can construct very large codesets by applying the function repeatedly from one or several seed codesets.

3.2   Construction of Stringent Codes from Shuffle Codes

We follow a similar procedure to the tensor product technique of (Garzon et al., 2004)[12] to improve the quality of the large shuffle codes by further processing. Strands are renoved for the sets that violate the requirements that (a) the Gibbs energies among strands in the set and complements of other strands in the set are high enough, and (b) the Gibbs energies among complements of strands in the set are also high enough. This method has the advantage of narrowing down the exponential size of the search space to a much smaller subset of all possible n-mers. As such, it is more efficient than exhaustive methods or heuristics that examine the whole search space. More specifically, first, a shuffle code is generated in such a way that the word-complement constraint is least violated (i.e. $h(u, v')$ being too large). This is done by partitioning the base codeset into disjoint sets and shuffle each of them individually as in (Phan and Garzon, 2004) [21]. The resulting codesets will be merged to produce a larger codeset, which has a minimal number of potential word-complement constraint violations. Second, the resulting codeset is processed to eliminate all strands that violate the word-word, word-complement and complement-complement constraint, i.e. eliminate all strands $u \in S$ such that for some strand $v \in S$, $\Delta G(u, v)$, $\Delta G(u, v')$ or $\Delta G(u', v')$ are too large. To do that, a *violation graph* G_S is constructed from S in such a way that vertices represent strands and edges represent pairs of strands that violate one of the three requirements. A *vertex cover* of such a graph G_S identifies the strands whose removal leaves no pair u and v violating any of the requirements. To minimize the number of strands needed to removed, we aim to find the *minimum vertex cover* of G_S, which unfortunately is an *NP*-hard problem. As such, we resort to approximations to find a close candidate (Cormen et al., 2001) [7].

We experimented with two ways of imposing thermodynamic requirements. The first way is similar to our notion of a codeset, namely an edge between two vertices in G_S is defined if at least one of three requirements fails: (1) $\Delta G(u, v) \geq t_1$, (2) $\Delta G(u, v') \geq t_2$ or (3) $\Delta G(u', v') \geq t_3$. The second way, proposed in (Tulpan et al., 2005) [25], measures the *energy gap* between the intended hybridization and cross-hybridization. Specifically, there's an edge between vertices u and v if

$$\min\{\Delta G(u, v), \Delta G(u, v'), \Delta G(u', v), \Delta G(u', v')\} - \Delta G(u, u') < \delta$$

Parameters t_1, t_2, t_3 and δ are used to produce codesets with the desired thermodynamic characteristics. Free energies are computed at the standard temperature 37°C using the PairFold package, which employs thermodynamic parameters (Santa Lucia, 1998)[23] to predict RNA and DNA secondary structures (Zuker

Table 1. A comparison of our codesets denoted as P1, P2, $\cdots$, P9 to those recently produced using other methods (Tulpan et al., 2005)[26]. The best result for each criterion is bold-faced; higher free energies between different strands are better. Unreported values are left blank. The comparison shows our codesets are of comparative quality and generally larger (in fact in some cases, much larger). Free energies are computed at the standard temperature 37°C, using the PairFold package, which employs thermodynamic parameters (Santa Lucia, 1998) [23].

Set	Length	Size	min $\Delta G(u,u')$	min $\Delta G(u,v)$	min $\Delta G(u,v')$	min $\Delta G(u',v')$	δ
P1	8	132	**-6.8**	**-3.3**	**-5.4**	**-3.9**	-0.1
P2	8	**173**	-7.1	-5.3	-6.5	-5.5	0.1
P3	8	135	-6.9	-5.3	-6.3	-5.4	0.2
S4-Frutos	8	108	-8.95		-6.83	-8.24	-0.21
S4-1	8	108					1.28
S4-2	8	173					**1.59**
P4	12	617	**-11.6**	**-5.9**	-9.9	-5.9	-0.1
P5	12	**1424**	-11.8	-9.2	-11.2	-10	0.2
P6	12	628	-11.8	-9	-11	-9.8	0.6
S7 Shortreed	12	64	-12.66		**-8.94**	**-5.06**	2.87
S7-1	12	64					3.72
S7-2	12	144					**3.01**
P7	15	42	**-15.4**	**-6**	-14.9	-7.3	-0.1
P8	15	96	-16.3	-12.3	-15.5	-12.3	0.2
P9	15	48	-16.3	-12.3	-15.5	-12.3	0.6
S1 Braich	15	40	-16.8		-9.29	**-4.4**	6.25
S1-1	15	40					7.57
S1-2	15	**114**					6.42
S3 Faulhammer	15	20	-17.56		**-7.98**	-5.02	6.13
S3-1	15	20					**8.59**
S3-2	15	110					6.25

et al., 1999)[27]. Table 1 shows the the resulting codesets are of comparable quality and generally larger sizes than those obtained by most other methods.

4   Conclusions and Future Work

We have obtained very large of very good noncrosshybridizing quality that are guaranteed to perform well under reaction conditions in wet test tubes by a *tour de force*, an exhaustive search that has lasted over 6 months (and yet has to bear fruits for 20−mers). As a result, we have in hand or can produce a number of code sets of n−mers for values of $n = 4 - 10, 13, 16, 20$. The search can be scaled for sets of longer oligonucleotides within feasible (but long) run times up to about 60−mers when the Gibbs energy model used begins to fail), although now optimality will be out of range. Arguments have been given that the size of these code sets may be in the order of magnitude of maximal sets. They provide, for example, a good practical estimate of the capacity of a DNA memory (Garzon et al, 2005; 2003)[16,10] based on Baum's construction [2]. These codes support virtually any kind of application of biomolecular computing that requires the use of DNA molecules in a controlled way.

The method is constructive, not in the sense that it produces the actual coding wet strands as the PCR Selection protocol does *in vitro* (Deaton et al., 2004) [5], but in the complementary sense that the *sequence* and *composition* of the actual codewords is known, thereby bypassing a costly or impossible sequencing procedure. As a result, careful analyses can be made of the structure and nature of the code and the abilities and limitations of storing information in DNA molecules. The results of some of these analyses have been presented, for example, the G/C-content, and a bias in the most frequent $k-$mer-blocks for $k = 2, 3, 4$. These results suggest further analyses that could be conducted with the products of the protocol *in vitro* to obtain further characterization of its products (so far only conducted *in vitro* for 20$-$mers (Chen et al., 2005) [5].)

We also presented a refinement of our shuffle algorithm (Phan and Garzon, 2004)[21] by post-filtering to produce sets where desriable more stringent non-crosshybridization criteria are met. The quality of the codes is competitive with those produced by recent works on codeword and structure-free design. Our codesets are typically larger and our method is more efficient since shuffle codes are constructed optimally in linear time (Phan and Garzon, 2004)[21][21] and the approximation algorithm for vertex cover in the post-filtering step is done in linear time in the number of strands and violating pairs. This is considerably more efficient than the other methods that employ local search and other search heuristics.

Acknowledgements

Partial support from the National Science Foundation grant QuBiC/EIA-0130385 is gratefully acknowledged. Special thanks to Russell Deaton (Arkansas) for making the C code for the Gibbs energy model available for the PCR simulations in Section 2.2 used in Section 2, as well as to Anne Condon and her team for making available the code for the Gibbs energy used in Section 3.

References

1. M. Arita and S. Kobayashi. Dna sequence design using templates. *New Gen. Comput.*, 20(3):263–277, 2002.
2. E. Baum. Building an associative memory vastly larger than the brain. *Science*, 268:583–585, 1995.
3. H. Bi, J. Chen, R. Deaton, M. Garzon, H. Rubin, and D. Wood. A pcr-based protocol for in vitro selection of non-crosshybridizing oligonucleotides. *J. of Natural Computing*, 2:4:417–426, 2003.
4. J. Chen, R. Deaton, M. Garzon, J.W. Kim, D.H. Wood, H. Bi, D. Carpenter, J.S. Le, and Y.Z. Wang. Sequence complexity of large libraries of dna oligonucleotides. In *11th International Conference on DNA Computing*, page in press, 2005.
5. J. Chen, R. Deaton, Max Garzon, D.H. Wood, H. Bi, D. Carpenter, and Y.Z. Wang. Characterization of non-crosshybridizing dna oligonucleotides manufactured in vitro. Proc. 8th Int Conf on DNA Computing DNA8.
6. J. Chen, R. Deaton, Max Garzon, D.H. Wood, H. Bi, D. Carpenter, and Y.Z. Wang. Characterization of non-crosshybridizing dna oligonucleotides manufactured in vitro. In L. Smith G.C. Mauri, editor, *10th International Workshop on DNA Computing*, pages 50–61, 2004.

7. T. Cormen, C. Leiserson, R. Rivest, and C. Stein. *Introduction to Algorithms, 2nd Edition.* The MIT Press, 2001.

8. R. Deaton, J. Chen, H. Bi, and J. Rose. A software tool for generating non-crosshybridizing libraries of dna oligonucleotides. pages 252–261, 2002. In: [17].

9. R.J. Deaton, J. Chen, H. Bi, M. Garzon, H.Rubin, and D.H. Wood. A pcr-based protocol for in vitro selection of non-crosshybridizing oligonucleotides. *In: (Hagiya & Ohuchi, 2002)*, pages 105–114, 2002a.

10. E. Cantu-Paz et al., editor. *Efficiency and Reliability of DNA-based Memories.*, 2003. The Genetic and Evolutionary Programming Conference. Springer-Verlag Lecture Notes in Computer Science 2723.

11. M. Garzon, D. Blain, K. Bobba, A. Neel, and M. West. Self-assembly of dna-like structures in silico. In M. Garzon, editor, *Biomolecular Machines and Artificial Evolution, Special Issue of the Journal of Genetic Programming and Evolvable Machines*, pages 185–200. Kluwer Academic Publishers, 2003.

12. M. Garzon, K. Bobba, and B. Hyde. Digital information encoding on dna. In *Aspects of Molecular Computing*, volume 20, pages 152–166, London, UK, 2004. Springer-Verlag.

13. M. Garzon and R. Deaton. Codeword design and information encoding in dna ensembles. *J. of Natural Computing*, 3:253–292, 2004.

14. M. Garzon, R. Deaton, P. Neathery, D. R. Franceschetti, and R. C. Murphy. A new metric for dna computing. In *Second Annual Genetic Programming Conference*, pages 472–478, 1997.

15. M. Garzon, R. Deaton, P. Neathery, R.C. Murphy, D.R. Franceschetti, and E. Stevens Jr. On the encoding problem for dna computing. In *The Third DIMACS Workshop on DNA-based Computing*, pages 230–237, 1997.

16. M. Garzon, V. Phan, K. Bobba, and R. Kontham. Sensitivity analysis of microarray data: A new approach. In *Proc. IBE Conference, Athens GA.*, 2005. Biotechnology Press.

17. M. Hagiya. Dnabasedcomputers. In A. Ohuchi, editor, Proc. 8th Int. Meeting on DNA-Based Computers, volume LNCS 2568. Springer-Verlag, 2002.

18. Garzon M, D. Blain, and A. Neel. Virtual test tubes for biomolecular computing. *J. of Natural Computing*, 3:4:460–477, 2004.

19. A. Marathe, A. Condon, and R. Corn. On combinatorial DNA word design. In Erik Winfree and David K. Gifford, editors, *Proceedings 5th DIMACS Workshop on DNA Based Computers*, pages 75–89. American Mathematical Society, 1999.

20. V. Phan. A method for constructing large dna codesets. In T. Pham, H Yan, and D.I. Crane, editors, *Advanced Computational Methods for Biocomputing and Bioimaging*. Nova Science Publishers, New York, 2006.

21. V. Phan and M. Garzon. The capacity of dna for information encoding. *Proc. 10th Int Conf on DNA Computing, LNCS*, 3384:281–292, 2005.

22. J. Roman. *The Theory of Error-Correcting Codes.* Springer-Verlag, 1995.

23. J. SantaLucia. A unified view of polymer, dumbbell, and oligonucleotide dna nearest-neighbor thermodynamics. *Proc Natl Acad Sci*, 95(4):1460–1465, 1998.

24. J. SantaLucia and D. Hicks. Thermodynamics of dna structural motifs. *Annu Rev Biophys Biomol Struct.*, 33:415–440, 2004.

25. M.R. Shortreed, S.B Chang, D. Hong, M. Phillips, B. Campion, D.C. Tulpan, M. Andronescu, A. Condon, H.H. Hoos, and L.M. Smith. A thermodynamic approach to designing structure-free combinatorial dna word sets. *Nucleic Acids Res.*, 33(15):4965–4977, 2005.

26. D. Tulpan, M. Andronescu, S.B. Chang, M.R. Shortreed, A. Condon, H.H. Hoos, and L.M. Smith. Thermodynamically based dna strand design. *Nucleic Acids Research*, 33(15):4951–4964, 2005.

27. M. Zuker, D.H. Mathews, and D.H. Turner. Algorithms and thermodynamics for rna secondary structure prediction: A practical guide in rna biochemistry and biotechnology. In J. Barciszewski and B.F.C. Clark, editors, *NATO ASI Series*. Kluwer Academic Publishers, 1999.

DNA Sequence Design
by Dynamic Neighborhood Searches*

Suguru Kawashimo, Hirotaka Ono,
Kunihiko Sadakane, and Masafumi Yamashita

Dept. of Computer Science and Communication Engineering, Kyushu University,
6-10-1 Hakozaki, Higashi-ku, Fukuoka, Fukuoka 812-8581, Japan
kawa@tcslab.csce.kyushu-u.ac.jp, {ono, sada, mak}@csce.kyushu-u.ac.jp

Abstract. We propose a local-search based algorithm to design DNA sequence sets that satisfy several combinatorial constraints about hamming-distance criteria. To deal with the constraints in the local search, we adopt elaborate (and dynamic) neighborhood search frameworks called the *Variable Neighborhood Search* (VNS) and the *Variable Depth Search* (VDS). Although our algorithm can deal with many types of hamming distance-based constraints and is easy to extend (e.g., also applicable for other constraints), in computational experiments, we succeeded in generating better sequence sets than the ones generated by exiting methods of more specified constraints.

Keywords: DNA Sequence Design, Local Search, Combinatorial Constraints, Variable Neighborhood Search, Variable Depth Search.

1 Introduction

Designing DNA sequence sets is a fundamental issue in the fields of nanotechnology and nanocomputing, e.g., Adleman's DNA solution for the Hamiltonian path [1], DNA tiling with its self-assemble [16], hairpin-based state machine [7] and so on. One point of DNA computing / technology is to control the DNA molecules reactions; to fulfill a robust "computation", it is important that DNA molecules react only in expected ways, because unexpected secondary structures of DNA sequences may cause error, for example. The sequence design is an approach of the control, which aims to design DNA sequences satisfying some constraints to avoid such unexpected molecular reactions. Since expected or unexpected reactions depend on the applications or the purposes, usually several representative constraints are considered as below mentioned. Another requirement for DNA sequence sets is that the sets should be large. This is because designed DNA sequences are used as elemental components of computation; the size of sequences is considered the size of the computational resource. In summary, our purpose is to design large sets of sequences that satisfy certain types of constraints.

---

* This research partly received financial support from Scientific research fund of Ministry of Education, Culture, Sports, Science and Technology.

C. Mao and T. Yokomori (Eds.): DNA12, LNCS 4287, pp. 157–171, 2006.

In the sequence design, many types of constraints are considered. Among them, combinatorial constraints and thermodynamics are well studied. Most common measures of combinatorial constraints are the *Hamming Distance constraint* (HD) and the *Reverse Complement Hamming Distance constraint* (RC). Also, the *overlapping measure*, which is more strict than the HD and RC, is well considered.

Based on these constraints (or measure), many researchers argue the sequence set design. Arita et al. apply techniques from the coding theory to design sequence sets [3,10]. Hamming distance measure has been discussed in the coding theory for long. In order to utilize the results, they propose the template method that projects a set of sequences satisfying HD (from the coding theory) into a set of sequences satisfying all the three measures based on Hamming distance. The template method achieves a fast design of sequence sets satisfying the Hamming distance-based constraints, though the distance parameter cannot be large. Tulpan et al. propose a Stochastic Local Search (SLS) method for the sequence sets design [13,14,15]. In [13,14], they treat the HD and RC constraint and another constraint called the GC-content constraint. Also, they treat thermodynamical constraints in [15]. Their method can be easily adapted to other constraints, since the local-search framework has a good flexibility. Other than these, many methods are proposed: a random generation and a genetic algorithm (Arita et al. [2]), lexicographic techniques and stochastic searches (Gaborit et al. [5]), traditional greedy methods (Asahiro [4]), a random generation with a local improvement (Kashiwamura et al. [8]), and so on.

In this paper, we focus on the combinatorial properties of sequence sets design problems. More concretely, we consider the problem from the view point of the combinatorial optimization, and propose a local-search based algorithm. The local-search framework is known to be useful to solve hard combinatorial optimization problems in practical sense. We adopt in our local search algorithm, the techniques called the Iterated Local Search (ILS), the Variable Neighborhood Search (VNS) and the Variable Depth Search (VDS) [17]. In local search, it is considered that controlling the range of searches well is important to achieve a good performance, and the idea is represented by *intensification* and *diversification*. Both of VNS and VDS dynamically control neighborhoods in our local search for this purpose. In passing, the VDS method is an abstraction of the Lin-Kernighan methods for TSP and the Ejection-Chain methods for GAP [6,9,11,18]).

In the design of our local search-based algorithm, we adopt these elaborate neighborhood search strategies, but the neighborhood definition itself is very simple; this realizes the flexibility of our algorithm, i.e., our algorithm is easy to apply other kinds of constraints. In spite of these, some preliminary computational experiments show that our method finds better sequence sets than ones by some existing methods. This implies that our approach is promising in designing favorable sequence sets.

2 Preliminaries

2.1 Constraints Using Hamming-Distance

Let S be the sequence set. In previous works, "hybridization" refers to "causing completely hydrogen-bonds between a sequence in S and its complement sequence", and "miss-hybridization" refers to "conformation changes which are not hybridization". We follow this manner in this paper. The constraints introduced in this section are proposed in order to avoid miss-hybridization.

Let s, s', s'' be DNA sequences of length n, then $s, s', s'' \in \{A, T, G, C\}^n$. Sequences are represented as $s = s_1 s_2 \cdots s_n$, $s' = s'_1 s'_2 \cdots s'_n$, and $s'' = s''_1 s''_2 \cdots s''_n$. In these representations, the left end of a sequence corresponds to 5' end of a DNA sequence. Let $s[i,j] = s_i s_{i+1} \cdots s_{j-1} s_j$ for $i \le j$. Concatenation of sequences s' and s'' is defined by $s's'' = s'_1 s'_2 \cdots s'_n s''_1 \cdots s''_n$. In addition, $wcc(s)$ denotes the Watson-Crick complement of DNA sequence s, here, $wcc(s)$ is the sequence which reverse s and replaced each A in s by T and vice versa, replaced each G in s by C and vice versa. Hamming-distance between s and s' is represented as $H(s, s')$ which denotes the summation of bases such as $s_i \ne s'_i$. Let $wcc(S) = \{wcc(s) | s \in S\}$.

Given a distance parameter d, we define the following constraints based on the Hamming distance measure:

Hamming Distance constraint(HD): for all pairs of distinct sequence s, s' in S, $HD(s, s') \ge d$. That is, $HD(S) \stackrel{\text{def}}{=} \min_{s,s' \in S, s \ne s'} \{H(s, s')\} \ge d$.

Reverse Complement Hamming Distance constraint(RC): for all pairs of sequence s in S and s' in $wcc(S)$, $H(s, s') \ge d$. That is, $RC(S) \stackrel{\text{def}}{=} \min_{s \in S, s' \in wcc(S)} \{H(s, s')\} \ge d$.

Overlapped Hamming Distance constraint(OL): for all combinations of sequence s in S, s', s'' in $S \cup wcc(S)$ (to attend that they are not distinct) and for $\forall i$, $H(s, s's''[i, i+n-1]) \ge d$. That is, $OL(S) \stackrel{\text{def}}{=} \min_{s \in S, s', s'' \in S \cup wcc(S), 1 < i \le n} \{H(s, s's''[i, i+n-1])\} \ge d$.

To avoid miss-hybridization, we consider to find S satisfying $HD(S) \ge d$, $RC(S) \ge d$ and $OL(S) \ge d$ for large d. The OL constraint is more effective to avoid miss-hybridization than the HD and RC constraints. In other words, the OL constraint is more strict than the other two.

In this paper, HD + RC and HD + RC + OL are adopted for constraints. Thus we design sets which satisfy $min\_hd(S) \ge d$, where $min\_hd(S) \stackrel{\text{def}}{=} \min \{HD(S), RC(S)\}$ or $min\_hd(S) \stackrel{\text{def}}{=} \min \{HD(S), RC(S), OL(S)\}$.

2.2 Local-Search

The local-search is known to be useful for solving hard combinatorial optimization problems in practical sense.

In general, combinatorial optimization problems are defined as follows: minimize (or maximize) $f(x)$ for the *solution* x and the *objective function* f. Let $N(x)$ denote the *neighborhood* of x (i.e., a set of solutions which are obtained by a slight perturbation to x). Operations which perturb certain solution x are called the *neighborhood operations*. When x satisfies $f(x) < f(x')$ (or $f(x) > f(x')$) for $\forall x' \in N(x)$, x is called the *local optima*. In general, solution spaces have a lot of local optima.

The local-search starts from a certain solution x and repeatedly moves to another solution x' in $N(x)$ where $f(x') < f(x)$ (or $f(x) > f(x')$) until a better solution can not be found; then it eventually obtains a local optimal solution. Thus, a standard strategy of local-search is as follows:

(1) Select an initial solution x.
(2) Search in $N(x)$.
(3) If an improved solution x' in (2) is found, the solution x is replaced by x', and go to (2). Otherwise, current x is local optima, then return x.

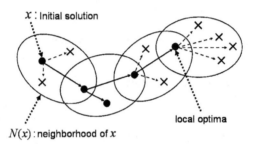

Fig. 1. Idea of local-search

The local-search frameworks have a trade-off between its accuracy and its running time. For example, larger neighborhood is searched, more accurate solution is found in general. However, to search in larger neighborhood requires more computational time. Therefore it is important to adopt neighborhoods which have appropriate size and structure. Another point is how we search. A traditional strategy of searches is the *first admissible move strategy* which moves immediately when a improved solution is found.

In addition, experiences show that "good solutions tend to have similar solution structures" in combinatorial optimization problems. Thus, it is considered effective to search solutions which are similar in structures to good ones, which is called the *intensification* of local-search frameworks. In contrast, superfluous intensification induces waste searches, for example, this leads to search the same solution repeatedly, to search repeatedly in spite that there exists no better solutions in the area where is searched intensively. Therefore, it is necessary to search solutions which have different solution structures, which is called the *diversification* of local-search frameworks. Thus, to balance conflicting ideas (the intensification and the diversification) is important. Figure 2 shows these ideas.

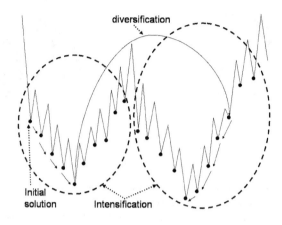

Fig. 2. Ideas of the intensification and the diversification

In summary, in order to design good local-search algorithms, it is necessary to carefully choose neighborhood operations which define appropriate size of neighborhoods, to consideration to adequate control of search strategies.

Based on these ideas, improved local-search frameworks are proposed. These are combinable and can be improved for specific problems, which shows the flexibility of local-search frameworks. We describe three ideas of improving local-search frameworks, used in this paper.

We first explain an *Iterated Local Search* (ILS) method. We consider the case of reaching a not-good local optimal solution during the search; it is not possible to move from local optima by the ordinary neighborhood operations. For such a case, we consider to perform a stronger perturbation to the local optima than one of ordinary neighborhood operations. Instead of that, we permit a corruption. As above, we move from the local optima, and search continuously. This operation is called the iterated local search (ILS) method. This method realizes the diversification of local-search frameworks. When an ILS method is adopted, searches can not finish, then it is necessary to define the condition of finish (e.g., time-limit, strict step number, and so on).

Secondly, we explain the *Variable Neighborhood Search* (VNS) method. Local-search frameworks use the determine neighborhood in general. However, to exchange neighborhoods depending on situations can lead effective search, since local-search frameworks have accuracy/time trade-off. Thus the method which exchanges neighborhoods depending on situations is called the Variable Neighborhood Search (VNS) method. This method realizes the intensification of local-search frameworks. One of the issue on a VNS method, it is necessary to define the rule how to exchange neighborhoods.

Finally, we explain a *Variable Depth Search* (VDS) method [6,9,11,17,18]. This is the method which defines neighborhoods as a set of solutions obtained by the chain of simple neighborhood operations. The chains of neighborhood operations are represented as a tree, and it is called the *search tree* (See Fig.3 and Fig.4).

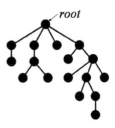

Fig. 3. Search tree in ordinary local search   **Fig. 4.** Search tree in a VDS method

In the search tree, the root corresponds to the current solution, the nodes correspond to the solutions, children of a certain node correspond to the solution obtained from the node by single neighborhood operation. A feature of the VDS method is that we can search the solutions which are not obtained by the single neighborhood operation. The single neighborhood operation does not move unimproved solutions, then it can not obtain improved one obtained by the move from unimproved solutions. However, the VDS method can find out these. In the case of the VDS method, structures of the search tree grow explosively without restricting it, then to restrict search tree is necessary. As a result, the depth of a search tree varies adaptively. The VDS method enables to balance the intensification and the diversification.

3   Sequence Design Based on Local-Search

In this section, we propose a local-search based algorithm for sequence sets design problems. Let the sequence set be S, the size of S be m, the length of sequences be n, the number of constraints be d.

For sequence sets, the maximum size of these depends on constraints. There are two optimization approaches for sequence design as follows:

approach(1) For fixed m and n, design S which has maximized d.
approach(2) For fixed d and n, design S which has maximized m.

In early studies of these problems, since there are not sufficient experiences, we could not know appropriate evaluations of d and m; they took approach(1). However, since we know evaluations of d and m in recent [4,5,8,13,14], it is possible to take approach (2) also. In this paper, we mainly consider the approach(2) by using approach (1) as a subprocedure.

More concretely, we take the following approach:

step 1. Find a set S that approximately maximizes $min\_hd(S)$ for fixed m and n (this part is smiler to approach(1)).
step 2. Increase m. Go back to step 1.

The outline of step 1, the main part of our method, is as follows: Apply a VNS method to find a local optima. Once it is found, escape from the local optima, and then search by the VNS method again. We use two varieties of escapes, one is the VDS, a small escape, and the other is the iterated local search (ILS), a big escape. The VNS part is oriented to the intensification, and the VDS part and the ILS part are oriented to the diversification. We describe the detailed ideas in section 3.2 and section 3.3.

In step 2, it is important to carefully choose a new sequence to be added. For this purpose, we adopt the idea introduced in [8], which will be explained in section 3.4.

3.1   Basic Properties of the Problem

Before explaining these details, we first give basic materials, such as the definition of neighborhood, which define our local search algorithm, in section 3.1.

First, we discuss the objective function of the problem. We then define the neighborhoods of the local search, and consider move strategies by focusing on the combinatorial aspects.

***$min\_count(S)$*: Evaluating Function of the Local Search.** This problem has the objective function $min\_hd(S)$. Although it is common to use the objective function for evaluation of the local search, adopting $min\_hd(S)$ as the evaluating function is not suitable in this problem; the value of $min\_hd(S)$ hardly varies in reasonable neighborhood definitions (e.g., the neighborhood defined in the next subsection). This implies that another measure is needed for neighborhood search.

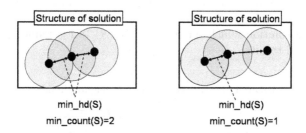

Fig. 5. Ideas of $min\_count(S)$, and $min\_related$
dots : sequences used in $min\_hd(S)$
radii of the circles : d
distance between two dots : hamming-distance between sequences
minimum distance between two dots : $min\_hd(S)$
If a dot exists in the circle regions of other dots, $min\_hd(S) \geq d$ is not satisfied.

See Figure 5, which shows an image of the solution space and the solution structures in the problem. In this figure, dots represent sequences used in $min\_hd(S)$. (For example, in the case of HD + RC, $\forall s \in S \cup wcc(S)$ are dots.) Intuitively, a neighborhood operation corresponds to a slight move of dots. The figure of left-side has the same $min\_hd(S)$ as the right-side. However, the left-side has more pairs of dots whose distances are $min\_hd(S)$ than the right-side. Therefore the left-side may need more neighborhood operations than the right-side in order to improve $min\_hd(S)$.

Here, we define $min\_count(S)$ as the number of pairs of sequences whose distance are $min\_hd(S)$. By using this, we evaluate the solution in the following way: For two sequence sets S and S', if $min\_hd(S) > min\_hd(S')$ (resp., $min\_hd(S) < min\_hd(S')$), $min\_hd(S)$ (resp., $min\_hd(S')$) is the better solution. For S and S' of $min\_hd(S) = min\_hd(S')$, the set having the smaller min_count value is the better solution.

Definition of Neighborhoods. We define neighborhoods in the local search. Naturally, two candidates of neighborhoods are considered:

$$N_i(S) = \{S' \mid \text{ sequence sets obtained by changing } i \text{ character(s)}$$
$$\text{of a sequence belonging to } S\} \cup N_{i-1}(S),$$
$$N'_i(S) = \{S' \mid \text{ sequence sets obtained by changing } i \text{ character(s)}$$
$$\text{in (some) sequence(s) belonging to } S\} \cup N'_{i-1}(S),$$

where $i = 1, 2, \ldots$. For convenience, we define $N_0(S) = \phi, N'_0(S) = \phi$. It should be noted that $N_i(S) \subseteq N'_i(S)$ for any i.

We performed a preliminary computational experiment to find out which neighborhood is better. In this experiment, we improve $min\_hd(S)$ with fixed m and n. Taking into account of section 3.4, we break off the search when $min\_hd(S)$ becomes greater or equal to d. We use constants $n = 10, m = 15$, $d = 4$. We performed 50 trials. The experimental environment is as follows: CPU - Pentium4 2.0GHz, MEM - 256MB, OS - VineLinux3.1, compiler - gcc3.3.2. We call success if a solution which satisfies $min\_hd(S) \geq d$ is found. Table 1 shows the experimental results.

Table 1. Results of comparing neighborhoods

	$N_2(S)$	$N_3(S)$	$N_2(S)'$	$N_3(S)'$
trial number of success	24	47	27	48
average CPU time of success trials(sec)	52.50	117.47	169.56	181.50
average CPU time of false trials(sec)	138.23	832.00	291.08	1088.00
average CPU time of all trials(sec)	97.08	160.34	230.82	217.42

The results show that $N'_i(S)$ are slightly better than $N_i(S)$ in the number of improvement, while $N_i(S)$ is much more preferable to $N'_i(S)$ in the running time. By these, we use $N_i(S)$ as the neighborhoods in the search.

Efficient Neighborhood Search. Here, we consider the effective order of checking sequences of S in our neighborhood searches. See the left-side of Fig.5 again. The contributions of dots for $min\_hd(S)$ vary. To clarify the variation, we define $min\_related$ as a counter for each sequence in S; the counter denotes the number of the contributions of the sequence for $min\_hd(S)$. Let us show examples. For $S = \{s_1, s_2, s_3\}$, in the case of $H(s_1, wcc(s_2)) = min\_hd(S)$, both s_1 and s_2 are contributing to min_hd, so $min\_relateds$ of s_1 and s_2 are at least 1. In the case of $H(s_2, s_2wcc(s_3)[3, n+2]) = min\_hd(S)$, $min\_related$ of s_2 is at least 2 and the one of s_3 is at least 1.

If a sequence has large $min\_related$, the sequence may be located in a wrong position. Therefore we change characters of sequences in descending order of $min\_related$ in the search. By the definition, changing character(s) of a sequence with $min\_related = 0$ is useless. We do not perform the neighborhood search for such sequences.

Another issue of efficient neighborhood searches is the relationship between a neighborhood search and its succeeding neighborhood search. The naive evaluation needs to calculate for all pairs of sequences. However, comparing the set after the neighborhood operation with before, only one sequence is changed, then some pairs are not changed. Therefore, we can skip re-calculation of pairs not changed, which makes the evaluation more efficient.

3.2   Iteration of VNS

In this section, we describe the iterated VNS part of our algorithm. We first explain VNS of our sequence design problem and how to control the size of neighborhood searches. We then explain the "iteration part" of our VNS, that is, the big perturbation to escape from local optima.

Algorithm Based on VNS. The idea of a VNS method is to dynamically change the range of neighboorhood. Our basic strategy is to enlarge the neighboorhood, $N_1(S)$, $N_2(S)$ and so on, in this order.

Here, the point is how we control the enlargement. In some cases, we hardly expect finding an improved solution in the $N_i(S)$ that we currently search. One easy case is that the size of $N_i(S)$ is too small. Let $m_R(S)$ denote the number of sequences whose $min\_related \neq 0$ in S, that is the size of searchable neighborhood, since changing characters of the sequences with $min\_related = 0$ is not performed in our neighborhood search as mentioned in the previous section. If $m_R(S)$ is too small, there is very little possibility of getting an improved solution in $N_i(S)$ with small i. For such S, we should search $N_i(S)$ with larger i in order to improve the solution. On the contrary, if $m_R(S)$ is large enough, there are many chances of improvements even in the search of $N_i(S)$ with small i; $N_i(S)$ with large i is not necessarily searched.

We decide the control by considering the relationship between $m_R(S)$ and the size of the searchable neighborhood. Let $f_i(S)$ denote the size of the searchable neighborhood in $N_i(S) \setminus N_{i-1}(S)$. Then,

$$f_i(S) = m_R(S) \binom{n}{i} 3^i.$$

If the searchable neighborhood of $N_i(S)$ is large enough, we search it, otherwise we do not. We quantify $f_i(S)$ by comparing with $N_1(S) = m\binom{n}{1}3 = 3mn$.

Let P_i be a parameter of the comparison, and let MAX_N be the maximum number of i. If $f_i(S) \leq 3mn \cdot P_i$ and $i \leq MAX\_N$, we search $N_i(S)$, otherwise, we do not. That is, we enlarge the search of $N_i(S)$ by

$$\max \left\{ i \ \middle| \ m_{\mathrm{R}}(S)\binom{n}{i}3^{i-1} \leq mn \cdot P_i \text{ and } i \leq MAX\_NUM \right\}.$$

Iteration Method. We achieve a perturbation as the "iterated" part of our search as follows:

(1) Delete a sequence in S and let the resulting set be S'.
(2) Add a sequence to S'.

In (1), we consider to delete a sequence that has negative effects on S, i.e., having large (actually, the largest) $min\_related$. This is because such a sequence is critical for min_hd in a sense and by eliminating it we can expect an improvement of the solution structure.

In (2), we consider to add a sequence that is farthest from all the sequences in the set. Here, "farthest" means a sequence that maximizes the distance from a nearest sequence in S. Such a farthest sequence is considered to have the least negative effect on the set (See Fig.6). However, since it may be time-consuming to compute such a farthest sequence, we use approximately farthest sequence instead. We find such a sequence in a greedy manner.

Here we use U as the set of sequences that appear explicitly or implicitly in S'. For example, in the case of HD + RC, $U = S' \cup wcc(S')$, and in the case of HD + RC + OL, $U = S' \cup wcc(S') \cup W$, where $W = \{w|w = ss'[i, i+n-1], s, s' \in S' \cup wcc(S'), 1 < i \leq n\}$. Let v be a sequence that approximates a "farthest" sequence from S'. Note that v and $wcc(v)$ approximately maximize the minimum distance between all sequences in U, and approximately minimize the number of pairs which have minimum distance.

We compute v in the following way: For all $i = 1, \ldots, n$, initialize $v_i = *$. Randomly choose v_i of $v_i = *$. We then replace v_i as one of $\{\mathtt{A},\mathtt{T},\mathtt{G},\mathtt{C}\}$ according to the following rule: the resulting v maximizes the minimum distance from all sequences in U, and minimizes the number of sequences whose distance from the resulting v is minimum. ($*$ is a wild card character and is considered any alphabet in $\{\mathtt{A},\mathtt{T},\mathtt{G},\mathtt{C}\}$). Continue the procedure until all $v_i \neq *$.

Our iterated VNS method is summarized as follows:

(1) Apply VNS for a initial solution S_0, and find a local optimal solution S.
(2) Delete a sequence whose $min\_related$ is the largest in S, and let the resulting set be S'.
(3) Find a sequence that approximates a farthest sequence from S'.
(4) Set new S_0 by adding the sequence obtained by (3) to S', and go to (1).

In this search, we expect that the intensification and the diversification are fulfilled by VNS and the iteration part, respectively.

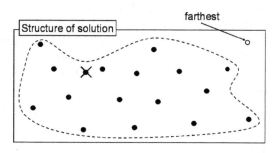

Fig. 6. Idea of *farthest*

3.3   Control of Iteration by VDS

We described the iterated VNS method in above. As I mentioned before, the intensification and the diversification are not always balanced in this method, then we adopt the idea of a VDS method in this operation for the balance of conflicting ideas.

A *Variable Depth Search* (VDS) method [6,9,11,17,18] is which defines neighborhoods as a set of solutions which obtained by a chain of simple neighborhood operations. The chains of neighborhood operations are represented as a search tree, then to define the structure of the search tree is necessary. In this paper, the nodes corresponds to the local optima obtained by the VNS method, the edges correspond to the iterations in section 3.2.

Since the search tree can be very large, we usually restrict its size by the degree of the tree and the depth of the tree, where we set these by parameters *MAX_WIDTH* and *MAX_DEPTH*, respectively (Fig.7).

On the search tree, VDS virtually traverses the nodes in the depth first manner. If an improving solution from the root solution is found, we set the improving solution as the new root immediately, and move on the search on the new search tree. Otherwise, we continue the search on the current search tree. If a node whose solution is as good as the current root is found, we expand the search tree by making the same tree from the node recursively. To avoid the explosion of the search tree, we do at most *REC_NUM* recursions from one root.

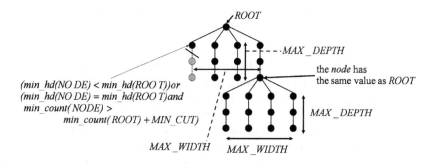

Fig. 7. Search tree of our approach

If we reach a node whose solution is rather worse than the current root, break off descending the tree. Figure 7 shows the threshold of breaking off.

As mentioned, we intend the intensification (with small diversification) by this VDS part. In spite of that, we sometimes fail to obtain better solutions than the iterated VNS method, after the VDS search. In such a case, there may be little possibility of existing improved solution around the local optima, which implies that it is better to search more different structures of search spaces. Hence, we add stronger perturbations than ones of the iterated VNS method as follows:

(1) Determine $JOIN\_NUM$.
(2) Design S' satisfying $|S'| = JOIN\_NUM$ and $min\_hd(S') \geq d$.
(3) Delete $JOIN\_NUM$ sequences in descending order of $min\_related$ from S.
(4) Add S' to S.

3.4   Extension of S

In this section, we explain how to increase m in step 2 of the beginning of Section 3. The idea is introduced in [8]: When the sequences (dots in the figure) are densely located, we can easily add a new sequence satisfying constraints (See Fig.8).

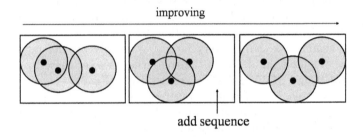

Fig. 8. Situation of increasing sequence
In all the figure, three dots are located but their balances are different. Since the left one does not satisfy the constraints, it may not be suitable to add a new sequence. The other two satisfy the constraints, but the center one is densely located than the right one. Intuitively, the center one is more adequate to put a new sequence than the right one, because the former has more free spaces.

However, the dots are densely located means that $min\_hd(S)$ is small and $min\_count(S)$ is large. On the other hand, in our search method presented in the previous subsections, an improvement means increasing $min\_hd(S)$ or reducing $min\_count(S)$; that is, as the search goes, the solution may get sparser. Therefore, the timing of the addition of a new sequence to S should be immediate after the sequence set satisfying $min\_hd(S) \geq d$ is found.

How to choose a sequence to be added is also an important factor. Here, we again adopt (approximately) farthest to S (in Section 3.2) to be added.

4   Computational Experiments

We implemented our algorithm, and performed computational experiments. The experimental environment is as follows: CPU - Pentium4 2.8GHz, MEM - 512MB, OS - Fedora, compiler - gcc3.3.2.

4.1   In the Case of HD + RC

The setting of the parameters is as follows: $MAX\_N = 3$, $P_2 = 1$, $P_3 = 1$, $MAX\_DEPTH = 3$, $MAX\_WIDTH = 5$, $REC\_NUM = 1$, $CUT\_NUM = 10$, $JOIN\_NUM = m/3$, and the initial set size is 1. Under this setting, we ran the algorithm with time-limit 10 minutes.

We compare our results with the traditional greedy methods [4] and the SLS method [13][1].

Table 2 shows the largest size of the sets obtained by our method when we performed five trials. The horizontal scale and the vertical scale represent the number of constraints and the length of sequences, respectively.

The results in bold faces means that we succeeded in finding larger sets than the existing works. The underlined results means that the same size as the existing works are found. The results with "!" means that they are missing in [4,13]. The results with "*" means that the results reach the upper bounds obtained in [12], i.e., optimal.

Table 2. Results of HD + RC

$n \setminus d$	2	3	4	5	6	7	8	9	10	11	12
4	32*	6	2*	-	-	-	-	-	-	-	-
5	116	32	4	2	-	-	-	-	-	-	-
6	407	60	28	4	2*	-	-	-	-	-	-
7	1268	186	42	12	2	2	-	-	-	-	-
8	2843	580	124	30	16	2	2*	-	-	-	-
9	2887!	1559	346	78	22	8	2	2	-	-	-
10	2786!	2702	965	201	54	17	8*	2	2	-	-
11	2677!	2054	1796	545	129	39	14	6	2	2	-
12	2734!	2609!	2592	1164	309	84	29	12	4	2	2

In the case of around $n - 3 \le d \le n$, our approach has the same size as the existing works. Most of the cases, they are shown to be optimal [12].

In the case when the number of constraints is smaller than above (around $n - 5 \le d \le n - 3$), we can obtain lager sizes than the existing works. This results show the effectiveness of our approach.

In the case when the number of constraints is much smaller (around $1 \le d \le n - 5$), our method fail to obtain better sequence sets than the existing works.

---

[1] Although SLS method [13] is extended to [14], [14] do not treat HD + RC, so we compare our result with the ones of [13].

However, in the case of around $d = n - 6$, the sizes of our results are just slightly smaller than the existing works. If we take the time-limit longer, there may be possibilities which our method can obtain better results.

4.2 In the Case of HD + RC + OL

The setting of the parameters is as follows: $MAX\_N = 3$, $P_2 = 3$, $P_3 = 6$, $MAX\_DEPTH = 3$, $MAX\_WIDTH = 5$, $CUT\_NUM = 10$, $REC\_NUM = \infty$, $JOIN\_NUM = 10$, and the initial size of sequence set is 1. Under this setting, we ran the algorithm with time-limit 24 hours for $n = 10$ and $d = 4$, and with time-limit 72 hours for $n = 15$ and $d = 6$.

We compare with the AG-template method [3] whose constraints are HD + RC + OL.

In the case of $n = 10$ and $d = 4$, we succeeded in finding the sequence set whose size is 39 in five trials. In the case of $n = 15$ and $d = 6$, we succeeded in finding the sequence set whose size is 80 in five trials. In the both cases, the results are better than ones obtained by AG-Template method [3] ($n = 10, d = 4, m = 36$ and $n = 15, d = 6, m = 70$).

AG-Template method cannot generate sets for large d, though our method can do. Furthermore, since our method is effective for strict constraints, our method has a good potential to generate good sets for large d.

These results may show that our approach has a good potential for the problem, although the experiments are not sufficient and there are many parts to be refined.

5 Conclusion

In this paper, we propose a local-search based algorithm of sequence sets design from the viewpoint of combinatorial optimizations. Taking into consideration the combinational properties of the problem, we proposed a dynamic neighborhood search strategy by combining VNS, ILS and VDS method based on natural neighborhood operations. In the computational experiments, we succeed in designing better sequence sets than the existing works in some cases.

As future works, we will consider to adjust our method to other constraints such as the GC-content constraint, and thermodynamical criteria.

References

1. L.Adleman, "Molecular Computation of Solutions to Combinatorial problems", Science 226, pp.1021–1024, 1994.
2. M.Arita, A.Nishikawa, M.Hagiya, K.Komiya, H.Gouzu,and K.Sakamoto, "Improving Sequence Design for DNA Computing", Proc. 5th Genetic and Evolutionary Computation Conference (GECCO-00), pp.875–882, 2000.
3. M.Arita, and S.Kobayashi, "DNA Sequence Design Using Templates", New Generation Computing, Vol.20(3), pp.263–273, 2002.
4. Y.Asahiro, "Simple Greedy Methods for DNA Word Design", Proc. 9th World Multi-Conference on Systemics, Cybernetics and Informatics, Vol.III, pp.186–191, 2005.

5. P.Gaborit, and O.D. King, "Linear Constructions for DNA Codes", Theoretical Computer Science, vol.334, no.1–3, pp.99–113, 2005.
6. F.Glover, "Ejection Chains, Reference Structures and Alternating Path Methods for Traveling Salesman Problem", Discrete Applied Mathematics, vol.65, pp.223–253, 1996.
7. A.Kameda, M.Yamamoto, H.Uejima, M.Hagiya, K.Sakamoto, and A.Ohuchi, "Hairpin-based State Machine and Conformational Addressing: Design and Experiment", Natural Computing, Vol.4, No.2, pp.103–126, 2005.
8. S.Kashiwamura, A.Kameda, M,Yamamoto, and A.Ouchi, "Two-Step Search for DNA Sequence Design", Proc. the 2003 International Technical Conference on Circuits/Systems, Computers and Communications (ITC-CSCC 2003), pp.1889–1892, 2003.
9. B.Kernighan, and S.Lin, "An Effective Heuristic Procedure for Partitioning Graphs", Bell System Technical J., Vol.49, pp.291–307, 1970.
10. S.Kobayashi, T.Kondo, and M.Arita, "On Template Method for DNA Sequence Design", Proc. 8th DNA Based Computers, LNCS(2568) pp.205–214, 2002.
11. S.Lin, and B.Kernighan, "An Effective Heuristic Algorithm for the Traveling Salesman Problem", Operation Research, Vol.21, pp.498–516, 1973.
12. A.Marathe, A.Condon, and R.Corn, "On Combinatorial DNA Word Design", Journal of Computational Biology, Vol.8(3), pp.201–220, 2001.
13. D.Tulpan, H.Hoos, and A.Condon, "Stochastic Local Search Algorithms for DNA Word Design", Proc. 8th DNA Based Computers, LNCS(2568), pp.229–241, 2002.
14. D.Tulpan, and H.Hoos, "Hybrid Randomized Neighborhoods Improve Stochastic Local Search for DNA Code Design", Proc. Advances in Artificial Intelligence, 16th Conference of the Canadian Society for Computational Studies of Intelligence, LNCS(2671), pp.418–433, 2003.
15. D.Tulpan, M.Andronescu, S.Chang, M.Shortreed, A.Condon, H.Hoos, and L.Smith, "Thermodynamically based DNA strand design", Nucleic Acids Res, 33(15), pp.4951–4964, 2005.
16. E.Winfree, F.Liu, L.Wenzler, and N.Seeman, "Design and Self-assembly of DNA Crystals", Nature 394, pp539–544, 1998.
17. M.Yagiura, T.Yamaguchi, and T.Ibaraki, "A Variable Depth Search Algorithm for the Generalized Assignment Problem", Meta-Heuristics: Advances and trends in Local Search Paradigms for Optimization, pp.459–471, 1999.
18. M.Yagiura, T.Ibaraki, and F.Glover, "An Ejection Chains Approach for the Generalized Assignment Problem", INFORMS Journal on Computing, Vol. 16, No. 2, pp. 133–151, 2004.

Sequence Design for Stable DNA Tiles

Naoki Iimura[1], Masahito Yamamoto[2], Fumiaki Tanaka[1], Atsushi Kameda[3],
and Azuma Ohuchi[2]

[1] Graduate School of Information Science and Technology, Hokkaido University
North 14, West 9, Kita-ku, Sapporo, Hokkaido 060-0814, Japan
{i~mu, fumi95}@complex.eng.hokudai.ac.jp
[2] CREST, Japan Science and Technology Agency (JST) and
Graduate School of Information Science and Technology, Hokkaido University
North 14, West 9, Kita-ku, Sapporo, Hokkaido 060-0814, Japan
{masahito, ohuchi}@complex.eng.hokudai.ac.jp
[3] Suyama Lab.Department of Life Sciences, The University of Tokyo
3-8-1, Komaba, Meguro-ku, Tokyo 153-8902, Japan
kameda@genta.c.u-tokyo.ac.jp
http://harmo.complex.eng.hokudai.ac.jp/

Abstract. DNA tile nanostructures have lately attracted a lot of attention as a new calculation technique and material on the nanometer scale. In forming DNA tiles, sequences need to bond in tile conformation. Conventional work can design sequences using overlapping subsequence. In this paper, we design tile sequences based on free energy. As a result of optimization, we show that we can design tile sequences as stable as conventional tiles. Moreover, we illustrate that the tile designed by the proposed method is perhaps more stable than conventional one. This method will be useful to design many tiles when forming large scale and complex DNA nanostructures.

1 Introduction

In recent years, the construction of molecular-scale structures has attracted attention in material engineering, medicine, and other fields. DNA tiles, which use DNA's nature that bonds by self-assembly, are studied as a parallel calculation technique and a nanostructure expected to apply nanodevices and nanocircuits [1][2][3]. Yan *et al.* visualized 4×4 DNA tiles, which array protein and produce silver nanowires, by atomic force microscope (AFM) [4]. In forming a tile, sequences need to bond (hybridize) accurately in tile conformation. Furthermore, the hybridization needs to be stable to prevent the tile from collapsing.

Conventional work SEQUIN can design tile sequences using overlapping subsequences (critons) and complementary sequences [5]. In designing DNA tile by SEQUIN, every critons in an individual single strand DNA molecule is proximate to be unique. However, it is possible that non-complementary subsequences (unique subsequence) mis-hybridize when many sequences exist in liquid solution, and ease of hybridization differs with the secondary structure of each sequence. Therefore, it is safe to say that SEQUIN can design tile sequences based

C. Mao and T. Yokomori (Eds.): DNA12, LNCS 4287, pp. 172–181, 2006.

on an approximate estimate of bond strength, not the secondary structure. We must consider the secondary structure in forming tiles because it affects hybridization between sequences. Sequence design regarding these points seems to form a stable and large scale DNA nanostructure. In DNA computing, sequence design based on the free energy is emphasized to form wanted structure and to design stable hybridization [6]. Free energy can evaluate a bond strength precisely because it considers kinds of base pair, loop structure and hairpin structure based on a thermodynamic nearest-neighbor parameters.

We previously proposed structural stability evaluation criteria for DNA tile and applied this method to 4×4 DNA tiles [7]. In this paper, we propose evaluation criteria based on the free energy and optimize tile sequences using the weighted sum of the ones as a measure of structural stability. Our sequence design considers the following points. First, single strand DNA does not have secondary structure. Second, needless hybridization must be avoided. Third, necessary hybridization must be tight. Furthermore, we show that we can design tile sequences as stable as or even more stable than conventional ones.

2 Design Method

2.1 Procedure

We propose stability evaluation criteria and define the weighed sum of the ones as a measure of tile stability. Our aim is to design tiles which maximize the evaluation value, that is, optimizing tile sequences to be stable. Although we can optimize tile sequences by any optimization method and need examine the most suited method, we employ the simplest hill-climbing algorithm as the optimization method in this paper.

First, we generate a tile so that each single strand DNA molecule has a certain GC rate (the ratio of G and C), and then this tile is set to a candidate tile. Second, the candidate tile is evaluated by evaluation criteria illustrated in the next section. In the optimization process, we change the candidate tile slightly by selecting a base pair in the tile randomly and flipping the bases of the selected base pair in order not to bias the GC rate : A and T flip and G and C flip. The changed tile is evaluated, and if the evaluation value of this tile is more than the previous candidate tile, this tile becomes a new candidate tile. If not, this changed tile is broken off. We select a base pair again and reexecute a similar operation. The above operation is repeated until the terminal conditions fulfilled, such as a predetermined number or a convergence of improvement (Fig. 1).

Although this method can be applied to any kind of tile, in this paper we explain the example of a 4 × 4 DNA tile, which consists of nine single strand DNA molecules, that is, one Core that has bulged T4 loops at each of the four corners, four Shells, and four Arms (Fig. 2) [4][8].

```
procedure Tile Sequence Design
    generate initial tile(x)
    evaluate(x)
    while terminal condition
        generate x' by flip two bases in x
    evaluate(x')
        if evaluate(x') is more stable
            than evaluate(x) then
                x = x'
        end if
    end while
    return x
end Tile Sequence Design
```

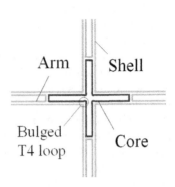

Fig. 1. Procedure of Sequence Design   **Fig. 2.** 4×4 DNA tile

2.2   Stability Evaluation

We propose a stability evaluation method. For more precise evaluations, this method uses the free energy for a single molecule [9][10] and pairs of molecules [11]. Free energy (ΔG) is well used as a measure of bond (hybridization) strength. For example, hybridization strength increases as the number of base pairs, especially GC pairs, increases. And ΔG decreases as the secondary structure or the hybridization strength is tightened. We use the minimum free energy (ΔG_{min}) of any secondary structure for one molecule or hybridization for two molecules.

Our evaluation method considers the following evaluation items, and the evaluation value is the sum of each weighted evaluation criterion by evaluation item. Each evaluation item rises as criterion value increases.

Item 1(I_1). $\Sigma \Delta G_{min}$ of each single strand DNA molecule
  Hybridization between single strand DNA molecules may be obstructed by the secondary structure. This criterion evaluates whether a single strand DNA molecule has secondary structure.

Item 2(I_2). $\Sigma \Delta G_{min}$ when two single strand DNA molecules mis-hybridize
  Needless hybridization may become competition that obstructs necessary hybridization. This criterion evaluates needless hybridization to form tiles. In the case of a 4×4 DNA tile, we evaluate between sequences which do not hybridize as a round robin : Core and Core, Shell and Shell, and Arm and Arm, including own sequence.

Item 3(I_3). $\Sigma(-\Delta G_{min})$ when two single strand DNA molecules hybridize
  Necessary hybridization must react more accurately and be tighter. This criterion evaluates necessary hybridization to form tiles. We use subsequences and evaluate the hybridization of the complementary subsequence in Shell and Arm (or Shell and Core). To align the magnitude relation of each evaluation item, this value is multiplied by -1.

We employ the weighted sum of each criterion as the stability of entire tile. The weighted sum for tile x $E(x)$ is defined by the following expression:

$$E(x) = I_1(x) + \alpha I_2(x) + \beta I_3(x) \qquad (\alpha = 0.086, \beta = 0.12). \qquad (1)$$

Although each weight needs to be set depending on the importance of each item, we use the frequency distributions of each evaluation value by calculating randomized tiles to set weights. We examined each value by randomly providing 100,000 tiles. Then, we weighed the I_2 and I_3 values such that the average of I_2 and I_3 equaled the average of I_1. As a consequence, α and β were 0.086, 0.12, respectively. The weight may be able to estimate by the number of sequence and the length of sequence, although this is not our concern in this paper.

3   Preliminary Experiment

We design tile sequences based on the above proposal. We use Yan's conventional tile [8], designed by SEQUIN as the initial tile here. We cut off the sticky ends to validate the stability of only a tile part without sticky ends. Hill-climbing steps were 2000.

Consequently, the transition of evaluation value E appears in Fig. 3. The figure shows that E improves, and the tile has the potential to gradually become stable. When we see optimized sequences, we find that there are places where the same base continues in a sequence except the bulged T4 loop. The reason for this is that the property of free energy which stabilizes the hybridization strength as the continuation of G or C is lengthened. Furthermore, the continuation of the same base causes mis-hybridization and a reduction of orthonormal sequences. By adding constraints for the initial tile, the above problem can be resolved. These constraints are introduced in the next section.

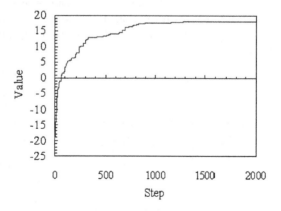

Fig. 3. Transition of E

4   Constraint for Reducing Sequence Similarity

4.1   GC Pair Template for Initial Tile

The continuation of GC pair in a sequence causes the difference of hybridization kinetics between sequences and the bad influence on forming DNA tile. Since the positions of the GC pair are not changed by flipping bases while optimization is being performed, an initial tile maintains their positions. Therefore, we consider that it is important to generate an initial tile. We adopt a template that decides the positions of the GC pair. Using this template, the positions of GC pair in an initial tile are designed as below.

- The number of GC pairs is in the range 55 to 66% of the total in a sequence. This percentage is ± 5% of Yan's tile.
- The GC pair or the AT pair does not permit more than 4 pairs in a row.

G or C and A or T are allocated based on the template at random. Then an initial tile limits the number of Gs in a sequence not to bias G in each Shell and each Arm. We set constraints that keep the limit of the number of Gs. In this study, the number of Gs in a shell sequence is 10 to 13 and the number of Gs in an arm sequence is 8 to 10. To reweigh items, we randomly generated 50,000 tiles based on templates and used the average of each evaluation value in the same way. The obtained α and β were 0.097, 0.14, respectively.

Figures 4-6 show the frequency of each value (I_1, I_2, I_3) of above random 50,000 tiles, respectively. As a result of calculation, Yan's tile designed by SE-QUIN allows improvement of the secondary structure (Fig. 4). In Fig. 5, Yan's tile marks a large value. Because SEQUIN uses overlapping subsequences and minimizes reuse of the subsequences, it is probable that sequences avoid mis-hybridization efficiently. In Fig. 6, the hybridization of local long GC pair continuations in Yan's tile causes this frequency. Although the meaning of I_2 and I_3 seems to be inversion and have relevance, the correlation coefficient between I_2 and I_3 was -0.21. Each has little relation. Fig. 7 shows the frequency of E. Yan's tile ranks in the top 1%. Therefore, it is difficult to design tiles at random.

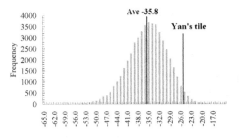

Fig. 4. Frequency of I_1

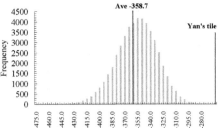

Fig. 5. Frequency of I_2

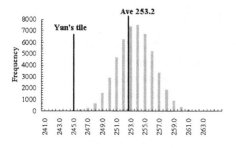

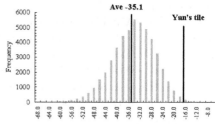

Fig. 6. Frequency of I_3 **Fig. 7.** Frequency of E

4.2   Experiment and Results

We executed sequence design 50 times with the GC pair templates and the above constraints. Templates were generated at random based on the constraints in section 4.1. Optimization steps, α and β were 2,000, 0.097, and 0.14, respectively. Our program was implemented using C language and a PC with 2.26GHz Pentium 4 processor, 512KB cache and 256MB RAM running Turbo Linux. The average computational time for each trial was about 92 minutes.

Table 1 shows the evaluation value after the optimization. Each value improved compared with Yan's tile. Any seed tile was optimized and the evaluation value was better than Yan's tile. An extensive improvement is found especially in I_1 and I_2. This implies that optimized sequences do not have the robust secondary structure and mis-hybridize easily. The improvement of the secondary

Table 1. Results of each value (50 hill-climbing executions)

	E	I_1	I_2	I_3
Yan's tile (reweighted)	-17.38	-25.0	-275.93	245.58
Best	10.57	-5.2	-211.03	258.86
Worst	-0.62	-10.9	-254.33	249.64
Average	5.63	-7.46	-230.52	253.43

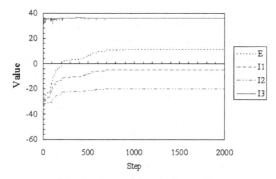

Fig. 8. Transition in best tile

structure appears to be due to introducing secondary structure prediction based on free energy. Value I_3 is not so much of a difference between SEQUIN algorithm and our evaluation item. We ought to have to discuss this evaluation item (I_3). We must also consider how to weigh each value, especially I_3, for example, using normal distribution. Moreover, we have to validate weights by chemical experiments. Fig. 8 shows the transition of each value in the best tile. Each value converges within 1,000 steps. Thus, the terminal condition also can be improved, for instance that the program stops at 1,000 steps. The best tile is demonstrated in Table 2. Each sequence is compatible with the sequence in Fig. 11. We validate the stability of this tile in the next section.

The optimization based on our evaluation function can thermodynamically design stabler tiles than Yan's tile.

Table 2. Best tile by Optimization (5'→3')

Name	Before (Initial tile)	After (Optimized tile)
Core (100mer)	ACGTGAGCAGGCTTTTTTACCACGC-TCAGAGGGTCCGTATTTTTGGTCTG-TCGGTCGTAGCGTCTTTTACGCTGC-TGTCGAGAGCCTTGTTTTCTCAGCG	TCGTCAGCAGGGTATTTTAGCTCGC-TCAGTCGGTGCCTATTTTTGGTGTG-TGGGTGCTTGCGTGTTTTTCCCTCC-AGTCGAGAGCCATGTTTTCACTGCG
Shell 1 (42mer)	CACCCACTCTCTGAGCGTGGTAAGC-CTGCTCCACGGACGTCA	GACCGACACACTGAGCGAGCTTACC-CTGCTGCACGGACCACA
Shell 2	CTCGGTCTCTACGTCGCTGAGCAAG-GCTCTCCTCCAGACGCT	CTGGGACTCAACGACGCAGTGCATG-GCTCTCCTCCTCACGCT
Shell 3	CTGTCGCTGTGACAGCAGCGTGACG-CTACGACTGCCACACCA	GACACGCACTGACTGGAGGGACACG-CAAGCACAGCCACACCT
Shell 4	CCGTCAGAGACCGACAGACCATACG-GACCCTCAGGGAGAGCT	GCGAGAGACACCCACACACCATAGG-CACCGACAGGCACACGA
Arm 1	AGCTCTCCCTGAGAGTGGGTG	TCGTGTGCCTGTGTGTCGGTC
Arm 2	TGACGTCCGTGAGAGACCGAG	TGTGGTCCGTGTGAGTCCCAG
Arm 3	AGCGTCTGGAGACAGCGACAG	AGCGTGAGGAGAGTGCGTGTC
Arm 4	TGGTGTGGCAGTCTCTGACGG	AGGTGTGGCTGTGTCTCTCGC

4.3   Review of Designed Sequences

We review sequences in terms of the free energy without chemical experiments. Designed sequences (Table 2) closely compare with Yan's tile in every evaluation item.

Evaluation Item 1 (I_1). Table 3 shows each free energy when a single strand DNA molecule has secondary structure. As the secondary structure is not stable and the tile forms successfully, the value grows. All values of our tile are greater than Yan's tile (Table 3). Every sequence is less likely to be able to have secondary structure which obstructs hybridization to form tile. For instance, Figs. 9 and 10 are the secondary structure prediction images of Yan's Core and the best tile's Core by Mfold [10]. The dot on the image expresses the bond of the base, and thus the secondary structure of Yan's Core is slightly more stable than the designed one. We see that secondary structure is improved by optimization.

Table 3. Breakdown of I_1

	Core	Shell 1	Shell 2	Shell 3	Shell 4	Arm 1	Arm 2	Arm 3	Arm 4
Yan's tile	-6.9	-2.8	-3.1	-1.1	-3.3	-1.3	-2.6	-2.4	-1.5
Designed tile	-2.2	-1.9	-0.3	-0.3	-0.5	0.0	0.0	0.0	0.0

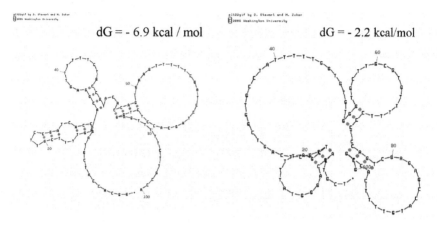

dG = - 6.9 kcal / mol   dG = - 2.2 kcal/mol

Fig. 9. Core of Yan's tile   **Fig. 10.** Core of designed tile

Evaluation Item 2 (I_2). Item 2 evaluates the bond strength of needless hybridization that does not form a tile. We use the free energy when Core and Core hybridize, Shell and Shell hybridize, and Arm and Arm hybridize. The value of I_2 increases as the tile optimizes because mis-hybridization has a larger value.

The free energy between Yan's Cores and between the optimized tile's Cores were -31.19 kcal/mol, -26.13 kcal/mol, respectively. Therefore, the optimized tile was harder to hybridize between Cores than Yan's one. Tables 4 and 5 show the free energy between Shells and between Arms. Although our optimization, which differed from SEQUIN that used overlapping sequences, did not use unique sequences, optimization could design sequences that did not mis-hybridize.

Evaluation Item 3 (I_3). Finally, we must evaluate the necessary hybridization strength to form tiles. Each sequence hybridizes with a few sequences, and hence we use complementary subsequences, not a full sequence. The evaluated hybridizations appear in Fig. 12. For example, S-A1 shows hybridization between the subsequence of Shell1 and the one of Arm1. Table 6 shows the free energy of the complementary subsequences. The complementary hybridization tightened in 13 out of 16.

We think that increasing the continuation of the GC pair or the ratio of GC leads to increasing I_3. On the other hand, increasing those permits decreasing unique sequences. Despite the same constraint with conventional tile, the designed tile is more stable than the one.

Table 4. Free energy between Shells (upper : Yan's tile, lower : designed tile)

	Shell 1	Shell 2	Shell 3	Shell 4
Shell 1	-18.29 (-14.16)	-18.55 (-14.13)	-14.42 (-12.59)	-16.54 (-10.44)
Shell 2	-	-20.71 (-12.80)	-14.40 (-14.65)	-17.35 (-14.77)
Shell 3	-	-	-12.19 (-8.87)	-15.72 (-11.78)
Shell 4	-	-	-	-16.72 (-8.71)

Table 5. Free energy between Arms (upper : Yan's tile, lower : designed tile)

	Arm 1	Arm 2	Arm 3	Arm 4
Arm 1	-9.61 (-4.73)	-8.38 (-5.67)	-6.99 (-6.56)	-7.10 (-5.60)
Arm 2	-	-10.04 (-5.20)	-8.16 (-5.10)	-7.82 (-5.50)
Arm 3	-	-	-8.40 (-4.88)	-7.10 (-7.67)
Arm 4	-	-	-	-6.25 (-5.38)

Table 6. Breakdown of I_3

	S-C1	S-C2	S-C3	S-C4	S-C5	S-C6	S-C7	S-C8
Yan's tile	-16.19	-13.61	-17.40	-13.79	-15.24	-13.62	-16.98	-14.84
Designed tile	-16.46	-14.93	-18.03	-13.71	-16.22	-16.02	-17.94	-15.77

	S-A1	S-A2	S-A3	S-A4	S-A5	S-A6	S-A7	S-A8
Yan's tile	-14.69	-18.88	-15.76	-13.85	-14.78	-15.60	-14.21	-16.14
Designed tile	-15.51	-17.97	-14.40	-16.53	-15.49	-17.21	-14.73	-17.94

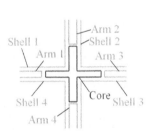

Fig. 11. Definition of sequence name

Fig. 12. Definition of subsequence pair

Our method designs, in short, sequences which have no secondary structure, mis-hybridize not to form tile and gain hybridization strength to form tile, by three evaluation items. Although this method can be applied to any kind of tile and sticky end, we need to customize program depending on kinds of tile, tile size, the length and constraint of sticky ends. We must normalize an evaluation value when the length of sticky end or tile size is different. Additionally, we need more computing power than SEQUIN because of calculating the free energy.

5   Conclusion

We designed DNA tile sequences by optimization based on our thermodynamical stability evaluation method. As a result of our design, we illustrated that the designed tile seems more stable than the conventional tile. Our sequence design method has a beneficial effect on designing many evenly stable tiles, because one can automatically design differently from semi-automatic SEQUIN.

Many very real problems remain: validity of weight, validity of evaluation criteria, and actual construction. These problems must be solved by chemical experiments and visualization by atomic force microscope (AFM).

Acknowledgment

Authors would like to thank to Prof.John H.Reif, Prof.Thomas H. LaBean and Dr.Sung Ha Park for valuable advice and discussions about 4 × 4 DNA tiles.

References

1. Erik Winfree, Furong Liu, Lisa A.Wenzler and Nadrian C.Seeman, "Design and self-assembly of two-dimensional DNA crystals", Nature, Vol.394, pp.539-544, 1998.
2. Erik Winfree, "DNA Computing by Self-assencly", in NAE's TheBredge 33(4): pp.31-38, 2003.
3. Hao Yan, Liping Feng, Thomas H.LaBean, John H.Reif,"Parallel Molecular Computations of Pairwise Exclusive-Or (XOR) Using DNA "String Tile" Self-Assembly", J.Am.Chem.Soc, Vol.125(47), pp.14246-14247, 2003.
4. Hao Yan, Sungg Ha Park, Gleb Finkelstein, John H.Reif, Thomas H.LaBean "DNA-Templated Self-Assembly of Protein Arrays and Highly Conductive Nanowires", Science, Vol. 301, pp. 1882-1884, 2003.
5. Nadrian C.Seeman, "De Nove Design of Sequence for Nucleic Acid Structual Engineering", Jornal of Biomolecular Structure & Dynamics, ISSNO. 739-1102, Vol.8, 1990.
6. Fumiaki Tanaka, Atsushi Kameda, Masahito Yamamoto, and Azuma Ohuchi, "Design of nucleic acid sequences for DNA computing based on a thermodynamic approach", Nucleic Acids Research, Vol. 33(3), pp. 903-911, 2005.
7. Naoki Iimura, Masahito Yamamoto, Fumiaki Tanaka, Atsuchi Kameda, Azuma Ohuchi, "Stability evaluation method of DNA tile Structure", Proceedings of the 11th International Symposium on Artificial Life and Robotics (AROB 11th '06), CD-ROM, 2006.
8. Suppoting Online Material, www.sciencemag.org/cgi/content/full/301/5641/1882/DC1
9. A.M.Zuker, B.D.H.Mathews, C.D.H.Turner, "ALGORITHMS AND THERMO-DYNAMICS FOR RNA SECONDARY STRUCTURE PREDICTION : A PRACTICAL GUIDEh, In RNA Biochemistryand Biotechnology, J. Barciszewski & B.F.C. Clark, eds., NATO ASI Series, Kluwer Academic Publisers, 1999
10. Michael Zuker, "Mfold web server for nulceic acid folding and hybridization prediction", Nucleic Acids Reserch, Vol.31, No.13, pp.3406-3415, 2003.
11. Mirela Andronescu, Rosalia Aguirre-Hernandez, Anne Condon, Hologer H.Hoos,"RNA soft : a suite of RNA secondary structure prediction and design software tools", Nucleic Acids Research, Vol.31, No.13, pp.3416-3422, 2003.

Hairpin Structures Defined by DNA Trajectories*

Michael Domaratzki

Department of Computer Science,
University of Manitoba,
Winnipeg, MB R3T 2N2 Canada
mdomarat@cs.umanitoba.ca

Abstract. We examine scattered hairpins, which are structures formed when a single strand folds into a partially hybridized stem and a loop. To specify different classes of hairpins, we use the concept of DNA trajectories, which allows precise descriptions of valid bonding patterns on the stem of the hairpin. DNA trajectories have previously been used to describe bonding between separate strands.

We are interested in the mathematical properties of scattered hairpins described by DNA trajectories. We examine the complexity of set of hairpin-free words described by a set of DNA trajectories. In particular, we consider the closure properties of language classes under sets of DNA trajectories of differing complexity. We address decidability of recognition problems for hairpin structures.

1   Introduction

A hairpin in a single strand of nucleotides is a structure formed by the bonding of two complementary regions, which form the *stem*, joined on one end by an intermediate, unbonded region. Together, the stem and the unbonded region are known as the hairpin. We illustrate this concept in Figure 1.

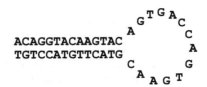

Fig. 1. A hairpin in a strand of nucleotides

As research into DNA computing applications and nanotechnology continues, the formal study of hairpins gains increasing significance. Kari *et al.* [2,3] survey the use of hairpins in various contexts. We also note the use of hairpins for visual

---

* For a full version, see [1]. Research supported in part by a grant from NSERC. Research conducted at the Jodrey School of Computer Science, Acadia University.

contrast in evaluating successful nanotechnological constructions, as described in, e.g., the work of Winfree *et al.* [4]. In some of these applications, hairpins are desirable, while in other applications, they are problematic and are to be avoided in sets of DNA strands. Further, hairpins serve as the basis for more complicated secondary structures such as pseudoknots.

Recently, Kari *et al.* [2,3] have studied hairpins using the tools of theoretical computer science. In particular, a single strand of nucleotides is viewed as a word over the alphabet $\Delta = \{A, C, G, T\}$. In this framework, a hairpin in a word z is a decomposition $z = uvwxy$ where v and x are complementary to each other, and form the stem of the hairpin. We characterize the complementarity of v and x using an antimorphism θ (for definitions, see Section 2). Among other results, Kari *et al.* characterize the complexity and decidability results for hairpin sets [3]. Further, Kari *et al.* [2] have also studied *scattered* hairpins, which represent hairpins in which the stem is not completely hybridized, i.e., where an arbitrary number of unbonded regions occur within the stem.

In this paper, we examine refinements of hairpins and scattered hairpins by incorporating an optional parameter—a set of *DNA trajectories*—to add increased capability in describing the set of hairpins which are of potential interest. The use of DNA trajectories has recently been employed to model bonding regions in separate strands, called *bond-free properties* [5]. One main benefit of DNA trajectories is that they enable constraints to be expressed as a formal language, rather than graphically or otherwise. Further, DNA trajectories are capable of adapting to minor structural changes: modifications such as enforcing a minimum length of a bond are easily introduced in DNA trajectories.

In our study of DNA trajectories and hairpins, we focus on closure properties, decidability and relations to problems from combinatorics on words. With respect to closure properties, we find that the situation is more complex than the case of hairpins and scatted hairpins studied by Kari *et al.*, and many interesting results have been obtained. Decidability problems are also more interesting, due to the fact that regularity of a set of DNA trajectories does not imply the regularity of the associated set of hairpins or the set of hairpin-free DNA words.

2   Definitions

For additional background in formal languages and automata theory, please see Rozenberg and Salomaa [6]. For an introduction to DNA computing, see Păun *et al.* [7]. Let Σ be a finite set of symbols, called *letters*. Then Σ^* is the set of all finite sequences of letters from Σ, which are called *words*. The empty word ϵ is the empty sequence of letters. The *length* of a word $w = w_1 w_2 \cdots w_n \in \Sigma^*$, where $w_i \in \Sigma$, is n, and is denoted $|w|$. Note that ϵ is the unique word of length 0. Given a word $w \in \Sigma^*$ and $a \in \Sigma$, $|w|_a$ is the number of occurrences of a in w. A *language* L is any subset of Σ^*. We use the notation $\prod_{i=1}^n L_i$ to denote $L_1 L_2 \cdots L_n$, and the notation $L^{\geq k}$ to denote $L^k L^*$. The reversal of a word $w = x_1 x_2 \cdots x_n$ $(x_i \in \Sigma)$, denoted w^R, is defined by $w^R = x_n \cdots x_2 x_1$. By extension, $L^R = \{x^R \ : \ x \in L\}$.

Let Σ, Δ be alphabets and $h : \Sigma \to \Delta$ be any function. Then h can be extended to a morphism $h : \Sigma^* \to \Delta^*$ via the condition that $h(uv) = h(u)h(v)$ for all $u, v \in \Sigma^*$. Similarly, h can be extended to an antimorphism via the condition that condition that $h(uv) = h(v)h(u)$ for all $u, v \in \Sigma^*$. An involution θ is any function $\theta : \Sigma \to \Sigma$ such that θ^2 is the identity function on Σ. Let μ denote the mirror involution (i.e., the identity function extended to an antimorphism). Let ι denote the identity morphism.

We assume that the reader is familiar with regular languages, linear context-free languages and context-free languages (CFLs). In particular, regular languages are accepted by deterministic or nondeterministic finite automata, while CFLs are generated by context-free grammars.

2.1   Trajectory-Based Operations

The shuffle on trajectories operation is a method for specifying the ways in which two input words may be interleaved to form a result. Each trajectory $t \in \{0,1\}^*$ with $|t|_0 = n$ and $|t|_1 = m$ specifies one particular way in which we can shuffle two words of length n (as the left input word) and m (as the right input word). The word resulting from the shuffle along t will have a letter from the left input word in position i if the i-th symbol of t is 0, and a letter from the right input word in position i if the i-th symbol of t is 1.

Formally [8], let x and y be words over an alphabet Σ and t, the *trajectory*, be a word over $\{0,1\}$. The shuffle of x and y on trajectory t is denoted by $x \sqcup_t y$. If $t = \prod_{i=1}^n 0^{j_i} 1^{k_i}$ for some $n \geq 0$ and $j_i, k_i \geq 0$ for all $1 \leq i \leq n$, then

$$x \sqcup_t y = \{\prod_{i=1}^n x_i y_i : x = \prod_{i=1}^n x_i, y = \prod_{i=1}^n y_i,$$

$$\text{with } |x_i| = j_i, |y_i| = k_i \text{ for all } 1 \leq i \leq n\}$$

if $|x| = |t|_0$ and $|y| = |t|_1$, and $x \sqcup_t y = \emptyset$ if $|x| \neq |t|_0$ or $|y| \neq |t|_1$. We extend the operation of shuffle on trajectories to sets of trajectories $T \subseteq \{0,1\}^*$ as follows:

$$x \sqcup_T y = \bigcup_{t \in T} x \sqcup_t y.$$

Further, if $L_1, L_2 \subseteq \Sigma^*$ are languages, then

$$L_1 \sqcup_T L_2 = \bigcup_{\substack{x \in L_1 \\ y \in L_2}} x \sqcup_T y.$$

As an example, note that if $T = 0^* 1^*$, then $\sqcup_T$ is the concatenation operation: $L_1 \sqcup_T L_2 = \{xy : x \in L_1, y \in L_2\}$.

We will also require the notion of the natural binary relation defined by shuffle on trajectories [9]. For $T \subseteq \{0,1\}^*$, define ω_T as follows: for all $x, y \in \Sigma^*$, $x \, \omega_T \, y \iff y \in x \sqcup_T \Sigma^*$. For example, if $T = 0^* 1^*$, then ω_T is the prefix order, defined by $x \, \omega_T \, y$ if and only if $y \in x \Sigma^*$. If $T = \{0,1\}^*$, then $x \, \omega_T \, y$ is the embedding order, defined by $x \, \omega_T \, y$ if and only if $y \in x \sqcup \Sigma^*$ (i.e., x can be obtained by from y by deleting zero or more letters). We denote the embedding order by $\leq_e$; note that if $x \leq_e y$ then x is a *scattered subword* of y.

2.2   DNA Trajectories and Hairpins

We now consider DNA trajectories, defined by Kari *et al.* [5]. A DNA trajectory is a word over the alphabet $V_D = \left\{ \binom{b}{b}, \binom{f}{f}, \binom{f}{\epsilon}, \binom{\epsilon}{f} \right\}$. The original use of a set of DNA trajectories was to define bonding between two separate single strands of DNA. The occurrence of $\binom{b}{b}$ implies a bond at a certain position, while $\binom{f}{f}$ (resp., $\binom{f}{\epsilon}$, $\binom{\epsilon}{f}$) denotes two bases which are free (resp., an extra unbonded nucleotide on the top strand, an extra unbonded nucleotide on the bottom strand). DNA trajectories are used to define so-called *bond-free properties* in DNA code word design [5], and we adopt them here for modelling the bonding of hairpins.

For hairpins, we can view words over V_D^* as designating where bonds can occur and cannot occur when viewing the strands with the loop at the right end. For instance, the DNA trajectory $t = \binom{f}{\epsilon}\binom{f}{f}^2\binom{b}{b}^3\binom{f}{f}^3\binom{f}{\epsilon}$ represents the bonding depicted in Figure 2. Note that the pairs x_4 and x_{16}, x_5 and x_{15}, as well as x_6 and x_{14} must be bonded together.

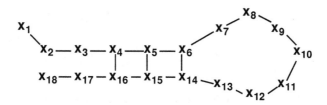

Fig. 2. A DNA bond specified by t. The letters x_i represent arbitrary letters from the alphabet.

Let $\varphi_u, \varphi_d : V_D^* \to \{0,1\}^*$ be morphisms defined by

$$\varphi_u(\binom{b}{b}) = 0, \; \varphi_u(\binom{f}{y}) = 1, \text{ for } y \in \{f, \epsilon\}, \; \varphi_u(\binom{\epsilon}{f}) = \epsilon,$$
$$\varphi_d(\binom{b}{b}) = 0, \; \varphi_d(\binom{y}{f}) = 1, \text{ for } y \in \{f, \epsilon\}, \; \varphi_d(\binom{f}{\epsilon}) = \epsilon.$$

We now give our main definition. Let Σ be an alphabet, $\theta : \Sigma \to \Sigma$ be an arbitrary involution, extended to a morphism or antimorphism, and $S \subseteq V_D^*$. Then a word w is said to be S-θ-hairpin-free, or simply $shp_\Sigma(S, \theta)$-free, if the following condition holds

$$\forall u, v, x, \in \Sigma^*, s \in S, (w = uv, x\, \omega_{\varphi_u(s)}\, u, \text{ and } \theta(x)\, \omega_{\varphi_d(s)^R}\, v) \Rightarrow x = \epsilon. \quad (1)$$

That is, if we can write w as $w = uv$ and there exists a word x—which represents the portions of u and v which are bonded together—such that x appears in u according to the bonding prescribed by $\varphi_u(s)$ and similarly for $\theta(x), v$ and $\varphi_d(s)^R$, then $x = \epsilon$. Note that $\varphi_d(s)$ is reversed since v runs backwards from the right-to-left in our hairpin. We say that a language L is $shp_\Sigma(S, \theta)$-free if w is $shp_\Sigma(S, \theta)$-free for all $w \in L$.

Let $shpf_\Sigma(S, \theta)$ denote the set of $shp_\Sigma(S, \theta)$-free words. Let $shp_\Sigma(S, \theta) = \Sigma^* - shpf_\Sigma(S, \theta)$. Clearly, L is $shp_\Sigma(S, \theta)$-free if and only if $L \subseteq shpf_\Sigma(S, \theta)$.

The definition of $shp_\Sigma(S, \theta)$-freeness is an extension of the notions of hairpin-freeness and scattered-hairpin-freeness, investigated by Kari et $al.$ [2,3].

Note that in the above definition θ can be an arbitrary involution, extended to either a morphism or antimorphism. This is similar to the work on bond-free properties [5] and hairpin-freeness [2,3]. In practice, an antimorphic involution yields hairpin and scattered-hairpin structures, while morphic involutions yield structures where the scattered stem is bonded in a parallel, rather than an anti-parallel, orientation. Of course, the antimorphic involution τ over the alphabet $\Delta = \{A, C, G, T\}$ defined by $\tau(A) = T, \tau(T) = A, \tau(C) = G$ and $\tau(G) = C$ is of particular interest in practice. This involution is called the Watson-Crick involution. In this paper, we study results applying to morphic involutions and antimoprhic involutions, as both give rise to interesting problems and results.

2.3   Examples of Hairpin Languages

Consider the following examples of hairpin languages:

(a) Let $k \geq 1$ and

$$S_k = \left\{ \binom{f}{\epsilon}^* \cup \binom{\epsilon}{f}^* \right\} \binom{f}{f}^* \binom{b}{b}^{\geq k} \binom{f}{f}^* \left\{ \epsilon, \binom{f}{\epsilon} \right\}. \tag{2}$$

The shape of this set of DNA trajectories is given by Figure 3(a). That is, only one bonded region (the $stem$) is formed in this simple hairpin structure, and the length of this stem is at least k. The set $shpf_\Sigma(S_k, \theta)$ is the set of all θ-k-hairpin-free words, studied by Kari et $al.$ [3].

(b) Let $k, m_1, m_2 \geq 1$. Jonoska et $al.$ [10,11] define $\theta(k, m_1, m_2)$-subword compliant languages, which are characterized by the following set of trajectories S_{k,m_1,m_2}:

$$S_{k,m_1,m_2} = \left(\binom{f}{\epsilon}^* \cup \binom{\epsilon}{f}^* \right) \binom{f}{f}^* \binom{b}{b}^{\geq k} \left(\bigcup_{m=m_1}^{m_2} \binom{f}{\epsilon}^m \right).$$

In particular, a language $L \subseteq \Sigma^*$ is $\theta(k, m_1, m_2)$-subword compliant for a morphic or antimorphic involution θ if $L \subseteq shpf_\Sigma(S_{k,m_1,m_2}, \theta)$.

(c) Let $k \geq 1$ and S_k be defined by

$$S_k = \left(\left(\binom{f}{\epsilon}^* \cup \binom{\epsilon}{f}^* \right) \binom{f}{f}^* \binom{b}{b}^{\geq k} \right) \binom{f}{f}^* \left\{ \epsilon, \binom{f}{\epsilon} \right\}. \tag{3}$$

The shape described by this set of trajectories is called $scattered$ $hairpins$ by Kari et $al.$ [2]. In particular, the condition is equivalent to the following: $x \leq_e u$ and $\theta(x) \leq_e v$ imply $|x| < k$. An example of the type of scattered hairpins described by S_k is given in Figure 3(b). The set $shpf_\Sigma(S_k, \theta)$ is denoted by $shpf(\theta, k)$ by Kari et $al.$ [2].

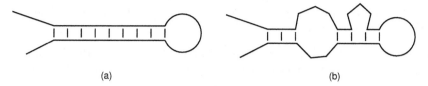

Fig. 3. (a) A simple hairpin structure. (b) A scattered hairpin structure.

By adding DNA trajectories to scattered hairpins, we can also define familiar languages which have been studied by researchers in formal language theory. We begin by demonstrating that the classical languages of palindromes (modulo short palindromes) and squares are definable by a trajectory-based hairpin condition:

Example 1. Let $S_p = \binom{b}{b}^* \{\epsilon, \binom{f}{\epsilon}\}$. Then $shp_\Sigma(S_p, \mu) = \{x \in \Sigma^* : |x| \geq 2, x = x^R\}$. Let $S_s = \binom{b}{b}^*$. Then $shp_\Sigma(S_s, \iota) = \{xx : x \in \Sigma^+\}$.

3   Preliminary Results

We first consider the implications of choosing alternate definitions for hairpin-freeness using DNA trajectories. In the first case, we show that, with DNA trajectories, there is no increase in power by adding a parameter $k \geq 1$ which enforces a minimum length of the (scattered) stem of the hairpin. In the second case, we show that if separate DNA trajectories are allowed to be chosen for the bonding on both sides of the stem, the result can destroy the structure described by the set of DNA trajectories.

In particular, let $k \geq 1$ and $S \subseteq V_D^*$. Say a word w is said to be θ-k-S-hairpin-free (or $shp_\Sigma(S, \theta, k)$-free) if the following condition holds

$$\forall u, v, x, \in \Sigma^*, s \in S, (w = uv, x \, \omega_{\varphi_u(s)} \, u, \theta(x) \, \omega_{\varphi_d(s)^R} \, v) \Rightarrow (|x| < k).$$

This definition more closely mirrors the definitions provided by Kari *et al.* [2,3]. Let $shpf_\Sigma(S, \theta, k)$ denote the set of $shp_\Sigma(S, \theta, k)$-free words. We now show that sets of DNA trajectories are sufficiently powerful to eliminate the need for considering S-θ-k-hairpin-free words.

Lemma 1. *Let $k \geq 1$ and $S \subseteq V_D^*$. Then there exists $S' \subseteq V_D^*$ such that $shpf_\Sigma(S, \theta, k) = shpf_\Sigma(S', \theta)$.*

For fixed k, the construction in Lemma 1 does not alter the complexity of S if S lies in a language class which is closed under finite modification[1]. Now we note the following relationship between inclusion of sets of DNA trajectories and the associated hairpin languages.

---

[1] A language class $\mathcal{C}$ is closed under finite modification if for all $L \in \mathcal{C}$ and all words x, $L \cup \{x\}, L - \{x\} \in \mathcal{C}$. Most common language classes are closed under finite modification; an example of a class that is not is the class of 0L languages.

Proposition 1. *Let $S_1, S_2 \subseteq V_D^*$ with $S_1 \subseteq S_2$. Then for all Σ and all θ, $shp_\Sigma(S_1, \theta) \subseteq shp_\Sigma(S_2, \theta)$.*

Finally, we note that distinct trajectories may represent the same bonding pattern. For instance, note that an occurrence of $\binom{f}{f}$ is equivalent to an occurrence of $\binom{f}{\epsilon}\binom{\epsilon}{f}$. Due to this equivalence, we show the existence of a normal form for sets of DNA trajectories which is sometimes useful.

Lemma 2. *For all sets of DNA trajectories $S \subseteq V_D^*$ there exists a set of DNA trajectories $S' \subseteq \left(\left(\binom{f}{\epsilon}^* \cup \binom{\epsilon}{f}^* \right) \binom{f}{f}^* \binom{b}{b}^* \right)^* \binom{f}{f}^* \left\{ \binom{f}{\epsilon}, \epsilon \right\}$ such that $shp_\Sigma(S, \theta) = shp_\Sigma(S', \theta)$.*

If S is in the form specified by Lemma 2, we say that S is in *normal form*. Further, if $S \subseteq V_D^*$, then by $[S]$ we mean the set of all DNA trajectories which can be rewritten to a DNA trajectory $s \in S$ by using the above rules.

4   Closure Properties

In this section we examine the closure properties of hairpin languages based on the complexity of S. Example 1 immediately yields the following lemma:

Lemma 3. *There exist a regular set of DNA trajectories S and an antimorphic involution θ (resp., morphic involution σ) such that $shp_\Sigma(S, \theta)$ is not a regular language (resp., $shp_\Sigma(S, \sigma)$ is not a CFL).*

Note that this is in contrast to the case of hairpin languages and scattered hairpin languages, studied by Kari *et al.* [2], where the associated languages are regular. Despite the fact that regularity is not preserved when using a set of DNA trajectories to describe hairpin trajectories, we can show that for all regular sets of trajectories S and all antimorphic involutions θ, the language $shp_\Sigma(S, \theta)$ is at worst context-free:

Theorem 1. *If θ is an antimorphic involution and S is a regular set of DNA trajectories, then $shp_\Sigma(S, \theta)$ is a linear context-free language.*

We note that if we relax the condition that S is regular, Theorem 1 does not hold.

Lemma 4. *Let Σ be an alphabet with $|\Sigma| \geq 3$. There exists a (linear) context-free set of DNA trajectories $S \subseteq V_D^*$ such that $shp_\Sigma(S, \mu)$ is not a CFL.*

Further, if we consider $shpf_\Sigma(S, \theta)$ for regular S and antimorphic θ the result may not be context-free. We expect this result, as the CFLs are not closed under complement.

Theorem 2. *Let Σ be an alphabet with $|\Sigma| \geq 3$. There exist a regular set of DNA trajectories $S \subseteq V_D^*$ and an antimorphic involution θ such that $shpf_\Sigma(S, \theta)$ is not a CFL.*

Thus, in general, $shpf_\Sigma(S, \theta)$ is not a CFL if S is regular and θ is an antimorphism. However, we can find conditions on S such that $shpf_\Sigma(S, \theta)$ is a CFL for all antimorphic involutions θ.

Call a language L *slender* if there exists a constant c such that L has at most c words of length n for all $n \geq 0$. A regular language R over Σ is slender if and only if there exist $k \geq 1$, and x_i, y_i, z_i for $1 \leq i \leq k$ such that $R = \bigcup_{i=1}^{k} x_i y_i^* z_i$. (see, e.g., Szilard *et al.* [12]). We can now demonstrate a nontrivial class of sets of DNA trajectories for which the set of hairpin-free words will be guaranteed to be a CFL:

Theorem 3. *Let $S \subseteq V_D^*$ be a slender regular set of DNA trajectories. Then for all antimorphic involutions θ, $shpf_\Sigma(S, \theta)$ is a CFL.*

Theorem 3 shows the power of using trajectories for characterizing hairpins. By using a well-studied property of languages—their density—and applying it to the set of DNA trajectories, we can observe important properties of the associated hairpin language. However, in this case, we find that in addition to the complexity of the set of DNA trajectories, it is also another measure of the complexity—the density of the language—that yields the result.

We can now turn to the complexity of $shpf_\Sigma(S, \theta)$ for morphic involutions θ. By Lemma 3, we know that $shp_\Sigma(S, \theta)$ can fail to be a CFL, even if S is regular. However, the example given (Example 2.1) yields a language whose complement $shpf_\Sigma(S, \theta)$ *is* a CFL. However, we can find an example of a regular set S such that $shpf_\Sigma(S, \theta)$ is not a CFL.

Theorem 4. *Let Σ be an alphabet with $|\Sigma| \geq 3$. There exist a regular set of DNA trajectories $S \subseteq V_D^*$ and an morphic involution θ such that $shpf_\Sigma(S, \theta)$ is not a CFL.*

4.1   Regularity of Hairpin Languages

In the previous section, we have seen that for some regular set of DNA trajectories S and antimorphic involution θ, the associated hairpin language $shpf_\Sigma(S, \theta)$ is not context-free. If we restrict S to be slender, then we can guarantee that $shpf_\Sigma(S, \theta)$ is context-free for all antimorphic involutions θ. In this section, we consider restrictions on S which will guarantee that $shpf_\Sigma(S, \theta)$ (and $shp_\Sigma(S, \theta)$) is regular. Instead of further constraining S by beginning with slender sets of DNA trajectories, we look at relations on S that ensure regularity of $shpf_\Sigma(S, \theta)$.

First, say that a set $S \subseteq V_D^*$ is *b-finite* if $\max\{|s|_{\binom{b}{b}} : s \in S\} < \infty$. The following lemma is easily proven:

Lemma 5. *Let $S \subseteq V_D^*$ be a b-finite regular set of DNA trajectories. Then for all morphic and antimorphic involutions θ, $shp_\Sigma(S, \theta)$ is regular.*

We now define a partial order $\prec$ on words over V_D^*. This partial order will help us define a significant class of sets of DNA trajectories where the language $shp_\Sigma(S, \theta)$ is regular. Let $s_1, s_2 \in V_D^*$ with

$$\varphi_u(s_1) = \prod_{i=1}^{n} 1^{j_i} 0^{k_i}, \text{ and } \varphi_d(s_1) = \prod_{i=1}^{n} 1^{\ell_i} 0^{k_i},$$

for $n \geq 0$ and $j_i, k_i, \ell_i \geq 0$ for all $1 \leq i \leq n$. Then $s_2 \prec s_1$ if there exist $\alpha_1, \ldots, \alpha_n \in \{0,1\}^*$ such that the following three conditions hold:

(i) $\varphi_u(s_2) = \prod_{i=1}^{n} 1^{j_i} \alpha_i$ and $\varphi_d(s_2) = \prod_{i=1}^{n} 1^{\ell_i} \alpha_i$;
(ii) $|\alpha_i| = k_i$ for all $1 \leq i \leq n$; and
(iii) $\prod_{i=1}^{n} \alpha_i \notin 1^*$.

We note that $\prec$ is also used to investigate bond-free properties between separate single strands of DNA [13].

The situation is illustrated in Figure 4. The figure illustrates that if $s_2 \prec s_1$, then we can get from s_1 to s_2 by replacing a bonding region of length k_i in s_1 with a region which is not completely bonded, but is of length k_i, in s_2.

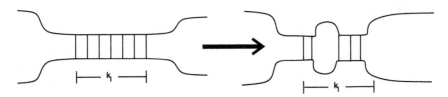

Fig. 4. A portion of s_1 is shown on the left, and a portion of s_2 is shown on the right

Example 2. Consider $s_1, s_2 \in V_D^*$ given by

$$s_1 = \binom{b}{b}\binom{b}{b}\binom{f}{\epsilon}\binom{b}{b}\binom{f}{f}, \quad s_2 = \binom{b}{b}\binom{f}{f}\binom{f}{\epsilon}\binom{f}{\epsilon}\binom{f}{f}\binom{\epsilon}{f}.$$

Note that $\varphi_u(s_1) = 00101, \varphi_u(s_2) = 01111, \varphi_d(s_1) = 0001$, and $\varphi_d(s_2) = 0111$. Thus, $s_2 \prec s_1$ holds with $\alpha_1 = 01$ and $\alpha_2 = 1$.

Note that Example 2 demonstrates that the relation $\prec$ is not simply defined by the idea "possibly replace $\binom{b}{b}$ with $\binom{f}{f}$". This is due to the equivalence of trajectories seen in the normal form of Lemma 2. However, the replacement intuition is formalized in the following result.

Proposition 2. *Let $\pi : V_D^* \to 2^{V_D^*}$ be the substitution defined by $\pi(x) = x$ if $x \neq \binom{b}{b}$ and $\pi(\binom{b}{b}) = \{\binom{b}{b}, \binom{f}{f}\}$. Then for all $S \subseteq V_D^*$,*

$$\left[\pi(S) \cap \left(V_D^* \binom{b}{b} V_D^*\right)\right] = \{x \in V_D^* : \exists s \in S, x \prec s\}.$$

We now define the minimal set of DNA trajectories with respect to $\prec$. For all $S \subseteq V_D^*$, let $\min(S) = \{s \in S : \forall t(\neq s) \in S, t \not\prec s\}$.

Example 3. Consider the k-hairpin languages (2):

$$S_k = \left\{ \left(\begin{smallmatrix} f \\ \epsilon \end{smallmatrix} \right)^* \cup \left(\begin{smallmatrix} \epsilon \\ f \end{smallmatrix} \right)^* \right\} \left(\begin{smallmatrix} f \\ f \end{smallmatrix} \right)^* \left(\begin{smallmatrix} b \\ b \end{smallmatrix} \right)^{\geq k} \left(\begin{smallmatrix} f \\ f \end{smallmatrix} \right)^* \left\{ \epsilon, \left(\begin{smallmatrix} f \\ \epsilon \end{smallmatrix} \right) \right\}.$$

Note that if we put $\min(S_k)$ in normal form, we get

$$\min(S_k) = \left\{ \left(\begin{smallmatrix} f \\ \epsilon \end{smallmatrix} \right)^* \cup \left(\begin{smallmatrix} \epsilon \\ f \end{smallmatrix} \right)^* \right\} \left(\begin{smallmatrix} f \\ f \end{smallmatrix} \right)^* \left(\begin{smallmatrix} b \\ b \end{smallmatrix} \right)^{k} \left(\begin{smallmatrix} f \\ f \end{smallmatrix} \right)^* \left\{ \epsilon, \left(\begin{smallmatrix} f \\ \epsilon \end{smallmatrix} \right) \right\}.$$

We now show that S and $\min(S)$ describe the same hairpin languages:

Theorem 5. *Let $S \subseteq V_D^*$. For all Σ and all morphic or antimorphic involutions θ, we have $shp_\Sigma(S, \theta) = shp_\Sigma(\min(S), \theta)$.*

Corollary 1. *Let $S \subseteq V_D^*$. If $\min(S)$ is a b-finite regular set of trajectories, $shp_\Sigma(S, \theta)$ is regular for all Σ and all morphic or antimorphic involutions θ.*

Example 4. Continuing with the previous example, we note that $shp_\Sigma(S_k, \theta)$ is regular for all morphic or antimorphic involutions θ. This was established by Kari *et al.* [3, Prop. 3].

Example 5. The k-scattered hairpin languages (see (3)) are given by

$$S_k = \left(\left\{ \left(\begin{smallmatrix} f \\ \epsilon \end{smallmatrix} \right)^* \cup \left(\begin{smallmatrix} \epsilon \\ f \end{smallmatrix} \right)^* \right\} \left(\begin{smallmatrix} f \\ f \end{smallmatrix} \right)^* \left(\begin{smallmatrix} b \\ b \end{smallmatrix} \right) \right)^{\geq k} \left(\begin{smallmatrix} f \\ f \end{smallmatrix} \right)^* \left\{ \epsilon, \left(\begin{smallmatrix} f \\ \epsilon \end{smallmatrix} \right) \right\}.$$

Note that if $\min(S_k) \subseteq S_k$ is put in normal form, we get

$$\min(S_k) = \left(\left\{ \left(\begin{smallmatrix} f \\ \epsilon \end{smallmatrix} \right)^* \cup \left(\begin{smallmatrix} \epsilon \\ f \end{smallmatrix} \right)^* \right\} \left(\begin{smallmatrix} f \\ f \end{smallmatrix} \right)^* \left(\begin{smallmatrix} b \\ b \end{smallmatrix} \right) \right)^{k} \left(\begin{smallmatrix} f \\ f \end{smallmatrix} \right)^* \left\{ \epsilon, \left(\begin{smallmatrix} f \\ \epsilon \end{smallmatrix} \right) \right\}.$$

Note that $\min(S_k)$ is b-finite and regular. Thus, $shp_\Sigma(S_k, \theta)$ is regular for all $k \geq 1$ (this was established by Kari *et al.* [2, Prop. 13(ii)]).

4.2 Finiteness of Hairpin Classes

We continue or investigation of conditions on S and θ that ensure the complexity of $shp_\Sigma(S, \theta)$ and $shpf_\Sigma(S, \theta)$ lie within a certain class of languages by considering conditions on S to ensure that $shpf_\Sigma(S, \theta)$ is finite. Kari *et al.* [3] have studied conditions which ensure finiteness of hairpin-free languages. (we note that Rampersad and Shallit [14] have independently established similar results).

Problems concerning finiteness of hairpin languages are sometimes related to problems in combinatorics on words. Using tools from the study of combinatorics on words, we can instantly conclude the finiteness of some scattered-hairpin languages by virtue of their coinciding with known unavoidable patterns. In particular, we use the results of Cassaigne [15], who gives a list of avoidability of patterns over 2- and 3-letter pattern alphabets, to derive finiteness results. These results are limited to the case where $\theta = \iota$, due to the emphasis on repetition of subwords in the study of combinatorics on words.

As an example, every sufficiently long word over any alphabet contains two occurrences of some letter. In terms of hairpins, we can phrase this equivalently as follows: the language $shpf_\Sigma(S, \iota)$ is finite for all Σ, where

$$ S = \left\{ \binom{f}{\epsilon}^* \cup \binom{\epsilon}{f}^* \right\} \binom{f}{f}^* \binom{b}{b} \binom{f}{f}^* \left\{ \binom{f}{\epsilon}, \epsilon \right\}. $$

Using the tools of avoidability, we can also conclude the following:

Lemma 6. *Let* $S_1 = \left(\binom{f}{\epsilon}^* \cup \binom{\epsilon}{f}^* \right) \binom{f}{f}^* \binom{b}{b}^+ \binom{f}{\epsilon}^+ \binom{b}{b}^+$. *The languages* $shpf_\Sigma(S_1, \iota)$ *are finite for all* Σ *with* $|\Sigma| \leq 2$.

Similarly, if $S_2 = \left(\binom{f}{\epsilon}^* \cup \binom{\epsilon}{f}^* \right) \binom{f}{f}^* \binom{b}{b}^+ \binom{\epsilon}{f}^+ \binom{b}{b}^+$, *the languages* $shpf_\Sigma(S_2, \iota)$ *are finite for all* Σ *with* $|\Sigma| \leq 2$.

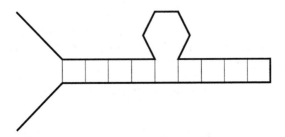

Fig. 5. Hairpins described by S_1 in Lemma 6

On the other hand, we can interpret the classic result of Entringer *et al.* [16] on avoidability of long squares in terms of hairpins:

Theorem 6. *Let* $S \subseteq V_D^*$ *be defined by* $S = \left\{ \binom{f}{\epsilon}^* \cup \binom{\epsilon}{f}^* \right\} \binom{f}{f}^* \binom{b}{b}^{\geq 3}$. *Then the language* $shpf_\Sigma(S, \iota)$ *is infinite if* $|\Sigma| \geq 2$.

Of course, there are both well-studied and novel problems in combinatorics on words and avoidability which cannot be expressed in terms of hairpins. However, the interaction between classical avoidability problems and hairpins is compelling, and the expressive power of hairpins suggests many problems, likely difficult, involving avoidability of patterns.

5   Decidability

Theorem 7. *Given an antimorphism* θ, *a regular set of DNA trajectories* S *and a regular language* L, *it is decidable whether* L *is* $shp_\Sigma(S, \theta)$-*free.*

For undecidability, we show that there exists a regular set of trajectories S such that determining whether context-free languages are $shp_\Sigma(S, \theta)$-free for morphic or antimorphic involutions.

Theorem 8. *There exists a fixed regular set of DNA trajectories S such that the following problem is undecidable: "Given an alphabet Σ, an antimorphic involution $\theta : \Sigma^* \to \Sigma^*$ and a CFL $L \subseteq \Sigma^*$ (resp., a morphic involution $\theta : \Sigma^* \to \Sigma^*$), is $L \subseteq shpf_\Sigma(S, \theta)$?"*

6   Conclusions

In this paper, we have considered modelling hairpin conditions on DNA words by using DNA trajectories. We have investigated the closure properties and decidability questions. In order to ensure positive closure properties, restrictions must be placed on the sets of DNA trajectories. In particular, if S is a slender regular set of DNA trajectories, then $shpf_\Sigma(S, \theta)$ is a context-free language for antimorphic involutions θ. On the other hand, for all regular sets of DNA trajectories S, the set $shp_\Sigma(S, \theta)$ is a context-free language for all antimorphic involutions θ.

To ensure regularity of scattered hairpin sets, we have considered a partial order $\prec$ and the minimal set of DNA trajectories with respect to $\prec$. Using the minimal set of DNA trajectories, we are able to ensure that $shp_\Sigma(S, \theta)$ is regular for all involutions for a large class of sets of trajectories, including those representing previously studied hairpin classes.

With respect to decidability, we have shown that hairpin-freeness of a regular language is decidable for regular set of trajectories and antimorphic involutions. However, there exists a fixed regular set of trajectories S such that it is undecidable, given an antimorphic involution and a context-free language L, whether or not L is $shp_\Sigma(S, \theta)$-free.

Acknowledgements. We thank the referees for their valuable comments.

References

1. Domaratzki, M.: Hairpin structures defined by dna trajectories. Technical Report TR-2006-001, Jodrey School of Computer Science, Acadia University (2006)
2. Kari, L., Konstantinidis, S., Losseva, E., Sosík, P., Thierrin, G.: Hairpin structures in DNA words. In Carbone, A., Daley, M., Kari, L., McQuillan, I., Pierce, N., eds.: The 11th International Meeting on DNA Computing: DNA 11, Preliminary Proceedings. (2005) 267–277
3. Kari, L., Konstantinidis, S., Sosík, P., Thierrin, G.: On hairpin-free words and languages. In Felice, C.D., Restivo, A., eds.: Developments in Language Theory: 9th International Conference. Volume 3572 of Lecture Notes in Computer Science., Springer (2005) 296–307
4. Rothemund, P., Papadakis, N., Winfree, E.: Algorithmic self-assembly of DNA Sierpinski triangles. PLoS Biol. **2**(12) (2004) e424
5. Kari, L., Konstantinidis, S., Sosík, P.: On properties of bond-free DNA languages. Theor. Comp. Sci. **334** (2005) 131–159
6. Rozenberg, G., Salomaa, A., eds.: Handbook of Formal Languages. Springer (1997)
7. Păun, G., Rozenberg, G., Salomaa, A.: DNA Computing: New Computing Paradigms. Springer (1998)

8. Mateescu, A., Rozenberg, G., Salomaa, A.: Shuffle on trajectories: Syntactic constraints. Theor. Comp. Sci. **197** (1998) 1–56
9. Domaratzki, M.: Trajectory-based embedding relations. Fund. Inf. **59**(4) (2004) 349–363
10. Jonoska, N., Kephart, D., Mahalingam, K.: Generating DNA code words. Congressus Numerantium **156** (2002) 99–110
11. Jonoska, N., Mahalingam, K.: Languages of DNA based code words. In Chen, J., Reif, J., eds.: DNA Computing, 9th International Workshop on DNA Based Computers. Volume 2943 of Lecture Notes in Computer Science, Springer (2004) 61–73
12. Szilard, A., Yu, S., Zhang, K., Shallit, J.: Characterizing regular languages with polynomial densities. In Havel, I., Koubek, V., eds.: Mathematical Foundations of Computer Science 1992. Volume 629 of Lecture Notes in Computer Science., Springer (1992) 494–503
13. Domaratzki, M.: Characterizing DNA bond shapes using trajectories. In Ibarra, O., Dang, Z., eds.: Developments in Language Theory. Volume 4036 of Lecture Notes in Computer Science, Springer (2006) 180–191
14. Rampersad, N., Shallit, J.: Words avoiding reversed subwords. J. Combin. Math. and Combin. Comput. **54** (2005) 157–164
15. Cassaigne, J.: Motifs évitables et régularités dans les mots. PhD thesis, Université Paris 6 (1994)
16. Entringer, R., Jackson, D., Schatz, J.: On nonrepetitive sequences. J. Combin. Theory. Ser. A **16** (1974) 159–164

Design and Simulation of Self-repairing DNA Lattices

Urmi Majumder, Sudheer Sahu, Thomas H LaBean, and John H Reif

Department of Computer Science, Duke University, Durham, NC, USA
{urmim, sudheer, thl, reif}@cs.duke.edu

Abstract. Self-repair is essential to all living systems, providing the ability to remain functional in spite of gradual damage. In the context of self-assembly of self-repairing synthetic biomolecular systems, recently Winfree developed a method for transforming a set of DNA tiles into its self-healing counterpart at the cost of increasing the lattice area by a factor of 25. The overall focus of this paper, however, is to develop *compact* designs for self-repairing tiling assemblies with reasonable constraints on crystal growth. Specifically, we use a special class of DNA tiling designs called *reversible* tiling which when carefully designed can provide inherent self-repairing capabilities to patterned DNA lattices. We further note that we can transform any irreversible computational DNA tile set to its reversible counterpart and hence improve the self-repairability of the computational lattice. But doing the transform with an optimal number of tiles, is still an open question.

1 Introduction

Currently, many scientists are in the process of substituting existing top-down techniques used in conventional manufacturing processes with bottom-up assembly techniques. This involves, among other things, developing self-assembly methods for patterning nano-materials as an alternative to using lithographic techniques. However, eventually, nanostructures can be damaged. What can we do when a self-assembled nanostructure is damaged?

1.1 The Challenge of Self Repairing Biomolecular Systems

This question leads us to realize that nature's capability to self-repair still far exceeds the self-healing capability of synthetic biochemical systems. As nanoscientists are building more complex systems at the molecular scale each day, this challenge of *Self-Repairing Biomolecular Systems* will become increasingly important. In fact, an interdisciplinary team at the University of Illinois at Urbana-Champagne has already developed a polymer composite that has the ability to self heal microcracks[24]. In the context of self-assembled nano-structures, such a system will provide a transition from the existing simple one-time assemblies to self-repairing systems, yielding capabilities that have broad impact to nano-engineering and provide numerous feasible practical applications.

C. Mao and T. Yokomori (Eds.): DNA12, LNCS 4287, pp. 195–214, 2006.

One interesting specific challenge in the area of self-repair to be addressed in this paper is to develop a molecular architecture for self-repairing memory. The interesting feature of this architecture is that, in spite of partial destruction of the nanostructure storing the memory, its bits can be restored.

1.2   Use of DNA Lattices to Demonstrate Self-repairing Processes

While the ultimate goal here is to build and experimentally demonstrate self-repairing capabilities for a variety of biomolecular systems, we need to begin first with well-understood chemistries. DNA has emerged as an ideal material for constructing self-assembled nanostructures because of its well defined structural properties, immense information encoding capacity and excellent Watson-Crick pairing. Exciting progress has been made on many frontiers of DNA self-assembled structures recently, especially in constructing *DNA Lattices* formed of DNA nanostructures known as *DNA Tiles*[4,3,2,6,1]. Thus we feel this provides an ideal platform on which self-repair at the molecular scale can be demonstrated.

1.3   Programmable Self-assembly and Self-repairability

One of the important goals of nanotechnology is to develop a method for assembling complex, aperiodic structures. Algorithmic self-assembly(AA) achieves this goal. AA has a strong theoretical foundation due to Winfree[4] but its experimental demonstration is quite limited by assembly errors, because while the theoretical model assumes a directional growth, crystals in reality can grow in all possible directions. This results in ambiguities at the binding sites. Consequently mismatch errors prevent further growth of the computational lattices. This is evident from the few experimental demonstrations of algorithmic assembly we have so far[17,18].

There have been several designs of error-resilient tile sets [9,7,10] that perform "proofreading" on redundantly encoded information [7] to decrease assembly errors. However, they too assume the notion of forward directional growth of the tiling lattice. Hence self-repair is not always feasible with such tile sets because of errors due to possible re-growth of the lattice in reverse direction.

Winfree[8], however, recently proposed an ingenious scheme that makes use of modified DNA tiles that force the repair reassembly to occur only in a forward manner. He converted an original tile set into a new set of self-healing tiles that perform the same construction at a much larger scale (5-fold larger scale in each direction and hence the new lattice requires a multiplicative factor of $5 \times 5 = 25$ more area) However, the much larger scale appears to make his construction more of theoretical interest than of practical use. The challenge is to limit the number of new tiles required, so that such a procedure can be applied in practice.

1.4   Our Paper's Results and Organization

The goal of this paper is to use a class of DNA tile sets with a certain property we call *reversibility* which will allow the reassembly of DNA tiles within a damaged lattice to occur in all possible directions without error with respect to at least two adjacent binding sites.

In section 2 of this paper we discuss how carefully designed reversible computations can improve self-repairing capability of the tiling with a specific instance called *Reversible XOR*. We observe that this lattice allows the first known molecular architecture for a self-repairing memory by storing bits in a two dimensional(2D) spatial domain which due to its self-healing properties is capable of restoring the bits in case of its partial destruction. We further introduce a new measure for computing the self-repairability of a tile set.

In section 3, we discuss two models for the extent of damage a DNA lattice suffers when acted upon by an external impulse in two different environments: 1)when it has a solid support and 2)when its free-floating in aqueous solution.

Section 4 of this paper provides DNA tile designs for RXOR tiling lattices. In the light of these designs, we discuss our experimental results with just the rule tiles. However, in order to have interesting patterns in the computational lattice we are currently designing a completely addressable nucleating structures for AA.

In section 5, we discuss techniques from theory of computation to transform irreversible CA to reversible CA that in theory improves the self-repairability of the corresponding computational DNA lattice. We also observe that doing the transformation with minimum number of tiles is still an unsolved problem.

2 Reversible Tiling Lattices and Their Self-repairing Properties

2.1 Reversible Computations and Reversible Tiling Lattices

A computation is said to be *reversible* if each step of the computation can be reversed, so that the computation can proceed in both forward or reverse manner. What is the relevance of reversible computation to the problem of self-repairing molecular assembly? Reversible computations have some unique properties, since they allow a partial computation to complete in a unique manner. A molecular assembly using DNA tiles can be viewed as a computation, where each step is executed by a given tile that is assembled adjacent to other previously assembled tiles. Each tile takes as its inputs the matching tiles of adjacent already placed tiles and provides as output the available free pads. Essentially, the tile computes an individual step mapping the values of its attached pads to the values of its still unattached pads. In general, forward-only tilings assume we are adding to a growing aggregate that started from within a concavity, and where further tiles can only be added to the lattice within the concavity. In contrast, some of the reversible tilings discussed here are also able to extend via tiles added to convex edges of the growing lattice. A careful tile design along with the reversibility property, allows a partially destroyed tiling lattice to be easily repaired, so long in the repair reassembly tiles are added with at least two adjacent matching binding sites. The reversible XOR tile set described just below is an interesting example of reversible self-assembly. It realizes a complex pattern that can achieve self-healing without increasing assembly time or number of tile types. In addition, such a self-healing assembly can act as a scaffold for other elements, for e.g. protein and would ensure self healing of the substance to which the self-assembled lattice acts as a scaffold. We now formally define self-repair in the context of self-assembly with square abstract tiles with four sticky ends on four sides, before discussing how reversibility can improve self-repairability.

Definition 1. *We call a tile set self-repairing, if any number of tiles are removed from a self-assembled aggregate to generate convex hole(s) such that all the remaining tiles still form a connected graph[1], then subsequent growth is guaranteed to restore every removed tile without error so long as repair reassembly happens with respect to at least two adjacent binding sites*

Note: This is a more restricted version of self-repairing tile set compared to the one that is described in [8]. Throughout the paper, we'll use this definition of self-repairing tile set and we also use the terms self-healing and self-repairing interchangeably.

2.2   The Reversible XOR Operation

We will now consider an interesting example of a reversible operation known as *Reversible XOR(RXOR)*.

The exclusive OR (known as XOR) operation takes as input two Boolean arguments (each can be *true(T)* or false (F)) and returns T if one, and only one, of the two inputs is true. Also, the XOR operation combined with one further simple Boolean operation (that does not alter values between input and output) makes the unit reversible and is known to provide a logical basis to do any Boolean computation. We call this *Reversible XOR(RXOR)*.

2.3   RXOR: A Family of Reversible Tiling Lattices

We describe a family of lattices called *RXOR lattices* that uses the XOR operation at each tile to form an interesting patterned lattice with reversible tiles. In the DNA tile implementation, each tile has two sticky ends on each side which is central to the lattice formation. Figure 1(a+b) gives the template and the set of rule tiles for one instance of reversible XOR.

Periodic and Nonperiodic Patterns RXOR Tiling Lattices. The figures in 1(c) illustrate some of the great variety of (periodic and nonperiodic) patterns that can be generated via RXOR operations at each tile. The rule tiles, the coloring scheme and the ideal lattice formed (when there are no errors) in each case is given in figure 1(c). (**Note:** *All the simulations assume a tile assembly model with $\tau = 2$, where τ is the number of binding sites for a tile to bind to the growing tiling lattice.*) The lattices with triangular patterns(figure 1(c):(ii)+(iv)) are interesting and complex, but it is difficult to determine errors in this lattice. The lattices

---

[1]   In the context of tile assembly, each tile is a vertex and the sticky end connections among the tiles denote the edges. An aggregate is connected if every tile can be reached from every other tile in the aggregate following the sticky end connections.

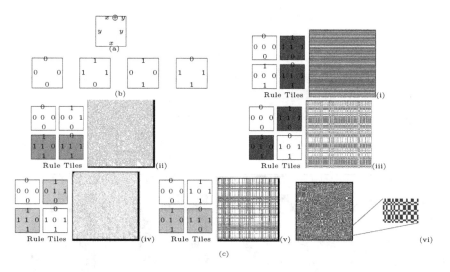

Fig. 1. (a)RXOR template, (b)Four Rule Tiles for RXOR, (c)Assembly of rule tiles *within a frame* defining the boundary of lattice growth according to aTAM:Rule tiles(left)+ resultant lattice(right):(i) propagation of the input y, (ii) propagation of the input x, (iii) propagation of input y but coloring of the tile based on the *xor* value in the tile, (iv) propagation of input x but coloring of the tile based on the *xor* value in the tile, (v)propagation of both inputs but coloring of the tile based on the *xor* value in the tile, (vi)Assembly of only the rule tiles, portion of error-free lattice(inset)

with band structure(figure 1(c):(i)+(iii)+(v)) are interesting since they can *redundantly store bits* in one/two dimension (by this we mean a bit is propagated through the linear lattice). In general such a $n \times n$ lattice can store n bits (by this we mean n bits are propagated through the $n \times n$ lattice)(figure 1(c):(i)+(iii)) and $2n$ bits as in figure 1(c):(v). Note that although figure 1(c):(ii)+(iv), demonstrate reversible computation they are not self-repairing. However, in figure 1(c):(i)+(iii) +(v) if some of the lattice tiles are removed, the self-repair will restore the lattice and preserve the bits. Error in the lattices from figure 1(c):(i)+(iii) occurs whenever there is a discontinuity in the horizontal bands. Error analysis of the lattice in figure 1(c):(v) is also quite simple(except when two blue bands intersect and the color reverses, any discontinuity in the band structure corresponds to a mismatch error).

Various RXOR Tiling Lattices with Errors. Although all the tiling lattices shown in figure 1 are reversible, we will use the tiling lattice given in Figure 1(figure 1(c):(v)) as the example RXOR lattice in our further discussions below of self-repair and experimental demonstrations.

As the AFM images of lattices in preliminary lab experiments did not have a frame, and self-assembly is error-prone, the lattice formed is not ideal. Hence, to estimate error rates, one can observe the largest portions of the lattice which are error-free[Figure 1(figure 1(c):(vi))].

(a)

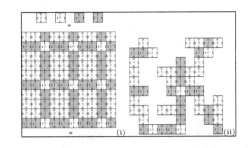

(b)

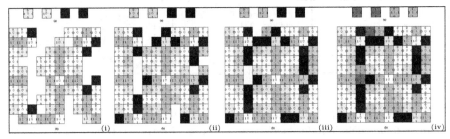

Fig. 2. a: Original self-assembled lattice(i) and damaged lattice(ii) **b**: A possible Self Healing Lattice Growth from (i) to (iv)

RXOR Tiling Lattices as an Example of Self-Healing Patterned Lattices. Previous work suggested that reversible self assembly can perform "proofreading" on redundantly encoded information[7]. Carefully designed RXOR is an interesting example of reversible self-assembly that achieves self-healing, without increasing assembly time or number of tile types as required by Winfree's Self-Healing construction [8]. For instance, consider the original 10×10 lattice in figure 2(a(i)). Suppose this lattice is damaged and the resulting structure looks like the one in figure 2(a(ii)). Since the tile set is self-healing, so one can recover the original lattice gradually. Ideally in the first step, all the tiles in the damaged lattice with at least two free binding site are available for attaching new tiles which are shown in different shades of the original color scheme[Figure 2(b(i))]. In the subsequent steps, the lattice grows further based on the newly incorporated tiles[Figure 2(b):(ii)+(iii)] and finally one obtains the original lattice[Figure 2(b):(iv)].

Use of RXOR Tiling Lattices to Redundantly Store/Copy Bits. Note that a row or a column in any rectangular window of the lattice in figure 1c(v) is entirely determined by a single cell. For instance, if a tile propagates 0 in the north-south direction and 1 in the east-west direction, then the corresponding column will have tiles with 0 in the north-south direction and the corresponding row will have tiles with 1 in the east-west direction. Thus a $m \times n$ lattice can store a total of $m + n$ bits and can be used as a self-healing memory because if damaged the $m \times n$ memory is capable of recovering all the $m + n$ bits as is shown in figure 2.

2.4   Reversibility Improves Self-repairability

In figure 3b we present three examples of computation in the increasing order of their reversibility. While computations for RXOR and Sierpinski Triangle pattern generation(ST) are self-explanatory, Binary Counter(BC) computation is reversible 50% of the time since out of the four possible combinations of the two inputs, we can retrieve them from the output sum and output carry only in two cases(when sum is zero and carry is either zero or one). However, in terms of self-repairability, RXOR tiling lattice completely self-heals, since for every possible open site in the latter lattice, there is a unique tile that can be bound to it given the constraints on crystal growth. But both BC and ST tiling lattices create ambiguity for tile attachment in a convex lattice site[Figure 3a]

Based on [27], we conclude that in general a tile set is self-healing, if the following constraints are satisfied. Let the inputs be x and y and the corresponding outputs be $f(x, y)$ and $g(x, y)$ with the arrangement of the input output ends in an abstract square tile starting from the north end in a clockwise direction as $g(x, y), x, y, f(x, y)$. The tile set is self-repairing if and only if

- if $f(x, y)$ is input sensitive to x if y is constant and $g(x, y)$ is input sensitive to y if x is constant and
- if both change then at least one of $f(x, y)$ or $g(x, y)$ also changes, the tile set is self-repairing

2.5   A Measure for Error-Resilience and Self-repairability

In general, when we design a tile set for algorithmic self-assembly it would be very useful if we can estimate its robustness against mismatch errors so long crystal growth occurs with respect to at least two adjacent binding sites. This also applies to the self-healing of a damaged lattice. Thus, we introduce a new measure for self-repairability of a tile set which we call "corner site ambiguity". This is inspired by [22] where the authors address the question of how the properties of a tile system are related to the periodicity of the resultant self-assembled nanostructures. We now define a *corner site* and *corner site ambiguity*

Definition 2. *A corner site is a pair of adjacent binding sites in a growing aggregate or in a convex hole in a damaged lattice.*

To reiterate, our abstract tiles are squares as in the original tile assembly model and the inputs are at the south and east ends. Thus the total number of possible corner sites with the number of tiles in the tile set T as w is $4w$.

Definition 3. *We define corner site ambiguity $C(T)$ as the average number of tiles in a tile set T that can bind to an available corner site.*

To measure $C(T)$, we compute the number of tiles that can bind to each corner site first and then compute the average.

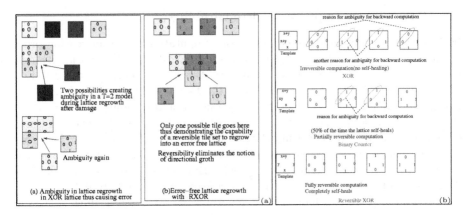

Fig. 3. (a)Concrete Self-healing comparison during lattice growth for completely reversible assembly and completely irreversible assembly, (b) Degree of reversibility comparison based on ambiguity of adjacent binding sites for the rule tiles of traditional assemblies

Corner Site Ambiguity and Self-Repairability. A tile set is self-repairing if and only if,

$$C(T) = \forall T, min\{C(T)\}$$

In other words, if all of the $4w$ corner sites are distinct, then exactly one tile can bind to it and hence $C(T) = 1$. In terms of concrete examples, the RXOR tile set has $C(T) = 1$ while each of ST and BC tile set has $C(T) = 1.25$. Obviously, the higher the value of $C(T)$, the more error-prone is the resultant assembly and re-assembly after lattice damage.

3   Models for Lattice Damage and Self-repair

Since we are yet to estimate the extent of damage that can happen in reality through concrete experiments, we developed a damage model to bridge the gap. Two different situations are considered: first when the lattice has a rigid support and second when its free floating in aqueous solution. Further we present a probabilistic model for self-repair given a damaged lattice.

3.1   Models for Lattice Damage

Probabilistic Model for Mechanical Damage on a Rigid Surface. Here we are concerned with the problem of estimating the extent of damage when the lattice with a rigid support is acted upon by an external impulse, for e.g damage created when the AFM tip hits the DNA lattice lying on mica during imaging . We model the lattice as a crystal as in [12]. Thus the force F_1 in a tile located at a distance of r from the tile receiving the impulse is proportional to $\frac{1}{\sqrt{r}}$. F_2 is the resistive force from the sticky end connections of the tiles. So long for any tile $F_1 > F_2$, the probability that a tile gets knocked off the lattice is greater than zero given that the shock wave has traveled to it from origin of damage. To estimate the fraction of the lattice damaged, we first compute the probability of a damage path of length i. The latter is defined as a path that originates in the tile which is directly hit by the tip say O, meanders outwards through its successors $< S_1, S_2, S_3, \ldots S_{i-1} >$ and stops at S_i. Here each of $O, S_1, S_2, S_3, \ldots S_{i-1}$ are knocked off the lattice except for S_i[Figure6(a) in the Appendix]. Let us denote the probability of the damage path that stops at S_i by $P(i)$. Now S_i is located at a Manhattan distance of i from O. As F_1 drops off as $\frac{1}{\sqrt{r}}$, so the probability that a tile will fall off, given that at least one of its neighbors is already knocked off, is given by $\frac{p}{\sqrt{r}}$ where $0 < p < 1$ is a function of F_1 and can be evaluated from the probability distribution for the damage path. Then,

$$P(i) = \frac{p^i}{\sqrt{i-1}!}(1 - \frac{p}{\sqrt{i}}) \tag{1}$$

Since $P(i)$ is a probability mass function, thus

$$\sum_{i=1}^{l} P(i) = 1 \tag{2}$$

Now we can estimate the expected fractional damage size $D(n, l)$ for a lattice L with n tiles, by summing the probabilities of a damage path from O to each S_i in the lattice. For ease of computation, we achieve this by calculating the number of tiles at a Manhattan distance of i from O, $\forall i, 1 \ldots l$.

$$D(n, l) = \frac{(1 - p) + \sum_{i=1}^{l} 4i \times P(i)}{n}$$

The first term account for the event when the damage path probabilistically stops at O and $4i$ is the number of tiles located at a Manhattan distance i from O. The simulation results are discussed in the appendix.

Model for Damage in a DNA Lattice in Aqueous Solution. Here we model the rectangular DNA lattice of size $m \times n$ as a simple mass spring system similar to cloth dynamics in computer graphics[23]. Here each tile is positioned at grid point (i, j), $i = 1, 2, \ldots, m$ and $j = 1, 2, \ldots n$. For simplicity, the external mechanical impulse F hits the lattice at a single tile location and the internal tension of the spring linking tile $T_{i,j}$ with each of its neighbor $T_{k,l}$ is given by $F\_int_{i,j} = -\sum_{(k,l) \in R} K_{i,j,k,l}[l_{i,j,k,l} - l^0_{i,j,k,l} \frac{l_{i,j,k,l}}{||l_{i,j,k,l}||}]$ where R is the set of $T_{i,j}$'s neighbors such that $T_{i,j}$ is linked to $T_{k,l}$ for all (k, l) in the neighbor set, $l_{i,j,k,l} = \overrightarrow{T_{i,j}T_{k,l}}$, $l^0_{i,j,k,l}$ is the natural length of the spring linking tiles $T_{i,j}$ and $T_{k,l}$ and $K_{i,j,k,l}$ is the stiffness of that spring. The value of $K_{i,j,k,l}$ is dependent on the type of bases that are involved at the sticky end match for the pair of tiles. We however, assume that the stiffness K and the unextended spring length l^0 are the same for every pair of adjacent tiles.

Since it is difficult to obtain a closed form solution for the expected number of tiles that gets knocked off the lattice because of the impulse, we outline a simple simulation algorithm for estimating the number of tiles removed when an impulse hits the lattice. The basic idea is as follows: Treat the lattice as a connected graph where the state of the tiles is updated every Δt time. To that extent, start a breadth first search from the tile receiving the original impulse. The tile under consideration moves in Δt extending the springs with its neighbors. If the extension is beyond some threshold value, the tile snaps off the lattice and there's no change of state for its neighbors. Otherwise, at the end of the Δt interval, a component of the spring force pulls each of the neighbor in the same direction as this tile while the same spring force tries to make this tile go in the opposite direction. Continue the breadth first search and update the state of all the tiles in the lattice. Repeat this whole set of events at every Δt time until the lattice stabilizes or has completely fallen apart.

Obviously, as $\Delta t \rightarrow 0$ we simulate the real situation. Some of the important assumptions we make are follows: Δt is negligibly small, such that the velocity of each tile remains constant in this interval. Further, it is difficult to model collisions with water molecules, so we assume a damping force on each tile with the damping constant of water α. The pseudocode for this algorithm is given in the appendix.

The model captures the overall behavior of a free floating lattice when acted upon by an external impulse, but it is rather sensitive to the parameter values such as the stiffness constant of the spring and the fracture threshold. Another important drawback is that our model does out consider tiles colliding spring extension/contraction. This changes their relative velocity and may affect the total number of tiles that actually get knocked off the lattice.

3.2   Probabilistic Model for Self-repair

Given a damaged lattice it is useful to estimate the likelihood of the lattice being self-healed. The probability that an aggregate of size n will be formed is given by $1 - 2ne^{-G_{se}}$[13] according to the kinetic Tile Assembly Model(kTAM). The same expression can be used for computing the extent of self-repair. Hence the probability that a damage of size n will be self-healed is $1 - 2ne^{-G_{se}}$.

4   DNA Implementation of Self-repairing Lattices

4.1   The Computational Tiles

We used *cross tile*, sometimes also called a 4 x 4 tile [6] for the physical implementation of our RXOR lattice. This tile contains 4-armed branch junction oriented in four directions in the lattice plane. [6] developed a method for further programming the assembly of these DNA tiles to form large extents of highly regular lattices which cancels out imperfections in individual tiles by a "corrugation" technique[Figure 4]. The sticky ends implementation is also given in figure 4. Two RXOR tiles, $RXOR_3$ and $RXOR_4$ were modified by incorporating a biotin group into a T4 loop at the center of the tile structure for visualizing an output of 1 bit. Streptavidin was then added to the solution of the self-assembled

RXOR lattice. Consequently, the interaction of streptavidin-biotin led to binary streptavidin patterns that can be readily imaged. Details of the experimental protocols is included in the appendix.

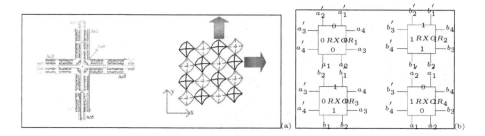

Fig. 4. (a)Corrugated cross Sequence Design and Lattice, (b)Corrugated RXOR Tiles with sticky ends. x' is the Watson Crick complement of x.

4.2   AFM Imaging of our Self-repairing Lattices

Our immediate goal was to verify the correct formation of the original lattice. We observed some micron sized lattices but they are probably formed by the self-association of either all zero tiles or all one tiles(both are without unary markers and are capable of forming nanogrids)[Figure 5](a+b). Although both patterns are valid in our computational lattice, as is evident from figure1c(v), we would prefer to observe some more interesting structures such as crisscrossing bands. This can be ensured through the use of a pre-assembled boundary for the computation[Figure 5c].

We also observed single tile mismatches. Our model assumes that tiles will attach with respect to at least two adjacent binding sites. Although in theory this can be achieved by manipulating the physical conditions, in reality, tiles still bind with respect to one matching binding site. One possible mechanism to deal with this will be the compact error-resilient design in [9], although it is likely to introduce new ambiguities.

4.3   Completely Addressable Nucleating Structure for Algorithmic Assembly

The ultimate goal of boundary formation in programmable assembly is to provide input for subsequent two dimensional growth. Until now we have been able to form specific structures but we have not been able to control their polymerization. Schulman, et al. demonstrated in [11] it is difficult to create large uniformly sized "V" and "X" boundaries with their DX tile[29] set. However, Pistol, et al. recently demonstrated [25,20] a $140nm \times 140nm$ fully synthetic and programmable 64-motif DNA nanostructure using cross tile. In the light of these findings, our current research focus involves meeting halfway between algorithmic assembly and addressable assembly with a fully programmable multi-tile seed and algorithmic rule tiles. Full addressability in a multi-tile system will guarantee it forms correctly and would also make our search for correct structures easier during AFM imaging. Using hierarchical assembly, a nanotrack of *AB* tile system[26] can further assemble onto the seed and provide an arbitrary combination of zero-one inputs to the computational tiles.

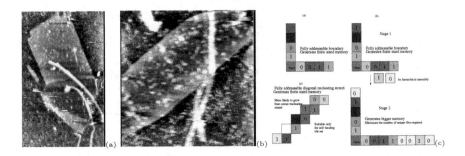

Fig. 5. a)Computational Lattice, $5\mu m \times 2.5\mu m$ formed by the association of zero tiles, b)Computational Lattice, $2\mu m \times 2\mu m$ formed again by the association of zero tiles. This reveals the beneficial effect of corrugation but the computational lattice lacks interesting band structure, c)Pre-assembled Nucleating Structure Design:Use of fully addressable lattice design

5   Self-repairing Transformation for any Tile Set

One of the major goals of algorithmic self-assembly is to provide compact designs for self-repairing error-resilient tile set. In this paper we demonstrated that a carefully designed tile set performing reversible computation can be self-repairing. Unfortunately, not all reversible tile sets are self-repairing. However, since reversibility ensures uniqueness for the adjacent pair of outputs, it definitely does improve the self-repairability of the tile set. So transforming an irreversible tile set into a reversible tile set improves its error-resilience. In fact we can show that reversible tiling is Turing Universal and thus any tile set will benefit from such a transformation.

It was shown by Winfree[14] that a two dimensional tiling lattice can be used to simulate any 1 dimensional Cellular Automata(CA). We can similarly prove that if the given CA is reversible, then the resulting tiling lattice is reversible.

Theorem 1. *A two dimensional DNA lattice with reversible tiles and size $n \times T$ can be used to simulate a one dimensional reversible CA with n cells running in time T.*

The proof of this theorem is given in the appendix.

Morita in [21], outlines a transformation technique for converting a 1D 3 neighbor CA to a 1D 3 neighbor partitioned CA. We can directly map this technique to DNA self-assembly.

If we restrict ourselves to one function instead of two, as in ordinary CA, one can have an abstract tile which has six sticky ends, three for inputs and three for outputs. In case of the ordinary CA, the three inputs denotes the values of the cell and its left and right neighbors while the three outputs contain the same output value which serve as inputs to the triplet directly above in the next computation step in a $\tau = 3$ Abstract Tile Assembly Model[13][Figure 9a in the appendix].

In case of a partitioned CA, the inputs are right output of the left cell, center output of the middle cell and the left output of the right cell directly below(corresponds to the previous computation step) and the three outputs are respectively the left, center and right outputs of the cell. The latter serve as the right input of the left cell, the center input of the middle cell and the left input of the right cell directly above(corresponds to the next computation step)[Figure 9b in the appendix].

Theorem 2. *A 2D DNA lattice performing reversible computation and size $(2n + 7)T$ can simulate a DNA lattice performing irreversible computation and size nT*

Proof. Morita in [21] further showed that the number of steps $T(n)$ to simulate one step of an ordinary 1D 3 neighbor CA A by a partitioned D 3 neighbor CA P is $2n + 7$ in the worst case. When combined with the previous theorem, this implies the result.

One observation here is that, although the size of the transformed tile set is still asymptotically the same as before(A CA with alphabet C and $|C|^3$ rules will have $O(|C|^3)$ for its reversible counterpart crystal growth in a $\tau = 3$ model), the blow up for all practical purposes is fairly high. For instance, if we consider a binary alphabet then a tile set with 8 tiles yield a set of 58 tiles in its reversible counterpart. Accommodating this seven fold increase in a biomolecular implementation is not very practical yet.

This motivates us to investigate redundancy based self-repairing schemes where redundancy is created by encodings in the pads of the tiles with no scale up of the assembly. Unfortunately this direction is no more promising.

Theorem 3. *There is no compact reversible transformation to generate a self-healing tile set for any irreversible computation using redundancy based scheme.*

Proof. Following the notation in [27], in order to make any tile set reversible we can incorporate the inputs at the corresponding output ends. In order to maintain consistency on the choice of inputs, we need to transform the original tile $V(i, j + 1), U(i, j), V(i, j), U(i + 1, j)$(from north in a clockwise fashion)[Figure 10a in the Appendix] to $V(i, j+1), V(i, j), V(i-1, j)$ at the north end, $U(i, j), U(i-1, j), U(i-1, j-1)$ at the east end, $V(i, j), V(i, j-1), V(i-1, j-1)$ at the south end and $U(i + 1, j), U(i, j), U(i, j - 1)$ at the west end[Figure 10b in the Appendix] . However, then the $U(i, j)$ and the $V(i, j)$ at the input ends become a function of the rest of the inputs and hence this tile does not represent a valid computational unit. Further, even if we accept the dependency the transformation incorporates 4 sets of computation in the new tile as opposed to one in the original tile and reversibility only helps for a single level[Figure 10c in the Appendix] . Thus for instance say keeping the values of $(U(i-1,j), V(i,j-1))$ constant, two choices for the tuple $(U(i-1,j-1), V(i-1,j-1))$ yield the same output $(U(i,j-1), V(i-1,j))$. Now the latter are input to the $(i-1,j)$ and $(i,j-1)$ original modules. Since the values of $(U(i-1,j), V(i,j-1))$ pair is constant, thus the outputs from the (i,j) original module will also be the same and hence such a construction will not be self-repairing.

Thus it still remains to be answered whether an irreversible tile set can be transformed to its reversible counterpart using a minimal tile set. However, reversible computation has its own merits in quantum computing, optical computing, nanotechnology and low power CMOS design. In fact, if we can have 3D DNA assembly, that would allow us to propagate the redundant bit in the third dimension, we can hope to improve the self-repairability of the resultant assembly given that crystal growth occurs with respect to at least *three* adjacent matching binding sites.

6   Discussion

Although molecular self-assembly appears to be a promising route to bottom-up fabrication of complex objects, to direct growth of lattices, error-free assembly cannot be assumed. Thus in this paper with our compact design of self-repairing tile sets, we addressed the basic issue of fault tolerant molecular computation by self-assembly. Our design exploited the reversibility property to provide inherent self-repairing capabilities with some constraints on crystal growth. We described the detailed design of DNA cross tiles for a particular reversible tiling lattice called the RXOR lattice. We observed that this lattice will allow the first known molecular architecture for self-repairing memory. We further discussed models for lattice damage when acted upon by external impulses and presented some experimental results with DNA tiles. Finally we observed that although in theory we can construct 2D reversible computational DNA lattices for 1D irreversible CAs and hence improve the self-healing capability of resultant computational lattice. Doing the transformation, however, with a minimal tile set is still an open question.

Acknowledgments

The authors are supported by NSF EMT Grants CCF-0523555 and CCF-0432038. We thank all our reviewers for their very useful comments and suggestions. We also thank Prof. Jie Liu, Chemistry Department, Duke University, for providing access to a Nanoscope IIIa AFM. Majumder also thanks Hanying Li for her help with the experimental protocols and Erik Halvorson for many stimulating discussions and immense support.

References

1. D. Liu, M. S. Wang, Z. X. Deng, R. Walulu, and C. D. Mao, *J. Am. Chem. Soc.*, 126:2324–2325, 2004.
2. T. H. LaBean, H. Yan, J. Kopatsch, F. Liu, E. Winfree, J. H. Reif, and N. C. Seeman, The construction, analysis, ligation and self-assembly of DNA triple crossover complexes, *J. Am. Chem. Soc.*, 122:1848–1860, 2000.
3. C. Mao, W. Sun, and N. C. Seeman, *J. Am. Chem. Soc.*, 121:5437–5443, 1999.
4. E. Winfree, F. Liu, L. A. Wenzler, and N. C. Seeman, Design and self-assembly of two-dimensional DNA crystals, *Nature*, 394(6693):539–544, 1998.
5. H. Yan, T. H. LaBean, L. Feng, and J. H. Reif, Directed nucleation assembly of DNA tile complexes for barcode patterned DNA lattices, *Proc. Natl. Acad. Sci. USA*, 100(14):8103–8108, 2003.
6. H. Yan, S. H. Park, G. Finkelstein, J. H. Reif, and T. H. LaBean, DNA-templated self-assembly of protein arrays and highly conductive nanowires, *Science*, 301(5641):1882–1884, 2003.
7. Erik Winfree and Renat Bekbolatov, Proofreading tile sets: Error correction for algorithmic self-assembly, *DNA Computing* , 2943:126-144
8. Erik Winfree Self Healing Tile Sets *Nanotechnology: Science and Computation, pages 3-21, 2006*
9. J H Reif, S Sahu and P Yin, Compact error-resilient computational DNA tiling assemblies, *Tenth International Meeting on DNA Based Computers (DNA10)*, 2004
10. HL Chen and A Goel, Error Free Self-Assembly using error-prone tiles, *DNA Computing 10*, 2004
11. R Schulman, E Winfree, Controlling nucleation rate in algorithmic self-assembly, *DNA Computing 10*, 2004
12. Hertzberg R, Deformation and Fracture Mechanics of Engineering Materials, John Wiley and Sons, NY 1996
13. Winfree E, Simulations of Computing by Self-Assembly, Caltech CS Tech Report 1998.22
14. Winfree E, On the Computational Power of DNA Annealing and Ligation, DNA Based Computers, pgs 199-221, 1996
15. Matthew Cook, Paul W.K. Rothemund and Erik Winfree, Self-Assembled circuit patterns, DNA Computers 9 LNCS volume 294:91-107, 2004
16. Toffoli T Reversible Computing, Automata, Languages and Programming, Springer Verlag, pp.632-644
17. Rob D. Barish, Paul W. K. Rothemund, and Erik Winfre,e Two Computational Primitives for Algorithmic Self-Assembly: Copying and Counting, Nano Letters 5(12): 2586-2592
18. Paul W.K. Rothemund, Nick Papadakis and Erik Winfree. Algorithmic Self-Assembly of DNA Sierpinski Triangles, PLoS Biology 2 (12) e424, 2004
19. Paul W. K. Rothemund, Folding DNA to create nanoscale shapes and patterns, Nature, 440:297-302,2006
20. S. H. Park, C. Pistol, S. J. Ahn, J. H. Reif, A. R. Lebeck, C. Dwyer and T. H. LaBean Finite-size, Fully-Addressable DNA Tile Lattices Formed by Hierarchical Assembly Procedures, Angewandte Chemie, 45:735-739, 2006
21. Morita K, Reversible simulation of one-dimensional irreversible cellular automata, Theoret. Comput. Sci., 148:157-163, 1995
22. Hönberg B and Olin H, Programmable Self-Assembly-Unique Structures and Bond Uniqueness, J. Comput. Theor. Nanosci. 2006, Vol 3, 1-7, 2006

23. Provot X, Deformation Constraints in a Mass Spring Model to Describe Rigid Cloth Behavior, In Proc. Graphics Interface 95, pp 147-154. 1995.
24. S.R. White, N.R. Sottos, P.H. Geubelle, J.S. Moore, M.R. Kessler, S.R. Sriram, E.N. Brown and S. Viswanathan, Autonomic healing of polymer composites, Nature, 409, 794-797, 2001
25. C. Pistol, A. R. Lebeck and C. Dwyer, Design Automation for DNA Self-Assembled Nanostructures, Proceedings of the 43rd Design Automation Conference (DAC), July 2006.
26. S-H. Park, P. Yin, Y. Liu, J.H. Reif, T.H. LaBean and Hao Yan, Programmable DNA Self-assemblies for Nanoscale Organization of Ligands and Proteins, Nano Letters 5, 729-733, 2005
27. Sudheer Sahu and John Reif, Capabilities and Limits of Compact Error Resilience Methods for Algorithmic Self-Assembly in Two and Three Dimensions, Twelfth International Meeting on DNA Based Computers (DNA12), Seoul, Korea, June 5-9, 2006.
28. Seeman, N. C, J. Biomol. Struct. Dyn. 1990, 8, 573.
29. E Winfree, F Liu, LA Wenzler and NC Seeman, Nature, Vol: 394, 1998

A   Appendix

A.1   Damage Model

Pseudocode of Algorithm for Computing Damage in a DNA Lattice Freely Floating in Aqueous Solution due to Mechanical Impulse. We estimate the number of tiles m that are knocked off the lattice of size n when an impulse F hits the lattice. The mass of each tile is μ, the velocity at time $\tau = 0$, $v_{i,j}$ of each tile $T_{i,j}$ except for the one which receives the impulse is zero and the threshold extension for fracture is d_{thres}. We'll use a queue Q for efficiency purposes and below is the pseudocode for the algorithm is given in section 3.1.
Program estimateDamage($F, n, \Delta t$)

> Choose with probability $\frac{1}{n}$, a tile $T_{p,q}$ for the initial impulse F
> $\forall i, j, color[T_{i,j}] = white$
> Compute velocity of $T_{p,q}$ as $v_{p,q} = \sqrt{\frac{2F\Delta t}{\mu}}$
> Enqueue($Q, T_{p,q}$)
> Repeat until $m = n$ or $v_{i,j}, \forall i, j$ is zero
> > While $Q \neq \phi$
> > > do $T_{a,b}$ =Dequeue($Q, T_{a,b}$)
> > > For each neighbor $T_{k,l}$ of $T_{a,b}$ and $color[T_{k,l}] = white$
> > > > $color[T_{k,l}] = gray$
> > > > Compute extended length $d_{a,b,k,l} = \sqrt{|v_{a,b}\Delta t|^2 + (l^0)^2}$
> > > > Enqueue($Q, T_{k,l}$)
> > > $color[T_{a,b}] = black$
> > > If $min_{k,l}\{d_{a,b,k,l}\} >= d_{thres}$
> > > > $m = m + 1$
> > > > Remove $T_{a,b}$
> > > else

Compute $v_{a,b} = v_{a,b} + \frac{(F\_int_{a,b,k,l} - \alpha v_{a,b})\Delta t}{\mu}$

For each neighbor $T_{k,l}$ of $T_{a,b}$ and $color[T_{k,l}] = gray$

Compute $F\_int_{a,b,k,l} = K[d_{a,b,k,l} - l^0]$

Compute velocity $v_{k,l}$ of each neighbor $T_{k,l}$ as $v_{k,l} = v_{k,l} +$
$\frac{(F_int_{a,b,k,l} - \alpha v_{k,l})\Delta t}{\mu}$

$\tau = \tau + \Delta t$

return m

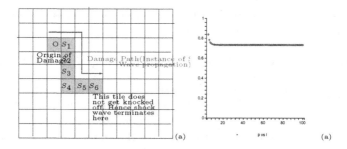

(a)   (a)

Fig. 6. (a)Instance of shock wave propagation and creation of damage path, (b)Estimation of p

Damage of a Lattice with Rigid Support due to a Mechanical Impulse

Simulation Results. Based on equation 1 and 2, we can solve for p and as the plot in figure6(b) reveals that the value of p stabilizes to 0.7316 beyond a relative hitting force of 10. Even the drop from an initial value of 0.87 to 0.75 occurs before the relative hitting force even reach a value of 5. So for all practical purposes we consider the value of p to be 0.73. Note that l is determined by the distance where F_1 and F_2 equalize. In essence, if $F_1 = \frac{c}{\sqrt{r}}$, where c is a constant,

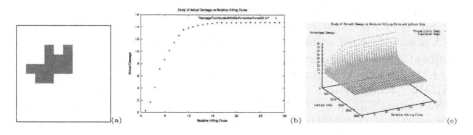

(a)   (b)   (c)

Fig. 7. a)Snapshot of a computer simulation of a damage(red region) by an AFM tip in a rectangular lattice(white region), b)Plot of Actual Damage vs relative hitting force of the AFM. The plot reveals the pseudo-geometric nature of the probability distribution of damage, c)Plot of percentage damage as a function of lattice size and relative hitting force.

then $l = (\frac{c}{F_2})^2$. So l is a measure of the relative hitting force and we will address it the same in all our plots.

A simulation snapshot for damage by an AFM tip in the lattice is shown in Figure 7a. We study the effect of relative hitting force on the lattice in isolation. Since the probability model is pseudo-geometric in nature, so the amount of actual damage reaches a constant value beyond a value of 10 for the former[Figure 7b]. We also compared the average fractional damage size from several simulation runs with the expected damage size as estimated from the probabilistic model as a function of relative hitting force and lattice size. They seem to agree well, as is evident from Figure 7c. The verification of the model by comparisons with experimental data is yet to be done.

A.2   Physical Implementation of RXOR Tiling with DNA Tiles

Experimental Protocols. Complex design, sample preparation, and stepwise assembly: The design of cross-tiles was based on the structure of immobile 4-arm branched junctions. The subsequence used for all bulged loops was four Ts, TTTT. Sequences were designed to minimize the chance of undesired complementary association and sequence symmetry[28]. Synthetic oligonucleotides were purchased from Integrated DNA Technologies (Coralville, IA) and purified by polyacrylamide gel electrophoresis (PAGE). Complexes were formed by mixing a stoichiometric quantity of each strand in physiological buffer, $1 \times TAEMg^{2+}$ (40 mM Tri acetate (pH 8.0), 2 mM EDTA, and 12.5 mM magnesium acetate). The final concentration of DNA was between 0.125 and 1.0μM. For the first step high-temperature annealing of unit tiles, equimolar mixtures of strands were cooled slowly from $95°C$ to $20°C$ by placing the tubes in 2 L of boiled water in a styrofoam box for at least 40 hours to facilitate hybridization. For the second step low-temperature annealing, DNA tiles mixtures were cooled slowly from $42°C$ to $20°C$ by placing the tubes in 1 L of water at room temperature for about 4 hours. After each step of annealing, samples were incubated overnight at $4°C$ before AFM imaging.

Streptavidin attachment: After 2nd-step DNA RXOR assembly, add streptavidin purchased from Rockland (www.rockland-inc.com, code no: $S000 - 01$, Lot no: 12088) of same volume of equal concentration of biotin in annealed DNA sample. Leave it an hour at room temperature, and then incubate $4°C$ for overnight before AFM experiment.

AFM imaging: AFM imaging was performed in tapping mode under $1 \times TAEMg^{2+}$ buffer. A $5\mu L$ annealed-sample was dropped on freshly cleaved mica for 3 minutes. $30\mu L$ of $1 \times TAEMg^{2+}$ buffer was then placed onto the mica and another $30\mu L$ of $1 \times TAEMg^{2+}$ buffer was placed onto the AFM tip. AFM images were obtained on a Digital Instruments Nanoscope IIIa with a multimode fluid cell head by tapping mode under buffer using a NP-S oxide-sharpened silicon nitride tips (Vecco).

A.3   Self-repairing Transformation for Any Tile Set

Proof of Theorem 1

Proof. Following Winfree[14]. we use a special kind of CA : *Blocked Cellular Automaton* where for each row, cells are read in pair and two symbols are written guided by the rule table. For each rule, $(x, y) \longrightarrow (u, v)$, we create a tile whose sticky ends on the input side(south and east sides of a cross tile) are $\overline{x}$ and $\overline{y}$ and that at the output ends(west and north sides of a cross tile) are u and v. We also have an initial assembly of tiles simulating the initial BCA tape. We add the rule tiles to the solution containing the initial tape. As figure 8 demonstrates, rule tiles anneal into position if and only if both sticky ends match. Thus we can simulate forward computation with DNA assembly. As described in [14] we access the output using a special "halting" tile gets incorporated in the lattice.

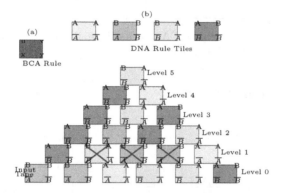

Fig. 8. (a)BCA rule in an abstract DNA tile,(b)DNA Rule Tiles for RXOR Computation, (c)Both forward and backward computation takes place starting from the initial tape

For a reversible CA, by definition, there's exactly one prior configuration for every current configuration. In terms of DNA assembly this implies that if we treat u and v as our inputs(north and west ends of a cross tile) and $\overline{x}$ and $\overline{y}$ as our outputs(south and east ends of the cross tile), then also the rule tiles will anneal into position abiding by the sticky ends match constraint. Thus we can simulate backward computation with reversible DNA assembly. For instance, if we remove the tiles which are crossed in Figure 8, since the two functions are invertible, so the correct rule tiles will reassemble, thus demonstrating reversible computation.

In particular, each horizontal row of n tiles of the tiling lattice simulates a given step of the CA, and the values from each automata step to step are communicated from a prior row of tiles to the next row of tiles above it. Thus, a two dimensional DNA lattice with reversible tiles and size $n \times T$ can be used to simulate a one dimensional reversible CA with n cells running in time T.

Figure for the reversible transformation according to [21]

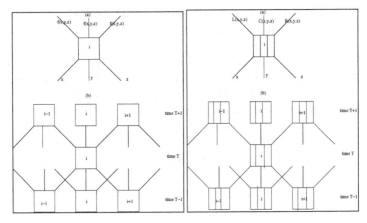

Fig. 9. a)Abstract Tile for original CA and its tiling simulation, b)Abstract Tile for corresponding PCA and its tiling simulation

Figure for Theorem 3

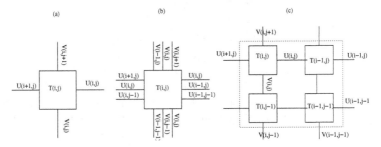

Fig. 10. a)Original Abstract Tile according to [27], b)Transformed Abstract Tile capable of reversible computation, c)A single computational unit in the transformed tile corresponds to four sets of computation in the original tile

On Times to Compute Shapes in 2D Tile Self-assembly

Yuliy Baryshnikov[1], Ed Coffman[2], and Boonsit Yimwadsana[2]

[1] Bell Labs, Lucent Technologies, Murray Hill, NJ 07974
`ymb@research.bell-labs.com`
[2] Department of Electrical Engineering, Columbia University, NY 10027
`{egc, teddy}@ee.columbia.edu`

Abstract. We study the times to grow structures within the tile self-assembly model proposed by Winfree, and the possible shapes that can be achieved during the self-assembly. Our earlier work was confined to the growth of rectangular structures, in which the border tiles are prefabricated. By varying the relative rates between the border-tile and rule-tile attachment, one can engineer interesting new shapes, which have been observed in the laboratory. We show that the results from an extension of our earlier stochastic models agree remarkably closely with experimental results. This is an important further demonstration of the validity and usefulness of our stochastic models, which have also been used successfully in studies of error correction in DNA self assembly.

1   The Tile Self-assembly Model

The focus of the work here is on mathematical foundations of self assembly based on Winfree's DNA tile model [1] which will be described shortly. More precisely, the emphasis is on the analysis of stochastic models. Although such insightful models and reference theories are ubiquitous in the physical sciences, they remain a fertile ground for self-assembly research in DNA-Based Computing, where stochastic analysis has only recently begun. The early work of Adleman [2] and colleagues, and that of the authors [3, 4, 5] sets the stage in this area, and serves as the point of departure for the analysis here.

The seminal mathematical *tile* model of DNA self-assembly, as developed by Winfree [1] and pursued by many others, has led to a much improved understanding of DNA self-assembly in two dimensions. At the physical layer being modeled, single-strand DNA molecules are manipulated to form complex DNA molecules (e.g., *double-crossover* molecules [6, 7]) which are designed to assemble (bond) with other such molecules in a two-dimensional crystal-growth process obeying bonding rules determined by the molecular motifs [1]. These building-block molecules are modeled as tiles.

There are three types of tiles involved in a self-assembly process: seed tiles, border tiles, and rule tiles. They participate in a growth process beginning at the origin of the positive lattice where the seed tile is located; the rest of the unit squares of the lattice are the potential sites occupied by tiles.

– The seed tile occupies the lower-left corner of the positive lattice (the origin) and is responsible for initiating the tile self-assembly process. Only border tiles, as described next, can stick to the two free (upper and right-hand) sides of the seed tiles.

C. Mao and T. Yokomori (Eds.): DNA12, LNCS 4287, pp. 215–222, 2006.

- A border tile can join the structure only by attaching to the seed tile or to another border tile along the horizontal and vertical boundaries of the positive lattice, each such attachment extending one of the borders of the structure assembled so far.
- A rule tile can attach to the growing structure at any available site which is adjacent to occupied sites both to the left and below the available site, where either site may contain another rule tile or, initially, a border tile.

In the probability model governing growth (cf. [4, 5]), as soon as a site becomes available to a rule tile (for attachment to the left and below) its waiting time for such a tile is exponentially distributed with mean 1. All such waiting times are independent. The attachment of new border tiles is also subject to independent exponential waiting times, but with a different rate parameter α. In the analysis of the first such model [4], the border (input) tiles were assumed to be prefabricated, occupying sites, let us say, from (0,0) to (M,0) and to (0,N), the so-called L input; the problem was to estimate the time $C_{M,N}$ necessary for a rule tile to attach to the site at (M,N).

One of the principal tools used in solving this type of problem has been the analysis of the TASEP (Totally Asymmetric Simple Exclusion Process) [8]; the main result is given below in terms of fluid limits where the discrete position variables M and N are replaced by continuous variables x and y. This gives us, with the obvious change in notation,

$$C_{(x,y)} \sim \left(\sqrt{x} + \sqrt{y}\right)^2 \tag{1}$$

as $x \to \infty$ with $x/y \to k$ where k is a constant. In [4], we show that the time it takes to place a tile at position (x, y) is the maximum of the times taken to place tiles along all possible paths from the origin to position (x, y). The path is the straight line connecting the origin, the location of the seed tile, to point (x, y) in the fluid limit case.

In the more general setting, where border tiles are not prefabricated, but take part in the self-assembly process, the problem becomes much more interesting and entails the assembly of a much larger class of shapes, particularly when $\alpha < 1$, i.e., the border-tile attachment rate is less than that of the rule tiles.[1] The more general extremal problem is illustrated in Figure 1. Since $\alpha < 1$, a path along which the sum of expected attachment times is maximum must first move along one of the axes; the slow growth rate of the borders is less than the rate of tiling the rectangles bounded by the borders. At point K, which is where these rates balance, the trajectory is determined by the rectangle tiling result of (1).

In Figure 1(b), the dotted line shows the shape of the self-assembled structure before the border reaches point K. When the crystal boundary is below point K, the rule tiles can completely fill in the rectangle covered by the length of the border before a new border tile arrives, since the border tiles' interarrival times are greater than the time that the rule tiles take to fill in small rectangles. This means that the time it takes to place a tile depends on the speed of border-tile attachment. When the boundary grows beyond point K, the rule tiles cannot complete the rectangle before the next arrival of a border

---

[1] K. Fujibayashi working in E. Winfree's lab at Cal Tech notes that $\alpha \approx 1/2$ was common in certain of his experiments, where α is a stoichiometric parameter giving the concentration of border tiles. In this case he found that the profile of the self-assembly structure is triangular, a fact that will also emerge in our purely mathematical framework.

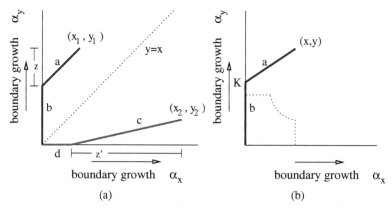

Fig. 1. (a) The paths that contribute to the time required for a tile to attach to the position (x, y). The path $\overline{ba}$ is followed when $x \leq y$, and the path $\overline{dc}$ is followed when $x \geq y$. (b) The dotted line illustrates the boundary of the shape of a self-assembled structure before the border tiles reach point K during the process. When the crystal boundary is below point K, the rule tiles (in the fluid limit) fill in the rectangle covered by the length of the border before a new border tile arrives. Beyond point K, the rule tiles cannot complete the rectangle before the next arrival of a border tile, so the attachment of rule tiles determines the speed of the approach to (x, y).

tile (in the fluid limit); hence the attachment of rule tiles contributes to the speed of the self-assembly process.

This argument holds true for the case where $x \geq y$ and $x \leq y$ (point K can be either on the vertical or horizontal axis). Thus, in the figure, to compute a path to (x, y) along which expected total attachment time is maximum, one must find this maximum among paths like $\overline{ba}$, if $x \leq y$; $\overline{dc}$, if $x \geq y$, and the path connecting the origin directly to (x, y) (for the case where the maximum time does not depend on the placement of border tiles).

With z as defined in the figure, the extremal path of tile attachments will start from the origin, go to position $(0, y - z)$ in expected time $(y - z)/\alpha$ for $x \leq y$ or go to position $(x - z, 0)$ in expected time $(x - z)/\alpha$ for $x \geq y$, and then go on to reach (x, y). We need only find the supremum over z to identify the extremal path. Thus, for finding the time to reach point (x, y) where $x \geq y$,

$$C_{x,y} \sim \sup_{z} \left(\left(\sqrt{x} + \sqrt{z} \right)^2 + \frac{y - z}{\alpha} \right) \tag{2}$$

as $(x, y) \to (\infty, \infty)$. A calculation for $C_{x,y}$ then shows that

$$C_{x,y} \sim x \left(1 + \frac{\alpha}{1 - \alpha} \right)^2 + \frac{y}{\alpha} - \frac{x\alpha}{(1 - \alpha)^2} = \frac{y}{\alpha} + \frac{x}{1 - \alpha} \tag{3}$$

as $(x, y) \to (\infty, \infty)$, where $z = x \left(\frac{\alpha}{\alpha - 1} \right)^2$, $x < y$ and $z < y$.

An analysis of the alternate path $\overline{dc}$ for $x \geq y$ follows the same arguments and yields the same result with x and y interchanged. By fixing the computing time $C_{x,y}$,

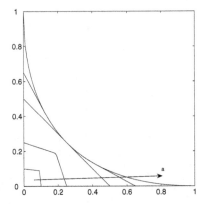

Fig. 2. Intermediate shapes of self-assembly structures for different border growth rates. Both horizontal and vertical lengths of the L-shape borders of the structures represent the border growth rates (a). The shapes of the structures are different for each value of the border growth rate.

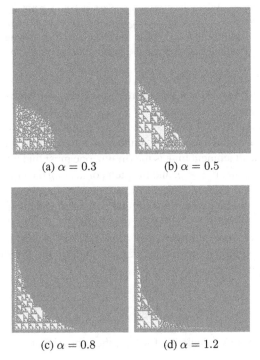

(a) $\alpha = 0.3$ (b) $\alpha = 0.5$

(c) $\alpha = 0.8$ (d) $\alpha = 1.2$

Fig. 3. Simulation of 2D structures with various values of α

we can plot the values of x and y corresponding to different values of α as shown in Figure 2.

Note that the lines produced by (3) are always tangent to the curve described by (1). As a result, for $0 \leq \alpha \leq 1/2$, the intermediate shape of the self-assembled crystal looks like those shown in Figure 2. When $\alpha = 1/2$, the shape of the self-assembled crystal is triangular. However, when $0.5 < \alpha < 1$, the shape of self assembly crystal has a linear shape near the border and a hyperbolic shape in the middle of the body since the time from (3) is the maximum time to place a tile near the border of the structure, and (1) is the maximum time to place a tile in the middle of the body of the structure.

Then when $\alpha \geq 1$, the intermediate shape of the self-assembled structure is no longer linear. It is described by the equation suggested in [4], which analyzed systems whose border tiles and seed tiles are prefabricated, or equivalently, systems where the rate of growth for the border tiles is greater than or equal to that for the rule tiles.

Remarkably, the simulations of self-assembled structures agree with our prediction for all values of $\alpha \leq 1$. Figure 3 was produced by Fujibayashi from simulation data based on kTAM model and shows results for various values of α.

We also simulated two dimensional self-assembled structures with one million tile attachments for $\alpha = 0.2, 0.5, 0.8, 1.0$. The shapes of the structures are plotted in Figure 4. The shapes of the structures are in remarkably good agreement with those described by equations (3) and (2).

It is possible to extend the scope of our discussion even further by allowing the rates of border growth to differ on the vertical and horizontal axes.

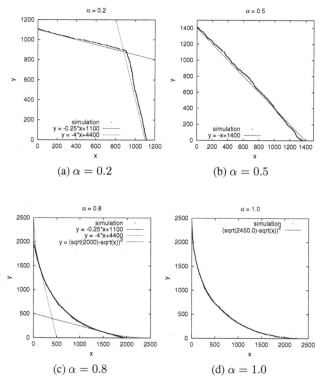

Fig. 4. Simulation of 2D structures with various values of α. The simulation was performed for one million attachments of tiles. The shapes of the structures are also plotted along with the equations obtained from (3) and (2).

2 Different Rates of Border Growth

Let α_1 be the ratio of the attachment rate of the vertical border tiles to that of the rule tiles, and let α_2 be the ratio of the attachment rate of the horizontal border tiles to that of the rule tiles. The paths that determine the time taken by a tile to arrive at position (x, y) with $x \leq y$ (or, respectively, with $x \geq y$) do not depend on the border tiles that are growing in the horizontal (or, respectively, vertical) direction because the attachment of the border tiles in the horizontal (respectively vertical) direction does not contribute to the maximum time it takes to place a tile at (x, y) where $x \leq y$. We arrive at

$$C_{x,y} = \frac{y}{\alpha_1} + \frac{x}{1 - \alpha_1} \qquad \text{for } x < y$$

$$C_{x,y} = \frac{y}{1 - \alpha_2} + \frac{x}{\alpha_2} \qquad \text{for } x > y$$

as $(x, y) \rightarrow (\infty, \infty)$. Figure 5 shows the shape of the self-assembly structures with different border growth rates.

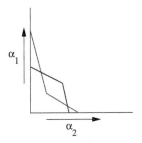

Fig. 5. Expected shapes of 2D self-assembly for different border growth rates: α_1 and α_2. The shapes are no longer symmetric along $y = x$.

Crystal self-assembly is typically designed to grow indefinitely, yielding congruent shapes at different times. A problem arises when we want a specific finite shape. Although we can design tile sets so that the growth process will stop at some pre-specified point, the complexity of the tile sets is high (cf. [9, 10, 11, 12]). In order to create shapes that are useful and economically efficient, a small number of simple tiles is desirable. A straightforward method of creating desired shapes is simply to remove the supply of tiles completely when we know that the desired crystal size has been reached. This method works since the shape (fluid limit) approximation illustrated in Figure 5 above allows one to estimate the size of the structure and the time it takes to create a specific structure.

3 Conclusions

As is well known, essentially arbitrary 2D shapes can be created by tile self assembly (cf. [12, 1]). For example, one technique is to create a domino pattern that serves as a skeleton on which to grow some given shape; the length L of a leg of the pattern can be determined by a counter structure of $O(\log L)$ width. Figure 6 shows an example which begins with a (blue) seed from which growth proceeds in each of the four directions. At the ends of legs, new (purple) seed structures can self-assemble to give the domino branching patterns. Once the domino pattern is in place, or as it self assembles, the growth, as described earlier, can flesh out the desired shape as shown in Figure 6(b). The domino patterns define independent, elementary growth regions as illustrated in Figure 6(a). In simple cases, expected times to compute such shapes may be expressible as the expected maximum of the times to grow the independent regions (cf. [4]). But in general, the computations of overall expected growth times are complicated.

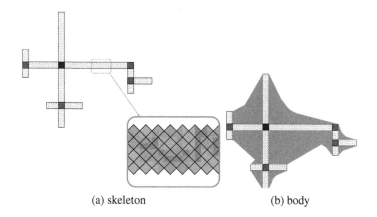

(a) skeleton   (b) body

Fig. 6. The shape of a tile self-assembly model built on a predesigned skeleton

Acknowledgement. We are grateful to K. Fujibayashi for pointing out and discussing the problem of (slowly) self assembling border tiles including the intermediate shapes of self-assembly of the Sierpinski triangle based on the kTAM model (see e.g., [1]) shown in Figure 3.

References

1. Winfree, E.: Algorithmic Self-Assembly of DNA. PhD thesis, California Institute of Technology, Pasadena, CA (1998)
2. Adleman, L., Cheng, Q., Goel, A., Huang, M.D., Wasserman, H.: Linear self-assemblies: Equilibria, entropy, and convergence rates. In Elaydi, Ladas, Aulbach, eds.: New progress in difference equations, Taylor and Francis, London (2004)
3. Baryshnikov, Y., Coffman, E., Momčilović, P.: Incremental self-assembly in the fluid limit. In: Proc. 38th Ann. Conf. Inf. Sys. Sci., Princeton, NJ (2004)
4. Baryshnikov, Y., Coffman, E., Momčilović, P.: DNA-based computation times. In: Proc. of the Tenth International Meeting on DNA Computing, Milan, Italy (2004)
5. Baryshnikov, Y., Coffman, E., Seeman, N., Yimwadsana, B.: Self correcting self assembly: Growth models and the hammersley process. In: Proc. of the Eleventh International Meeting on DNA Computing, London, Ontario (2005)

6. Ding, B., Sha, R., Seeman, N.: Pseudohexagonal 2D DNA crystals from double crossover cohesion. J. Am. Chem. Soc. **126** (2004) 10230–10231

7. Fu, T.J., Seeman, N.: DNA double crossover structures. Biochemistry **32** (1993) 3211–3220

8. Liggett, T.M.: Interacting Particle Systems. Springer-Verlag, New York (1985)

9. Adleman, L., Cheng, Q., Goel, A., Huang, M.D., Kempe, D., de Espanés, P.M., Rothemund, P.: Combinatorial optimization problems in self-assembly. In: Proc. ACM Symp. Th. Comput., Montreal, Canada (2002) 23–32

10. Adleman, L., Cheng, Q., Goel, A., Huang, M.D.: Running time and program size for self-assembled squares. In: Proc. ACM Symp. Th. Comput. (2001) 740–748

11. Rothemund, P., Winfree, E.: The program-size complexity of self-assembled squares. In: Proc. ACM Symp. Th. Comput. (2001) 459–468

12. Soloveichik, D., Winfree, E.: Complexity of self-assembled shapes. In: Proc. of the Tenth International Meeting on DNA Computing. (2004)

Capabilities and Limits of Compact Error Resilience Methods for Algorithmic Self-assembly in Two and Three Dimensions[*]

Sudheer Sahu and John H. Reif

Department of Computer Science, Duke University
Box 90129, Durham, NC 27708-0129, USA
{sudheer, reif}@cs.duke.edu

Abstract. Winfree's pioneering work led the foundations in the area of error-reduction in algorithmic self-assembly [26], but the construction resulted in increase of the size of assembly. Reif et. al. contributed further in this area with compact error-resilient schemes [15] that maintained the original size of the assemblies, but required certain restrictions on the Boolean functions to be used in the algorithmic self-assembly. It is a critical challenge to improve these compact error resilient schemes to incorporate arbitrary Boolean functions, and to determine how far these prior results can be extended under different degrees of restrictions on the Boolean functions. In this work we present a considerably more complete theory of compact error-resilient schemes for algorithmic self-assembly in two and three dimensions. First we consider two-dimensional algorithmic self-assembly. We present an error correction scheme for reduction of errors from ϵ to ϵ^2 for arbitrary Boolean functions in two dimensional algorithmic self-assembly. Then we characterize the class of Boolean functions for which the error reduction can be done from ϵ to ϵ^3, and present an error correction scheme that achieves this reduction. Then we prove ultimate limits on certain classes of compact error resilient schemes: in particular we show that they can not provide reduction of errors from ϵ to ϵ^4 is for any Boolean functions. Further, we develop the first provable compact error resilience schemes for three dimensional tiling self-assemblies. We also extend the work of Winfree on self-healing in two-dimensional self-assembly [25] to obtain a self-healing tile-set for three-dimensional self-assembly.

1 Introduction

Self-assembly is the ubiquitous process in which smaller objects combine together to form larger complex objects. Recently, it has been demonstrated as an efficient mechanism for bottom-up construction of nanostructures in nanotechnology [19, 27, 11, 8, 32, 31, 3, 10]. The potential of self-assembly is not limited to nanofabrication. The ability of two-dimensional and three-dimensional assemblies to perform parallel universal computations has been explored in development of self-assembly of DNA tiles as a tool for nanocomputation [9, 14, 23, 28, 30]. Self-assembly has been demonstrated at

[*] The work is supported by NSF EMT Grants CCF-0523555 and CCF-0432038.

C. Mao and T. Yokomori (Eds.): DNA12, LNCS 4287, pp. 223–238, 2006.

larger scales (meso-scale) using capillary forces for interactions between meso-scale tiles [2, 16]. However, major hurdle in harnessing the capabilities of algorithmic self-assembly are the errors that occur during the assembly. Incorrect tiles are incorporated in the growing structure with error rate ranging from 1% to 5% [26]. There are two approaches to combat the errors. The first is to reduce the inherent error rate by optimizing the physical conditions [24] or using newer molecular mechanisms [4], while the other approach is to improve the tile design so that the total number of errors in the final structure is reduced in spite of the intrinsic error-rate remaining the same [26, 15, 5].

Winfree's pioneering work in error-correction [26] laid the foundations towards improving the tile-design to reduce the errors in assembly. Though it resulted in the total size of assembly to be 2×2 times for error reduction to ϵ^2 and 3×3 times for error reduction to ϵ^3, it paved the way for further work in error-reduction using the concept of redundancy. The basic idea was that an error in the assembly of a tile forced more errors in the immediate neighborhood of that tile, making it extremely prone to detachment, and hence reducing the error. Later, the snaked proof-reading scheme that could correct both growth and nucleation errors in the self-assembly was built upon this construction [5]. However, it required replacing a tile by a $k \times k$ block of tiles. Later a method was proposed to control nucleation errors programmably [18]. However, each of these schemes significantly scaled up the overall size of assembly. In applications like molecular fabrication tasks where the scale of final pattern is of critical importance, this scaling up is undesirable. Reif et al. [15] proposed a *compact error-resilient tiling schemes* in which errors could be reduced to ϵ^2 (2-way overlay redundancy) and ϵ^3 (3-way overlay redundancy) without increasing the size of the assembly. The analysis of error was done in the equilibrium state of the assembly. Another distinction of this scheme was that it considered the error resilience in the whole pattern and not only in the output row. It means that this scheme had a tendency to remove any incorrectly placed tile from the assembly even if the ongoing computation was not affected by that tile. This is important in the assembly of a nanostructure of desired pattern, where any incorrect placement of any tile is a defect (even though it might not have interfered with the subsequent growth of assembly). But it had its limitations on the Boolean functions that could be used for the error-resilient algorithmic assembly. In particular, it required one of the function to be *XOR*, and for reduction to ϵ^3 the additional requirement was that the other function should be input-sensitive to one of the inputs. A Boolean function $f(x)$ is called *input-sensitive* to a Boolean variable x if whenever x changes $f(x)$ also changes. It is thus a critical challenge to improve these compact error-correction schemes to incorporate any arbitrary Boolean functions. In case that is not possible, it is important to characterize the class of Boolean functions to which these error-correction schemes can be extended. Recently Winfree [20] presented a *compact error resilient scheme* based on Chen et al [5]. They also overlooked the errors that did not affect the ongoing computation.

Self-assembly in three dimensions is extremely promising in the field of microelectronics assembly, where independent manipulation of each component is required. It is already being seen as promising candidate for heterogeneous three-dimensional integration of next-generation microsystems [29, 12, 6, 22]. In light of the inherent parallelism, three-dimensional nature and larger range (nanoscale to mesoscale) of application of self-assembly, it has a great potential as tool for building complex systems from microscaled templates. Apart from this, the utility of three-dimensional structures for computing has been known for a long time [7]. Simple examples of algorithmic computation in three dimensions includes the generalization of Pascal triangle to 3D [1] and three dimensional multiplexers (the latter would provide a mechanism for 3D memory addressing with the appropriate affixed molecular electronic components). Analogous to the simulation of a finite state automata through two-dimensional self-assembly, three dimensional self-assembly can be used to simulate a two-dimensional cellular automata, where the third spatial dimension of the 3D tiling is the time step of the cellular automata. The tiles in a horizontal plane will represent the current state of all the cells of a two-dimensional cellular automata, then the tiles assembled in horizontal plane on top of it will be states at next time instance. This allows one to derive 3D tiling assemblies from a wide variety of known two-dimensional cellular automata designs, including matrix multiplication, integer multipliers, context free language recognition, etc. Recently crystal structure of three-dimensional DNA lattices formed by self-assembly was demonstrated [13]. The question of fault-tolerance naturally arises with the increasing popularity of self-assembly for construction of three dimensional self-assembled structures. It will be critical to determine how successfully can the error-correction techniques used for two-dimensional assemblies be extended to three-dimensions.

Self-healing is a very important process in nature. The damage to the living cells can be caused by an external intruder or some mechanical impulse or unfavorable physical conditions. The one property of biological systems that make them robust is their ability to self-heal in case of damages. It would be really interesting to design the DNA tiles that forms the lattices having the ability to self-heal, thereby imparting them the much desired robustness. Winfree [25] gave a construction in which he replaced a single tile with 3×3 (for simple assemblies like Sierpinski triangles) , 5×5 (for general assemblies) and 7×7 (for additional robustness to nucleation errors) block of tiles for self-healing in a two-dimensional assembly. It would be interesting to know if compact self-healing tilesets can be formed and whether the techniques given by Winfree can be extended to three dimensions.

In this paper, we follow the notion of *compactness* as presented in [15], which requires the new error-resilient tiling assembly to be of no larger size than the original assembly. Like [15] we consider any incorrect placement of a tile anywhere in the assembly as an error and aim at reducing them as well, even though these errors might not affect the ongoing computation. As mentioned earlier, this is important for construction of nanostructures of desired pattern. In this paper, the analysis of the error in the assembly is done in the equilibrium state of the assembly. Throughout this paper *redundancy based compact error resilient scheme* refers to any error resilient scheme that does not scale up the assembly and in which the encodings on the pads of the tiles are used to create redundancy. In the event of an error this redundancy forces more errors, which makes the incorrectly placed tiles and their neighborhoods more unstable and prone to removal from assembly, thereby reducing the error. Also we refer to *k-expansive error resilient schemes* as the error correction schemes that work by replacement of a tile by a block of multiple tiles. In case of three dimensional tiling, we carry forward this notion of *redundancy based compact error resilient schemes*.

In this paper, we present a comprehensive theory of redundancy based compact error resilient tiling schemes and examine the prospects of constructing compact self-healing tile sets in two and three-dimensions. The error analysis throughout this paper is in the equilibrium state of the assembly. In Section 2, first we present a compact error correction schemes in two dimensional self-assembly that reduces the error from ϵ to ϵ^2 for arbitrary Boolean functions. Then we characterize the class of Boolean functions for which error reduction from ϵ to ϵ^3 is possible using redundancy based compact error resilient schemes. Also we prove that error reduction from ϵ to ϵ^4 is impossible using redundancy based compact error resilient schemes. Next in Section 3 we examine three-dimensional self-assembly. First we present a compact error resilient scheme that reduces error to ϵ^2 for arbitrary Boolean functions and ϵ^3 for a restricted class of input-sensitive Boolean functions. We also prove that error reduction to ϵ^4 can not be obtained for arbitrary Boolean functions using redundancy based compact error resilient schemes. In Section 4 we extend the idea of Winfree's construction for self-healing in two-dimensions [25] to three-dimensional assembly. In the conclusion, we review our results and state various open problems and conjectures. We conjecture stronger results that error reduction to ϵ^3 in three dimensions can not be achieved outside the previously characterized class, and error reduction to ϵ^4 is impossible to achieve for any Boolean functions using these error resilient techniques.

2   Error Correction in Self-assembly in Two Dimensions

2.1   Assembly in Two Dimensions

We will consider a general assembly problem in two dimensions consisting of the assembly of a two-dimensional Boolean array of size $N \times M$, where the elements of each column are indexed from 0 to $N - 1$ from right to left and rows are indexed from 0 to $M - 1$ from bottom to top. The bottom row and the rightmost column provide the inputs to the assembly. Let $V(i, j)$ be the value of the ith column (from the right) in the jth row(from the bottom). Let $V(i, j + 1)$ be the value communicated to the position $(i, j+1)$ and $U(i+1, j)$ be the value com-

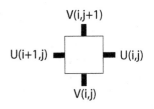

Fig. 1. Two dimensional algorithmic self-assembly

municated to the position $(i + 1, j)$. We define $U(i + 1, j) = U(i, j)OP_1V(i, j)$ and $V(i, j + 1) = U(i, j)OP_2V(i, j)$ for two Boolean functions OP_1 and OP_2.

Figure 1 shows a computational tile that can be used for constructing two dimensional self-assembly. Bottom and right pads are the input pads, while the pads on top and left are output pads. A pad *matches* with the neighbor's contiguous pad if the values communicated by these pads are the same. $U(i, j)$ and $V(i, j)$ are the right and bottom input pads, respectively, to the ith column from right and jth row from bottom. Then $U(i + 1, j)$ the left output pad is given by $U(i + 1, j) = U(i, j)OP_1V(i, j)$, while $V(i, j + 1)$ the top output pad is given by $V(i, j + 1) = U(i, j)OP_2V(i, j)$. Examples of simple two dimensional assemblies: sierpinski triangle and binary counter, are given in [15]. Highly complex two-dimensional assemblies are possible due to the universal computability of two-dimensional self-assembly [21, 28].

2.2 The Error Model

We assume that error probability ϵ is defined as the probability that there is mismatch between two tiles and they still stay together in the equilibrium. This probability is independent of any other match or mismatch and hence we term this probabilistic model the *independent error model*. We also want to put emphasis on the correct assembly of all the tiles in the assembly (and hence on the correctness of complete pattern), and not just on the correctness of final output only. There might be wrong placement(s) of tile(s), that do not affect the ongoing computation. But in our error model, we count them as errors and need the error correction schemes to reduce such errors as well. In this way we differ from [20], who overlooked the errors that did not affect the ongoing computation.

Consider a tile $T(i, j)$ in a $N \times M$ tiling assembly where $0 < i < N - 1, 0 < j < M - 1$. We define the *immediate neighborhood* of a tile $T(i, j)$ as 8 tiles surrounding it, whose coordinates differ from (i, j) by at most 1. Formally speaking, $\{T(i', j') : |i' - i| \leq 1, |j' - j| \leq 1\} \setminus \{T(i, j)\}$. Tile $T(i', j')$ is said to be *a-dependent* (for assembly dependent) on tile $T(i, j)$ if $i' \geq i$ and $j' \geq j$ and *a-independent* otherwise. Next we examine the schemes to reduce the errors in self-assembly. To reiterate, throughout this paper, we refer to redundancy based compact error resilient scheme as error reduction scheme, where redundancy is created by encodings in the pads with absolutely no scale up of the assembly.

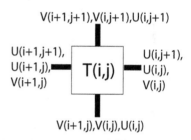

Fig. 2. Construction for error reduction to ϵ^2

Proposition 1. *Under our independent error model, if an error in a pad in a tile enforces k further mismatches in the assembly in the immediate neighborhood of that tile, then error probability is reduced to ϵ^{k+1}.*

Proof. If one error guarantees k more errors, then the probability that the tile and its neighborhood in the assembly will stay together in the equilibrium in spite of these $k + 1$ errors is ϵ^{k+1}. And hence the error reduction.

2.3 Error Reduction to ϵ^2

It is known that if an error in a tile can guarantee another error in immediate neighborhood, then it reduces the rate of errors from ϵ to ϵ^2 [26, 15]. Next we describe our construction to achieve this goal in the form of Theorem 1.

Theorem 1. *There exists a compact error correction scheme that will reduce the error from ϵ to ϵ^2 for two-dimensional algorithmic self-assembly for any arbitrary Boolean functions OP_1 and OP_2.*

Proof. **Construction.** Before we begin the proof we would like to emphasize the wholeness of the pad. Each side of the tile has one pad in Figure 2, and it encodes the triplet shown in the Figure. Disagreement between corresponding elements of two such triplets in any two pads results in the total mismatch between those two pads. Consider the tile with input $U(i, j)$ and $V(i, j)$ at the right and bottom pads respectively. Our goal is to guarantee one more error in the immediate vicinity of this tile if there is one error. For that, we construct an error checking portion $(V(i, j))$ in the right side pad and one error checking portion $(U(i, j))$ in the bottom pad. We will need corresponding parts in the pads on the top $(U(i, j + 1))$ and the left side $(V(i + 1, j))$ also, which will match with the error checking parts in the bottom pad of the top neighbor $T(i, j + 1)$ and right pad of the left neighbor $T(i + 1, j)$ respectively. Now since top output pad

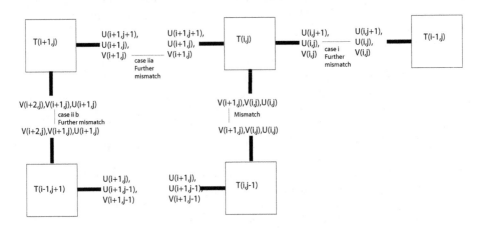

Fig. 3. Case 1 b) A further mismatch is caused by an error in the input pads

depends on the value of $U(i, j + 1)$ (which is the right input of the top neighbor) we need to incorporate it in our input pads. It is necessary otherwise there will be multiple type of tiles for any given set of input pads. But for successful functioning of algorithmic self-assembly it is required that there should be only one possible tile-type for every set of input pads. So, we need one more portion in the right input pad $(U(i, j + 1))$ and hence a corresponding part in the left output pad $(U(i + 1, j + 1))$. Similarly, the need for another portion in bottom input pad $(V(i + 1, j))$ and subsequently, in top output pad $(V(i + 1, j + 1))$ can be explained.

This completes our description of a tile in the required tile-set. It should be noted that the number of different tile types in this tile-set will be 4 times as compared to number of tiles in a tileset without any error-correction. It can be attributed to the two possible values for each of $U(i, j + 1)$ and $V(i + 1, j)$, for every value of the inputs $U(i, j)$ and $V(i, j)$.

Error-Analysis: We show that if the neighborhood tiles a-independent of $T(i, j)$ are assembled correctly then a pad binding error in any of the input pads in $T(i, j)$ causes an additional mismatch error in its neighborhood in equilibrium. We need to consider only the cases where the pad binding error occurs in either the bottom or the right pad of tile $T(i,j)$. Otherwise, if the error occurs in left (or top) pad of $T(i,j)$ then we can consider the right pad of $T(i+1,j)$ (or bottom pad of $T(i,j+1)$) for the analysis. The following case analysis provides the required proof.

1. If the bottom pad of $T(i,j)$ has a mismatch:
    (a) If $V(i,j)$ on the bottom pad has a mismatch, then $V(i,j)$ on right pad is incorrect, which causes an additional mismatch.
    (b) If $V(i,j)$ on the bottom pad is correct and $V(i+1,j)$ on right pad has a mismatch, $V(i+1,j)$ on left pad is incorrect. Now we will prove that it causes a further mismatch by exactly same technique as used by Reif et al [15]. We have assumed that all the rows and columns that are a-independent of tile $T(i,j)$ are correctly assembled so $T(i+1,j-1)$ is correctly assembled and has correct values of its top output pad. Hence $T(i,j)$'s left neighbor $T(i+1,j)$ is dependent upon the incorrect value communicated by the left pad of $T(i,j)$ and correct values communicated by top pad of $T(i+1,j-1)$. Now consider the pads of $T(i+1,j)$. The right pad includes $U(i+1,j+1), U(i+1,j), V(i+1,j)$ and bottom pads include $V(i+2,j), V(i+1,j), U(i+1,j)$. Since the value $V(i+1,j)$ communicated by $T(i+1,j-1)$ is correct and the value $V(i+1,j)$ communicated by $T(i,j)$ is wrong, this implies there will be a mismatch at the right or bottom pad of Tile $T(i+1,j)$.
2. If there is no error in bottom pad, but the right pad of $T(i,j)$ has mismatch:
    (a) If $U(i,j)$ on the right pad has a mismatch, then $U(i,j)$ on bottom pad is incorrect, which causes an additional mismatch.
    (b) If $U(i,j)$ on right pad is correct but $U(i,j+1)$ on right pad is incorrect, then $U(i,j+1)$ on top output pad is incorrect. Now we will show that it causes a further mismatch as argued above. Since we assume that all the rows and columns that are a-independent of tile $T(i,j)$ are correctly assembled $T(i-1,j+1)$ is correctly assembled and has correct values of its left output pad. Hence $T(i,j)$'s top neighbor is dependent upon the incorrect value communicated by the top pad of $T(i,j)$ and correct values communicated by left pad of $T(i-1,j+1)$. Now consider the pads of $T(i,j+1)$. The right pad includes $U(i,j+2), U(i,j+1), V(i,j+1)$ and bottom pads include $V(i+1,j+1), V(i,j+1), U(i,j+1)$. Since $V(i+1,j)$ communicated by $T(i-1,j+1)$ is correct and the value $V(i+1,j)$ communicated by $T(i,j)$ is wrong, this implies there will be a mismatch at the right or bottom pad of Tile $T(i,j+1)$.

Hence any mismatch on the right or bottom pad of tile $T(i,j)$ causes one more mismatch in the vicinity of the tile. Together with the Proposition 1 this implies that this scheme can reduce the pad mismatch errors from ϵ to ϵ^2.

2.4 Error Reduction to ϵ^3

At this point we would like to reiterate that *redundancy based compact error resilient scheme* refers to any error resilient scheme that does not scale up the assembly and in which only the encodings on the pads of the tiles are used to create redundancy. Also, a

Boolean function $f(x)$ is said to be *input-sensitive* to Boolean input x if it changes for every change in the value of x.

Theorem 2. *For arbitrary OP_1 and OP_2, there does not exist any redundancy based compact error resilient scheme for two-dimensional self-assembly that can reduce the error from ϵ to ϵ^3.*

Proof.

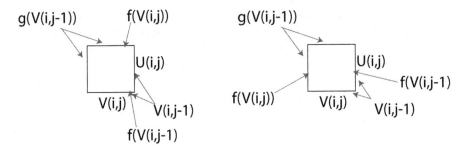

Fig. 4.

For errors to reduce from ϵ to ϵ^3, an error in any input pad, say $V(i,j)$ should cause two further mismatches in the immediate neighborhood. At least one of those mismatches should be caused because of an error on one of the output pads. It should be noted that if OP_1 and OP_2 are arbitrary Boolean functions then the output $U(i+1,j)$ or $V(i,j+1)$ cannot be guaranteed to be wrong for incorrect value of $V(i,j)$. Hence, in at least one of the output pads an additional error checking portion $f(V(i,j))$ (that is input-sensitive to $V(i,j)$ and hence can reflect the error in $V(i,j)$) is required. It can be located on the top or left output pad.

- Assume that $f(V(i,j))$ is located on top side, which implies $f(V(i,j-1))$ is located on the bottom side.
  1. If $V(i,j-1)$ does not exist within the input pads, then we need to consider the case when $f(V(i,j-1))$ has a mismatch. Since we require two further errors in the neighborhood of $T(i,j)$, as argued above it requires an additional error checking function $g(f(V(i,j-1)))$ (that is input-sensitive to $f(V(i,j-1))$) on at least one of the top or left output pad.
  2. If $V(i,j-1)$ exists in the input pads, then in case when $V(i,j-1)$ is mismatched, and two further errors in the neighborhood of $T(i,j)$ are required, it needs an additional error checking function $g'(V(i,j-1))$ (that is input-sensitive to $V(i,j-1)$) on at least one of the top or left output pad.
- Assume that $f(V(i,j))$ is located on left side, which implies $f(V(i-1,j))$ is located on the right side.
  1. If $V(i-1,j)$ does not exist within the input pads, we need to consider the case when $f(V(i-1,j))$ is mismatched. Since two further errors are required, as argued above it requires an additional error checking function $h(f(V(i-1,j)))$ (that is input-sensitive to $f(V(i-1,j))$) to be located on at least one of the top or left output pad.

2. If $V(i-1,j)$ exists in the input pads, then in case when $V(i-1,j)$ is mismatched, and two further errors are required, it requires an additional error checking function $h'(V(i-1,j))$ (that is input-sensitive to $V(i-1,j)$) to be present on at least one of the top or left output pads.

Hence, an additional error checking pad $(g(f(V(i,j-1))), g'(V(i,j-1))$ or $h(f(V(i-1,j)))$ or $h(V(i-1,j)))$ is required on at least one of the output pads. Arguing in the same manner as above it can concluded that this cycle will keep on repeating. Hence, it is not possible to construct tile with a bounded number of parameters in the pads and we conclude that redundancy based compact error resilient schemes can not reduce error from ϵ to ϵ^3.

However, it will be proved that for a rather restricted class of Boolean functions OP_1 and OP_2, error can be reduced to ϵ^3 by using the construction of Figure 2, which is stated as Theorem 3.

Before we proceed with the error-analysis, it will be useful to understand the function class characterized in the Theorem 3. OP_1 and OP_2 are such that:(1) $U(i,j)$ OP_1 $V(i,j)$ is input-sensitive to $U(i,j)$, if $V(i,j)$ is kept constant and $U(i,j)$ OP_2 $V(i,j)$ is input-sensitive to $V(i,j)$ if $U(i,j)$ is kept constant. (2) When both of them change at least one of the $U(i,j)$ OP_1 $V(i,j)$ or $U(i,j)$ OP_2 $V(i,j)$ should also change. For $U(i,j)=0$, there are 2 possible assignments to $U(i,j)OP_1V(i,j)$ maintaining its input-sensitivity to $V(i,j)$. Similarly, for $U=1$ there are 2 possible assignments to $U(i,j)OP_1V(i,j)$ conditioned to its input-sensitivity to $V(i,j)$. Similarly for $V(i,j)=0$ and $V(i,j)=1$ there are 2 independent assignments each. But among these half of the assignments do not satisfy the second condition. Hence the total number of Boolean functions in this class are 8. An example of such a function is given in the Table 1.

Table 1. An example of the OP_1 and OP_2

U	V	UOP_1V	UOP_2V
0	0	1	0
0	1	1	1
1	0	0	0
1	1	0	1

Theorem 3. *For restricted class of Boolean functions OP_1 and OP_2 such that at least one of the $U(i+1,j)$ or $V(i,j+1)$ changes for any change in $U(i,j)$ or $V(i,j)$, there exists a redundancy based compact error resilience scheme that can reduce the error to ϵ^3, and one such scheme is as shown in Figure 2.*

Proof. If OP_1 and OP_2 are restricted to be as described, and if the neighborhood tiles that are a-independent of $T(i,j)$ are assembled correctly, then a pad binding error in any of the input pads in $T(i,j)$ causes two additional mismatch errors in its neighborhood. As explained earlier, we need to consider only the cases where the pad binding error occurs in either the bottom or the right pad of tile T(i,j). The following case analysis provides the required proof.

1. If the bottom pad of $T(i,j)$ has a mismatch:
    (a) If $V(i,j)$ in bottom pad of $T(i,j)$ has a mismatch, then the $V(i,j)$ in the right pad of $T(i,j)$ is incorrect. This causes a mismatch because according to our assumption, all the tiles a-independent of $T(i,j)$ are assembled correctly. Also:

     i. If $U(i,j)$ on right pad is correct, $V(i, j + 1)$ on top pad is incorrectly computed because of restrictions on OP_1 and OP_2. This will cause further mismatch at the right or bottom pad of the top neighbor $T(i, j + 1)$, as argued in the proof of Theorem 1.

     ii. If $U(i,j)$ on right pad has a pad-mismatch, then at least one of the $V(i, j + 1)$ on top pad or $U(i + 1, j)$ on left pad is incorrectly computed, because of the restrictions on OP_1 and OP_2. This will cause a further mismatch at right or bottom pad of the left neighbor $(T(i + 1, j))$ or top neighbor$(T(i, j + 1))$ in the same way as argued earlier.

(b) If $V(i,j)$ on bottom pad is correct and $V(i+1, j)$ on bottom pad has mismatch, then $V(i + 1, j)$ on the left pad is incorrect, which causes a further mismatch in the right or bottom pad of the left neighbor $T(i + 1, j)$. Also:

     i. If $U(i,j)$ on right pad is incorrect, then this causes a mismatch on the right pad of $T(i, j)$, because according to our assumption, all the tiles a-independent of $T(i, j)$ are assembled correctly.

     ii. If $U(i,j)$ on right pad is correct, then $U(i + 1, j)$ on left output pad is correct. But since $V(i + 1, j)$ has a mismatch, $V(i + 1, j + 1)$ on the top pad is incorrectly computed, because of the restriction on OP_1 and OP_2. This causes a further mismatch on the bottom or the right pad of the top neighbor tile $T(i, j + 1)$.

2. If there is no error in the bottom pad and there is mismatch in right pad:

(a) If $U(i,j)$ on right pad has a pad-mismatch, then at bottom $U(i, j)$ is incorrect, and causes a mismatch. However since $V(i, j)$ on the bottom pad is correct so $U(i + 1, j)$ on left pad is incorrectly computed because of the restriction on OP_1 and OP_2. This causes a further mismatch on right or bottom pad of left neighbor as explained earlier.

(b) If $U(i,j)$ on right pad is correct and $U(i, j+1)$ has a mismatch, then $U(i, j+1)$ on top pad is incorrect., which causes a further mismatch in right or bottom pad of the top neighbor tile $T(i, j + 1)$. Also since $V(i, j)$ is correct, $V(i, j + 1)$ is also correct, and hence $U(i + 1, j + 1)$ on left pad is incorrectly computed because of restriction on OP_1 and OP_2. This causes a further mismatch in the right or bottom pad of the left neighboring tile $T(i + 1, j)$.

Hence any mismatch on the right or bottom side of the tile $T(i, j)$ causes two further mismatches in the vicinity of tile $T(i, j)$ and this results in error reduction from ϵ to ϵ^3.

For any other combination of Boolean functions OP_1 and OP_2, which do not satisfy the conditions of Theorem 3, the redundancy based compact error resilient schemes fail to achieve the reduction from ϵ to ϵ^3. It can be proven along the similar lines of reasoning as the proof of Theorem 2. Therefore we state it without proof.

Theorem 4. *For any combination of Boolean functions OP_1 and OP_2 outside the restricted class of Theorem 3, there exists no redundancy based compact error correction schemes that can reduce the error from ϵ to ϵ^3 in two-dimensional self-assembly.*

2.5 Error Reduction to ϵ^4

Theorem 5. *For any Boolean functions OP_1 and OP_2, there exists no redundancy based compact error correction scheme that can reduce error from ϵ to ϵ^4 in two-dimensional self-assembly.*

Proof. For the reduction of error from ϵ to ϵ^4, a mismatch in any input pad should cause 3 more mismatches. It means that for any error in one of the input pads both the output pads should have errors. In case an output pad requires any additional error checking portion to detect an error in an input, then by arguments similar to the proof of Theorem 2, it can be shown that such a tile cannot be constructed.

Hence, the only possibility is when, the left and top outputs $U(i+1, j)$ and $V(i, j+1)$ both change for any change in the input $U(i, j)$ or $V(i, j)$. This means that we have different values for each of $U(i + 1, j)$ and $V(i, j + 1)$ for 4 different values of input pair, which is not possible as $U(i + 1, j)$ and $V(i, j + 1)$ are Booleans.

3 Error Correction in Self-assemblies in Three Dimensions

Three dimensional self-assembly is being described as the most promising tool for heterogeneous integration of next generation microsystems. Its potential to build complex systems from microscale templates can not be overlooked [29, 12, 6, 22]. Besides the assembled three-dimensional structures can be extremely useful in computations [7]. It is possible to simulate a two-dimensional cellular automata, using three-dimensional self-assembly, which then paves way to perform a rich class of computations including matrix multiplication, integer multiplications, context-free language recognition etc.

3.1 Assembly in Three Dimensions

The assembly problem in three-dimensions can be generalized from the two-dimensional assembly as the assembly of a three-dimensional Boolean array of size $N \times M \times P$, where the elements are indexed from 0 to $N - 1$ from right to left, 0 to $M - 1$ from bottom to top, and 0 to $P - 1$ from front to back. The rightmost plane, bottommost plane and frontmost plane provide the inputs to the assembly.

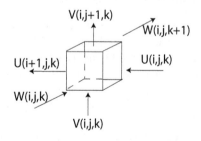

Fig. 5. Three dimensional algorithmic self-assembly

Let $V(i, j, k)$ be the i-th value from right, j-th from bottom, and k-th from front. Let $U(i, j, k)$ be the value communicated to the position $(i+1, j, k)$, $V(i, j, k)$ be communicated to the position $(i, j+1, k)$, and $W(i, j, k)$ be communicated to the position $(i, j, k + 1)$. Figure 5 shows a computational tile that can be used for construction of three-dimensional assembly. $U(i, j, k)$, $V(i, j, k)$ and $W(i, j, k)$ are inputs at right pad, bottom pad and front pad respectively, to the tile located at position (i, j, k). Then $U(i + 1, j, k)$, $V(i, j + 1, k)$ and $W(i, j, k + 1)$ are the output values at left, top and back pads, respectively. Also, $U(i, j, k) = f_1(U(i-1, j, k),$

$V(i, j-1, k), W(i, j, k-1)), V(i, j, k) = f_2(U(i-1, j, k), V(i, j-1, k), W(i, j, k-1)), W(i, j, k) = f_3(U(i-1, j, k), V(i, j-1, k), W(i, j, k-1))$ where f_1, f_2 and f_3 are the ternary Boolean functions that take as input three Boolean values and give a Boolean output. It is assumed that initially a frame is assembled, with $M \times P$ tiles in rightmost plane, $N \times P$ tiles in bottommost plane and $N \times P$ tiles in frontmost plane. Next we examine the error resilience in three-dimensional self-assembly.

3.2   The Error Model

We extend the error model in two-dimensions to three-dimensional assembly in an obvious way. We follow the *independent error model* for three dimensional assembly. We also want to emphasize on the correct assembly of all the tiles in the assembly (and hence on the correctness of complete pattern), and not just on the correctness of final output only. We want to emphasize that the error analysis is done in the equilibrium state of the assembly. Consider a tile $T(i, j, k)$ in a $N \times M \times P$ tiling assembly where $0 < i < N - 1, 0 < j < M - 1, 0 < k < P - 1$. We define the *immediate neighborhood* of a tile $T(i, j, k)$ as 26 tiles surrounding it, whose coordinates differ from (i, j, k) by at most 1. Formally speaking, $\{T(i', j', k') : |i' - i| \leq 1, |j' - j| \leq 1, |k' - k| \leq 1\} \setminus \{T(i, j, k)\}$. Tile $T(i', j', k')$ is said to be *a-dependent* on tile $T(i, j, k)$ if $i' \geq i$, $j' \geq j$, and $k' \geq k$ and *a-independent* otherwise. Next we examine the schemes to reduce the errors in self-assembly. As mentioned earlier *redundancy based compact error resilient scheme* refers to an error resilient scheme that does not scale up the assembly and in which the encodings on the pads of the tiles are used to create redundancy.

3.3   Error Reduction to ϵ^2

Theorem 6. *There exists a redundancy based compact error resilient tiling scheme in three dimensional assembly which can reduce the error from ϵ to ϵ^2 for any arbitrary Boolean functions f_1, f_2, and f_3, and it is shown in Figure 6.*

Construction. Before we describe the construction, we would like to emphasize on the wholeness of pad. Each side of the tile has one pad in Figure 6, that encodes a 5-tuple as shown in the Figure. Disagreement between corresponding elements of two such 5-tuples in any two pads results in the total mismatch between those two pads. Consider the tile $T(i, j, k)$ with inputs $U(i, j, k)$, $V(i, j, k)$ and $W(i, j, k)$ on the right, bottom and front pads respectively. Our goal is to guarantee one more error in the vicinity of this tile if there is one error in any of the input pads.

We add error checking parts to the right, bottom and front pads as shown in the Figure 6: $V(i, j, k)$ and $W(i, j, k)$ on right pad, $W(i, j, k)$ and $U(i, j, k)$ on bottom pad and $U(i, j, k)$ and $V(i, j, k)$ on front pad. Corresponding to these, we need to add $V(i + 1, j, k)$ and $W(i + 1, j, k)$ on left pad, $W(i, j + 1, k)$ and $U(i, j + 1, k)$ on top pad and $U(i, j, k + 1)$ and $V(i, j, k + 1)$ on back pad, as explained in the case of two-dimensional tile.

As described in two-dimensional assembly, every value in the output pads should be uniquely derivable from the values on the input pads. For $V(i + 1, j, k)$ and $W(i + 1, j, k)$ on the left pad we add $V(i+1, j, k)$ on the bottom pad, and $W(i+1, j, k)$ on the front pad. For $U(i, j + 1, k)$ and $W(i, j + 1, k)$ on the top pad, we add $U(i, j + 1, k)$ to

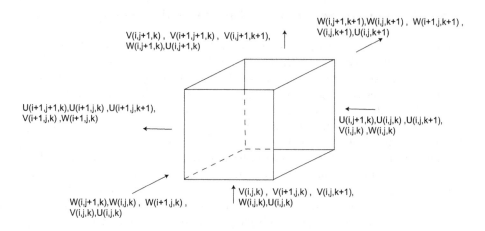

Fig. 6. Construction for error reduction to ϵ^2

the right pad and $W(i, j+1, k)$ to the front pad. For $U(i, j, k+1)$ and $V(i, j, k+1)$ on the back pad, we add $U(i, j, k+1)$ to the right pad and $V(i, j, k+1)$ to the bottom pad. The construction is complete with addition of $U(i+1, j+1, k)$ and $U(i+1, j, k+1)$ to left pad, $V(i+1, j+1, k)$ and $V(i, j+1, k+1)$ to top pad, and $W(i+1, j, k+1)$ and $W(i, j+1, k+1)$ to back pad.

This completes our description of a tile in the required tile-set. It should be noted that the number of different tile types in this tile set will be 64 times as compared to number of tiles in a tileset without any error-correction. It can be attributed to the two values for each of the $U(i, j+1, k)$, $U(i, j, k+1)$, $V(i+1, j, k)$, $V(i, j, k+1)$, $W(i+1, j, k)$ and $W(i, j+1, k)$, for every value of the inputs $U(i, j, k)$, $V(i, j, k)$ and $W(i, j, k)$.

Refer to [17] for detailed error analysis.

3.4   Error Reduction to ϵ^3

Theorem 7. *If Boolean functions f_1, f_2, and f_3 satisfy the following conditions:*

- *for fixed $V(i, j, k)$ and $W(i, j, k)$, $f_1(U, V, W)$ is input-sensitive to $U(i, j, k)$.*
- *for fixed $U(i, j, k)$ and $W(i, j, k)$, $f_2(U, V, W)$ is input-sensitive to $V(i, j, k)$.*
- *for fixed $U(i, j, k)$ and $V(i, j, k)$, $f_3(U, V, W)$ is input-sensitive to $W(i, j, k)$.*

Then there exists a compact error resilient scheme to reduce error from ϵ to ϵ^3 for three-dimensional self-assembly, and it is shown in Figure 6.

Refer to [17] for detailed proof.

3.5   Error Reduction to ϵ^4

Theorem 8. *For arbitrary Boolean functions f_1, f_2, and f_3, there exists no redundancy based compact error resilient scheme that can reduce error from ϵ to ϵ^4 in three-dimensional self-assembly.*

Refer [17] for proof.

4 Self-healing Tile Set for Three Dimensional Assembly

Winfree [25] provided the basis for studying self-healing in the self-assembly in a rigorous manner. We need to consider the repairability of a self-assembled structure in the face of a damage. A tile-set is called self-healing, if at any point during error-free growth, when n tiles are removed, subsequent error free growth will repair the damage rapidly [25]. Winfree's scheme of correctly repairing the damage (hole) is by ensuring that the holes are filled in the original forward direction of the algorithmic assembly and there is no backward growth in the holes.

Winfree proposed constructions of self-healing tile-sets for two dimensional algorithmic self-assembly by replacing a single tile by a 3×3 (for simple assemblies like sierpinsky triangles) , 5×5 (for general assemblies) and 7×7 (for additional robustness to nucleation errors) block. We can extend his construction to three-dimensions. We have discussed the construction of self-healing tile set by replacing a tile by $3 \times 3 \times 3$ block of tiles in [17].

5 Discussion

In this paper, we presented a theoretical analysis of redundancy based compact error resilient tiling in two and three dimensions. We conjecture the following stronger results for three-dimensional assemblies. Currently these conjectures are open questions. We state them without proofs as follows:

Conjecture 1. For arbitrary Boolean functions f_1, f_2, and f_3, there exists no redundancy based compact error correction scheme that will reduce error from ϵ to ϵ^3 in three-dimensional self-assembly.

Conjecture 2. For any functions f_1, f_2, and f_3 that are outside the restricted class of the functions defined in Theorem 7 there exists no redundancy based compact error correction scheme that will reduce error from ϵ to ϵ^3 in three-dimensional self-assembly.

Conjecture 3. For any Boolean functions f_1, f_2, and f_3, , there exists no redundancy based compact error resilient scheme that can reduce error from ϵ to ϵ^4 in three-dimensional self-assembly.

The immediate future work will be to prove or disprove these conjectures. We have presented a three-dimensional extension to Winfree's self-healing tile set in two-dimensions. It remains an open question if it is possible to design a compact self-healing tile set for two and three-dimensional self-assembly.

References

1. B.A. Bondarenko. *Generalized Pascal Triangles and Pyramids, Their Fractals, Graphs and Applications*. The Fibonacci Association, 1993. Translated from Russion and edited by R.C. Bollinger.
2. N. Bowden, A. Terfort, J. Carbeck, and G.M. Whitesides. Self-assembly of mesoscale objects into ordered two-dimensional arrays. *Science*, 276(11):233–235, 1997.

3. N. Chelyapov, Y. Brun, M. Gopalkrishnan, D. Reishus, B. Shaw, and L. Adleman. DNA triangles and self-assembled hexagonal tilings. *J. Am. Chem. Soc.*, 126:13924–13925, 2004.
4. H.L. Chen, Q. Cheng, A. Goel, M.D. Huang, and P.M. de Espanes. Invadable self-assembly: Combining robustness with efficiency. In *Proceedings of the 15th annual ACM-SIAM Symposium on Discrete Algorithms (SODA)*, pages 890–899, 2004.
5. H.L. Chen and A. Goel. Error free self-assembly using error prone tiles. In *DNA Based Computers 10*, pages 274–283, 2004.
6. T. D Clark, R Ferrigno, J Tien, K E Paul, and G M Whitesides. Template-directed self-assembly of 10-microm-sized hexagonal plates. *J Am Chem Soc.*, 124(19):5419–26, 2002.
7. N. Jonoska, S.A. Karl, and M. Saito. Three dimensional DNA structures in computing. *BioSystems*, 52:143–153, 1999.
8. T.H. LaBean, H. Yan, J. Kopatsch, F. Liu, E. Winfree, J.H. Reif, and N.C. Seeman. The construction, analysis, ligation and self-assembly of DNA triple crossover complexes. *J. Am. Chem. Soc.*, 122:1848–1860, 2000.
9. M.G. Lagoudakis and T.H. LaBean. 2-D DNA self-assembly for satisfiability. In *DNA Based Computers V*, volume 54 of *DIMACS*, pages 141–154. American Mathematical Society, 2000.
10. D. Liu, M. Wang, Z. Deng, R. Walulu, and C. Mao. Tensegrity: Construction of rigid DNA triangles with flexible four-arm dna junctions. *J. Am. Chem. Soc.*, 126:2324–2325, 2004.
11. C. Mao, W. Sun, and N.C. Seeman. Designed two-dimensional DNA holliday junction arrays visualized by atomic force microscopy. *J. Am. Chem. Soc.*, 121:5437–5443, 1999.
12. B. R. Martin, D. C. Furnange, T. N. Jackson, T. E. Mallouk, and T. S. Mayer. Self-alignment of patterned wafers using capillary forces at a water-air interface. *Advanced Functional Materials*, 11:381–386, 2001.
13. P.J. Paukstelis, J. Nowakowski, J.J. Birktoft, and N.C. Seeman. Crystal structure of a continuous three-dimensional DNA lattice. *Chemistry and Biology*, 11:1119–1126, 2004.
14. J.H. Reif. Local parallel biomolecular computation. In H. Rubin and D.H. Wood, editors, *DNA-Based Computers 3*, volume 48 of *DIMACS*, pages 217–254. American Mathematical Society, 1999.
15. J.H. Reif, S. Sahu, and P. Yin. Compact error-resilient computational DNA tiling assemblies. In *Proc. 10th International Meeting on DNA Computing*, pages 248–260, 2004.
16. P.W.K. Rothemund. Using lateral capillary forces to compute by self-assembly. *Proc. Natl. Acad. Sci. USA*, 97(3):984–989, 2000.
17. S. Sahu and J.H.Reif. Capabilities and limits of compact error resilience methods for algorithmic self-assembly in two and three dimensions. Technical Report CS-2006-04, Duke University, 2006.
18. R. Schulman and E. Winfree. Programmable control of nucleation for algorithmic self-assembly. In *DNA Based Computers 10*, LNCS, 2005.
19. N.C. Seeman. DNA in a material world. *Nature*, 421:427–431, 2003.
20. D. Soloveichik and E. Winfree. Complexity of self-assembled shapes. In *DNA Based Computers 10*, LNCS, 2005.
21. H. Wang. Proving theorems by pattern recognition ii. *Bell Systems Technical Journal*, 40:1–41, 1961.
22. George M. Whitesides and Bartosz Grzybowski. Self-assembly at all scales. *Science*, 295:2418 – 242, 2002.
23. E. Winfree. Complexity of restricted and unrestricted models of molecular computation. In R. J. Lipton and E.B. Baum, editors, *DNA Based Computers 1*, volume 27 of *DIMACS*, pages 187–198. American Mathematical Society, 1996.
24. E. Winfree. Simulation of computing by self-assembly. Technical Report 1998.22, Caltech, 1998.

25. E. Winfree. Self-healing tile sets. *Nanotechnology: Science and Computation*, 2006. Preprint. One-page abstract in proceedings of FNANO 2005.
26. E. Winfree and R. Bekbolatov. Proofreading tile sets: Error correction for algorithmic self-assembly. In *DNA Based Computers 9*, volume 2943 of *LNCS*, pages 126–144, 2004.
27. E. Winfree, F. Liu, L.A. Wenzler, and N.C. Seeman. Design and self-assembly of two-dimensional DNA crystals. *Nature*, 394(6693):539–544, 1998.
28. E. Winfree, X. Yang, and N.C. Seeman. Universal computation via self-assembly of DNA: Some theory and experiments. In L.F. Landweber and E.B. Baum, editors, *DNA Based Computers II*, volume 44 of *DIMACS*, pages 191–213. American Mathematical Society, 1999.
29. X. Xiong, Y. Hanein, J. Fang, Y. Wang, W. Wang, D. Schwartz, and K. Bohringer. Controlled multibatch self-assembly of microdevices. *Journal Of Microelectromechanical Systems*, 12:117–127, 2003.
30. H. Yan, L. Feng, T.H. LaBean, and J.H. Reif. Parallel molecular computation of pair-wise xor using DNA string tile. *J. Am. Chem. Soc.*, 125(47), 2003.
31. H. Yan, T.H. LaBean, L. Feng, and J.H. Reif. Directed nucleation assembly of DNA tile complexes for barcode patterned DNA lattices. *Proc. Natl. Acad. Sci. USA*, 100(14):8103–8108, 2003.
32. H. Yan, S.H. Park, G. Finkelstein, J.H. Reif, and T.H. LaBean. DNA-templated self-assembly of protein arrays and highly conductive nanowires. *Science*, 301(5641):1882–1884, 2003.

A Mathematical Approach to Cross-Linked Structures in Viral Capsids: Predicting the Architecture of Novel Containers for Drug Delivery

Thomas Keef

Department of Mathematics
Univerity of York
tk506@york.ac.uk

Abstract. Understanding the structure of viruses is an important first step in terms of many applications in virology, including the protein engineering of containers to enable more effective drug delivery. In particular, the viral capsids, i.e. the protective shells on the exterior of viruses containing the important genetic code, play an important role in the context of gene therapy, where small amounts of therapeutic DNA is packaged into a capsid which then penetrates the cell membrane and delivers its payload. Cross-linking structures are particular additional covalent bonds that can occur in addition to the already present hydrophobic interactions and hydrogen bonds between the proteins. Their importance lies in the fact that they render the capsid particularly stable. Here we shall introduce a mathematical method to predict possible locations for these additional bonds of cross-linking. We will give examples of failed cases as well as of cases where cross-linking structures are possible. These results serve as a pointer for experimentalists as to which types of cross-linking structures may possibly be engineered and exploited in the framework of drug delivery.

1 Introduction

Viruses are fascinating micro-organisms, consisting of a very compact genome and a protective protein shell that hijack host cells typically between one hundred or one thousand times their size. Viral capsids are shells constructed from many copies of one, or a few, identical protein subunits. In order to encode the locations of these proteins, viruses use symmetry so that the subunits are located in as few as possible equivalent locations. The most common symmetry observed in viral capsids is icosahedral, a structure with five-, three-, and two-fold rotational symmetries encoding 60 equivalent regions called fundamental domains, three per triangular face of an icosahedron. This naturally predicts the locations of the subunits in a capsid, with 60 copies of each protein in the corners of each of the triangular faces, clustered in groups of five around the vertices where the clusters are known as capsomeres. For viruses with more than 60 subunits in

C. Mao and T. Yokomori (Eds.): DNA12, LNCS 4287, pp. 239–249, 2006.

their capsids it is neccessary to subdivide each of the 60 fundamental domains into smaller areas. The first model which satisfactorally predicted a way to do this and also the possible numbers of protein subunits allowed in a capsid was suggested by Caspar and Klug [1] and concerns the division, or triangulation, of the 20 triangular faces of the icosahedral capsid into smaller triangular facets with protein subunits at the corners of each facet, and is known as Caspar-Klug Theory. This leads to a series of polyhedra, called the Caspar Klug series, that encode the surface structures of viruses in this way.

The allowed subdivisions are based on the possible ways of superimposing a smaller regular hexagonal lattice onto the larger lattice used for the overall icosahedron subdividing the triangular faces of an icosahedron into smaller triangular facets. The resulting members of the Caspar-Klug series are $T = h^2 + hk + k^2$ where $h, k \in \mathbb{Z}^{\geq 0}$ with the first three shown diagramatically in Fig. 1. The resulting shell encodes a virus with twelve capsomeres of five protein subunits, called pentamers, and $10(T - 1)$ capsomeres with six protein subunits, called hexamers, located on the vertices as predicted by the above lattice.

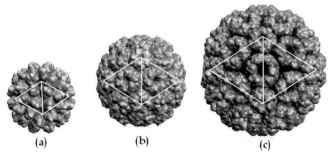

Fig. 1. The first three members of the Caspar-Klug series are (a) $T = 1$, (b) $T = 3$ and (c) $T = 4$ with two of the icosahedral triangles sketched to show the introduction of more protein subunits. All figures have been adapted from [2].

This model, widely accepted and used for classification and for three-dimensional reconstructions of viral capsids based on experimental results, cannot explain all viral structures, for example Papovaviridae fall out of the scope of this theory (see Rayment *et al.*[3] and Liddington *et al.*[4] for examples).

Viral Tiling Theory (VTT) has been introduced to close this gap and provides a more general framework to model the structures of viral capsids in terms of tessellations [5]. The dividing of a surface into a number of shapes is called a tiling with each individual shape known as a tile. These tiles are not only mathematical objects, but are geometric representations for *inter-subunit bonds* where the tilings encode the bonding structure between subunits. This suggests that also other types of bonds, such as *covalent bonds*, where two atoms share an electron, should be representable in a similar way. In particular, relaxing the assumptions on the construction of the tilings as discussed in Section 2 one obtains new tilings that encode the chainmail formed by the covalent bonds between capsid proteins. An example of a viral capsid with a covalently linked chainmail structure described in this way is HK97 [6] and Fig. 2(a and b) show a mature HK97 capsid and the rhomb tile which are used to describe its structure. In Fig. 2 the chainmail structure is shown along with the higher level tiling producing the interlinked covalently bonded rings over the capsid surface. These cross-linked bonds provide added stability to the capsid and could present an opportunity for protein engineering to create more stable, robust containers for drug delivery. It is necessary to produce a classification of the viruses that have structures consistent with the requirements for cross-linking in order to predict suitable candidates as a pointer for experimentalists.

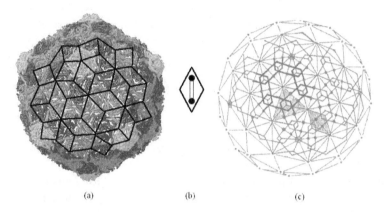

(a)   (b)   (c)

Fig. 2. (a) The surface structure of HK97 can be described by rhomb tiles with inter-subunit bonds between capsomeres as shown in (b). (c) The interlinked cova-lently bonded rings are shown along with the rhomb tiles and also the triangular tiles used to encode the chainmail structure. (a) has been adapted from [7] and (c) from [8].

2 A New Method for the Prediction of Cross-Linking Structures in Viral Structures

Here we will show how to create the tilings required to encode crosslinked structures from the tilings which describe the structure of the viral capsid. The technique is to superimpose a further set of tiles over those which encode the inter-subunit bonding structure of the viral capsid. To produce a successful tiling, one must follow the rules below as set out in [8]:

1. The decorations of the original tiling and all higher-level tilings superimpose. This is required due to the fact that the decorations indicate the locations of the protein subunits in the capsid.
2. Higher-level tilings have decorations on the edges of the tiles and each edge has at least one decoration.

3. As with all tiles in Viral Tiling Theory, the tiles in higher-level tilings represent inter-subunit interactions, in this case covalent bonds, and are given by connections between the decorations. Each subunit participates in precisely one bond per tile, and a line representing a bond connects decorations located on different edges on the tiles such that the overall symmetries of the tiles are not changed.

To create the higher level tiling we first need the tiling structure for the capsid. The new tiles should have vertices at the same locations as the decorated vertices for the capsid tiling. The edges should then pass through the decorations (or proteins in the capsid), bisecting the centres of the first tiling. For a model to have a successful crosslinked structure we require interlinked rings which do not self intersect and whose overall length is not such that resulting rings are so long that they result in a tangled web of bonds over the surface of the capsid.

We will first investigate the higher level tilings for Caspar-Klug type viruses and show why cross-linked structures cannot be observed. We will then provide an example of a non cross-linked virus with possibilities for the locations of covalent bonds resulting in a cross-linked structure that could be useful for the engineering of more robust virus-like particles.

2.1   Application to Caspar-Klug Type Viruses

Viruses with 12 pentamers and otherwise hexamers with trimer interactions between capsomeres can be described by triangular tilings as shown in Fig. 3(a). The higher level tilings for these viruses consist of rhombs with one decoration on each edge (see Fig. 3(b)).

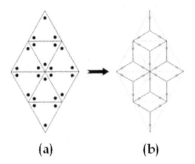

(a)   (b)

Fig. 3. Caspar-Klug triangular tilings admit rhomb higher level tilings with one decoration on each edge

The options for the lines encoding the possible covalent bonds between protein subunits that obey rule 3 above are shown in Fig. 4. None of these options produce cross-linking, two providing no new bonds and the other, Fig. 4(a), simply producing long bond lines across the capsid surface which suggests cross-linking in Caspar-Klug Theory is not possible.

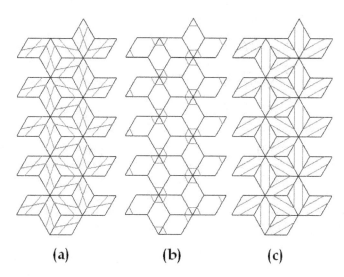

(a)   (b)   (c)

Fig. 4. Possibilities for cross-linking structures for a $T = 1$ Caspar-Klug type virus with rhomb shaped higher-level tilings. None produce cross-linking structures. (a) This rhomb produces long lines around the viral capsid. (b) Here we have intra-capsomere bonds, i.e. those bonds within the pentamers. (c) Here the covalent bonds would follow the inter-capsomere bonds. We remark theat larger (higher T number) viruses have the same general structures.

2.2   Application to a Particle of MS2 Type

MS2 is an icosahedral RNA bacteriophage classified as $T = 3$ with a capsid formed by 180 copies of the coat protein arranged in clusters of three proteins, or trimers, with dimer interactions between them. It is represented in VTT by a rhomb tiling with decorations on the vertices representing the locations of the proteins a patch of which is shown in Fig. 5(a). The corresponding higher level tiling, a tiling with hexagons and pentagons, each with two decorations along each edge, is shown in Fig. 5(b) along with the options for covalent bonds which retain the symmetries of the underlying hexagonal and pentagonal tiles respectively. We shall refer to the tiles as **P1** to **P4** for the four pentagonal tiles and **H1** to **H5** for the five hexagonal tiles. The lines have been drawn as straight lines, although any line between the two points could be drawn as long as the resulting symmetry of the tiles is the same as for the tile with no decorations. The case where this is easiest to show is the tile **P3** where all lines cross over in the centre of the tile: it is unrealistic, because the lines represent polypeptide chains with associated volumes, so it would be unphysical to have all of the chains stacked one on top of another. However the lines may be chosen in a way compatible with symmetry and avoiding cross-overs.

For a $T = 3$ structure such as MS2 it is necessary to look at the twenty combinations of hexagonal and pentagonal tiles and investigate the possible cross-linked structures resulting from them. The possibilities include cross-linked

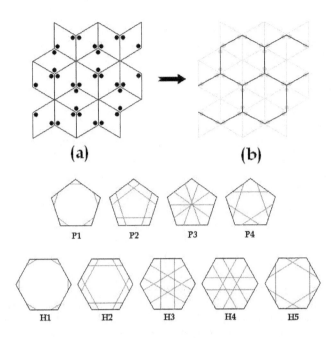

Fig. 5. Trimers with dimer interactions between them admit higher level tilings consisting of hexagons and pentagons, each with two decorations on each edge. The higher level tiles are shown with all types of decorations corresponding to the locations of covalent bonds which retain the rotational symmetry of the tile. The lines in **P3** have been left as straight lines, but could be drawn in any way between the points as long as the five-fold rotational symmetry in the tile is retained.

covalent models, i.e. those where the closed covalent rings intersect each other in a 'chainmail' structure, and also models which do not produce cross-linked structures for a variety of reasons. One failed cross-linking model is produced by the tiles **H1** and **P1** because instead of a cross-linking structure, intra-capsomere bonds are generated. We have also ruled out models where there is no cross-linking, where the covalent bondings overlap themselves, referred to as self-intersecting chains, and also where the chains are extremely long, called tangled chains, that result in a cobweb of chains over the whole surface of the virus, because these are not expected to contribute significantly to the stability of the capsid.

These results rule out thirteen of our options shown in Table 1. Some of the seven remaining options have furthermore been excluded for different reasons which we shall now indicate: For capsids whose possible cross-linked structures have long or overly involved chains we feel that it would be much more difficult to engineer them. This suggests that although the structures are in principle possible, practically they would not be a viable choice for protein engineering to create new, more stable structures. To decide which of these should be discarded we have to look at the individual characteristics of each possible structure with

Table 1. Table showing possible options for cross-linked structures for a $T = 3$ virus of MS2 type formed from hexagonal and pentagonal higher-level tiles. "None" indicates that there is no crossing of chains present, "SI" indicates that the resultant chains are self-intersecting, "T" indicates long tangled chains across the capsid surface, and "CL" the occurrence of cross-linking.

	P1	P2	P3	P4
H1	None	SI	SI	SI
H2	CL	CL	T	CL
H3	SI	T	T	SI
H4	CL	T	CL	CL
H5	SI	SI	SI	CL

the view of ease of creating new virus-like particles with the covalent bonds in the model. The cases using the tiles **P3, H3** and **H4**, have a large number of crossovers and are therefore difficult to engineer. One such example is **P4H4**, i.e. the tiling corresponding to a combination of **P4** and **H4** tiles shown in Fig. 6(a). As the covalent bonds are between proteins already present in the capsid, the distance between proteins on the edges of the tiles is too great for one proteins polypeptide chain to reach the other protein.

Of the remaining four possible cross-linked structures indicated in the table (**P1H2, P2H2, P4H2** and **P4H5**) only **P1H2** provides a possible reason for exclusion. This is because the bonds encoded by the tile **P1** are the intra-capsomere bonds within the trimer itself. The resulting 'chainmail' has two types of chains, one with four edges of two distinct lengths and another

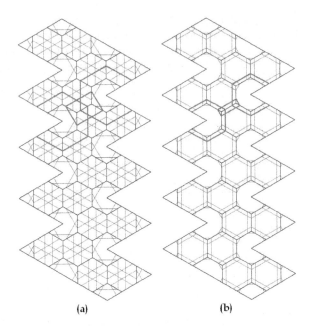

Fig. 6. (a) **P4H4** is not a suitable model for cross-linked structures due to the long edge lengths in the five sided rings surrounding the five fold vertices. (b) **P1H2** has two very distinct rings, one with four edges having two edge lengths and one with a five-membered ring having equal length sides. The models are sections of an icosahedral lattice, so the hexagons with sections cut out correspond to pentagons when the overall lattice is folded up. The resulting icosahedron is a qualitative model and as such bond angles and lengths are not accurate.

with five edges each of the same length. The two different edge lengths and the tight curves required for the polypeptide chains at the ends with shorter edges would suggest that **P1H2** with its four-edged chains is not a suitable model for the creation of more stable engineered capsid-like particles. The short distance between the long chains would make it difficult to form interlinked catenanes from the capsid proteins. A diagramatic example of this is shown in Fig. 6(b). The resulting three possibilities for the creation of cross-linked virus-like particles for a $T = 3$ type viral capsid whose capsomeres are trimers with dimer interactions between them are shown in Fig. 7. These are: **P2H2** which has five- and six-membered rings around the pentamers and hexamers respectively, **P4H2** which has ten-membered rings centred on a five-fold rotational symmetry axis, and **P4H5** which consists of three-membered rings everywhere centred over the trimers, or local three fold axes. We feel that of these three options **P2H2** would be the model which provides the highest stability for the capsid. This is due to the fact that three loops intersect locally, as opposed to only two loops for **P4H5**. Moreover, **P2H2** is preferable to **P2H2** because it has smaller sized loops. It could, however, be the case that the model shown in

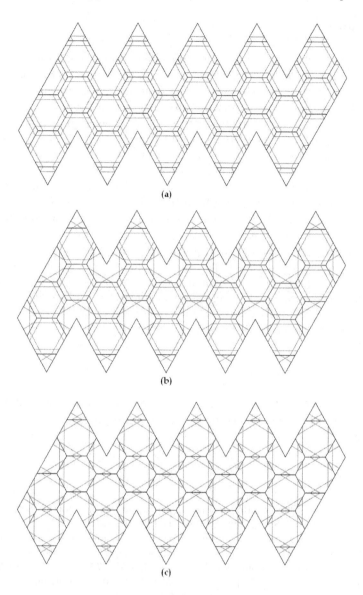

Fig. 7. The three possibilities for successfully engineered cross-linked structures for a $T = 3$ virus with trimeric capsomeres such as MS2. (a) **P2H2** (b) **P4H2** (c) **P4H5**. The lattices represent nets for icosahedra and can be thought of as qualitative cartoons showing the locations of the bonds, but not necessarily the exact arcs of the polypeptide chains, as stated earlier.

Fig. 7(c) is easier to engineer due to the relatively uncomplicated cross-linking structures with the loops only having three edges around the outside of each trimer, and each loop only crosslinking with one of its neighbouring loops at a time.

3 Outlook

Based on Viral Tiling Theory we have provided an explanation for the lack of cross-linking in Caspar-Klug type viral capsids, i.e. those consisting of hexamers and pentamers with trimer interactions between them. We have also shown that for a $T = 3$ particle, whose capsomeres are trimers with dimer interactions such as MS2, there are twenty different structures with covalent bonding patterns over the capsid. We have reduced these down to three models which we consider to be viable choices for the creation of cross-linked virus-like particles. All of these have closed interlinked rings centered on local symmetry axes, i.e. symmetry axes that have a local environment invariant under rotations. The number of intersections along these rings has been restricted in order to focus on models that are both easy to realise from a practical, protein engineering, point of view, and correspond to more stable structures.

Work in progress considers the classification of cross-linking structures for viruses and virus-like particles in general [9]. The use of virus-like particles for drug delivery has been well documented, see for example a review by Garcea and Gissmann [10]. We hope that our classification paves the way for the engineering of such particles with higher stability targeted at the production of new capsid-based vectors for the medical community.

Acknowledgments

I would like to thank Reidun Twarock, my PhD supervisor, for helpful discussions. This work has been funded by the EPSRC LSI programme, grant number GR/T26979/01.

References

1. Caspar, D.L.D., Klug, A.: Physical principles in the construction of regular viruses. Cold Spring Harbor Symp. Quant. Biol. **27** (1962) 1–24
2. Reddy, V.S., Natarajan, P., Okerberg, B., Li, K., Damodaran, K., Morton, R., Brooks, C.r., Johnson, J.: Virus particle explorer (viper), a website for virus capsid structures and their computational analyses. J. Virol. **75**(24) (2001) 11943–11947
3. Rayment, I., Baker, T., Caspar, D.L.D., Murakami, W.T.: Polyoma virus capsid structure at 22.5Å resolution. Nature **295** (1982) 110–115
4. Liddington, R.C., Yan, Y., Moulai, J., Sahli, R., Benjamin, T.L., Harrison, S.C.: Structure of simian virus 40 at 3.8-Å resolution. Nature **354** (1991) 278–284
5. Twarock, R.: A tiling approach to virus capsid assembly explaining a structural puzzle in virology. J. Theor. Biol. **226**(4) (2004) 477–482
6. Ross, P., Cheng, N., Conway, J., Firek, B., Hendrix, R., Duda, R., Steven, A.: Crosslinking renders bacteriophage hk97 capsid maturation irreversible and effects an essential stabilization. The EMBO Journal **24**(7) (2005) 1352 – 63
7. Conway, J.F., Wikoff, W.R., Cheng, N., Duda, R.L., Hendrix, R.W., Johnson, J.E., Steven, A.C.: Virus maturation involving large subunit rotations and local refolding. Science (292) (2001) 744–48

8. Twarock, R., Hendrix, R.W.: Crosslinking in viral capsids via tiling theory. J. Theor. Biol. (2005)
9. Keef, T., Twarock, R.: Classification for crosslinking in viral capsids based on tiling theory. ((*Work in progress*))
10. Garcea, R.L., Gissmann, L.: Virus-like particles as vaccines and vessels for the delivery of small molecules. Curr. Opin. Biotechnolgy **6**(15) (2004) 513–7 Review.

A Framework for Modeling DNA Based Molecular Systems*

Sudheer Sahu, Bei Wang, and John H. Reif

Department of Computer Science, Duke University
Box 90129, Durham, NC 27708-0129, USA.
{sudheer, beiwang, reif}@cs.duke.edu

Abstract. In this paper, we propose a framework for a discrete event simulator for simulating the DNA based nano-robotical systems. We describe a physical model that captures the conformational changes of the solute molecules. We also present methods to simulate various chemical reactions due to the molecular collisions, including hybridization, dehybridization and strand displacement. The feasibility of such a framework is demonstrated by some preliminary results.

1 Introduction and Related Work

Recent research has explored DNA as a material for self-assembly of nanoscale objects [19,47,58,75,96,100,101], for performing computation [1,11,9,10,55,54,57,94,95,97], and for the construction of nano-mechanical devices [2,20,21,28,53,59,86,70,76,77, 78,79,89,88,102,106,107]. One key application of an autonomous unidirectional DNA device is to perform computation. Recently Yin proposed the design of an autonomous universal turing machine and cellular automata [105, 104]. One potential application beyond computation is the design of a controllable moving device which can be integrated into a DNA lattice for efficient transportation. One major challenge in the design of DNA based devices is the cost and time required for the experiments. Computer simulations can be performed to capture the essential physical and chemical properties, and serve as an effective tool in the design process.

Our method of simulation is different from the commonly used Gillespi algorithm [33,42,34,90,32,69]. In the system of our interests, the geometry of the nano-structures plays an important role apart from the concentrations of the reactants and the reaction rates. Physical simulations are performed to model the molecular conformations and the chemical reactions are monitored explicitly.

Sales-Pardo et. al. modeled a ssDNA as a bead-pin rotational polymer chain and used a modified Monte Carlo simulation to investigate the dynamics of a single-stranded DNA and its associated hybridization events [72]. The geometric constraints of the nucleic chain was handled by a lattice model [72]. Isambert and Siggia modeled RNA helices as rods and single stranded RNA as Gaussian chains [38]. Kinetic Monte Carlo method was used to sample RNA conformational changes [38]. They also used the short-scale and the large-scale conformation descriptors, i.e. *nets* and *crosslinked gel*, to model geometric constrains related to complex RNA folding conformations. Bois

---

* The work is supported by NSF EMT Grants CCF-0523555 and CCF-0432038.

et. al. investigated the possible effects of topological constraints in DNA hybridization kinetics [13]. Recently Dirks et. al. developed an algorithm aiming at analyzing the thermodynamics of unpseudo-knotted multiple interacting DNA strands in a dilute solution [27].

In this paper, we describe a framework for the design of a discrete event simulator, which simulates DNA based nano-robotical devices. Section 2 gives an overview of the system. Section 3 describes the physical simulation of the molecules. Section 4 discusses the event simulation based on the kinetic and thermodynamic studies. Section 5 describes the adaptive time-steps to optimize the physical simulation, and Section 6 describes the analysis of the complete algorithm. Section 7 presents some preliminary results to support such a framewok. Discussions and future work is described in Section 8. It should be noted that in this paper, we present the framework for building such a simulator and not the simulator itself.

2 Discrete Event Simulation

The simulator performs the molecular-level simulations and provides an useful tool to study DNA based nano-mechanical devices. It has two major components. The first component is the physical simulation of the molecule conformations. The second component is the event simulation (hybridization, dehybridization and strand displacement events) which depends on the kinetics, thermodynamics and geometry of the molecules. Due to the large number of molecules in a given solution, we sample and simulate molecules within a small cell volume, assuming the solution is well mixed.

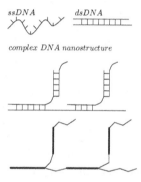

Fig. 1. Schematic view of the molecules in the modeled system. Bold solid lines represent the WLC model used for ds-DNA segments while thin solid lines represent the WLC model used for ssDNA segments.

The modeled system consists of three types of molecules, single-stranded DNA (ssDNA), double-stranded DNA (dsDNA) and complex DNA nano-structure with both single-stranded and double stranded segments, as shown in Figure 1. We assume no self-hybridization and no pseudo-knots formation for the complex DNA nano-structures. Therefore, to the first approximation, the complex DNA nano-structure is reducible to a collections of WLC segments with different parameters, i. e. persistence length. For more complicated DNA nano-structures, we can adapt the geometric descriptors used in [38, 13], as discussed in section 8.

During the simulation, three types of reaction take place in the solution: the hybridization between a pair of ssDNA segments with complementary base-pairing, the dehybridization of the dsDNA portion of a nano-structure and the strand displacement. The DNA molecule contains potential hybridization sites at its free-end (sticky ends). During the simulation, when two molecule come into contact (reactive collision), a potential hybridization event is reported. The corresponding free-end base-pairs are investigated to determine the probability of its actual occurrence. Strand displacement is a reaction in which two strands compete against each other to hybridize with a common strand as shown in Figure 2. Strand B and C compete against each other to hybridize with strand A. At a time instance, B (or C) makes one more bond with A and removes one bond of C (or B).

The required discrete event simulation with Δt as the time-interval is described as follows. Algorithm 1 describes the major steps of the simulation. m_i is a data structure that stores individual molecular configurations, including sequence and secondary structure. MQ stores all m_i in the system. T is the total simulation time. Δt is the simulation time per step. $Initialize$ is a function that initialize the MQ based on the user input. The detailed algorithms are described in the subsequent sections.

Algorithm 2 describes steps involved in generating random conformations for all molecules in the system. $Enqueue$ and $Dequeue$ are standard queueing operations that insert and delete an element in the queue. $MCSimulation(m)$ generates new conformation for the molecule m.

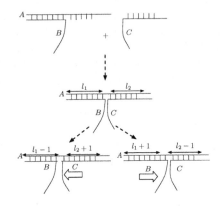

Fig. 2. Strand displacement: molecule B and C compete against each other to hybridize with molecule A

Algorithm 3 describes reactive collision detection which leads to potential hybridization events. $Collide(m_i, m_j)$ returns true if the sticky ends of molecule m_i and m_j collide. e is a data structure that stores an event (hybridization, dehybridization or strand displacement), including all the molecular configurations involved in the event and inter-molecule relations. For example, in the case of hybridization, it stores the molecular configurations and the information of the hybridization sites. $HEvent(m_i, m_j)$ creates a potential hybridization event based on colliding molecule m_i and m_j. HQ stores all potential hybridization events.

Algorithm 4 presents algorithm involved in hybridization. $Hybridize(e)$ probabilistically determines the hybridization product based on the change in free energy as described in Section 4. $PotentialSD(e)$ returns true if event e is a potential strand-displacement event. SDQ stores all potential strand-displacement events. $Update(MQ, e)$ updates the configurations of the molecule in the system based on the occurred event e.

Algorithm 5 describes dehybridization event. $PotentialD(m)$ returns true if molecule m could potentially dehybridize. $Dehybridization(m)$ probabilistically dehybridizes molecule m.

Algorithm 6 shows the steps involved in the strand displacement event. $StrandDisplacement(e, \Delta t)$ probabilistically proceeds with the strand displacement event e within time frame Δt. $IncompleteSD(e)$ returns true if the strand displacement event has not completed within the given time frame.

Algorithm 1. Discrete Event Simulation

1: Initialize(MQ)
2: **while** $t \leq$ T **do**
3:     $t = t + \Delta t$
        {PHYSICAL SIMULATION}
4:     Physical simulation
5:     Collision detection
        {EVENT SIMULATION}
6:     Hybridization
7:     Dehybridization
8:     Strand displacement
9: **end while**

Algorithm 5. Dehybridization

1: **for** $\forall m_i \in MQ$ **do**
2:     **if** PotentialD(m_i) **then**
3:         $e = $ Dehybridization(m_i)
4:         Update(MQ, e)
5:     **end if**
6: **end for**

Algorithm 2. Physical Simulation

1: **for** $\forall m_i \in MQ$ **do**
2:     MCSimulation(m_i)
3: **end for**

Algorithm 6. Strand Displacement

1: **while** SDQ is NOT empty **do**
2:     $e = $ Dequeue(SDQ)
3:     $e^* = $ StrandDisplacement($e, \Delta t$)
4:     **if** IncompleteSD(e^*) **then**
5:         Enqueue(SDQ^*, e^*)
6:     **end if**
7:     Update(MQ, e^*)
8: **end while**
9: $SDQ = SDQ^*$

Algorithm 3. Collision Detection

1: **for** $\forall m_i, m_j \in MQ, i \neq j$ **do**
2:     **if** collide(m_i, m_j) **then**
3:         $e = $HEvent($m_i, m_j$)
4:         Enqueue(HQ, e)
5:     **end if**
6: **end for**

Algorithm 7. MCSimulation (m), $m \in MQ$

1: $m^* = $RandomConformation($m$)
2: **if** SelfCollision(m^*) **then**
3:     continue to next iteration
4: **end if**
5: $\Delta E = E(m^*) - E(m)$
6: **if** ($\Delta E > 0$) **then**
7:     $x \in_{var} [0, 1]$
8:     **if** ($x > \exp - \frac{\Delta E}{K_B T}$) **then**
9:         continue to next iteration
10:     **end if**
11: **end if**
12: $m = m^*$

Algorithm 4. Hybridization

1: **while** HQ is NOT empty **do**
2:     $e = $ Dequeue(HQ)
3:     $e^* = $ Hybridize(e)
4:     **if** PotentialSD(e^*) **then**
5:         Enqueue(SDQ, e^*)
6:     **end if**
7:     Update(MQ, e^*)
8: **end while**

3 Physical Simulation

The discrete worm-like chain model (WLC) is used to model the polymer-like DNA molecules in solution. Monte Carlo (MC) computer simulations are used to determine their conformations.

3.1 Discrete Wormlike Chain Model

The advancement of single molecule dynamics offers experimental validations of various DNA polymer models, among which Gaussian Chain Model, Freely-Jointed Chain (FJC) and Worm-Like Chain (WLC) are widely investigated [67,45,39,81,65,29,44,5, 99,80,48,14,15,46]. The choice of a polymer model depends on the physical property of the DNA chain, affordable computation and molecular-details of interest [26].

Our simulation is constructed using the discrete wormlike chain model. Marko and Siggia used the model to derive the elastic theory suitable for DNA and further completed the model to include bending and twisting elasticity of DNA and the free energy required for deformation [60, 61]. Bustamante el al. proposed an interpolation of the Marko-Siggia model for fitting and experimental elasticity curve of single DNA molecules [15]. Klenin et al. modeled linear and circular DNA where the DNA polymers are represented by a WLC of stiff segments connected by bending torsion and stretching potentials [43]. Tinnoco et. al. used WLC as their polymer chain conformation to investigate force effect on thermodynamics and kinetics of single molecule reaction [87]. Larson et al used a similar model to predict the behavior of tethered dsDNA in a constant-velocity flow [51, 25]. Experimental data has shown some reasonably good agreement with the model [64].

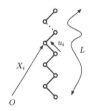

Fig. 3. WLC model

The DNA molecule (Figure 3) is initialized as $N + 1$ beads $(0, 1..N)$ connected by N mass-less extendable segments (springs) of the same length [25,31,52]. The contour length of the chain is L. The position of the bead i is denoted as $\mathbf{x}_i$. The segment vectors are given by

$$\mathbf{u}_i = \mathbf{x}_i - \mathbf{x}_{i-1} \tag{1}$$

Therefore the chain is represented by a set of $N + 1$ vectors $\mathbf{x}_0, \mathbf{x}_1, \mathbf{x}_2, ..., \mathbf{x}_N$ [18]. We use WLC to model ssDNA, dsDNA and complex DNA nanostructure. Specifically for complex DNA nanostructure, different parameters are applied to different segments of the chain depending on whether the segment is double-stranded or single-stranded.

3.2 Monte Carlo Simulation

The molecules are simulated through Monte Carlo simulation for a desired number of time steps as Algorithm 7. According to the Metropolis algorithm used in the simulation, $E(m)$ is the energy associated with conformation of molecule m. The computation of $E(m)$ will be discussed in a later section. ΔE is defined as the energy change of the system due to the new conformation. K_B is the Boltzman constant, and T is the absolute temperature. MQ is the set of all molecules in the simulation. $RandomConformation$ is a function that achieves a new conformation of the molecule through random walk in three dimension. *SelfCollision* detects and excludes the self-crossing conformations. The detail algorithm is shown in Algorithm 7. Similar methods have been used in [108,7,56]. To achieve random conformation of the molecules, more permutations can be used such as random rotation of an interval chain or bond-length change, which are described in [108].

3.3 Collision Detection: Cylinder Model

To simulate the motion of the molecule, each segment occupies a finite volume. Instead of using repulsive forces (weak and short-ranged) to maintain the excluded volume of the individual segments of the chain [17], for two disjoint segments, we assume a minimum distance D between them in three-dimension. In other words, we assume each segment is a cylinder with a certain radius R, when two cylinders contact ($2R \geq D$), a collision occurs (Figure 4). If the two cylinders belong to the same chain, the self-avoiding criteria is violated. If the two cylinders belong to neighboring DNA molecules, a potential hybridization event occurs.

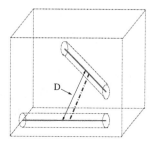

Fig. 4. Collision detection in 3D

3.4 Random Conformation

The random conformation of the DNA molecule is generated by a random walk in three dimension. Based on [6],

$$\Delta \mathbf{x}_i = \mathbf{R}_i \qquad (2)$$

where $\Delta \mathbf{x}_i$ is the change of $\mathbf{x}_i$ in time step Δt, $\mathbf{R}_i$ is the random displacement. Let D be the diffusion coefficient, we assume $\mathbf{R}_i$ as a Gaussian random variable which is distributed according to

$$W(\mathbf{R}_i) = (4A\pi)^{-3/2} \exp(-\mathbf{R}_i/4A) \qquad (3)$$

where $A = D\Delta t$. The diffusion coefficient D of a macromolecule in an ideal dilute solution is computed according to $D = K_B T/f$, where f is the hydrodynamic frictional coefficient of the macromolecule [83]. f of a rigid, rod-like molecule can be written as $f = 3\pi \eta L/(\ln \rho + \gamma)$, where η is the viscosity of the solution, L is the length of the DNA molecule, ρ is the axial ratio and γ is a correction for end effects [83].

3.5 Energy

Now we describe how we calculate $E(m)$ as stated in Algorithm 7. Our current simplified model neglects the following energies though more accurate model should take them into consideration [108, 24]: pairing potential between complementary bases, stacking energy from the vertical interactions between neighboring base pairs and hydrodynamic interaction energy with the solvent. We shall consider the torsional rigidity in the forms of bending torque and twisting torque for the DNA molecules in a more sophisticated model. The total energy of a DNA conformation is given as the sum of stretching, bending, twisting and electrostatic interaction energy among negatively charged phosphate groups along the chain [43, 108, 49], which are denoted as E^s, E^b, E^t and E^e, respectively.

$$E^{total} = E^s + E^b + E^t + E^e \tag{4}$$

Stretching Energy. The stretching energy is defined as

$$E^s = \frac{1}{2} Y \sum_{i=1}^{N} (u_i - l_0)^2 \tag{5}$$

where l_0 is the segment equilibrium length, Y is the stiffness parameter defined previously [108].

Please refer to [74] for description of the *bending energy, twisting energy, electrostatic energy*, and other physical models.

3.6   Parameters

We use WLC model for both ssDNA and dsDNA for modeling consistency, it is important to notice that there are different set of parameters used for each of them.

Parameters for ssDNA. Let L be the contour length of the ssDNA, $L = l_{bp}N_{bp} = l_0N$. l_{bp} is the length of the ssDNA per base pair. N_{bp} is the number of bases. N is the number of beads (monomer) in our WLC model. l_0 is the length per segment. The average length of ssDNA in the system is approximately $25 - 30\ bp$. According to [103], $l_{bp} = 0.7\ nm$. Many groups have obtained the force/extension data for ssDNA in different salt environment [108, 81, 71, 8, 16]. Parameters used in our model is obtained from [108], where $l_0 = 1.5\ nm$ and $Y = 120\ K_BT/nm^2$. The persistence length $P = 0.7\ nm$ [81]. The diffusion coefficient D of ssDNA is obtained from [83] as approximately $1.52 \times 10^{-6}\ cm^2 s^{-1}$ for a 20 bp strand. The diameter of the ssDNA backbone is $1\ nm$ [23].

Parameters for dsDNA. For dsDNA, the parameters associated with the equations are difference, i.e. $l_0 = 100\ nm$ [43, 22, 62], $P = 50\ nm$, $Y = 3K_BT/2P$ [22, 84], $l_{bp} = 0.34\ nm$ [103], and $D = 1.07 \times 10^{-6}\ cm^2 s^{-1}$ [83]. For short dsDNA segment (20 bp), WLC model can be simplified as the straight, rigid cylinder model with reasonable adequacy [3, 62]. WLC models are used for simulation consistency.

3.7   Motion of the Complex Nano-structure

The MC simulation described previously can can applied to the the complex nanostructure. Since it is reducible to a collection of ssDNA and dsDNA WLC segements, perturbations of each segment is done independently. The total energy is computed as a summation of the energies associated with individual segments. For more accurate model, loop energy and the energy associated with each branching point should also be considered.

3.8   Physical Model for Hybridization

Though extensive research has been done for RNA folding simulation [30, 98], to the best of our knowledge, there is no empirical results that describe: 1) the location of contact that initializes the hybridization; 2) the motions of each individual strands during the hybridizations; nor 3) the actual physical location of the hybridized products relative to other molecules in the system.

We make the following hypothesis: 1) location of contact is not explicitly modeled in the simulation; 2) upon collision that leads to potential hybridization, two strands immediately align their bases involved in the formation of duplex with the right orientation; 3) during the hybridization process, the displacement of the two strands is inversely proportional to their mass (or number of bases in the structure). The model can be subsequently improved as the empirical evidence become available. Figure 5 illustrates one schematic to depict our hypothesis.

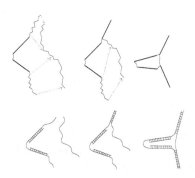

Fig. 5. Figure illustrates various steps wrt the physical motion of the strands during hybridization

4   Event Simulation

In the event simulation module, we use thermodynamics and kinetics principles to calculate the probabilities of various events. Possible events in our systems are hybridization, dehybridization (melting/dissociation) and strand displacement.

4.1   Hybridization

The nearest-neighbor (NN) model is used to model the hybridization event [41]. The model assumes that the stability of a given base-pair depends on the identity and orientation of neighboring base pairs [41]. Empirical data are used to determine parameters for all possible alignments of base pairs. The model has been shown to describe the thermodynamics of DNA structure that involves mismatches and neighboring base pairs beyond the Watson-Crick pairs [68,73]. When a potential hybridization event that involves molecules m_1 and m_2 is detected due to a collision, the simulator examines all possible alignments of m_1 and m_2. For hybridization according to alignment i, its free energy ΔG_i° is computed using the NN model. Let $m_1 m_2{}^i$ be its hybridization product. Let p_i be the stability measurement of $m_1 m_2{}^i$, $p_i \propto \exp(-\Delta G_i^\circ / RT)$. Let P_h^i be the probability of hybridization according to alignment i, for all p_j that exceeds a given threshold, we have

$$P_h^i = \frac{p_i}{\sum_j p_j} \qquad (6)$$

4.2 Dehybridization

Let P_d be the dehybridization probability of a molecule $m_1 m_2$ and $[m_1 m_2]$ be its concentration. Let k_r be the reverse rate constant, R_r be the reverse rate, where $R_r = k_r[m_1 m_2]$. The number of molecules dehybridized in time Δt is $R_t \Delta t$. Therefore the probability that the molecule $m_1 m_2$ dehybridizes in Δt can be approximated as

$$P_d = \frac{k_r[m_1 m_2]\Delta t}{[m_1 m_2]} = k_r \Delta t \qquad (7)$$

Refer to [74] for more details.

4.3 Strand Displacement

Strand displacement is modeled as a random walk in which the direction of movement of the branch point along the DNA is chosen probabilistically and is independent of its previous movements. It has been shown that the branch migration and strand displacement is a biased random walk due to mismatches [12]. In other words, migration probability towards the direction with mismatches are substantially decreased. Based on Figure 2, molecule ABC is denoted as the DNA nanostructure involving molecule A, B and C before the strand displacement. Let G°_{ABC} be its free energy. Denote G°_{rABC} and G°_{lABC} as the free energy of ABC after 1 base pair migration towards right, and left, respectively. Let $\Delta G^\circ_r = G^\circ_{rABC} - G^\circ_{ABC}$ and $\Delta G^\circ_l = G^\circ_{lABC} - G^\circ_{ABC}$. Let p_r be the probability of the right-directional migration and p_l be the probability of the left-directional migration. It has been shown in [12] that $p_r \propto \exp(-\Delta G^\circ_r / RT)$, similarly $p_l \propto \exp(-\Delta G^\circ_l / RT)$, where the change of free energies can be computed by the NN model described previously.

5 Adaptive Time Step

We use adaptive time steps in our simulation. The simulation captures various processes at different time-scales. Ideally, the smallest time unit should be chosen as the time step $\delta t \sim 10^{-6}$ to resolve the conformations and trajectory of each individual molecule using the WLC model and MC simulation. Inspired by ideas in the kinetic Monte Carlo method [92], long-time system dynamics of the system consists of diffusive jumps from state to state. There are series of simulation steps where no collisions take place and molecules remain far apart. We attempt to overcome the limitations of such a short time-scale approach. In other words, we differ the time intervals between long-ranged molecules and short-ranged molecules.

If all the strands are far apart, we can guarantee that within a particular time-interval δT there will not be any collisions. We treat each molecule in the system as a unit of rigid body and assign a random momentum to each unit. We apply this large-scale time step δT as the simulation step at that instance. δT moves the entire system from state to state, which is computational efficient. We store the distance between the closest pair of potential reactive molecules. As the distance reaches a given threshold where the conformations of molecules can no longer being ignored, we change to a smaller-scale time step δt.

6   Algorithm Analysis

The major portion of the time taken by the algorithm is in the physical simulation, so it suffices to analyze the time-complexity of the physical simulation of the molecules in the system. The discrete WLC model is used to replace the continuos WLC for computational simplification. As the (discrete) WLC consists of N segments, it is an approximation that improves as N increase. For a WLC simulation of a single chain (dsDNA or ssDNA), due to the self-collision detections, it runs in $O(N^2)$ time for a single simulation step. Similar analysis applies to complex DNA nano-structures where N is equal in this case to the total number of double-stranded or single-stranded segments in the structure. Let M be the total number of the molecules in the small cell volume. For a long-ranged simulation period, each molecule in the system is treated as a rigid unit. Therefore the complexity per simulation step is $O(M)$. If M' is the number of molecule pairs that reach the short-range simulation threshold, then the simulation time per step is $O(M'N^2)$.

7   Preliminary Results

Our preliminary results demonstrate the feasibility of such a framework in modeling DNA based molecular systems.

7.1   Physical Simulation

The results presented here are obtained using the less computer-intensive Monte Carlo simulation of a discrete WLC model. The physical simulation module is demonstrated through the simulation of a tethered ssDNA. The same module applies to the modeling of other DNA molecules in the system. For demonstration purpose, we neglect twisting energy and focus primarily on

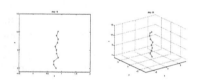

Fig. 6. 2D and 3D snapshots of the simulation for a single tethered DNA

the stretching energy and optional bending energy of the tethered DNA. Ideally, relatively long runs are carried out to generate initial conditions for simulations of the tethered-DNA chains, allowing the chains to reach their equilibrium configurations [52]. Then these configurations are saved for the actual simulation. The figures shown here are snapshots of a simulation during different time steps, from both 2D and 3D (Figure 6) perspective, visualized by Matlab. The scales for the x-axis and the y-axis are enlarged to show the details of the conformational changes relative to the horizontal plane. The simulations are preliminary but promising.

7.2   Event Simulation

We present here a snapshot of a hybridization event in simulation based on our framework in Figure 7. Bold black lines represent the double stranded DNA regions, while the thinner lines are single-stranded. The ssDNA we display in the above snapshots are $20 - 30$ bp.

Fig. 7. Simulation of a hybridization event

8   Discussion and Future Work

We presented a comprehensive framework for building a software tool for simulating DNA based molecular system, and not the actual software tool itself. It is important to note that, as a framework, the physical simulation component and event simulation component can be decoupled as we improve each component individually. As we improve the accuracy of the physical simulation, i.e. to reflect topological constraints by modeling more complicated DNA nano-structures such as pseudo-knots [38, 13]; to provide more biophysical sound behavior of DNA strands by considering stacking energy and electrostatic energy; or to achieve the molecular details by replacing the MC simulation with a BD simulation once computational resources are available, we can validate its correctness against polymer theory and experimental data, i.e. average radius of gyration and the diffusion constant. We can constantly update the physical simulation component to result in more realistic simulation.

During the physical simulation the random perturbations often lead to a configuration that can be achieved only with a low probability. Can we optimize the simulation so that we sample a larger space of configurations to avoid these with low probability, therefore making the simulation more computational efficient?

The first extension to our framework is to consider more complicated interactions, i.e. the enzyme restriction event and the hairpin formation. The second extension is to incorporate sequence design capabilities. We would like to design and optimize sequences based on the given nano-structure conformations. Furthermore, a conformation change of a nano-device can be decomposed into units of local deformations to ease the sequence design.

We believe that the methods presented here make a good framework for designing the simulator for DNA based molecular systems. The preliminary results in this paper support the feasibility of the approach. We describe that it is possible to capture geometric constraints of the molecules with the polymer theory and MC simulation. We also described the approximations and limitations in this framework and the ways of improving them.

References

1. L. Adleman. Molecular computation of solutions to combinatorial problems. *Science*, 266:1021–1024, 1994.
2. P. Alberti and J.L. Mergny. DNA duplex-quadruplex exchange as the basis for a nanomolecular machine. *Proc. Natl. Acad. Sci. USA*, 100:1569–1573, 2003.
3. S.A. Allison and S. Mazur. Modeling the free solution electrophoretic mobility of short dna fragments. *Biopolymers*, 46:359–373, 1998.

4. S.A. Allison and J.A. McCammon. Multistep brownian dynamics: application to short wormlike chains. *Biopolymers*, 23:363–375, 1984.

5. S. R. Aragon and R. Pecora. Dynamics of wormlike chains. *Macromolecules*, 18:1868, 1985.

6. R.G.C. Arridge. An introduction to polymer mechanics. 1985.

7. G.A. Arteca, T. Edvinsson, and C. Elvingson. Compaction of grafted wormlike chains under variable confinement. *Phys. Chem. Chem. Phys.*, 3:3737–3741, 2001.

8. B. Maier B, D. Bensimon, and V. Croquette. Replication by a single dna polymerase of a stretched single-stranded dna. *Proc. Natl. Acad. Sci. U.S.A.*, 97(22):12002–7, October 2000.

9. Y. Benenson, R. Adar, T. Paz-Elizur, Z. Livneh, and E. Shapiro. DNA molecule provides a computing machine with both data and fuel. *Proc. Natl. Acad. Sci. USA*, 100:2191–2196, 2003.

10. Y. Benenson, B. Gil, U. Ben-Dor, R. Adar, and E. Shapiro. An autonomous molecular computer for logical control of gene expression. *Nature*, 429:423–429, 2004.

11. Y. Benenson, T. Paz-Elizur, R. Adar, E. Keinan, Z. Livneh, and E. Shapiro. Programmable and autonomous computing machine made of biomolecules. *Nature*, 414:430–434, 2001.

12. I. Biswas, A. Yamamoto, and P. Hsieh. Branch migration through dna sequence heterology. *J. Mol. Bio*, 1998.

13. Justin S. Bois, Suvir Venkataraman1, Harry M. T. Choi1, Andrew J. Spakowitz, Zhen-Gang Wang and Niles A. Pierce1,2,* Topological constraints in nucleic acid hybridization kinetics. *Nucleic Acids Research*, 33(13):4090–4095, 2005.

14. C. Bouchiat, M.D. Wang, J. Allemand, T. Strick, S. M. Block, and V. Croquette. Estimating the persistence length of a worm-like chain molecules from force-extension measurements. *Biophys. J.*, 76:409, January 1999.

15. C. Bustamante, J. F. Marko, E. D. Siggia, and S. Smith. Entropic elasticity of lambda-phage dna mechanics. *Science*, 265:1599, 1994.

16. C. Bustamante, S. Smith, J. Liphardt, and D. Smith. Single-molecule studies of dna mechanics. *Current Opinion in Structural Biology*, 10:279, 2000.

17. J.E. Butler and E.S.G. Shaqfeh. Brownian dynamics simulations of a flexible polymer chain which includes continuous resistance and multi-body hydrodynamic interaction. *Journal of Chemical Physics*, 122(014901), 2005.

18. G. A. Carri and M. Marucho. Statistical mechanics of worm-like polymers from a new generating function. *J. Chem. Phys.*, 121(12):6064–6077, 2004.

19. N. Chelyapov, Y. Brun, M. Gopalkrishnan, D. Reishus, B. Shaw, and L. Adleman. DNA triangles and self-assembled hexagonal tilings. *J. Am. Chem. Soc.*, 126:13924–13925, 2004.

20. Y. Chen and C. Mao. Putting a brake on an autonomous DNA nanomotor. *J. Am. Chem. Soc.*, 126:8626–8627, 2004.

21. Y. Chen, M. Wang, and C. Mao. An autonomous DNA nanomotor powered by a DNA enzyme. *Angew. Chem. Int. Ed.*, 43:3554–3557, 2004.

22. S. Cocco, J. F. Marko, and R. Monasson. Theoretical models for single-molucule dna and rna experiments: from elasticity to unzipping. *to appear in CRAS, special issue dedicated to Single Molecule Experiments*, 2002.

23. C. Desruisseaux, D. Long, G. Drouin, and G. W. Slater. Electrophoresis of composite molecular objects. 1. relation between friction, charge and ionic strength in free solution. *Macromolecules*, 34:44–59, 2001.

24. M. N. Dessinges, B. Maier, Y. Zhang, M. Peliti, D. Bensimon, and V. Croquette. Stretching single stranded dna, a model polyelectrolyte. *Phys. Rev. Lett.*, 89:248102, 2002.

25. P. Dimitrakopoulos. Stress and configuration relaxation of an initially straight flexible polymer. *J. Fluid Mech.*, 513:265–286, 2004.

26. P.S. Doyle and P.T. Underhill. Brownian dynamics simulations of polymers and soft matter. *S. Yip, (ed.), Handbook of Materials Modeling*, pages 2619–2630, 2005.

27. R. M. Dirks, J. S. Bois, J. M. Schaeffer, E. Winfree and N. A. Pierce. Thermodynamic analysis of interacting nucleic acid strands. *SIAM Rev*, in press.

28. L. Feng, S.H. Park, J.H. Reif, and H. Yan. A two-state DNA lattice switched by DNA nanoactuator. *Angew. Chem. Int. Ed.*, 42:4342–4346, 2003.

29. M. Fixman and J. Kovac. Polymer conformation statistics iii: Modified gaussian models of the stiff chains. *J. Chem. Phys.*, 58:1564–1568, 1973.

30. C. Flamm, W. Fontana, I. L. Hofacker, and P. Schuster. RNA folding at elementary step resolution. *RNA*, 6(3):325-38, 2000.

31. J.B. Fournier. Wormlike chain or tense string? a question of resolution. *Continuum Mechanical Thermodynamics*, 14:241, 2002.

32. M.D. Frank-Kamenetskii. Biophysics of dna molecule. *Phys. Rep.*, 288:13 – 60, 1997.

33. D. T. Gillespie. Exact stochastic simulation of coupled chemical reactions. *J. Phys. Chem.*, 81:2340–2361, 1977.

34. D. T. Gillespie. Approximate accelerated stochastic simulation of chemically reacting systems. *J. Chem. Phys.*, 115:1716–1733, 2001.

35. A. J. Hartemink and D. K. Gifford. Thermodynamics simulation of deoxyoligonucleotide hybridization for dna computation. 1997.

36. P. J. Heath, J. A. Gebe, S. A. Allison, and J. M. Schurr. Comparison of analytical theory with brownian dynamics simulations for small linear and circular dnas. *Macromolecules*, 29:3583, 1996.

37. J. S. Hur and E. S. G. Shaqfeh. Brownian dynamics simulations of single dna molecule in shear flow. *J. Rheol.*, 44(4):713–742, July-August 2000.

38. H. Isambert and E. D. Siggia. Modeling RNA folding paths with pseudoknots: application to hepatitis delta virus ribozyme. *Proc Natl Acad Sci U S A.*, 97(12):6515–20, 2000.

39. H. M. James and E. Guth. Theory of the elastic properties of rubber. *Journal of Chemical Physics*, 10:455–481, 1943.

40. R. M. Jendrejack, J.J. Pablo, and M. D. Graham. Stochastic simulations of dna in flow: Dynamics and the effects of hydrodynamic interactions. *Journal of Chemical Physics*, 116(17):7752, 2002.

41. J. Santalucia Jr. A unified view of polymer, dumbbell and oligonucleotide dna nearest-neighbor thermodynamics. *PNAS*, 95:1460–1465, 1998.

42. A. M. Kierzek. Stocks: Stochastic kinetic simulations of biochemical systems with gillespie algorithm. *Bioinformatics*, 18:470–481, 2002.

43. K. Klenin, H. Merlitz, and J. Langowski. A brownian dynamics program for the simulation of linear and circular dna and other wormlike chain polyelectrolytes. *Biophys J*, 74(2):780–788, February 1998.

44. J. Kovac and C. Crabb. Modified gaussian model for rubber elasticity. 2. the wormlike chain. *Macromolecules*, 15(2):537, 1982.

45. M. Kuhn and F. Grun. Relationships between elastic constants and stretching double refraction of highly elastic substances. *Kolloid-Z*, 101:294, 1942.

46. S. Kutter. Elasticity of polymers with internal topological constraints. *PhD Thesis*, August 2002.

47. T.H. LaBean, H. Yan, J. Kopatsch, F. Liu, E. Winfree, J.H. Reif, and N.C. Seeman. The construction, analysis, ligation and self-assembly of DNA triple crossover complexes. *J. Am. Chem. Soc.*, 122:1848–1860, 2000.

48. B. Ladoux, J. P. Quivy, P. S. Doyle, G. Almouzni, and J. L. Viovy. Direct imaging of single-molecules: from dynamics of a single dna chain to the study of complex dna-protein interactions. *Sci. Prog.*, 84:267, 2001.

49. J. Langowski. Polymer chain models of dna and chromatin. *Manuscript*, 2006.

50. R. G. Larson, H. Hu, D. E. Smith, and S. Chu. Brownian dynamics simulation of a dna molecule in an extensional flow field. *J. Rheol.*, 43(2):267–304, March-April 1999.

51. R.G. Larson, T. Perkins, D. Smith, and S. Chu. Hydrodynamics of a dna molecule in a flow field. *Phys. Rev. E.*, 55:1794–1797, 1997.

52. R.G. Larson, T.T. Perkins, D.E. Smith, and S. Chu. Brownian dynamics simulations of a dna molecule in an extensional flow field. *J. Rheol.*, 43:267, 1999.

53. J. Li and W. Tan. A single DNA molecule nanomotor. *Nano Lett.*, 2:315–318, 2002.

54. D. Liu, M. Wang, Z. Deng, R. Walulu, and C. Mao. Tensegrity: Construction of rigid DNA triangles with flexible four-arm dna junctions. *J. Am. Chem. Soc.*, 126:2324–2325, 2004.

55. Q. Liu, L. Wang, A.G. Frutos, A.E. Condon, R.M. Corn, and L.M. Smith. DNA computing on surfaces. *Nature*, 403:175–179, 2000.

56. A. Malevanets and J. M. Yoemans. Dynamics of short polymer chains in solution. *Europhysics Letters*, 52(2):231, 2000.

57. C. Mao, T.H. LaBean, J.H. Reif, and N.C. Seeman. Logical computation using algorithmic self-assembly of DNA triple-crossover molecules. *Nature*, 407:493–496, 2000.

58. C. Mao, W. Sun, and N.C. Seeman. Designed two-dimensional DNA holliday junction arrays visualized by atomic force microscopy. *J. Am. Chem. Soc.*, 121:5437–5443, 1999.

59. C. Mao, W. Sun, Z. Shen, and N.C. Seeman. A DNA nanomechanical device based on the B-Z transition. *Nature*, 397:144–146, 1999.

60. J. Marko and E. D. Siggia. Bending and twisting elasticity of dna. *Macromolecules*, 27:981, 1994.

61. J. F. Marko and E. D. Siggia. Stretching dna. *Macromolecules*, 28:8759, 1995.

62. R. J. Meagher, J. Won, L. C McCormick, S. Nedelcu, M. M. Bertrand, J. L. Bertarm, G. Drouin, A. E. Barron, and G. W. Slaters. End-labeled free-solution electrophoresis of dna. *Electrophoresis*, 26:331–350, 2005.

63. J. Mercier and G. W. Slater. Solid phase dna amplification: a brownian dynamics study of crowding effects. *Biophysical Journal*, 89:32–42, July 2005.

64. M. C. Murphy, I. Rasnik, W. Cheng, T. M. Lohman, and T. Ha. Probing single-stranded dna conformation flexibility using fluorescence spectroscopy. *Biophysical Journal*, 86:2530–2537, April 2004.

65. T. Odijk. Stiff chains and filaments under tension. *Macromolecule*, 28:7016–7018, 1995.

66. I.G. Panyutin and P. Hsieh. The kinetics of spontaneous dna branch migration. *Proc Natl Acad Sci U S A.*, 91(6):2021–5, 1994 Mar 15.

67. J.S. Pedersen, M. Laso, and P. Schurtenberger. Monte carlo study of excluded volume effects in wormlike micelles and semiflexible polymers. *Phys Rev E.*, 54(6):5917–5920, December 1996.

68. N. Peyret, P. A. Seneviratne, H. T. Allawi, and J. Santalucia. Nearest-neighbor thermodynamics and nmr of dna sequences with internal aa,cc,gg and tt mismatches. *Biochemistry*, 38:3468, 1999.

69. C. Rao and A. Arkin. Stochastic chemical kinetics and the quasi-steady-state assumption: application to the gillespie algorithm,. *J. of Chem. Phys.*, 118:4999–5010, 2003.

70. J.H. Reif. The design of autonomous DNA nanomechanical devices: Walking and rolling DNA. *The 8th International Meeting on DNA Based Computers (DNA 8)*, 2002.

71. M. Rief, H. Clausen-Schaumann, and H. E. Gaub. Sequence-dependent mechanics of single dna molecules. *Nature Structural Biology*, 6:346 – 349, 1999.

72. M. Sales-Pardo, R. Guimera, A. A. Moreira, J. Widom, and L. A. Amaral. Mesoscopic modeling for nucleic acid chain dynamics. *Phys Rev E Stat Nonlin Soft Matter Phys.*, 71:051902, 2005.

73. J. Santalucia and D Hicks. The thermodynamics of dna structural motifs. *Annu. Rev. Biophys. Biomol. Struct.*, 33:415, 2004.

74. S. Sahu, B. Wang, and J. H. Reif. A Framework for Modeling DNA Based Molecular Systems. *Technical Report*, Duke University, 2006.

75. R. Sha, R. Liu, D.P. Millar, and N.C. Seeman. Atomic force microscopy of parallel DNA branched junction arrays. *Chemistry and Biology*, 7:743–751, 2000.

76. W.B. Sherman and N.C. Seeman. A precisely controlled DNA biped walking device. *Nano Lett.*, 4:1203–1207, 2004.

77. J.S. Shin and N.A. Pierce. A synthetic DNA walker for molecular transport. *J. Am. Chem. Soc.*, 126:10834–10835, 2004.

78. F.C. Simmel and B. Yurke. Using DNA to construct and power a nanoactuator. *Phys. Rev. E*, 63:041913, 2001.

79. F.C. Simmel and B. Yurke. A DNA-based molecular device switchable between three distinct mechanical states. *Appl. Phys. Lett.*, 80:883–885, 2002.

80. S. B. Smith, L Finzi, and B. Bustamante. Direct mechanical measurements of the elasticity of single dna molecules by using magnetic beads. *Science*, 258:1122, 1992.

81. S.B. Smith, Y. Cui, and C. Bustamante. Overstretching b-dna: the elastic response of individual double-stranded and single-stranded dna molecules. *Science*, 271:795–799, Feb 1996.

82. M. Somasi, B. Khomami, N. J. Woo, J. S. Hur, and E. S. G. Shaqfeh. Brownian dynamics simulations of bead-rod and bead-spring chains: numerical algorithms and coarse-graining issues. *J. Non-Newtonian Fluid Mech.*, 108:227–255, 2002.

83. E. Stellwagen and N. C. Stellwagen. Determining the electrophoretic mobility and translational diffusion coefficients of dna molecules in free solution. *Electrophoresis*, 23(16):2794–2803, 2002.

84. C. Storm and P. C. Nelson. Theory of high-force dna stretching and overstretching. *Physical Review E.*, 67:051906, 2003.

85. B. J. Thompson, M. N. Camien, and R.C.Warner. Kinetics of branch migration in double-stranded dna. *Proc Natl Acad Sci U S A*, 73(7):2299–303, 1976 Jul.

86. Y. Tian, Y. He, Y. Chen, P. Yin, and C. Mao. Molecular devices - a DNAzyme that walks processively and autonomously along a one-dimensional track. *Angew. Chem. Intl. Ed.*, 44:4355–4358, 2005.

87. I. Tinoco and C. Bustamante. The effect of force on thermodynamics and kinetics of single molecule reactions. *Biophys Chem.*, 101-102:513, December 2002.

88. A.J. Turberfield, J.C. Mitchell, B. Yurke, Jr. A.P. Mills, M.I. Blakey, and F.C. Simmel. DNA fuel for free-running nanomachines. *Phys. Rev. Lett.*, 90:118102, 2003.

89. A.J. Turberfield, B. Yurke, and Jr. A.P. Mills. DNA hybridization catalysts and molecular tweezers. *DNA5*, 2000.

90. T. E. Turner, S. Schnell, and K. Burrage. Stochastic approaches for modelling in vivo reactions. *Computational Biology and Chemistry*, 2004.

91. A.V. Vologodskii. Monte carlo simulation of dna topological properties. *Preprint*, 2004.

92. A.F. Voter. Introduction to kinetic monte carlo method. *Springer*, NATO publishing unit, 2005.

93. J. G. Wetmur and N. Davidson. Kinetics of renaturation of dna. *J. Mol. Biol.*, 31:349–370, 1968.

94. E. Winfree. Complexity of restricted and unrestricted models of molecular computation. In R. J. Lipton and E.B. Baum, editors, *DNA Based Computers 1*, volume 27 of *DIMACS*, pages 187–198. American Mathematical Society, 1996.

95. E. Winfree. Simulation of computing by self-assembly. Technical Report 1998.22, Caltech, 1998.

96. E. Winfree, F. Liu, L.A. Wenzler, and N.C. Seeman. Design and self-assembly of two-dimensional DNA crystals. *Nature*, 394(6693):539–544, 1998.

97. E. Winfree, X. Yang, and N.C. Seeman. Universal computation via self-assembly of DNA: Some theory and experiments. In L.F. Landweber and E.B. Baum, editors, *DNA Based Computers II*, volume 44 of *DIMACS*, pages 191–213. American Mathematical Society, 1999.

98. M. T. Wolfinger, W. A. Svrcek-Seiler, C. Flamm, I. L. Hofacker, and P. F. Stadler. Exact Folding Dynamics of RNA Secondary Structures. *J.Phys.A: Math.Gen.*, 37:4731-4741, 2004.

99. H. Yamakawa and T. Yoshizaki. Dynamics of helical wormlike chains. i. dynamic model and diffusion equation. *Journal of Chemical Physics*, 75(2):1016, July 1981.

100. H. Yan, T.H. LaBean, L. Feng, and J.H. Reif. Directed nucleation assembly of DNA tile complexes for barcode patterned DNA lattices. *Proc. Natl. Acad. Sci. USA*, 100(14):8103–8108, 2003.

101. H. Yan, S.H. Park, G. Finkelstein, J.H. Reif, and T.H. LaBean. DNA-templated self-assembly of protein arrays and highly conductive nanowires. *Science*, 301(5641):1882–1884, 2003.

102. H. Yan, X. Zhang, Z. Shen, and N.C. Seeman. A robust DNA mechanical device controlled by hybridization topology. *Nature*, 415:62–65, 2002.

103. J. Yan and J. F. Marko. Localized single-stranded bubble mechanism for cyclization of short double helix dna. *Phys. Rev. Lett.*, 93(10):108108, September 2004.

104. P. Yin, S. Sahu, A.J. Turberfield, and J.H. Reif. Design of autonomous DNA cellular automata. In *Proc. 11th International Meeting on DNA Computing*, pages 376–387, 2005.

105. P. Yin, A.J. Turberfield, S. Sahu, and J.H. Reif. Design of an autonomous DNA nanomechanical device capable of universal computation and universal translational motion. In *Proc. 10th International Meeting on DNA Computing*, pages 344–356, 2004.

106. B. Yurke, A.P. Mills, and A.J. Turberfield. A molecular machine made of and powdered by DNA. *Biophysics*, 78:2629, 2000.

107. B. Yurke, A.J. Turberfield, Jr. A.P. Mills, F.C. Simmel, and J.L. Neumann. A DNA-fuelled molecular machine made of DNA. *Nature*, 406:605–608, 2000.

108. Y. Zhang, H. Zhou, and Z. Ou-Yang. Stretching single-stranded dna: Interplay of electrostatic, base-pairing, and base-pair stacking interactions. *Biophys J.*, 81(2):1133–1143, August 2001.

Uniquimer: A *de Novo* DNA Sequence Generation Computer Software for DNA Self-assembly

Bryan Wei, Zhengyu Wang, and Yongli Mi[*]

Department of Chemical Engineering
Hong Kong University of Science and Technology
Clear Water Bay, Kowloon, Hong Kong

Abstract. We developed a computer-software with graphic interfaces for generating *de novo* DNA sequences of various DNA motifs for DNA nanotechnology research. The software is free of charge for academic and non-profit organizations.

1 Introduction

The powerful molecular recognition system of DNA base pairing can be used in nanotechnology to direct self-assembly of DNA architectures. DNA self-assembly has been found to be an innovative methodology for preparing nano-scaled patterns [1]. DNA architecture involves the following: firstly, designing synthetic DNA oligonucleotides for creating specific topologies, shapes and arrangements of secondary and tertiary structures; secondly, enabling oligonucleotides to self-assemble into a desired structure (scaffold); and thirdly, modifying the DNA scaffold for specific applications [2]. In DNA nanotechnology, a variety of rigid DNA motifs have been obtained, such as the DNA double crossover (DX) [3], the DNA triple crossover (TX) [4], and the DNA paranemic crossover (PX) [5-7]; from which, 1-D and 2-D arrays with topographic features visible in atomic force microscopy (AFM) and transmission electron microscopy (TEM) can be attained [3,4,8-11]. 3-D objects can also be constructed by careful design of the branched DNA motifs, including the DNA cube [12] and the octahedron [13]. Other motifs, such as Four Holliday junction analogues parallelogram [14], 4×4 tile [15], DNA triangle [16] and six-helix bundle [17], three-helix bundle [18] and TXT [19] have also come to the horizon recently.

From an engineering point of view, the DNA research is now permeating into two fast growing areas, namely, DNA computing and DNA nanotechnology. In DNA computing, programmed DNA 1-D, 2-D and possibly 3-D assemblies can be used for both areas [20, 21]. With this regard, DNA sequences for self-assembly are to be generated with computer software, so that the secondary structures of simple DNA strands can be obtained.

After the motif for a specific DNA self-assembly is well designed (length of the helices, crossover points joining the helices, sticky end matching strategy, etc.), the bases, A, T, G, and C for DNA strands for the specific architecture need to be well specified. There are two major principles for specifying the bases. First, a certain

---

[*] Corresponding author. keymix@ust.hk

C. Mao and T. Yokomori (Eds.): DNA12, LNCS 4287, pp. 266–273, 2006.
© Springer-Verlag Berlin Heidelberg 2006

fragment of a certain strand must be complementary to another fragment of another stand to follow the motif design (i.e. length of the helices, crossover points joining the helices, sticky ends matching strategy, etc.); second, in order to prevent from mismatching, the length of repeating sequences should be minimized among all the strands of the motif for the specific DNA architecture. It would be really labor intensive work to get satisfactory strand sequence design by hand. However, if computer software of appropriate algorithm is developed, specifying A, T, G, and C will be done with extremely high speed and accuracy.

2 Algorithm of the Program

After the structure of a certain motif is determined, the next step for the DNA architecture is to fill each base (A, T, G, C) for a number of oligonucleotides of certain lengths. One out of these four letters, A, T, G and C will be chosen for each base of a certain oligonucleotide and random seeds are used for the initiation. The segment to be paired with another segment, whose sequence is already generated by random seeds, can be simply determined by the base-pairing rule. As long as the bases of oligonucleotides are filled, the checkup of the limiting condition is applied to the sequences generated. If there is no violation, the sequence generation keeps going on until all sequences of oligonucleotides for the motif are generated. However, if there is a violation of the limiting condition, the violated base will be canceled and a new base will be chosen using random seeds. Sine only one possible solution is needed for the generation, the depth first searching is chosen. In the flow of base-by-base checkup, a branch-and-cut algorithm is used. Algorithm flow diagram is show as in next page.

3 Depth First Searching (DFS)

The DNA sequence generation is a typical problem of trial and error; therefore, a proper searching algorithm should be applied. Depth first searching (DFS), which aim at giving

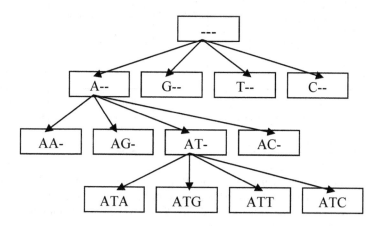

Fig. 1. Diagram of the algorithm tree

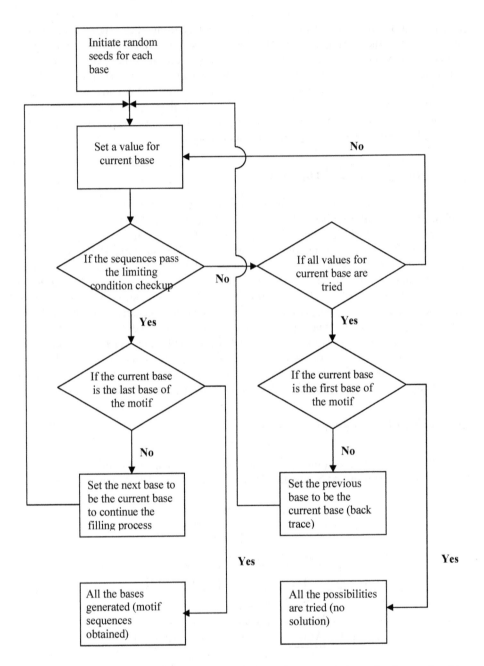

a possible solution rather than finding out all the possible solutions, is chosen for this software since only one of the possible solutions for the problem is needed for a special architecture. We can see the process of solution as a traversal of a certain tree structure. In Figure 1, a solution tree for filling a tri-nucleotide sequence is shown. Each level of the tree represents the filling of a certain base. Since there are 4 choices for each filling

(A, T, G or C), each unfinished status (non-leaf node) has 4 child nodes and each child node is further filled based on the status of its parental node.

4 Random Seeds

The sequence generated should also be as random as possible. However, using the DFS algorithm, the result is predictable (i.e., there will be the same solution for the same input). If different result for different round of running is expected, random seeds need to be introduced to the algorithm. Random seed(s) is an integer between 0 and 3 and each base is specified to a random seed, by which the filling of the certain base is determined. Before each round of running of the algorithm, random seed is applied to the process of the filling for each base. Therefore, the results for different round of running of the algorithm will be different (i.e. different transversal of the solution tree).

5 Branch-and-Cut Algorithm

Restrictions in the searching process are not concerned so far. However, some limiting conditions must be included in each type of searching algorithm. Otherwise, it will degenerate to an exhaustive searching algorithm. There are two methods to include limiting conditions. One is not to make restrictions in the searching process until a possible solution is obtained. When a leaf node is met (i.e. possible solution), the solution is checked by the limiting condition. The other way is to check each node in the

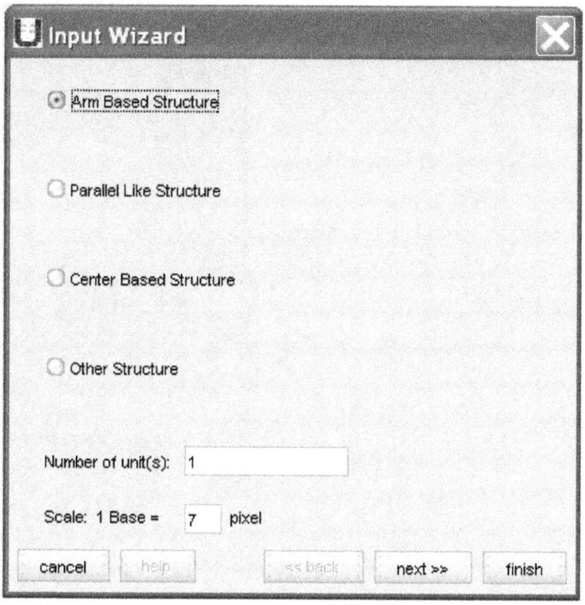

Fig. 2. The input wizard for DNA sequence generation

searching process. If the node is found not to pass the limiting conditions. For instance, if melting temperature for a specific node is above a certain value, all the offspring nodes of it will have higher melting temperature and will be "less" supposed to meet the limiting condition. Therefore, the node and all its offspring nodes will be discarded and won't be visited further. The second method, branch-and-cut algorithm, is used in our software.

6 Features

Graphic interfaces are available for almost all motifs with known structures for DNA nanotechnology research, which can be classified into three categories of DNA structures, namely, arm-based structures (e.g., junction motifs), center-based (e.g., 4×4 tiles), and parallel-like structures (e.g., DX, TX) in this software. Input wizard is provided for a stepwise sequence generation. Real-time highlight effects are available

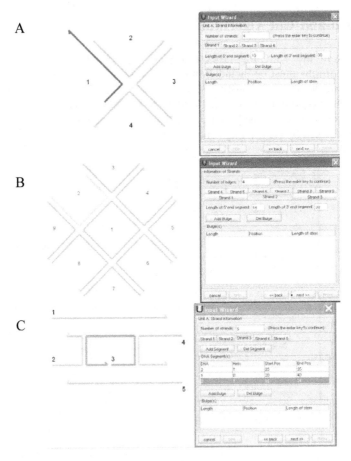

Fig. 3. Graphic interfaces for three kinds of DNA motifs. (A) an arm-based structure, (B) a center-based structure, and (C) a parallel-like structure

for most of the operations such as length specification for certain strands, sticky end matching setting.

Bulges and pre-specified segments (named as fixed segments in the software) are treated as non-extra sequences in the motifs, which means that they still can be included in the overall algorithm for sequence generation.

Fig. 4. Bulge and fixed segments addition

Control of the additional limiting conditions, such as the melting temperature for complementary segments and GC percentage besides the maximum length of repeating sequences, are available in the software.

7 Prospect

Currently, there are programs used in other groups for the research of DNA nanotechnology for sequence generation. There is not yet a program available in the public domain with graphic interface. This computer software, *Uniquimer*, is aimed at providing an easy access to researchers who want to do research in DNA nanotechnology. The software can be used to generate DNA sequences for DNA self-assembly with graphic interface. Fixed segments, extra bulges and melting temperature of certain segments can be assigned by users. Future development will include, arbitrary motif generation (like single strand DNA/RNA folding prediction program MFOLD, energy minimization is expected to be involved for the tertiary structure of a motif to be constructed) and 3-D display of motifs with related arrays. Moreover, some useful databases such as restriction enzyme sites and aptamers are expected to be integrated to the software.

Acknowledgement. Research grant from the Hong Kong Government of the University Grant Council, RGC 602603, is greatly acknowledged.

References

1. Seeman, N. C., *Nature* **2003**, 421, 427-431.
2. Seeman, N. C., *Trends Biotechnol.* **1999**, 17, 437-442.
3. Li, X.; Yang, X.; Qi, J.; Seeman, N. C. *J. Am. Chem. Soc.* **1996**, 118, 6131-6140.
4. LaBean, T. H.; Yan, H; Kopatsch, J.; Liu, F; Winfree, E; Reif, J. H.; Seeman, N. C. *J. Am. Chem. Soc.* **2000**, 122, 1848-1860.
5. Seeman, N. C. *Nano Lett.* **2001**, 1, 22-26.
6. Zhang, X.; Yan, H.; Shen, Z.; Seeman, N. C. *J. Am. Chem. Soc.* **2002**, 124, 12940-12941.
7. Shen, Z.; Yan, H.; Wang, T.; Seeman, N. C. *J. Am. Chem. Soc.* **2004**, 126, 1666-1674.
8. Winfree, E.; Liu, F.; Wenzler, L. A.; and Seeman, N. C.; *Nature* **1998**, 394, 539-544.
9. Liu, F.; Sha, R.; Seeman, N. C. *J. Am. Chem. Soc.* **1999**, 121, 917-922.
10. Liu, D.; Park, S. H.; Reif, J. H.; LaBean, T. H. *PNAS* **2004**, 101, 717-722.
11. Li, H.; Park, S. H.; Reif, J. H.; LaBean, T. H.; Yan, H. *J. Am. Chem. Soc.* **2004**, 126, 418-419.
12. Chen, J.; Seeman, N. C. *Nature* **1991**, 350, 631-633.
13. Shih, W. M.; Quispe, J. D.; Joyce, G. F. *Nature* **2004**, 427, 618-621.
14. Mao, C.; Sun, W.; Seeman, N. C. *J. Am. Chem. Soc.* **1999**, 121, 5437-5443.
15. Yan, H.; Park, S. H.; Finkelstein, G.; Reif, J. H.; LaBean, T. H. *Science* **2003**, 301, 1882-1884.
16. Chelyapov, N.; Brun, Y.; Gopalkrishnan, M.; Reishus, D.; Shaw, B.; Adleman, L. *J. Am. Chem. Soc.* **2004**, 126, 13924-13925.
17. Mathieu, F.; Liao, S.; Kopatsch, J.; Wang, T.; Mao, C.; Seeman, N. C. *Nano Lett.* **2005**, 5, 661-665.
18. Park, S. H.; Barish, R.; Li, H.; Reif, J. H.; Finkelstein, G.; Yan, H.; LaBean, T. H. *Nano Lett.* **2005**, 5, 693-696.
19. Wei, B.; Mi, Y. *Biomacromolecules* **2005**, 6, 2528-2532.
20. Yan, H.; Zhang, X.; Shen, Z.; Seeman, N. C. *Nature* **2002**, 415, 62-65.
21. Winfree, E. In *DNA Based Computing*; Lipton, E. J., Baum, E. B., Eds.; Am. Math. Soc.: Providence; 1996; pp 199-219.

Appendix-I

System Specification

Platform --Windows, Mac OS X, etc.
Input interface -- Java Wizard Framework (JWF)
Input file format -- Extensible Markup Language (XML)
Output file format -- Scalable Vector Graphics (SVG)

Function of the Software

The main function of the software is sequence generation of DNA motifs. Graphic interface is available for the existing motifs classified as arm-based structures, center-based structures, and parallel-like structures.

Extra Functions

1. Designation of certain segment sequences.
2. Designation of bulges for certain strands of DNA.
3. Designation of melting temperature (T_m) for complementary segments of DNA.
4. Designation of GC percentage for DNA motifs.
5. Exportation of motif pictures as JPG, PNG, TIFF format and the motif sequences in TXT format.

Appendix-II

The software and its operation manual are available in the following web address: http://ihome.ust.hk/~keymix/uniquimer.htm

A Probabilistic Model of the DNA Conformational Change[*]

Masashi Shiozaki, Hirotaka Ono, Kunihiko Sadakane, and Masafumi Yamashita

Dept. of Computer Science and Communication Engineering, Kyushu University
{masashio, ono, sada, mak}@tcslab.csce.kyushu-u.ac.jp

Abstract. Predicting the behavior of DNA molecules in vitro is one of the most fundamental issues on DNA computing, but is also known to be quite difficult. Shiozaki et al. proposed a probabilistic model that can simulate many features of biochemical experiments in terms of the reaction rate [7], although there are several differences between the biochemical experiments and the computational simulations on the model.

In this paper, we extend the model to support base pairs construction among k DNA sequences, which plays an essential role in realizing branch migrations. The simulation results have much more similarities to the biochemical experiments results than ones on the previous model, which implies that the analysis of the model may give some insight about the reaction rate. Through the analysis, we conclude this paper by giving a guideline for designing DNA sequences that can quickly react.

1 Introduction

Predicting the behavior of DNA molecules in vitro is one of the most fundamental issues on DNA computing, but also known to be quite difficult. Many researchers try to understand the principle of DNA molecule reactions from several viewpoints. The aim of this paper is to understand the mechanism of the DNA conformational change.

Over the past few decades, modelling the behavior of RNA (not DNA) has been intensively investigated [1,2,8,10]. Most of the models are essentially based on Markov process under the assumption that an RNA sequence changes its own structure according to the transition probability defined by the free energy. Some models succeeded; they can simulate the folding behavior of RNA well. These models are based on the simplicity of the RNA reaction; a base pair of RNA is made only in one sequence [1]. On the other hand, the DNA reactions are usually more complicated, because base pairs of DNA are made not only on one sequence but also on two or more sequences. That is, the behavior of DNA molecules is supposed to be quite different from that of RNA, and the

---

[*] This research partly received financial support from Scientific research fund of Ministry of Education, Culture, Sports, Science and Technology.

[1] This is actually a bit exaggerating and in RNA base pairs among more than one sequence are made. What we want to say here is that in RNA reaction a base pair construction in one sequence is known to be essential.

C. Mao and T. Yokomori (Eds.): DNA12, LNCS 4287, pp. 274–285, 2006.
© Springer-Verlag Berlin Heidelberg 2006

same technique is not applicable directly. In fact, there had been no study on the analysis of the DNA behavior, as far as the authors know.

Among them, Shiozaki et al., the authors of this paper, proposed a probabilistic model based on Markov process for simulating the behavior of DNA, as a first step of the DNA behavior's analysis [7]. Considering the hybridization of multiple DNA sequences, we introduced a model with two macro states and the coupling factor p that can control the state transition between the two macro states. The model can treat the structures formed with one or two sequences, which are not so essential in the RNA conformation but are crucial in the DNA conformation. We conducted computational simulations to compare with biochemical experiments, whose results are provided by Prof. Suyama's Group [2]. That biochemical experiments observe the reaction rate in which two types of DNA sequences (one type of sequences and its complement type of sequences) hybridize in vitro under a constant temperature. Through the comparison studies, we see that our computational simulations could imitate the biochemical experiments on some of their typical features, though some features are still left to be realized.

In this paper, we propose an advanced model, which is an extension of the previous model, so as to explain the unrealized features. The previous model considers the structures formed with only at most two sequences. This implies that some conformational changes in which more than two sequences react, such as branch migrations, are not supported in the model. It is a problem, because branch migrations are considered to play a very important role in DNA reactions. A point of the new model is that it permits any DNA sequence to have base pairs among k sequences, which make it possible to simulate the mutual influences of multiple DNA sequences. For example, a double stranded structure makes a base pair with other structures; the branch migrations also may occur in the model.

We then conduct computational simulations on the new model. Simulation results show more similarities to biochemical experiments than the previous one. Through the computational simulations, we propose a guideline for designing DNA sequences that quickly react, and further simulation studies support the availability of the guideline.

These simulation studies on our model correspond to the biochemical experiments explained above. Simulation studies for another kind of DNA reaction (biochemical experiments) in Prof. Hagiya's group [3], which is about the branch migrations, also show that they can nicely imitate the bioexperimental results. The similarities between the simulation results on our model and the biochemical experiments imply that our model may be a good approximation of the DNA conformational system in vitro; if our model approximates essential parts of DNA conformation indeed, we can estimate the DNA reactions without any biochemical experiments, which means that we can save many costs of biochemical experiments, i.e., a lot of money, time, labor, and so on. Of course, more computational simulations and more biochemical experiments should be

---

[2] http://dna.c.u-tokyo.ac.jp/
[3] http://hagi.is.s.u-tokyo.ac.jp/

performed to see the validity of the model, if yes, our results may be a giant leap for DNA computing.

2   Preliminaries

In this section, we first explain the results of biochemical experiments that motivate us to design a new model of DNA conformational change, then give a basic simulation model of RNA conformation.

2.1   Motivating Biochemical Experiments

First we explain biochemical experiments which motivate us to design a simulation model of DNA conformational changes.

Basic DNA reactions. We consider the situation that one type of DNA sequences (we call them primary sequences) and the same number of its complements are put in vitro at the same time under a constant temperature. At the initial state all DNA sequences have no base pair. Then DNA sequences start to change their own structures randomly and after a while they form the perfect double stranded structures. This is considered one of the most basic DNA reactions, and actually Suyama et.al. also concentrate to analyze this basic situation.

In the biochemical experiments conducted by Suyama et al., they use six types of DNA sequences that have the same length and almost the same minimum free energy (MFE) of the structure that is formed with a primary sequence and its complement. Table 1 presents the primary sequences of six instances and the MFEs that are calculated with a Vienna-RNA1.4 package [3,4] which was yet improved by Uejima [9]. In this table, MFEs are minimum relative free energies with respect to the structure with no base pair.

These chemical experiments were done in the following environment; the absolute temperature is 298 K and the concentration of DNA is 10 nmol/l. For comparison, we use three instances #60, #171 and #176, each of which has obvious characteristics in terms of the reaction rate.

Table 1. The primary sequences and MFEs of the six instances

ID	Primary Sequence	MFE (kcal/mol)
60	TTCGCTGATTGTAGTGTTGCACA	-36.16
171	CGCGATTCCTATTGATTGATCCC	-34.78
176	GGGATCAATCAATAGGAATCGCG	-37.10
117	CGCGATTCCTATTGATTGATCCC	-35.60
227	TAGCACCCGTTAAAACGGAAATG	-35.27
247	GCCTCCTTAGTAAGAACTGGTGC	-36.14

Figure 1 shows the experimental results in biochemistry[4] of #60, #171 and #176. The horizontal and vertical axes represent real time and fluorescence

---

[4] This figure was offered by Prof. Suyama of the University of Tokyo.

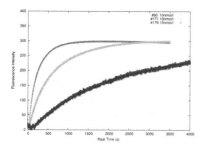

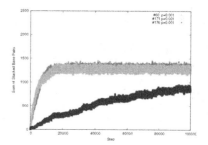

Fig. 1. Experimental results   **Fig. 2.** Simulation results

intensity, respectively. In these experiments, fluorescence intensity is proportional to the number of stacked base pairs between two sequences.

Figure 2 shows the simulation result of Shiozaki et al. [7]. The horizontal and vertical axes represent the number of steps and the sum of the number of stacked base pairs, respectively. We can see that the biochemical experiments and the simulations have some common tendencies. For example, both of the figures show that the reactions of #176 are slow. This is because #176 tends to make base pairs in one sequence frequently. Table 2 presents the MFEs in one of the primary sequences and the complements of #60, #171 and #176. It is known that the lower the free energy is, the more stable the structure is. Hence that means both of the primary sequence of #176 and its complement can form hairpin structures whose free energy are lower than the structure with no base pair, while those of #60 and #171 do not form such structures in one sequence. Figure 3 shows the structures corresponding to the MFEs in Table 2. Forming a hairpin structure prevents a strand from hybridizing with another to make a duplex. This may be the reason that the reaction of #176 is much slower than the other two.

These are the features of the biochemical experiments that our simulation can imitate. On the other hand, there are some differences between the results of our basic model and the biochemical experiments. For example, the results of the biochemical experiments show that the DNA sequences of #60 react faster than those of #171, while our simulation model shows the opposite results.

Branch migrations. In the previous subsection, we explain a basic DNA reaction, its biochemical experiments and the simulation results of our old model. In this section, we introduce two new reactions about about branch migration, whose biochemical experiments are provided by Prof. Hagiya group. We call them BM1 and BM2, respectively.

BM1 is described in Figure 4. We prepare three types of DNA sequences; these are sequences consisting of 40 bases (normal color one in the figure), called the primary sequences, their complement sequences (shaded one) and the half sequences of the complement (normal color one but the length is half). A primary sequence has a black hole quencher (BHQ) at the 3' end of the sequence. On the other hand, a half of its complement has a fluorescent material, FAM at the

Table 2. MFE in one sequence of #60, #171 and #176

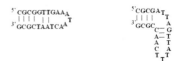

	MFE in one sequence (kcal/mol)	
ID	Primary Seq	Complement
60	0.0	0.0
171	0.0	0.0
176	-4.92	-3.88

Fig. 3. Structures of the MFE in one sequence of the primary sequence of #176 (left) and its complement (right)

5′ end of the sequence. At the initial state, these two types of sequences form double stranded DNA structures (Figure 4 (1)); that is, BHQs neighbors FAMs and the light of FAMs is absorbed by BHQs. Therefore the molecules do not glow at this moment. As time advances, shaded sequences start to make base pairs with the primary sequences (Figure 4 (2)). Then branch migrations occur and the half of their complements with a FAM start to dissociate from the primary sequence with a BHQ. Finally the primary sequences and their complements form perfect double stranded structures (Figure 4 (3)).

BM2 is similar to BM1, but four types of sequences participate. Figure 5 explains BM2. We prepare three types of DNA sequences as BM1 and another type of sequences; We add the half sequences of the primary sequence, that is the length is 20. These sequences also make double stranded DNA structure with the complements of the primary sequences at the initial state (Figure 5 (1)). Hence there exist two kinds of double stranded DNA structures. If the reactions perfectly occur, then two types of perfect double stranded structures are formed in the equilibrium state (Figure 5 (3)).

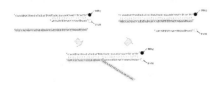

Fig. 4. The reaction of BM1 (1,2,3)   **Fig. 5.** The reaction of BM2 (1,2,3)

In these two experiments, molecules glow when BHQs and FAMs stand off to each other, which means that fluorescence intensity becomes strong after the branch migrations completely finishes. Hence the fluorescence intensity shows the progress of the reaction.

Figure 6 shows the results of the biochemical experiments[5]. The horizontal and the vertical axes represent real time and fluorescence intensity respectively. These experiments are done in the environment where the absolute temperature is 318 K and the concentration of DNA is 100 nmol/l.

As shown in this figure, BM1 reacts in a very short time while BM2 takes a lot of time to finish branch migrations completely.

---

[5] This result is offered by Prof. Hagiya of the University of Tokyo.

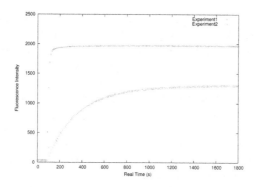

Fig. 6. Results of biochemical experiments

2.2   The Simulation Model of RNA

There exist many studies on analyzing and simulating the folding behavior of
RNA [1,2,5,8,10,11]. The basic assumptions of their models are the following:

- Each RNA molecule changes its structure randomly and independently of
  other molecules.
- At the equilibrium state, the probability $p_s(x)$ that an RNA molecule s has
  a structure $x \in S$ is

$$p_s(x) = \frac{\exp(-E(x)/RT)}{Z}$$

where $E(x)$ is the free energy of the molecule that form the structure x, S
is the set of all the different structures of s, T is the temperature, R is a
constant, and $Z = \sum_{s \in S} \exp(-E(x)/RT)$.

Note that this model explains only the equilibrium state and does not explain
the transition period. Wolfinger et al. propose a model for simulating the folding
behavior of RNA, which is based on Markov process [10]. In their study, each
state in the Markov process corresponds to each RNA structure. They assume
that RNA sequences change their own structures according to the state transition
probability defined by the free energy of each structure. In their simulation
model, the neighborhood of the state is defined as the set of structures that are
formed from the present structure with a single base pair variation. The state
transition probability is as follows:

$$P(X_{t+1} = y | X_t = x) = \frac{r_{xy}}{\sum_{z \in N(x)} r_{xz}}, \tag{1}$$

where X_t is a random variable indicating the structure of a molecule at time t,
$N(x)$ is the neighborhood of the state x and r_{xy} is the transition rate from the
state x to y defined as follows:

$$r_{xy} = \exp\left\{-\frac{E_{yx}^{\neq} - E(x)}{RT}\right\}, \qquad E_{yx}^{\neq} = \max\{E(x), E(y)\}.$$

The transition state energies $E_{yx}^{\neq}$ assures the detailed balance, $p_s(x) \cdot P(X_{t+1} = y | X_t = x) = p_s(y) \cdot P(X_{t+1} = x | X_t = y)$. As a result, this model could imitate the folding behavior of RNA quite well. The model is based on the independence of the past history and the random behavior of RNA conformational change. Since these also hold for DNA conformation, we expect that the behavior of DNA can be modelled as a Markov process, but straightforward application is not possible because of the following reasons:

- DNA reactions among multiple sequences are not considered.
- The probability $p_s(x)$ depends on only the structure x and the temperature T, whereas the results of the above biochemical experiments indicate that the probability depends on the concentration of molecules.

3   New Simulation Model

As mentioned in Section 2.1, old results [7] have some differences from those of the biochemical experiments in terms of the reaction rate of #60 and #171. That is, the old model may extract some part of the DNA reaction, but that extraction is rough and not good enough. In this section, we propose a more polished model, which can simulate making base pairs among k DNA sequences, and then show the simulation results.

3.1   The Simulation Model with the Extended Neighborhood

This model considers any structure created by k DNA sequences, so that the branch migration may be realized. Although the branch migration is considered essential for DNA reactions, the old model does not support; if it is the reason that the old model is not good enough, we may expect good improvements for the new model.

In the design of Markov process system as a simulation model, there are several computational problems to be considered, because our objective, DNA conformation, has an enormous number of states (DNA structures). The first problem is how to define the transition. Given a state x (DNA structure), We call a set of states (DNA structures) that can be changed from x, *neighborhood* of x. To simulate a Markov process, we compute all the transition probability from x to the neighborhood. That is, it is necessary to enumerate the neighborhood of x. However, enumerating all the neighborhood is too time-consuming if we adopt very general models of DNA reaction. On the other hand, it is considered that the transitions from x to most of neighbor states of x in such a model seldom occur. By these observation, we consider a model that has a restriction where k DNA sequences are not affected by any other sequences and all of them are neighbored at any time. In our model, we define the neighborhood as the set of structures that is formed with k DNA sequences from the present structure with a single base pair variation; it includes the present structure. By this, the size of the neighborhood is $O(k^2 l^2)$ at any time.

Preparing k DNA sequences, we move to the next state selected from all the neighborhood according to the state transition probability at each step; we use the same transition probability with [7] which is defined as follows:

$$P(X_{t+1} = y | X_t = x) = \frac{R_{xy}}{\sum_{z \in N(x)} R_{xz}}, \qquad (2)$$

where $N(x)$ is the neighborhood of a state x and R_{xy} is the transition rate defined as follows:

$$R_{xy} = \begin{cases} p \cdot r_{xy} & \text{(a first base pair between two sequences)} \\ r_{xy} & \text{(otherwise)}, \end{cases}$$

However, enumerating all the neighborhood in this model is still too complicated when k is large. This is because we need to change the order of the k sequences to get the free energy of each structure in the neighborhood. We implement this model as $k = 3$ and $k = 4$ with computer so far.

In order to simulate the basic DNA reactions we use this model as $k = 4$. Preparing two primary sequences and the same number of their complements, we can simulate any pairs of them. In order to realize branch migrations, we use this model as $k = 3$ and $k = 4$. In fact, in the BM1 and BM2, three and four types of DNA sequences participate in the reaction.

3.2   Computer Simulations of Basic DNA Reactions

We use three types of DNA sequences (#60, #171 and #176) in Table 1 as instances and do Monte Carlo simulations 50 times. Then the total number of sequences is 200, which is the same to the previous simulation [7]. Considering the results of the previous simulation model, we set $RT = 2.0$ and $p = 0.0001$. These parameters seem to depend on the concentrations.

Figure 7 shows the results of 0 to 10^6 steps when $p = 0.0001$. The horizontal and vertical axes represent the number of steps and the sum of the number of stacked base pairs of 50 times Monte Carlo simulations. In this figure, the rank of the reaction rate of #60, #171 and #176 is the same as the results of biochemical experiments, while not in the previous simulations. Therefore this model can simulate the DNA reactions well.

Discussion. It is natural to consider that these results come from the realization of the branch migration. Intuitively if the MFEs with two primary sequences and with their complements are small, the reaction becomes slow. In fact these MFEs of #60 are smaller than those of #171. Therefore, in terms of the free energy, #171 is supposed to react faster than #60. However the results of the biochemical experiments are the opposite to this prediction. This fact implies that we can not judge what DNA sequences react fast only from the free energies.

We then focus on the structures of the MFEs with two primary sequences and their complements. Table 3 represents these MFEs and the number of bases of *free 5' end* and *free 3' end* of #60, #171 and #176; the free 5' (3') end is

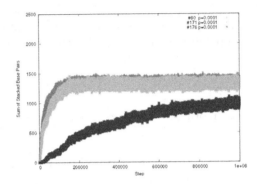

Fig. 7. The results of our advanced simulation model ($p = 0.0001$)

the number of bases that form no base pair at the 5' (3') end of the structure. Since two DNA sequences form these structures, we show two numbers of bases of free 3' end and 5' end for each. The more the number of bases of free 5' (3') end is, the more easily the other DNA sequences can make base pairs with the structure. Especially if both of free 5' (3') end of two primary sequences and that of free 3' (5') end of their complements are large, the branch migration occurs with ease and the reaction can be accelerated.

As presented in Table 3, the structure of two primary sequences of #60 and their complements have many free bases that do not make base pairs at the 5' and 3' end respectively; it implies that the branch migration occurs with ease when the DNA sequences form such structures. On the other hand, that of #171 and its complement have less free bases at both of the edges. We consider that this is the reason that #60 reacts faster than #171 in biochemical experiments.

These discussions also explain the slow reaction of #176. As shown in Table 3, #176 has much lower MFEs. In addition, both of these structures have no free base that does not make base pairs at the edges. Therefore the branch migration hardly occurs and the reaction is not accelerated.

Now we consider the reaction rate of six DNA sequences including #117, #227 and #247. Table 4 presents the MFEs with two primary sequences and their compliments and the number of bases of free 5' end and free 3' end of #117, #227 and #247. From the above discussions and Tables 3 and 4, we predict the following:

- The reactions of #117 and #60 are much faster than the others.
- The reactions of #171 and #227 are a bit slower.
- The reaction of #176 is the slowest.

Table 5 shows the rank of the reaction rate of the six types of DNA sequences in the biochemical experiments for each concentration. The predictions seem to explain the results of the biochemical experiments especially when the concentration of DNA is high. The higher the concentration is, the more frequently the DNA sequences encounter with each other; it means that the branch migrations

Table 3. The characteristics of the structures of MFEs with two primary sequences and their complements of #60, #171 and #176

	Primary Seq			Complement		
ID	MFE(kcal/mol)	free 5' end	free 3' end	MFE(kcal/mol)	free 5' end	free 3' end
60	-8.54	13, 13	1, 1	-6.60	0, 1	13, 14
171	-4.98	3, 3	7, 7	-6.22	7, 7	3, 3
176	-14.59	0, 0	0, 0	-10.42	0. 0	0, 0

Table 4. The characteristics of the structures of MFE with two primary sequences and their complements of #117, #227 and #247

	Primary Seq			Complement		
ID	MFE(kcal/mol)	free 5' end	free 3' end	MFE(kcal/mol)	free 5' end	free 3' end
117	-8.49	0, 0	19, 19	-8.25	19, 19	0, 0
227	-14.84	6, 6	5, 5	-14.62	5, 5	6, 6
247	-9.27	5, 5	9, 9	-6.85	0. 17	8, 0

Table 5. Ranking of strands by reaction rates in the biochemical experiments

	Concentration		
Rank	5 nmol/l	10 nmol/l	20 nmol/l
1	60	247	117
2	247	60	60
3	117	117	247
4	227	227	227
5	171	171	171
6	-	176	176

easily occur. Therefore these predictions match the experimental results better when the concentration is high.

There still remain some problems. The comparison between the results of the biochemical experiments and our simulation model including #117, #227 and #247 shows some differences. These may be due to the lack of the dynamical elements. It is known that DNA sequences start to make base pairs from an endpoint and they sequentially bond like zippers. Nevertheless our model can not simulate such reactions.

We conclude this section by the following.

1. We have almost succeeded in simulating the DNA reactions by the probabilistic model based on the free energy of each structure.
2. However, only the free energy information may not to be enough to simulate the precise behavior of DNA.
3. A new model reflecting the dynamical elements might approximate DNA reactions much better.

3.3   Computer Simulations of Branch Migrations

We apply our advanced model as $k = 3$ and $k = 4$ to simulate BM1 and BM2, respectively. We set $RT = 1.0$ and do Monte Carlo simulations 200 times.

Figure 8 shows the results. The horizontal and vertical axes represent the number of steps and the number of structures in that a BHQ and a FAM are away from each other (we call them the fluorescent structures), respectively. We can see many common tendencies between the simulation results and the biochemical experiments also in these reactions; for example, BM1 reacts in a very short time while BM2 takes a lot of time to be in the steady state. These results support that our model is useful not only for the basic DNA reactions but also for more complicated DNA reactions.

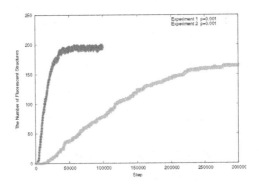

Fig. 8. The comparison between BM1 and BM2 ($p = 0.001$)

4   Conclusion

In this paper, we propose a new probabilistic model of DNA conformational change. The simulation studies for two basic DNA reactions show that the model may be a good approximation of (actual) biochemical reaction. Although further improvements may be possible, the authors think our model has succeeded in extracting some essence of the DNA reactions. To confirm this, much more simulation and biochemical experiments are needed. Also some (computational) theoretical studies might be useful and interesting to analyze thermo-dynamical properties of DNA reactions.

As a byproduct of the simulation studies, we obtain a guideline for designing DNA sequences that can react fast in terms of the basic DNA reactions; these are summarized as follows:

1. Both the primary sequence and its complement should not form hairpin structures (from the results of [7]).
2. Both two primary sequences and their complements also should not form the stable structure; in the case that they form such structures, the numbers of bases of free 3' end and 5' end of those should be large (Subsection 3.2).

Acknowledgments

The authors sincerely appreciate Professor Suyama and Professor Hagiya of the university of Tokyo who offered experimental data in biochemistry to us. We also thank the anonymous reviewers for their insightful comments to improve the presentation of the paper.

References

1. C. Flamm, *Kinetic Folding of RNA*, Dissertation, 1998.
2. C. Flamm, W. Fontana, I. L. Hofacker and P. Schuster, *RNA Folding at Elementary Step Resolution*, Santa Fe Institute 12-078, 1999.
3. I. L. Hofacker, *The Vienna RNA Secondary Structure Server*, Nucleic. Acids Res. 31 3429-31, 2003.
4. I. L. Hofacker, W. Fontana, P. F. Stadler, L. S. Bonhoeffer M. Tacker and P. Schuster, *Fast Folding and comparison of RNA Secondary Structures*, Monatsh. Chem. 125 167-88, 2003.
5. A. J. Hartemink and D. K. Fifford, *Thermodynamic Simulation of Deoxyligonucleotide Hybridization for DNA Computation*, DIMACS Series in Discrete Mathematics and Theoretical Computer Science.
6. S. Kobayashi, *Strand Design for Molecular Computation*, Sciences, 2003.
7. M. Shiozaki, H. Ono, K. Sadakane and M. Yamashita, *Modeling DNA Conformation Change and Theoretical Analysis on the Reaction Rate*, Preproceedings of the 11th International Meeting on DNA Computing (DNA11), 408, 2005.
8. M. Tacker, W. Fontana, P. F. Stadler and P. Schuster, *Statistics of RNA Melting Kinetics*, Eur. Biophys. J., 23:29–38, 1993.
9. H. Uejima, M. Hagiya, *Analyzing Secondary Structure Transition Paths of DNA/RNA molecules*, PreProceedings of the 9th International Meeting on DNA Computing (DNA9) 4731-4741, 2004.
10. M. T. Wolfinger, W. A. Svrcek-Seiler, C. Flamm, I. L. Hofacker and P. F. Stadler, *Efficient Computation of RNA Folding Dynamics*, J. Phys. A: Math. Gen. 37: 4731-4741, 2004.
11. M. Zucker, *The Equilibrium Partition Function*, http://www.bioinfo.rpi.edu/ zukerm/lectures/RNAfold-html/node3.html, 2003.

Simulations of Microreactors:
The Order of Things

Joseph Ibershoff[1], Jerzy W. Jaromczyk[1], and Danny van Noort[2]

[1] Department of Computer Science, University of Kentucky,
Lexington, KY 40506, USA
jwiber0@cs.uky.edu, jurek@cs.uky.edu
[2] Biointelligence Lab., School of Computer Science and Engineering,
Seoul National University, San 56-1, Sinlim-dong, Gwanak-gu, Seoul 151-742, Korea
present address: IBN, the Nanos, 31 Biopolis Way, Singapore
dvannoort@ibn.a-star.edu.sg

Abstract. Simulations are needed to predict various parameters for chemical reactions and error propagation in microfluidic networks. This paper studies the impact of the order of microreactors implementing a fluidic network on the error in solutions for Boolean expressions. Additionally, we present a computer program that augments the software toolkit introduced in our previous work. The program is useful for simulating microfluidics; we present an example from DNA computing. It monitors the concentration of every molecule throughout the fluidic network and assists in predicting how the layout of the network contributes to the error in the DNA computation.

1 Introduction

Microfluidics is the system of choice for performing computations using biological material. It can be incorporated as an information carrier in a DNA computing scheme [6] [10]. The advantages of microfluidics are the small volumes (in the picoliter range) and the speed of reactions. When using fluidic valves and micro pumps, the flow can be (re-)directed [9]. The channels are like the wires in an electronic circuit, transporting the information from one operator to another, to fluidic flip-flops, i.e. logical operators. For that reason, a microreactor network can be viewed as a realization of a Boolean expression.

There is need for simulation to allow researchers to determine various parameters essential to the costly computations without repeatedly having to expend resources. For example, simulations can be used to monitor how many strands are left after each Boolean operation and how many of these strands are errors; by running various simulations on a Boolean expression and examining the resulting simulated error, the researcher can gain insight into appropriate parameter values to maximize the effectiveness of real DNA computations.

In our previous paper on networks of microreactors [8], we investigated strand concentrations in microreactors using dataflow techniques characteristic of

C. Mao and T. Yokomori (Eds.): DNA12, LNCS 4287, pp. 286–297, 2006.

syntax-directed translation (see [1]). In particular, we discussed how to determine properties and attributes of a microreactor network, such as maximum serial length of the network or a subnetwork, or maximum simultaneous parallel paths of the network or a subnetwork; we also showed how that information can be used in the calculation of strand loss, a factor relevant to the overall effectiveness of a microreactor network. To illustrate the usefulness of these calculations, we compared two logically equivalent Boolean expressions and the strand loss resulting from them. We also proposed that the combination of this information with that from simulation could lead to further insight into microreactor network structures that help or hinder overall effectiveness.

In this paper, we discuss and present software that implements simulation and visualization of a microreactor network. In addition to looking at summary information on groups of strands as discussed in [8], the software presented here tracks the concentration for every distinct strand of the solution library individually. This allows more subtle analysis of the behavior of microreactor networks. For example, in the comparison of two logically equivalent Boolean expressions presented in [8], we chose one as preferable on the basis of strand loss. Here, we combine the strand loss information with incomplete hybridization rates from the simulation to show that the chosen Boolean expression from that comparison which exhibits less strand loss also exhibits more incomplete hybridization, and therefore may actually be less desirable. Taking this one step further, we present a third expression which retains the best characteristics of both the previously compared expressions. In addition, to illustrate observations we have made through the use of the simulation software, we examine a pair of Boolean expressions which have identical strand loss (and are therefore equivalent using the techniques presented in [8]) but with vastly different incomplete hybridization rates.

In the following sections, the principle of Boolean operations in microfluidics will be reviewed, after which the model of operation and the software will be introduced. Finally, the results of the simulation of simple Boolean expressions in different configurations are presented and discussed.

2 Computation of a Boolean Expression by a Microreactor Network

A potential solution to a Boolean expression is an assignment of truth values (TRUE or FALSE) to each variable in the expression which results in a value of TRUE for the expression as a whole. If sequences of nucleotides (which we will refer to as *words*) are used to represent assignments of values to variables (e.g. *GTTTACGTAGC* could mean "A==TRUE"), then a molecule of ssDNA consisting of these words can represent a potential solution to a Boolean expression. This is the essence of the DNA computation model, and its attractiveness is in the potential for efficient computations for problems that are currently computationally prohibitive.

A library of ssDNA can be built based upon these nucleotide words such that every potential assignment is in the library. Computation of a Boolean expression then means determining which potential solutions are correct solutions (solutions such that the value of the expression as a whole is TRUE), or perhaps simply determining whether there exist any correct solutions to the Boolean expression. The latter problem is known as the Boolean Satisfiability problem (abbreviated SAT) and it is a prototypical NP-Complete problem, one with no known efficient solution. In the context of DNA computing, computation of a Boolean expression generally refers to some process which begins with a full library of potential solutions, performs biochemical operations on the library, and ideally results in a set of ssDNA containing only those molecules which represent correct solutions. Realistically, this sort of computation will not be perfect, and will leave a (hopefully small) number of ssDNA which correspond to incorrect solutions. These trace leftovers may be referred to as *error strands* or *incorrect strands*, and it is desirable to reduce the level of error strands both absolutely and relative to the level of correct strands in the result. The ratio of error strands to *correct strands* (those which correspond to correct solutions) is referred to as the *error ratio*.

2.1   Computation on the Individual Microreactor Level

The function of a microreactor is to divide the ssDNA into two sets, based on their value assignments for a single variable in the Boolean expression (e.g. to separate strands with the word meaning "A==TRUE" from those with the word meaning "A==FALSE"). In theory, both groups could then be retained for use in different parts of the calculation, but usually this is not the case – a microreactor is used to select one of the sets of strands.

Selection is accomplished using biotinylated complementary strands (strands which are the Watson-Crick complement of one of the words used in the library), which are immobilized to streptavidin functionalized beads and packed into a microreactor. Hybridization between strands in the library and strands on the beads (abbreviated "CP" for "capture probes") is selection of one of the two groups of strands; this can be positive selection (which hybridizes the strands desired for later computation) or negative selection (which hybridizes the unwanted strands) [5]. In this paper, we only use negative selection.

2.2   Computation on the Network Level

Each microreactor represents a variable in the Boolean expression. Operators are represented in the structure of the fluidics network [9]. An OR operator is two reactors in parallel, while an AND operator is two reactors in series. In this manner a Boolean expression can be translated into a microfluidic structure. For example, the expression (**A** OR **B**) AND (**C** OR **D**) can be translated as shown in the left side of Figure 2. Only one thing is missing: a general NOT operation is not easily implemented in the same way as AND and OR. This does not cause a problem, however. It is trivial to implement NOT on a particular variable, simply by hybridizing the opposite set of strands. For example, since a microreactor using negative selection to represent **A** hybridizes strands with "A==FALSE", a microreactor using negative selection to represent NOT **A** could hybridize strands

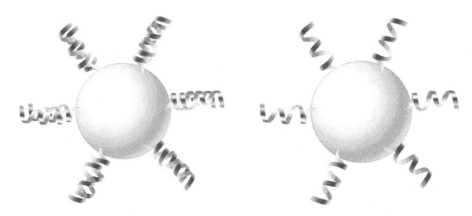

Fig. 1. A bead with bound ssDNA (*left*), and a bead without bound ssDNA (*right*)

with "**A**==TRUE". de Morgans Law then allows normalization of NOT operations such that they are always applied to variables, for instance by changing (NOT (**A** OR **B**)) to the equivalent (NOT **A**) AND (NOT **B**). A microreactor network realization of a more complex Boolean expression which utilizes NOT normalization is shown in the right side of Figure 2.

Unfortunately, the OR operator is prone to loss of information when used in conjunction with negative selection, because the flow is split between two paths; 50% of the total strands go one way while the other 50% go the other. This means that a significant portion (approaching 50%) of correct strands can be deleted from the flow during this operation, e.g. if they will hybridize in one path but not the other [8]. Viable solution strands are lost, which results in an overall increase of the error relative to the remaining solution strands. Thus it is pertinent to minimize the number of OR operators.

3   Model of Operation

Hybridization depends on the following factors [7]:

1. the size of the reactor.
2. the size of the beads.
3. the concentration and volume of ssDNA.
4. the concentration of CP.
5. flow velocity.
6. environmental conditions (pH, salt concentration, temperature).

However, in the program presented here we will only consider the reactor size, flow velocity, concentration of ssDNA and CP strands, and ssDNA solution volume as input parameters. To include the environmental conditions requires a more complex model, which will not be considered here.

The simulation of a single selection module (i.e. microreactor) is based on compartmentalization. The module is divided into n compartments of volume V.

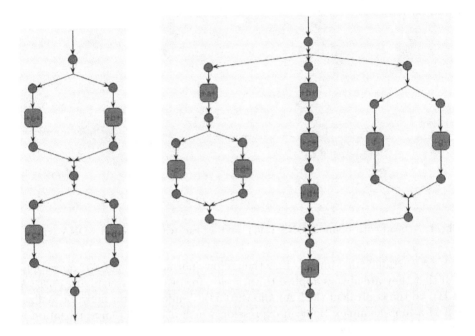

Fig. 2. The expressions (**A** OR **B**) AND (**C** OR **D**) (*left*) and ((**A** AND ((NOT **C**) OR **D**)) OR (**B** AND **C** AND **D**) OR (NOT (**F** AND **G**))) AND NOT **H** (*right*) represented by fluidic networks

Larger values of n will result in smaller compartments, and so the model will be closer to a continuous model. The DNA solution is presumed to flow in a plug, so no diffusion at the interface of DNA and carrier solution is assumed. The plug is then divided into compartments with the same volume V as the modules (see Figure 3). Time and total volume decide the accuracy of the selection. The uptake of the ssDNA is calculated in a static fashion, which in principle means that the flow is stopped and started with time intervals depending on the flow velocity and the size of the compartment.

The objective of negative selection in a microreactor is to hybridize (and thus filter out) all the ssDNA *not* wanted. In every compartment the uptake of ssDNA is calculated, and that information is passed on to the next, in the direction of the flow. Since the number of molecules involved is sufficiently large ($\sim 10^{10}$), it is possible to use deterministic and continuous, rather than stochastic, expressions. It is assumed that annealing of two ssDNA molecules follows a second-order reaction kinetics [5]:

$$[W] + [C] \xrightarrow{k} [WC] \tag{1}$$

where $[W]$ is the concentration of the ssDNA in solution and $[C]$ is the concentration of the CP. From Equation 1 we can derive the following about the decrease of the concentration of ssDNA and concentration of the CP, i.e. the binding capacity.

$$W_{i+1} = \frac{\varepsilon W_i(W_i - C_i)}{\varepsilon W_i - C_i} \tag{2}$$

$$C_{i+1} = \frac{C_i(W_i - C_i)}{\varepsilon W_i - C_i} \tag{3}$$

$$\varepsilon = e^{kt(W_i - C_i)} \tag{4}$$

where i is the compartment, k is the second-order rate constant and t is the duration of the incubation. An n-bit Boolean problem gives rise to a $2n$ library of all possible solutions. The concentration of each member in the library will be monitored individually so a comparison can be made between the correct solutions and errors after each selection module. This will show the impact of the erroneous members on the problem solving capabilities of this system and ways in which the system can be improved.

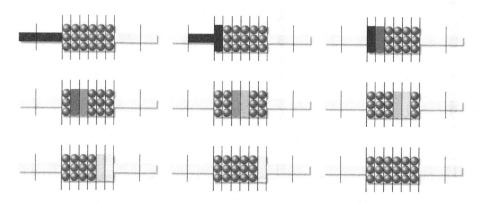

Fig. 3. A schematic of the flow of a plug with ssDNA though a compartmentalized microreactor filled with beads. The sequence of the figures is from left to right and top to bottom. In this example the volume of the plug corresponds to two compartments.

Error can be grouped into two basic categories: loss of correct strands (ones that satisfy the equation being computed), and misclassification/retention of incorrect strands (ones that do not satisfy the equation) [2]. In a system based on negative selection, misclassification of incorrect strands is typically caused by incomplete hybridization; the effect of microreactor parameters on incomplete hybridization in a single negative selection is examined in [4]. Strand loss has two sources: non-specific hybridization within a microreactor, and loss resulting from parallel selection in an OR structure. Non-specific hybridization is said to occur when ssDNA hybridizes with non-complementary strands on the beads, binds to the walls of the microreactor, or in any other way binds to something other than complementary strands. In this paper, the level of non-specific hybridization is assumed to be zero.

4   Software

The program that performs the simulation based on the model discussed above has a structure typical of the Model-View-Controller design pattern [3]. A screenshot of the current interface (View) is shown in Figure 4. As can be seen in the screenshot, the window is divided into three major components: the first part is responsible for controls and input parameters; the part below that contains numerical output of the simulation results; and the area to the right displays the schematic diagram of the microreactor network constructed.

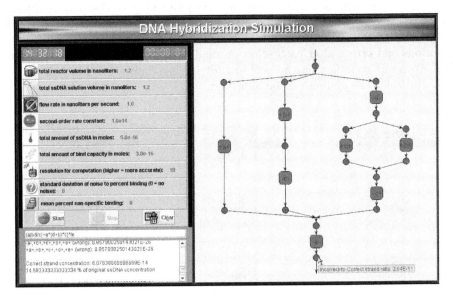

Fig. 4. A screenshot of the program, showing the network layout for the equation (**A** OR (**B** AND (NOT **C**)) OR ((NOT **A**) AND (**D** OR **B**) AND **C**)) AND (NOT **E**))

4.1   Simulation Control

There are seven parameters governing the simulation of the network, in addition to a text field for accepting the Boolean equation to compute.

1. *Reactor volume.* This is the total free space available for flow of ssDNA solution (i.e. it does not include the volume consumed by beads), given in nanoliters.
2. *ssDNA volume.* This is the volume of ssDNA solution that will flow through the reactor, given in nanoliters.
3. *Reactor flow rate.* This is the rate at which ssDNA solution flows into and out of the reactor, given in nanoliters per second.
4. *Rate constant.* This is the second-order rate constant k used in Equation 1. This value specifies how quickly hybridization occurs when ssDNA is in the presence of CP in the microreactor. Since this value may depend on the

design of the microreactors being used as well as other factors, it is necessary to allow the user to specify it.

5. *ssDNA amount.* This specifies the total amount of ssDNA present in the solution, given in moles.

6. *Bind capacity.* This specifies the total amount of ssDNA that can theoretically be bound within the microreactor, given in moles.

7. *Resolution.* This specifies the granularity of calculation for the simulation; higher values will require more computation, but will yield more accurate results. Unless many significant figures are necessary in the result, a value of about 10 is appropriate.

4.2   Microreactor Network Model

The microreactor network is automatically laid-out based on its structure, so that all important structural features are easily discernible. As a side remark, this layout is an example of an algorithm that belongs to the group of so-called graph drawing algorithms; in this case the drawing is directed by the syntactical structure of the Boolean expression. The AND and OR operators are treated as n-ary rather than binary; for example, a+b+c+d is understood directly as a single operation on four expressions, rather than being interpreted in the usual left-associative, binary way as (((a+b)+c)+d) (which has a very different, and less desirable, network representation). The model shows individual microreactors as red rectangles, each labeled with the variable that microreactor is selecting; the label uses the plus sign ('+') to indicate that the microreactor should allow strands to pass which have the value of TRUE for that variable, while it uses the minus sign ('-') to indicate that the microreactor should allow strands to pass which have the value of FALSE for that variable. Note that since this simulation uses negative selection, a microreactor with the label "+a+" actually hybridizes strands which have "a==FALSE" (thus allowing strands which have "a==TRUE" to pass). Holding the mouse cursor over a microreactor rectangle causes a ToolTip to be displayed which shows the error rate of that microreactor's selection.

The model also has numerous gray colored nodes on the connections between microreactors; these *tap points* are locations at which the flow of ssDNA can be inspected to examine the error ratio within the context of the entire network (as opposed to the error shown for a microreactor, which only gives error within the context of a single selection). A tap point is included in the model every time the concentrations in the flow would be altered, and may be inspected via ToolTip (a tap point ToolTip is shown in the screenshot in Figure 4).

4.3   Output View

The current output view area displays textual and numerical data for the out-flow of the network. A normalized version of the input Boolean expression is displayed, followed by a concentration readout for each distinct ssDNA strand in the simulation. Then, summary information such as the overall concentration of correct strands and incorrect strands is given, along with the error ratio (concentration of incorrect strands divided by concentration of correct strands).

An improved output view can accompany this textual information with a visualization of the strand concentrations. In addition, all the information (both the textual and visual) can be made available not just for the network out-flow, but for any tap point in the model, selectable by clicking on the appropriate tap point. This would allow the researcher to use the simulation to investigate not just the overall result of a particular microreactor network, but also to investigate substructures within the network (e.g. to locate the source within a given network of a large amount of strand loss).

5   Results and Discussion

To demonstrate the importance of careful use of the OR operator, three logically equivalent Boolean expressions (i.e. expressions having the same truth table values) were compared. Furthermore, to show the importance of the placement of the OR operator within and AND operation, we tested the associative law. The results obtained in these two experiments illustrate the usefulness of the simulation software as a tool in the DNA computing researcher's toolkit.

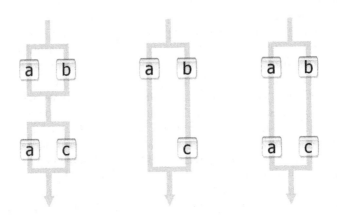

Fig. 5. A microfluidic representation of the Boolean expressions: (**A** OR **B**) AND (**A** OR **C**) (*left*), **A** OR (**B** AND **C**) (*center*), and (**A** AND **A**) OR (**B** AND **C**) (*right*)

5.1   OR Operator Minimization

To show the impact of the OR operator on strand loss and the resulting error ratio, we have compared three logically equivalent Boolean expressions, each with a difference in the number and/or structure of OR operators: (**A** OR **B**) AND (**A** OR **C**) vs **A** OR (**B** AND **C**) vs (**A** AND **A**) OR (**B** AND **C**) (see Figure 5).

The first network in Figure 5, representing (**A** OR **B**) AND (**A** OR **C**), has two ORs placed in series, with strand mixing in between. The fact that the two ORs each have an identical selection (on the variable a) leads to significant strand loss, as detailed in [8]. The first chart in Figure 6 compares the resulting concentration of correct strands (strands that satisfy the Boolean expression) from

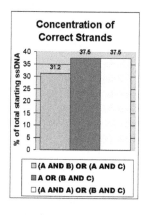
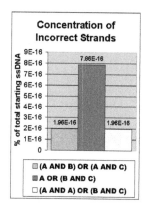
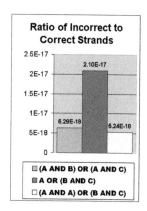

Fig. 6. Comparison of strand concentrations in three different but logically equivalent expressions for a simulation using default running parameters. (*left*) shows the concentration of strands that satisfy the Boolean expression, (*center*) shows the concentration of strands that do not satisfy the Boolean expression, and (*right*) shows the ratio of the previous two values.

simulating each of the expressions using default parameters. As has been shown previously [8], the first expression resulted in less correct strands retained than the second expression.

The second network in Figure 5, representing **A** OR (**B** AND **C**), eliminates the problems of the first network by using a simplified Boolean expression which eliminates the second OR. Unfortunately, it has more retained incorrect strands than the first expression, as can be seen in the second chart of Figure 6. This is because the redundant selection of a, although responsible for additional correct strand loss in the first expression, also removed additional incorrect strands. The ratio of incorrect strand concentration to correct strand concentration (shown in the third chart in Figure 6) indicates that this second expression may actually be worse for many purposes, as it has a higher error ratio.

The third network in Figure 5, representing (**A** AND **A**) OR (**B** AND **C**), keeps the best of both the previous expressions by using only a single OR, but adds back in the redundant selection on a. It has the same incorrect strand concentration as the first expression, and the same correct strand concentration as the second expression, resulting in a better error ratio than either.

As the OR operator is a major source of strand loss, there is a need to minimize its use. Incorrect strand concentration can be reduced without requiring additional uses of OR.

5.2   Order and Associative Law

To show that the order of serial operations (stemming from an AND) in the microfluidic system is important, we demonstrate by simulation that error in microfluidic systems does not obey the associative law. Figure 7 shows the networks that were used for comparison.

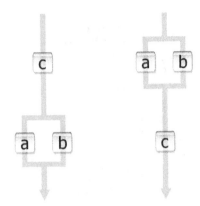

Fig. 7. Even changing the evaluation order of operations in series, such as in the two networks above, results in changes to the end concentrations

Simulation of the first network using default parameters results in an incorrect strand concentration of 8.3e-3%, compared to a tiny 5.7e-8% for the second network, a factor of 10^5 difference. This occurs because the flow entering an OR is split, leaving only a fraction as much volume available to each of its branches; when the concentration of ssDNA entering the OR is reduced by the previous selection on c, and then also split in volume, it causes the second operation in the series to be substantially less effective. This is an indirect result of the decaying-exponential behavior of ssDNA concentration within a single microreactor; see [7] for a more detailed analysis of behavior within a single microreactor. On the other hand, the AND is less sensitive to a previous reduction in concentration of ssDNA, since it does not have to split its volume. Thus, it appears that OR operations should be performed first within an AND operation, at least for simple cases.

It should be reiterated that in the simulations just described, non-specific binding of ssDNA is set to zero. This results in the two different networks each having the same concentration of correct strands remaining at the out-flow of the network. However, a more complicated and accurate model that calculates non-specific binding with attention to the overall ssDNA concentration and remaining binding capacity would most likely show different concentrations of correct strands for these two networks.

6   Conclusions

Maximizing correct strand concentration while also minimizing incorrect strand concentration in a network of microreactors is a subtle and difficult task. The OR operator is particularly sensitive, due to the necessary reduction in ssDNA volume that is inherent to its calculation when using negative selection – it has been shown that the number and structure of OR operators, as well as the placement of ORs within an AND series, can have a significant impact on the result.

Investigation of the circumstances which increase or decrease the effectiveness of a particular realization of a Boolean problem is tedious and difficult by hand, but is relatively simple with simulation software. Eventually, the insight gained from experimenting with simulation might lead to observations, and ultimately to proven algorithms that can automatically restructure a microreactor network to increase its effectiveness. The results obtained in the simulations presented in the paper illustrate the usefulness of the simulation software as a tool in the DNA computing researcher's toolkit.

Acknowledgments

D. van Noort would like to acknowledge the support from the Molecular Evolutionary Computing (MEC) project of the Korean Ministry of Commerce, Industry and Energy, and the National Research Laboratory (NRL) Program from the Korean Ministry of Science and Technology. J. W. Jaromczyk acknowledges the support by the Kentucky Biomedical Research Infrastructure Network, funded by grant 2P20RR016481-04 from the National Center for Research Resources and awarded to Nigel Cooper from the University of Louisville.

References

1. A.V. Aho, R. Sethi, and J.D. Ullman, Compilers: Principles, Techniques, and Tools, Addison-Wesley, 1986.
2. K. Chen and E. Winfree: Error Correction in DNA Computing: Misclassification and Strand Loss, DIMACS 5, (1999).
3. E. Gamma, R. Helm, R. Johnson, and J. Vlissides: Design Patterns: Elements of Reusable Object-Oriented Software. Addison-Wesley, (1995).
4. J. Ibershoff, J. W. Jaromczyk, D. van Noort: Simulation and Visualization of Hybridization Events for DNA Computing in Microreactors. Journal of the Japan Society for Simulation Technology; 2005 December, Vol.24, No. 4, pp277–284 (2005).
5. M. S. Livstone and L. F. Landweber: Mathematical considerations in the design of microreactor-based DNA computers. LNCS 2943, pp180–189, (2004).
6. J. S. McCaskill: Optically programming DNA computing in microflow reactors. BioSystems 59 pp125–138, (2001).
7. D. van Noort, Y. Hong, J. Ibershoff, and J. W. Jaromczyk: Simulation and Visualization for DNA Computing in Microreactors. Proceedings of the International Conference on Natural Computing ICNC 2005; LNCS 3611 Part II, (2005).
8. D. van Noort, J. Ibershoff, J. W. Jaromczyk: Dataflow-Based Simulation of DNA Computing in Cascading Microreactors. Proceedings of 3rd International conference on Computational Intelligence, Robotics and Autonomous Systems (CIRAS 2005). Elsevier, ISSN: 0219-6131, (in press).
9. D. van Noort, Z.-L. Tang, and L. F. Landweber: Fully controllable microfluidics for molecular computers. JALA 9, (2004).
10. D. van Noort, P. Wagler, and J. S. McCaskill: The role of microreactors in molecular computing. Smart Mater. Struct. 11, pp756–760, (2002).

DNA Hypernetworks for Information Storage and Retrieval

Byoung-Tak Zhang and Joo-Kyung Kim

Biointelligence Laboratory, School of Computer Science and Engineering
Seoul National University, Seoul 151-742, Korea
{btzhang, jkkim}@bi.snu.ac.kr
http://bi.snu.ac.kr/

Abstract. Content-addressability is a fundamental feature of human memory underlying many associative information retrieval tasks. In contrast to location-based memory devices, content-addressable memories require complex interactions between memory elements, which makes conventional computation paradigms difficult. Here we present a molecular computational model of content-addressable information storage and retrieval which makes use of the massive interaction capability of DNA molecules in a reaction chamber. This model is based on the "hypernetwork" architecture which is an undirected hypergraph of weighted edges. We describe the theoretical basis of the hypernetwork model of associative memory and its realization in DNA-based computing. A molecular algorithm is derived for automatic storage of data into the hypernetwork, and its performance is examined on an image data set. In particular, we study the effect of the hyperedge cardinality and error tolerance on the associative recall performance. Our simulation results demonstrate that short DNA strands in a vast number can be effective in some pattern information processing tasks whose implementation is within reach of current DNA nanotechnology.

1 Introduction

Content-addressable memories or associative memories are storage devices which return stored contents from partial contents. These are contrasted to typical location-based storage devices where addresses are to be provided rather than contents. Content-addressable memories are useful in search intensive applications such as information retrieval, data compression, and database search [9]. It has long been known that human memory is based on content-addressing [8].

The property of massive parallelism along with associative search capability of DNA computing can be very useful in realizing content-addressable memory. This has been pointed out by several researchers in DNA computing community [1,10] and there are some experimental works going on in this line of research, for example [3] and [5]. However, there is lack of theoretical studies on developing systematic models of associative memory based on molecular computing or DNA nanotechnology.

C. Mao and T. Yokomori (Eds.): DNA12, LNCS 4287, pp. 298–307, 2006.

Here we propose a graphical model of associative memory, called hypernetworks, which is naturally implemented as a library of interacting DNA nanostructures. A hypernetwork is a weighted hypergraph, i.e., graphs having "hyperedges". Thus, a hypergraph can have edges connecting three or more vertices while in an ordinary graph the edges can connect maximum two vertices. As we shall see, the use of these hyperedges allows for additional degrees of freedom in representing memory elements in a network representation while preserving the mathematical tools provided by the graph theory.

The purpose of this paper is to introduce the hypernetwork model of associative memory from a theoretical point of view, and study its essential properties such as the tolerance of associative recall against errors in input and/or chemical reaction. We report on simulation results on pattern completion tasks where the corrupted, input images are to be reconstructed into the original or clean patterns.

The paper is organized as follows. Section 2 introduces the hypernetwork model of data storage. Section 3 presents a method for encoding the hypernetworks in DNA molecules. Section 4 describes a method for automatic storage of patterns on the hypernetwork along with its theoretical background. Section 5 shows the simulation results on the hand-written image data set. Section 6 draws conclusions.

2   The Hypernetwork Model

A hypergraph is an undirected graph G whose edges connect a non-null number of vertices [2], i.e. $G = (V, E)$, where $V = \{v_1, v_2, ..., v_n\}$, $E = \{E_1, E_2, ..., E_m\}$, and $E_i = \{v_{i1}, v_{i2}, ..., v_{ik}\}$. E_i is called hyperedges. Mathematically, E_i is a set and its cardinality is $k \geq 1$, i.e., the hyperedges can connect more than two vertices while in ordinary graphs the edges connect up to two vertices, i.e., $k \leq 2$. A hyperedge of cardinality k will be referred to as a k-hyperedge.

Figure 1 shows a hypergraph consisting of seven vertices $V = \{v_1, v_2, ..., v_7\}$ and five hyperedges $E = \{E_1, E_2, E_3, E_4, E_5\}$. A hypergraph can be represented as an incidence matrix. The incidence matrix of a hypergraph $G = (V, E)$ is a matrix $((a_j^i))$ with m rows that represent the hyperedges of G and n columns that represent the vertices of G, such that $a_j^i = 1$ if $v_j \in E_i$ and $a_j^i = 0$ if $v_j \notin E_i$. Each $(0, 1)$-matrix is the incidence matrix of a hypergraph if no row or column contains only zeros. Figure 1 also shows the incidence matrix corresponding to the hypergraph.

We now generalize the hypergraph into hypernetworks by assigning the weight values to the hyperedges. Formally, we define a hypernetwork as a triple $H = (V, E, W)$, where

$$V = \{v_1, v_2, ..., v_n\} \tag{1}$$
$$E = \{E_1, E_2, ..., E_m\} \tag{2}$$
$$W = \{w_1, w_2, ..., w_m\}, \tag{3}$$

where $E_i = \{v_{i1}, v_{i2}, ..., v_{im}\}$. An m-hypernetwork consists of a set V of vertices, a subset E of $V[m]$, and a set W of hyperedge weights, i.e. $H = (V, E, W)$, where

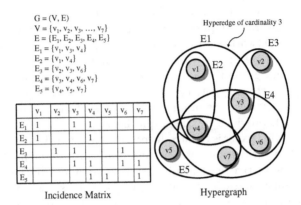

Fig. 1. An example hypergraph consisting of seven vertices and five hyperedges of variable cardinality. The hypernetwork can be represented a matrix, called an incidence matrix, of m rows of hyperedges and n columns of vertices.

$E = V[m]$ is a set of subsets of V whose elements have precisely m members. A hypernetwork H is said to be k-uniform if every edge E_i in E has cardinality k. A hypernetwork H is k-regular if every vertex has degree k. Note that an ordinary graph is a 2-uniform hypernetwork with $w_i = 1$.

We wish to store a data set $D = \{\mathbf{x}^{(n)}\}_{n=1}^{N}$ in a hypernetwork so that they can be retrieved later by content. $\mathbf{x}^{(n)}$ denotes the n-th pattern to store. To do this we require the hypernetwork to represent the probabilistic distribution of the data. We define the energy of the hypernetwork

$$
\begin{aligned}
E(\mathbf{x}^{(n)}; W) = &-\sum_{i_1} w_{i_1}^{(1)} x_{i_1}^{(n)} - \frac{1}{2} \sum_{i_1, i_2} w_{i_1 i_2}^{(2)} x_{i_1}^{(n)} x_{i_2}^{(n)} \\
&- \frac{1}{6} \sum_{i_1, i_2, i_3} w_{i_1 i_2 i_3}^{(3)} x_{i_1}^{(n)} x_{i_2}^{(n)} x_{i_3}^{(n)} - \dots \\
= &-\sum_{k=1}^{K} \frac{1}{k!} \sum_{i_1, i_2, \dots, i_k} w_{i_1 i_2 \dots i_k}^{(k)} x_{i_1}^{(n)} x_{i_2}^{(n)} \dots x_{i_k}^{(n)},
\end{aligned}
\tag{4}
$$

where W represents the parameters (hyperedge weights) for the hypernetwork model. Note that $x_{i_1}^{(n)} x_{i_2}^{(n)} \dots x_{i_k}^{(n)}$ is a combination of k components of the data item $\mathbf{x}^{(n)}$ which is represented as a k-hyperedge in the network. The probability of the data being generated from the hypernetwork is then expressed as

$$
\begin{aligned}
P(\mathbf{x}^{(n)}|W) &= \frac{1}{Z(W)} \exp\left[-E(\mathbf{x}^{(n)}; W)\right] \\
&= \frac{1}{Z(W)} \exp\left[\sum_{k=1}^{K} \frac{1}{k!} \sum_{i_1, i_2, \dots, i_k} w_{i_1 i_2 \dots i_k}^{(k)} x_{i_1}^{(n)} x_{i_2}^{(n)} \dots x_{i_k}^{(n)}\right],
\end{aligned}
\tag{5}
$$

where the normalizing term (known as the partition function in statistical physics) is given as

$$Z(W) = \sum_{\mathbf{x}^{(m)}} \sum_{k=1}^{K} \frac{1}{k!} \sum_{i_1,i_2,...,i_k} w_{i_1 i_2 ... i_k}^{(k)} x_{i_1}^{(m)} x_{i_2}^{(m)} ... x_{i_k}^{(m)}. \tag{6}$$

In effect, the hypernetwork represents a probabilistic model of the data set using a collection of hyperedges and their weights.

3   Representing a Hypernetwork with DNA Molecules

Given two sets of data items, $D = X = \{\mathbf{x}^{(n)}|n = 1, ..., N\}$ and $Y = \{\mathbf{y}^{(n)}|n = 1, ..., N\}$, where Y is a corrupted version of X. The goal is to store X in a hypernetwork H in such a way that, given a corrupted data $\mathbf{y}^{(n)}$, the original data $\mathbf{x}^{(n)}$ is recovered or a clean version of it is reconstructed.

This task is known as pattern completion or pattern restoration for which content-addressing and associative capability is required, such as in the self-organizing systems [8,13]. The Hopfield network is another model of content-addressable memory [6]. Boltzmann machines and the Helmholtz machines [7] are generalizations of the Hopfield model by introducing the hidden variables in addition to the observable variables. All these models are based on the second-order correlations of the data. Higher-order correlations are captured by intro-ducing hidden variables, and no explicit use of higher-order terms are made. There is, however, evidence that higher-order correlation terms are useful. The hypernetwork model has the advantage that the higher-order correlation terms can be directly represented by the hyperedges. For example, a 3-hypernetwork encodes the memory using the third-order correlation terms made of combina-tions of the input variables.

We now explain the method for representing the hypernetwork using DNA molecules so that the networks can be built and maintained by molecular com-putational operators. The idea is based on the observation that the hypernetwork is a collection of hyperedges with duplicates allowed. Basically, the original data are fragmented into a set of vertices, i.e. hyperedges, and maintained as a col-lection of hyperedges or, equivalently, an incidence matrix. Then the hyperedges are encoded as DNA strands. In effect, a hypernetwork is represented as a library of DNA strands where duplicates are allowed. The procedure is schematically illustrated in Figure 2.

To be more concrete, let us assume that we opted for a 3-hypernetwork model. We generate all possible hyperedges $V[3] = \{E1 = \{x_1, x_2, x_3\}, E2 = \{x_1, x_2, x_4\}, ..., E_{|V[3]|} = \{x_{n-2}, x_{n-1}, x_n\}\}$ or some subset of it to initialize the library. This results in a hypernetwork represented as a collection of hyperedges. The number of possible hyperedges increases by $2^k \times {}_nC_k$ in the number n of variables and the cardinality k of hyperedges. There should be some mechanism to choose the right hyperedges or to penalize the growth of the model complexity and we will study some of these issues in a later section.

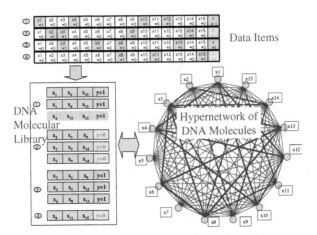

Fig. 2. General procedure for building a hypernetwork from a data set. From the data items, higher-order correlation terms are extracted and represented as hyperedges (with duplication allowed) which are then encoded as DNA strands. This library of DNA molecules represents the hypernetwork where the weights are encoded as the number of copies of the DNA molecules for hyperedges.

We use a similar method described in [12] to encode the hyperedge using a DNA strand. Each vertex (i.e. the variable and its value) in the hyperedge is encoded as a DNA sequence. For example, $x_1 = 0$ is assigned a DNA codeword 'TACAGG', where 'TACA' is for variable x_1 and 'GG' is for value '0'. In this scheme, a hyperedge $(x_1 = 0, x_3 = 1, x_4 = 0)$ is represented as 'TACAGG CTACAA GCATGG' assuming that x_3 = 'CTAC', x_4 = 'GCAT', and the value 1 is encoded as 'AA'. Then the collection of DNA-encoded hyperedges represent a hypernetwork of DNA molecules or a molecular hypernetwork.

4   Constructing a DNA Hypernetwork from Data

We now describe the procedure for building a molecular hypernetwork that fits a given data set. The basic idea is, starting with a random network, to let the network self-organize to learn the data as they are observed. The procedure is illustrated in Figure 2. The hypernetwork is represented as a collection of hyperedges, and each hyperedge is encoded as a DNA molecule, as described in the preceding section. The random k-uniform hypernetwork is then represented as a collection (or library) L of hyperedges of cardinality k where the component variables of the hyperedge and the number of copies of the hyperedges are initialized at random or according to some prior knowledge in the problem domain.

The procedure is summarized as follows:

- 1. Generate a library L of random hyperedges of cardinality k.
- 2. Get a pattern $\mathbf{x}$. Generate the hyperedges of cardinality k from $\mathbf{x}$ (with duplication permitted) into K.

- 3. (Retrieval) Find the hyperedges of L matching to those of K with error tolerance τ into M. Optionally (in case of multi-cycle retrieval), repeat this step using M as K.
- 4. (Storage) Update L by $L \leftarrow L + M + Copy(u)$ for hyperedge $u \in M$.
- 5. Go to step 2 if not terminated.

The library starts with a random collection of hyperedges of cardinality k (Step 1). As a training pattern $\mathbf{x} = (x_1, x_2, ..., x_n)$ is observed, we sample a collection K of hyperedges from $\mathbf{x}$ (Step 2). The hyperedges are generated in multiple copies where the component variables and the number of copies are chosen at random. The library L and the hyperedge collection K from the example are merged to find the matching hyperedges in L (Step 3). This can be done by hybridizing the DNA encoded hyperedges in the two collections (this to be effective, we encode the example hyperedges in complementary DNA sequences to those for the library hyperedges). The matching hyperedges are then copied by some rate, i.e $Copy(u)$ for $u \in M$, and merged with the current library L to update it (Step 4). The whole procedure is repeated for the next training pattern (Step 5). Note that the procedure makes use of relatively simple molecular operators such as selection, separation, and replication of DNA strands.

It is important to note that we allow error tolerance in this matching process where the tolerance level τ is an algorithmic parameter. The error tolerance parameter has several implications. First, it is useful to model the degree of unreliability of DNA hybridization reaction. Second, it is useful to control the generalization ability of the molecular hypernetwork memory, since the mismatches have some effect of reducing the noises in raw data. A low error tolerance (allowing only a small number of mismatches) might lead to overfitting while a high error tolerance might result in unstable learning.

It can be shown that the storage process performs gradient search to find maximum-likelihood parameters for the training data set. To see this, given a set $D = \{\mathbf{x}^{(n)}\}_{n=1}^{N}$ of n independently and identically distributed examples, we consider the likelihood of the parameters W:

$$P(D|W) = \prod_{n=1}^{N} P(\mathbf{x}^{(n)}|W), \tag{7}$$

where W consists of the weights or the number of copies of the hyperedges of order k. Taking the logarithm of the likelihood we get

$$\ln P(D|W) = \ln \prod_{n=1}^{N} P(\mathbf{x}^{(n)}|W) \tag{8}$$

$$= \sum_{n=1}^{N} \left\{ \left[\sum_{k=1}^{K} \frac{1}{k!} \sum_{i_1, i_2, ..., i_k} w_{i_1 i_2 ... i_k}^{(k)} x_{i_1}^{(n)} x_{i_2}^{(n)} ... x_{i_k}^{(n)} \right] - \ln Z(W) \right\},$$

where Eqn. (5) is used for $P(\mathbf{x}^{(n)}|W)$. We take the derivative of the log-likelihood

$$\frac{\nabla}{\nabla w_{i_1,i_2,\ldots,i_k}^{(k)}} \ln \prod_{n=1}^{N} P(\mathbf{x}^{(n)}|W) \tag{9}$$

$$= \frac{\nabla}{\nabla w_{i_1,i_2,\ldots,i_k}^{(k)}} \sum_{n=1}^{N} \left\{ \left[\sum_{k=1}^{K} \frac{1}{k!} \sum_{i_1,i_2,\ldots,i_k} w_{i_1 i_2 \ldots i_k}^{(k)} x_{i_1}^{(n)} x_{i_2}^{(n)} \ldots x_{i_k}^{(n)} \right] - \ln Z(W) \right\}$$

$$= \sum_{n=1}^{N} \left\{ \frac{\nabla}{\nabla w_{i_1,i_2,\ldots,i_k}^{(k)}} \left[\sum_{k=1}^{K} \frac{1}{k!} \sum_{i_1,i_2,\ldots,i_k} w_{i_1 i_2 \ldots i_k}^{(k)} x_{i_1}^{(n)} x_{i_2}^{(n)} \ldots x_{i_k}^{(n)} \right] - \frac{\nabla}{\nabla w_{i_1,i_2,\ldots,i_k}^{(k)}} \ln Z(W) \right\}$$

$$= \sum_{n=1}^{N} \left\{ x_{i_1}^{(n)} x_{i_2}^{(n)} \ldots x_{i_k}^{(n)} - \langle x_{i_1} x_{i_2} \ldots x_{i_k} \rangle_{P(\mathbf{x}|W)} \right\}$$

$$= N \left\{ \langle x_{i_1} x_{i_2} \ldots x_{i_k} \rangle_{Data} - \langle x_{i_1} x_{i_2} \ldots x_{i_k} \rangle_{P(\mathbf{x}|W)} \right\}, \tag{10}$$

where the two terms in the last line are defined as

$$\langle x_{i_1} x_{i_2} \ldots x_{i_k} \rangle_{Data} = \frac{1}{N} \sum_{n=1}^{N} \left[x_{i_1}^{(n)} x_{i_2}^{(n)} \ldots x_{i_k}^{(n)} \right] \tag{11}$$

$$\langle x_{i_1} x_{i_2} \ldots x_{i_k} \rangle_{P(\mathbf{x}|W)} = \sum_{\mathbf{x}} \left[x_{i_1} x_{i_2} \ldots x_{i_k} P(\mathbf{x}|W) \right]. \tag{12}$$

The learning rule (10) suggests that maximum-likelihood is achieved by reducing the difference between the average frequencies of the hyperedges in the data set and in the hypernetwork model, as was described above. The next section studies the empirical behavior of this procedure under various experimental set-ups.

5 Simulation Results

We are interested in examining the system property of the hypernetworks in storing and retrieving patterns. In particular, we ask the following questions:

- 1. What is the effect of cardinality of hyperedges on the retrieval performance? Does increasing the cardinality help the correct retrieval or deteriorate it?
- 2. How tolerant is the hypernetwork model of associative memory against the noise or corruption of data.

To study the above questions, we used a data set consisting of handwritten numeral images. The original data came from Optical Recognition of Handwritten Digits in UCI machine learning repository[1] and we preprocessed it into a training set of 3760 examples of 8×8 bitmap. The process of image storage into the hypernetwork proceeded as described in the preceding section. To see the noise effect, another set of "corrupted" images was made by randomly toggling

---

[1] ftp://ftp.ics.uci.edu/pub/machine-learning-databases/optdigits/

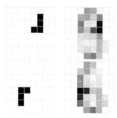

Fig. 3. Pattern completion by the hypernetwork model of associative memory. (Left column) The partial images given as input cues. (Right column) The corresponding output images reconstructed from the inputs.

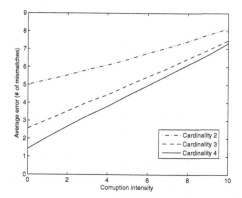

Fig. 4. The effect of cardinality k of hyperedges. Shown is the average error in image reconstruction as a function of the corruption intensity for $k = 2, 3, 4$. The hypernetworks with higher-cardinality hyperedges obtain better restoration performance than the low-cardinality networks.

the pixels of the original images. We compared the image reconstruction performances of k-uniform hypernetworks with varying $k = 2, 3, 4$, i.e., those consisting of the hyperedges of cardinality k with associated weights. We also have run the experiments by varying the level of error tolerance in matching between the hyperedges from the input image and those maintained in the library.

Figure 3 shows the images completed from the partial input. In this task, images of numeral '6' and '9' have been stored into the hypernetwork. The network recovers the appropriate patterns given partial contents as input cues. Figure 4 shows the effect of the cardinality parameter k on the associative recall of the image. The restoration error was measured as the average number of mismatches between the original image and the restored image. The results show that the restoration errors for the hypernetworks of higher k were smaller than those of lower k. It can be clearly seen that by increasing the cardinality of the hyperedges, the reconstruction error tends to decrease. This suggests the importance of higher-order correlation terms in associative recall. In this experiment, we also

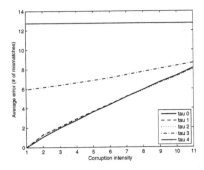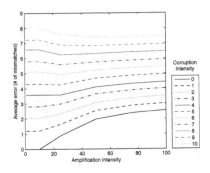

Fig. 5. The effects of the error tolerance level (left) and the amplification intensity (right). The left panel shows the stability of associative recall in the hypernetwork model against mismatches in reaction. The right panel suggests that lower amplification intensities are more appropriate to achieve a good restoration performance for slightly-corrupted data, while higher amplification intensities are not necessarily harmful for highly corrupted data.

changed the corruption levels of the images, which is depicted on the x-axis. The graph shows that the higher-order effect is especially clear when the corruption level is low.

The effect of error tolerance level τ is shown in Figure 5(left). When the number of mismatches between an input pattern and a library element is 2 or less, the library element was assumed to be matched and duplicated as if it were perfectly matched. This result shows that the performance is relatively unaffected up to some critical tolerance level (in this experiment, $\tau = 2$). However, as error tolerance increases, the average error of recovery increases because of overgeneralization. Since the error tolerance parameter indirectly reflects the unreliability in molecular reaction, this shows a stability of the hypernetwork memory in this setting of experimental parameters.

We also studied the effect of amplification intensity, i.e., the strength of learning for an observed image. The curves in Figure 5(right) show that keeping the amplification rate small helps reduce the reconstruction error, especially when the images have a low-level of corruption. However, when the images are highly corrupted, a higher rate of amplification does not necessarily hurt the performance.

6 Conclusion

We have presented a hypernetwork-based molecular architecture which allows for content-addressable storage and retrieval of patterns. The realization of this architecture using DNA molecules is described, and an algorithm is presented that automatically store data into and retrieve them from this architecture using massively parallel molecular operations. Simulation results demonstrate the possibility of using this network for pattern completion and reconstruction, i.e.

as associative memory devices. Due to lack of computing power for simulation, we were not able to perform simulations on k-hypernetworks for hyperedge cardinality $k \geq 5$. However, realized in DNA computers, we expect the molecular computational method to scale up better than in silicon computers. Another implication of the hypernetwork model is that it suggests an interesting new application of DNA-based molecular computing where a vast number of DNA molecules with short, not necessarily long, strands is useful.

Acknowledgements

This research was supported by the Ministry of Science and Technology (NRL), the Science and Engineering Foundation (Korean-German Researcher Exchange Program), and the Ministry of Industry and Commerce (MEC).

References

1. Baum, E. B., "Building an associative memory vastly larger than the brain," *Science*, 268:583-585, 1995.
2. Berge, C. *Graphs and Hypergraphs*, North-Holland Publishing, Amsterdam, 1973.
3. Chen, J. Deaton, R. and Wang, Y.-Z., "A DNA-based memory with in vitro learning and associative recall," DNA9, *LNCS* 2943:145-156, 2004.
4. Chisvin, L. and Duckworth, R. J., "Content-addressable and associative memory: Alternatives to the ubiquitous RAM," *IEEE Computer*, 22(7): 51-64, 1989.
5. Garzon, M. Bobba, K. and Neel, A., "Efficiency and reliability of semantic retrieval in DNA-based memories," DNA9, *LNCS* 2943:157-169, 2004.
6. Hopfield, J., "Neurons with graded response have collective computational properties like those of two-state neurons," *Proc. Nat. Acad. Sci.*, 81:3088-3092, 1984.
7. Hinton, G. E. Dayan, P. Frey, B. J. and Neal, R. M. "The wake-sleep algorithm for unsupervised neural networks," *Science*, 268:1158-1161, 1995.
8. Kohonen, T. *Content-Addressable Memories*, Springer-Verlag, Berlin, 1980.
9. Paziamtzis, K. and Sheikholeslami, A., "A low-power content-addressable memory (CAM) using pipelined hierarchical search scheme," *IEEE Journal of Solid-State Circuits*, 39(9):1512-1519, 2004.
10. Reif, J.H., LaBean, T.H., Pirrung, M., Rana, V.S., Guo, B., Kingsford, C., Wickham, G.S., "Experimental construction of very large scale DNA databases with associative search capability," DNA7, *LNCS* 2340:231-247, 2002.
11. Thurber, K. J. and Wald, L. D. "Associative and parallel processors," *ACM Computing Surveys*, 7(4): 215-225, 1975.
12. Zhang, B.-T. and Jang, H.-Y., "A Bayesian algorithm for in vitro molecular evolution of pattern classifiers," DNA10, *LNCS* 3384:458-467, 2005.
13. Zhang, B.-T. Yang, J.-S., and Chi, S.-W., "Self-organizing latent lattice models for temporal gene expression profiling," *Machine Learning*, 52(1/2):67-89, 2003.

Abstraction Layers for
Scalable Microfluidic Biocomputers

William Thies[1], John Paul Urbanski[2],
Todd Thorsen[2], and Saman Amarasinghe[1]

[1] Computer Science and Artificial Intelligence Laboratory
[2] Hatsopoulos Microfluids Laboratory
Massachusetts Institute of Technology
{thies, urbanski, thorsen, saman}@mit.edu

Abstract. Microfluidic devices are emerging as an attractive technology for automatically orchestrating the reactions needed in a biological computer. Thousands of microfluidic primitives have already been integrated on a single chip, and recent trends indicate that the hardware complexity is increasing at rates comparable to Moore's Law. As in the case of silicon, it will be critical to develop abstraction layers—such as programming languages and Instruction Set Architectures (ISAs)—that decouple software development from changes in the underlying device technology.

Towards this end, this paper presents BioStream, a portable language for describing biology protocols, and the Fluidic ISA, a stable interface for microfluidic chip designers. A novel algorithm translates microfluidic mixing operations from the BioStream layer to the Fluidic ISA. To demonstrate the benefits of these abstraction layers, we build two microfluidic chips that can both execute BioStream code despite significant differences at the device level. We consider this to be an important step towards building scalable biological computers.

1   Introduction

One of the challenges in biological computing is that the laboratory protocols needed to carry out a computation can be very time consuming. For example, a 20-variable 3-SAT problem required 96 hours to complete [1], not counting the considerable time needed for setup and evaluation. To automate and optimize this process, researchers have turned to microfluidic devices [2,3,4,5,6,7,8,9,10]. Microfluidics offers the promise of a "lab on a chip" system that can individually control picoliter-scale quantities of fluids, with integrated support for operations such as mixing, storage, PCR, heating/cooling, cell lysis, electrophoresis, and others [11,12,13]. Apart from being amenable to computer control, microfluidics drastically reduces the volumes of samples, thereby reducing costs and improving capture kinetics. Using microfluidics, DNA hybridization times can be reduced from 24 hours to 4 minutes [10] and the number of bases needed to encode information can be decreased from 15 bases per bit to 1 base per bit [1,8].

C. Mao and T. Yokomori (Eds.): DNA12, LNCS 4287, pp. 308–323, 2006.
© Springer-Verlag Berlin Heidelberg 2006

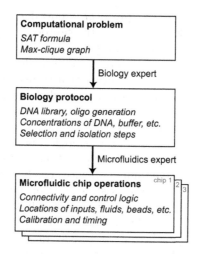

Fig. 1. Abstraction layers for DNA computing

Thus has emerged a vision for creating a hybrid DNA computer: one that uses microfluidics for the plumbing (the control paths) and biological primitives for the computations (the ALUs). On the hardware side, this vision is becoming scalable: microfluidic chips have integrated up to 3,574 valves with 1,000 individually-addressable storage chambers [14]. Moreover, recent trends indicate that microfluidics is following a path similar to Moore's law, with the number of soft-lithography valves per unit area doubling every 4.5 months [15].

On the software side, however, the microfluidic realm is lagging far behind its silicon counterpart. For silicon computers, the complexity and scale of the underlying hardware is masked by a set of well-defined abstraction layers. For example, transistors are organized into gates, which combine to form functional units, which together can implement an Instruction Set Architecture (ISA). The user operates at an even higher level of abstraction (e.g., C++), which is automatically translated into the ISA. These abstraction layers have proven critical for managing complexity. Without them, the computing field would have stagnated as every researcher tried to gain a transistor-level understanding of his machine.

Unfortunately, the current practice in experimental microfluidics is to expose all of the hardware resources directly to the experimentalist. Using a graphical system such as Labview, the user orchestrates the individual behavior of each valve in the microfluidic device. While this practice is merely tedious for today's devices, it will soon become completely intractable—akin to programming a modern microprocessor by directly toggling each of a million gates.

In this paper, we present a system and methodology that uses new abstraction layers for scalable biological computing. As illustrated in Figure 1, our system consists of three layers. At the highest level, the programmer indicates the abstract computation to be performed—for example, in the form of a SAT formula. With some expertise in DNA computing and experimental biology, the computation can be transformed to the next layer: a portable biological protocol for performing the computation. The protocol is portable in that it does not depend on the physical implementation of the protocol; for example, it specifies fluid concentrations but not fluid volumes. Finally, the bottom layer specifies the operations needed to execute the protocol on a specific microfluidic chip. Each microfluidic chip designer provides a library that translates an abstract protocol into the specific sequence of valve actuations needed to execute that protocol on a specific chip.

These abstraction layers provide many benefits. Primarily, by using an architecture-independent description of the biological protocol (the middle layer), the application development can be decoupled from advances in the underlying device technology. Thus, as microfluidic devices come to support additional inputs, mixers, storage cells, etc., the existing suite of protocols can run without modification (much as C programs run without modification on successive generations of microprocessors). In addition, the protocol layer serves as a division of labor. Rather than requiring a heroic and brittle translation from a SAT formula directly to a microfluidic chip, a biologist provides a mapping to the abstract protocol while a microfluidics expert maps the protocol to the underlying device. The abstract protocol is also perfectly suited to simulation, thereby allowing the logical operations to be verified without relying on any physical implementation. Further, a portable protocol description could serve the role of pseudocode in technical publications, providing a precise account of the experimental methods used. Third-party protocols could be downloaded and executed (or called as sub-routines) on one's own microfluidic device.

In the long term, the protocol description language will support all of the operations needed for biological computing. However, as there does not yet exist a single microfluidic device that can encompass all the functionality (preparation of DNA libraries, selection, readout, etc.), this paper focuses on three fundamental primitives: fluid mixing, fluid transport, and fluid storage. We describe a programming system called BioStream (Section 2) that provides an architecture-independent interface for these operations. To show that BioStream is portable, we execute BioStream code on two fundamentally different microfluidic architectures (Section 3). We also present a novel algorithm for mixing fluids to a given concentration using the minimal number of simple on-chip mixing steps (Section 4). Our system represents a fully-functional, end-to-end demonstration of portable software on microfluidic hardware.

2   BioStream Protocol Language

We have developed a software system called BioStream for portable microfluidics protocols. BioStream is a Java library that virtualizes many aspects of the underlying hardware resources. While BioStream can be targeted by a compiler (for example, a DNA computing compiler that converts a mathematical problem into a biological protocol), it is also suitable for direct programming and experimentation by biologists. As such, the language provides several high-level abstractions to improve readability and programmer productivity.

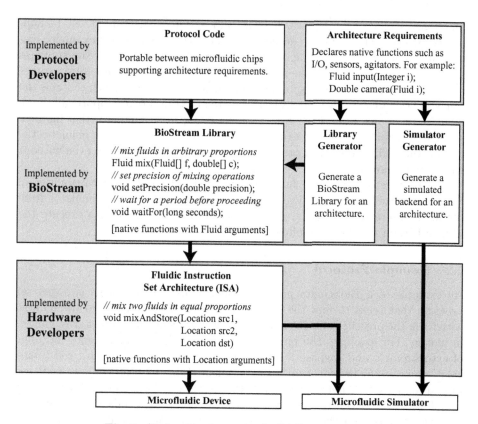

Fig. 2. Abstraction layers in the BioStream system

2.1   Providing Portability

As shown in Figure 2, BioStream offers two levels of abstraction underneath the protocol developer. The first abstraction layer is the BioStream library, which provides first-class Fluid objects to represent the physical fluids on the chip. The programmer deals only with Fluid variables, while the runtime system automatically assigns and tracks the location of the corresponding fluids on the device. The library also supports a general mix operation for combining Fluids in arbitrary proportions and with adjustable precision.

The second abstraction layer, the Fluidic Instruction Set Architecture (ISA), interfaces with the underlying hardware. The fundamental operation is mixAnd-Store, which mixes two fluids in equal proportions and stores the result in a destination cell. (We describe how to translate the flexible mix operations in BioStream to a series of equal-proportion mixes in Section 4.) As all storage cells on the chip have unit volume, only one unit of mixture is stored in the destination; any leftover mixture may be discarded. As detailed in Section 3, this allows for a flexible implementation of mixAndStore on diverse architectures.

In addition to the abstractions for mixing, there are some architecture-specific features that need to be made available to the programmer. These "native functions" include I/O devices, sensors, and agitators that might not be supported by every chip, but are needed to execute the program; for example, special input lines, cameras, or heaters. As shown in Figure 2, BioStream supports this functionality by having the programmer declare a set of architecture requirements. BioStream uses the requirements to generate a library which contains the same functionality; it also checks that the architecture target supports all of the required functions. Finally, BioStream includes a generic simulator that inputs a set of architecture requirements and outputs a virtual machine that emulates the architecture. This allows full protocol development and validation even without hardware resources.

The BioStream system is fully implemented. The reflection capabilities of Java are utilized to automatically generate the library and the simulator from the architecture requirements. As described in Section 3, we also execute the Fluidic ISA on two real microfluidic chips.

2.2 Example Protocol

An example of a BioStream protocol appears in Figure 3. This is a general program that seeks to find the ratio of two reagents that leads to the highest activity in the presence of a given indicator. Experiments of this sort are common in biology. For example, the program could be applied to investigate the roles of cytochrome-c and caspase 8 in activating apoptosis (cell death); cell lysate would serve as the indicator in this experiment [16]. The protocol uses feedback from a luminescence detector to guide the search for the highest activity. After sampling some concentrations in the given range, it descends recursively and narrows the range for the next round of sampling. Using self-directed mixing, a high precision can be obtained after only a few rounds.

The recursive descent program declares a `SimpleLibrary` interface (see bottom of Figure 3) describing the functionality required on the target architecture. In this case, a camera is needed to detect luminescence. While we have not mounted a camera on our current device, it would be straightforward to do so.

2.3 Automatic Fluid Regeneration

A distinguishing feature of BioStream code is the use of Fluid variables to represent samples on the device. The challenge in implementing this functionality is that physical fluids can be used only once, as they are consumed in mixtures and reactions. However, the programmer might reference a Fluid variable multiple times (e.g., variables A and B in the recursive descent example). BioStream supports this behavior by keeping track of how each Fluid was generated and automatically regenerating Fluids that are reused. This process assumes that the original steps employed to generate a Fluid (input, mixing, agitation, etc.) will produce an equivalent Fluid if repeated. While this assumption is a natural fit for protocols depending only on the concentrations of reagents, there are also non-deterministic systems (such as directed evolution of cells) to which it does not apply. We leave full consideration of such systems for future work.

```
import biostream.library.*;

public class RecursiveDescent {

    public static void main(String[] args) {
        String backend = args[0];

        SimpleLibrary lib =
            (SimpleLibrary)LibraryFactory.
            buildLibrary("SimpleLibrary", args[0]);
        run(lib);
    }

    private static void run(SimpleLibrary lib) {
        int ROUNDS = 10; int SAMPLES = 5;

        Fluid A = lib.input(new Integer(0));
        Fluid B = lib.input(new Integer(1));
        Fluid indicator = lib.input(new Integer(2));

        double center = 0.5, radius = 0.5;

        for (int i=0; i<ROUNDS; i++) {
            lib.setPrecision(0.1*(2*radius)/ SAMPLES);

            double bestActivity = -1; int bestJ = -1;
            for (int j=1; j<SAMPLES; j++) {

                double target = center+radius*
                    (1-2*(double)j/SAMPLES);
                Fluid sample = lib.mix(A, target, B, 1-target);

                Fluid test = lib.mix(indicator, 0.9, sample, 0.1);
                lib.wait(30);
                double act = lib.luminescence(test).doubleValue();

                if (act > bestActivity)
                    bestActivity = act; bestJ = j;
            }

            center = center+radius*(1-2*(double)bestJ/SAMPLES);
            radius = radius / 2;

            if (center < radius) center = radius;
            if (center > 1-radius) center = 1-radius;
        }
    }
    System.out.println("Best activity: " + center);
}

interface SimpleLibrary extends FluidLibrary {
    Fluid input(Integer i);
    Double luminescence(Fluid f);
}
```

// *The Recursive Descent protocol* recursively
// zooms in on the ratio of fluids A and B that
// has the highest activity. It requires the
// following setup in the laboratory:
// - input(0) -- fluid A
// - input(1) -- fluid B
// - input(2) -- luminescent activity indicator

// *Initialize the backend to use* (for example,
// an actual chip or a microfluidic simulator)
// based on command-line input.

// Create an interface to the backend using the
// native functions declared in SimpleLibrary.

// *Perform the protocol:*
// Set number of rounds and samples per round.

// Assign names to the input fluids.

// Initialize center, radius of concentration range.

// Repeat for a number of rounds:
//   Set absolute mixing precision to 10X
//   more than the granularity of sampling.

//   Repeat across concentrations in range:

//   Obtain sample of the
//   target concentration.

//   Mix sample with indicator,
//   wait, and measure activity.

//   Remember highest activity.

// Zoom in by factor of 2 around best activity.

// If needed, move center away from boundary.

// Print concentration yielding highest activity.

// *Declare devices needed by RecursiveDescent:*
// Require array of fluid inputs.
// Require luminescence camera.

Fig. 3. Recursive descent search in BioStream

Table 1. Key properties of the microfluidic chips developed. Chip 1 provides better isolation and retention of samples, while Chip 2 offers faster and simpler operation.

	Driving fluid	Wash fluid	Mixing	Sample size	Inputs	Storage cells	Valves	Control lines
Chip 1	oil	N/A	rotary mixer	half of mixer	2	8	46	26
Chip 2	air	water	during transport	full mixer	4	32	140	21

The regeneration mechanism works by associating each Fluid object with the name and arguments of the function that created it. The creating function must be a mix operation or a native function, both of which are visible to BioStream (the Fluid constructor is not exposed). BioStream maintains a valid bit for each Fluid, which indicates whether or not the Fluid is stored in a storage chamber on the chip. By default, the bit is true when the Fluid is first created, and it is invalidated when the Fluid is used as an argument to a BioStream function. If a BioStream function is called with an invalid Fluid, that Fluid is regenerated using its history. Note that this regeneration mechanism is fully dynamic (no analysis of the source code is needed) and is accurate even in the presence of pointers and aliasing.

The computation history created for Fluids can be viewed as a dependence tree with several interesting applications. For example, the library can execute a program in a demand-driven fashion by initializing each Fluid to an invalid state and only generating it when it is used by a native function. Dynamic optimizations such as these are especially promising for microfluidics, as the silicon-based control processors operate much faster than their microfluidic counterparts.

3   Microfluidic Implementation

To demonstrate an end-to-end system, we have designed and fabricated two microfluidic chips using a standard multi-layer soft lithography process [13]. While there are fundamental differences between the chips (see Table 1), both provide support for programmable mixing, storage, and transport of fluid samples. More specifically, both chips implement the `mixAndStore` operation in the Fluidic ISA: they can load two samples from storage, mix them together, and store the result. Thus, despite their differences, code written in BioStream will be portable between the chips.

The first chip (see Figure 4) isolates fluid samples by suspending them in oil [17]. To implement `mixAndStore`, each input sample is transported from a storage bin to one side of the mixer. The mixer uses rotary flow, driven by peristaltic pumps, to mix the samples to uniformity [18]. Following mixing, one half of the mixer is drained and stored in the target location. While the second half could also be stored, it is currently discarded, as the basic `mixAndStore` abstraction produces only one unit of output.

The second chip (see Figure 5) isolates fluid samples using air instead of oil. Because fluid transport is very rapid in the absence of oil, a dedicated mixing element is not needed. Instead, the input samples are loaded from storage and

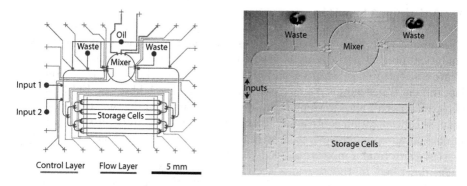

Fig. 4. Layout and photo of Chip 1 (driven by oil)

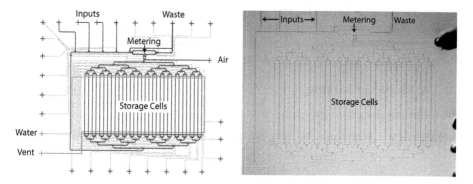

Fig. 5. Layout and photo of Chip 2 (driven by air)

aligned in a metering element; when the element is drained, the samples are mixed during transport to storage. Because the samples are in direct contact with the walls of the flow channels, a small fraction of the sample is lost during transport. This introduces the need for a wash phase, to clean the channel walls between operations. Also, to maintain sample volumes, the entire result of mixing is stored. Any excess volume is discarded in future mixing operations, as the metering element has fixed capacity.

To demonstrate BioStream's portability between these two chips, consider the following code, which generates a gradient of concentrations:

```
Fluid blue = input(1);
Fluid yellow = input(2);
Fluid[] gradient = new Fluid[5];
for (int i=0; i<=4; i++) {
  gradient[i] = mix(blue, yellow, i/4.0, 1-i/4.0);
}
```

This code was used to generate the gradient pictured in Figure 4 and produces an identical result on both microfluidic devices. (The gradient shown in Figure 5 is different and was generated by a different program.)

4   Mixing Algorithms

The mixing and dilution of fluids plays a fundamental role in almost all bio-analytical procedures. Mixing is used to prepare input samples for analysis, to dilute concentrated substances, and to control reagent volumes. In DNA computing, mixing is needed for reagent preparation (e.g., DNA libraries, PCR buffers, detection assays) and, in some techniques, for restriction digests [19,20] or fine-grained concentration control [21]. It is critical to provide integrated support for mixing on microfluidic devices, as otherwise the samples would have to leave the system every time a mixture is needed.

As described in the previous sections, our microfluidic chips support the `mixAndStore` instruction from the Fluidic ISA. This operation simply mixes two fluids in equal proportions. However, the mix command in BioStream allows the programmer to specify complex mixtures involving multiple fluids in various concentrations. To bridge the gap between these abstractions, this section describes how to obtain a complex mixture using a series of simple steps. We describe an abstract model for mixing, an algorithm for minimizing the number of steps required, and how to deal with error tolerances.

4.1   A Model of Mixing

The following definition gives our notation for mixtures.

Definition 1. *A mixture $\mathcal{M}$ is a set of substances S_i at given concentrations c_i:*

$$\mathcal{M} = \{\langle S_1 \rangle c_1 \ldots \langle S_k \rangle c_k\}$$
$$\sum_{i=1}^{k} c_i = 1$$

For example, a mixture of 3/4 buffer and 1/4 reagent is denoted as $\{\langle \text{buffer} \rangle 3/4,$ $\langle \text{reagent} \rangle 1/4\}$. We further define a ***sample*** to be a mixture with only one substance ($|\mathcal{M}| = 1$). For example, a sample of buffer is denoted $\{\langle \text{buffer} \rangle 1\}$, or just $\langle \text{buffer} \rangle$.

To obtain a given mixture on a microfluidic chip, one performs a series of mixes using an on-chip mixing primitive. While the capabilities of this mixer might vary from one chip to another, a simple ***1-to-1 mixing model*** can be implemented on both continuous flow and droplet-based architectures [18,22]. In this model, all fluids are stored in uniform chambers of unit volume. The mix operation combines two fluids in equal proportions, producing two units of the mixture. However, since there may be some amount of fluid loss with every operation, the result of the mixture might not be able to completely fill the contents of two storage cells. Thus, the result is stored in only one storage cell, and the extra mixture is discarded.

The 1-to-1 mixing process can be visualized using a "mixing tree". As depicted in Figure 6, each leaf node of a mixing tree represents a sample, while each internal node represents the mixture resulting from the combination of its children. Figure 7 illustrates that the mixture at an internal node can be calculated as the arithmetic mean of the components in child mixtures. In the 1-to-1

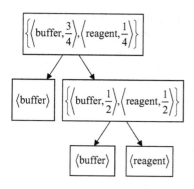

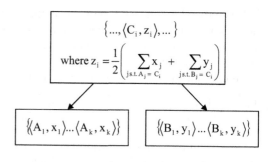

Fig. 6. Mixing tree yielding 3/4 buffer and 1/4 reagent

Fig. 7. Calculation of a parent mixture from child mixtures using a 1-to-1 mixer. For each substance, the resulting concentration is the average of the concentrations in the children.

model, mixing trees are binary trees because each mix operation has two inputs. Evaluation of the tree proceeds from the leaf nodes upwards; the mixture for a given node can be produced once the child mixtures are available. The overall result of the operation is the mixture specified at the root node.

The following theorem is useful for reasoning about mixing trees. It describes the concentration of a substance in the overall mixture based on the depths of leaf nodes containing samples of the substance. The *depth* of a node n in a binary tree is the length of the path from the root node to n.

Theorem 1. *Consider a mixing tree and a substance S. Let m_d denote the number of leaf nodes with sample $\langle S \rangle$ appearing at depth d of the tree. Then the concentration of S contained in the root mixture is given by $\sum_d m_d * 2^{-d}$.*

Proof. A sample at depth d is diluted d times in the mixing process, each time by a factor of two. Thus it contributes 2^{-d} to the root mixture. Since each mix operation sums the concentrations from child nodes, the overall contribution is the sum across the leaf nodes at all depths: $\sum_d m_d * 2^{-d}$. $\qquad\square$

The following theorem describes the set of mixtures that can be obtained using a 1-to-1 mixer. Informally, it states that a mixture is reachable if and only if the concentration of each substance can be written as an integral fraction $k/2^d$.

Theorem 2. *(1-to-1 Mixing Reachability) Consider a finite set of substances $\{S_1 \ldots S_k\}$ with an unlimited supply of samples $\langle S_i \rangle$. Let $\mathcal{R}$ denote the set of mixtures that can be obtained via any sequence of 1-to-1 mixes. Then:*

$$\mathcal{R} = \left\{ \begin{array}{c} \{\langle S_1 \rangle c_1 \ldots \langle S_k \rangle c_k\} \text{ s.t. } \exists\, p_i, q_i, d \in \mathcal{Z} : \\ LCM(q_1 \ldots q_k) = 2^d \ \wedge\ \forall i \in [1, k] : c_i = \dfrac{p_i}{q_i} \end{array} \right\}$$

Proof. Available in an extended version of this paper [23].

It is natural to suggest a number of optimization problems for mixing. Of particular interest are the number of mixes and the number of samples consumed, as these directly impact the running time and resource requirements of a laboratory experiment. The following theorem shows that (under the 1-to-1 model) these two optimization problems are equivalent.

Theorem 3. *In any 1-to-1 mixing sequence, the number of samples consumed is exactly one greater than the number of mixes.*

Proof. By induction on the number of nodes, there is always exactly one more leaf node than internal node in a binary tree. The mixing tree is a binary tree in which each internal node represents a mix and each leaf node represents a sample. Thus there is always exactly one more sample consumed than there are mixes. □

Note that this theorem only holds under the 1-to-1 mixing model, in which two units of volume are mixed but only one unit of the mixture is retained. For microfluidic chips that attempt to retain both units of mixture (such as droplet-based architectures or our oil-driven chip), it might be possible to decrease the number of samples consumed by increasing the number of mix operations.

4.2   Algorithm for Optimal Mixing

In this section, we give an efficient algorithm for finding a mixing tree that requires the minimal number of mixes to obtain a given concentration. For clarity, we frame the problem as follows:

Problem 1. (Minimal Mixing) Consider a finite set of substances $\{S_1 \ldots S_k\}$ with an unlimited supply of samples $\langle S_i \rangle$. Given a reachable mixture $\{\langle S_1 \rangle p_1/n \ldots \langle S_k \rangle p_k/n\}$, what is the mixing tree with the minimal number of leaves?

Our algorithm runs in $O(k \lg n)$ time[1] and produces an optimal mixing tree (with respect to this metric). The tree produced has no more than $k \lg n$ internal nodes.

The idea behind the algorithm, which we refer to as MIN-MIX, is to place a leaf node with sample $\langle S \rangle$ at depth d in the mixing tree if and only if the target concentration for S has a 1 in bit $\lg n - d$ of its binary representation. Theorem 1 then ensures that all substances have the desired concentrations, while fewer than $\lg n$ samples are used for each one.

Psuedocode for MIN-MIX appears in Figure 8. We illustrate its operation for the example mixture of $\{\langle A \rangle 5/16, \langle B \rangle 4/16, \langle C \rangle 7/16\}$. As shown in Figure 9, the algorithm begins with a pre-processing stage that allocates substances to bins according to the binary representation of the target concentrations. It then builds the mixing tree via calls to MIN-MIX-HELPER, which descends through the bins. When a bin is empty, an internal node is created in the graph and the procedure recurses into the next bin. When a bin has a substance identifier in it, the substance is removed from the bin and a corresponding sample is added

---

[1] $\lg n$ denotes $\log_2 n$.

node **Min-Mix**
    (mixture { $\langle S_1, p_1/n \rangle$, ...,$\langle S_k, p_k/n \rangle$ }) {
    depth = lg(n)
    bins = **new** stack[depth+1]
    // pre-processing: build a stack of the
    // bitwise components of each concentration
    for i = 1 to k
      mask = 1
      for j = 0 to depth-1
        if (mask & $p_i \neq 0$) **then**
          bins[j].**push**(S_i)
        endif
        mask = mask << 1
      endfor
    endfor
    return buildMixingHelper(bins, depth)
}

node **Min-Mix-Helper**(stack[] bins, int pow) {
    if bins[pow].**empty**() **then**
      node child1 = Min-Mix-Helper(bins,
                 pow-1)
      node child2 = Min-Mix-Helper(bins,
                 pow-1)
      return $\langle$child1, child2$\rangle$ **as** internal node
    else
      return bins[pow].**pop**() **as** leaf node
    endif
}

Fig. 8. Min-Mix algorithm

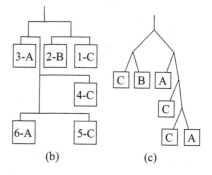

bin	2^{bin}	5A	4B	7C
4	16			
3	8			
2	4	A	B	C
1	2			C
0	1	A		C

(a)

(b)   (c)

Fig. 9. Example operation of Min-Mix for the mixture $\{\langle A\rangle 5/16, \langle B\rangle 4/16, \langle C\rangle 7/16\}$. Part (a) illustrates the algorithm's allocation of substances to bins. The bin layout directly translates to a valid mixing tree, which appears in (b) with numbers indicating the order in which nodes are added to the tree. The mixing tree is redrawn in (c) for clarity.

as a leaf node to the graph. Figure 9 labels the order in which the nodes in the final mixing tree are created by the algorithm.

The following lemma is key to proving the correctness of Min-Mix. We denote the nth least significant bit of x by $LSB(x, n)$. That is, $LSB(x, n) \equiv (x \ll n)$ & 1.

Lemma 1. *Consider the mixing tree t produced by* Min-Mix$(\{\langle S_1\rangle p_1/n \ldots \langle S_k\rangle p_k/n\})$. *A substance S_i appears at a depth d in t if and only if $LSB(p_i, \lg n - d) = 1$.*

Proof. **If:** It suffices to show that there is a substance added to the mixing tree for each LSB of 1 drawn from the p_i (that the substance appears at depth d is given by the only if direction.) Further, since bins[j] is constructed to contain all substances i for which $LSB(p_i, j) = 1$, it suffices to show that a) all bins are empty at the end of the procedure, and b) the procedure does not try to pop from an empty bin. To show (a), use the invariant that each call to Min-Mix-Helper adds a total of 2^{-d} to the mixing tree, where d is the current depth;

either a leaf node is added (which contributes 2^{-d} by Theorem 1) or two child nodes are added, contributing $2 * 2^{-(d+1)} = 2^{-d}$. But since the initial depth is 0, the external call results in $2^0 = 1$ unit of mixture being generated. Since the bins represent exactly one unit of mixture (i.e., $\sum_j bins[j] * 2^{-j} = 1$), all bins will be used. To show (b), observe that MIN-MIX references the bins in order, testing if each is empty before proceeding. Thus no empty bin will ever be dereferenced.

Only if: When a substance is added to the tree from bins[j], it appears at depth $\lg n - j$ in the tree. This is evident from the recursive call in MIN-MIX-HELPER: it initially draws from bins[$\lg n$] and then works down when the upper bins are empty. By construction, bins[j] contains only substances S_i with $LSB(p_i, j) = 1$. Thus, if S_i appears at depth d in the mixing tree, it was added from bins[$\lg n - d$] which has $LSB(p_i, \lg n - d) = 1$. □

The following theorem asserts the correctness of MIN-MIX.

Theorem 4. *The mixing tree given by* MIN-MIX *gives the correct concentration for each substance in the target mixture.*

Proof. Consider a component $\langle S \rangle p/n$ of the mixture passed to MIN-MIX. Let m_d denote the number of leaf nodes with sample S at depth d of the resulting mixing tree. By Lemma 1, $m_d = LSB(p, lg(n) - d)$. Using Theorem 1, this implies that the concentration for S in the root mixture is given by:

$$
\begin{aligned}
c &= \sum_d LSB(p, lg(n) - d) * 2^{-d} \\
&= \sum_x LSB(p, x) * 2^{-(lg(n) - x)} \\
&= \sum_x LSB(p, x) * 2^x / n \\
&= p/n
\end{aligned}
$$

Thus the concentration in the root node of the mixing tree is the same as that passed to MIN-MIX. □

The MIN-MIX algorithm requires $O(k \lg n)$ time to find a mixing tree for mixture ($\{\langle S_1 \rangle p_1/n \ldots \langle S_k \rangle p_k/n\}$). The resulting mixing tree is optimal in that there does not exist a mixing tree that yields the same concentration using fewer mixes. Proofs of these properties are available in an extended version of this paper [23].

4.3 Supporting Error Tolerances

Thus far the presentation has been in terms of mixtures that can be obtained exactly with a 1-to-1 mixer, i.e., those with target concentrations in the form of $k/2^d$. However, the programmer should not be concerned with the reachability of a given mixture.

In the BioStream system, the programmer specifies a concentration range $[c_{min}, c_{max}]$ and the system ensures that the mixture produced will fall within the given range[2]. Such error tolerances are already a natural aspect of scientific

---

[2] Alternately, BioStream supports a global error tolerance ϵ that applies to all concentrations.

experiments, as all measuring equipment has a finite precision that is carefully noted as part of the procedure. Given a concentration range, the system increases the internal precision d until some concentration $k/2^d$ (which can be obtained exactly) falls within the range.

5   Related Work

Several researchers have pursued the goal of automating the control systems for microfluidic chips. Gascoyne et al. describe a graphical user interface for controlling chips that manipulate droplets over a two-dimensional grid [24]. By varying parameters in the interface, the software can target grids with varying dimensions, speeds, etc. However, portability is limited to grid-based droplet processors. While the BioStream protocol language could target their chips, their software is not suitable for targeting ours.

Su et al. represent protocols as acyclic sequence graphs and map them to droplet-based processors using automatic scheduling [25] and module placement [26]. While the sequence graph is portable, it lacks the expressiveness of a programming language and cannot represent feedback loops (as in our recursive descent example). King et al. demonstrate a "robot scientist" that directs laboratory experiments using a high-level programming language [27], but lacks the abstraction layers needed to target other devices. Gu et al. have controlled microfluidic chips using programmable Braille displays [28], but protocols are mapped to the chip by hand.

Johnson demonstrates a special-purpose robotic system (controlled by Labview) that automatically solves 3-SAT problems using DNA computing [29]. Miniaturizing his benchtop devices could result in a fully-automatic microfluidic biocomputer. Livstone et al. compile an abstract SAT problem into a sequence of DNA-computing steps [5]. The output of their system would be a good match for BioStream and the abstraction layers proposed in this paper.

There are other microfluidic chips that support flexible gradient generation [30,31,32] and programmable mixing on a droplet array[33]. To the best of our knowledge, our chips are the only ones that provide arbitrary mixing of discrete samples in a soft lithography medium. A more detailed comparison of the devices is published elsewhere [17].

Fair et al. also suggest a mixing algorithm for diluting a single reagent by a given factor [34]. It seems that their algorithm performs a binary search for the target concentration, progressively approximating the target by a factor of two. However, since intermediate reagents must be regenerated in the search, this algorithm requires $O(n)$ mixes to obtain a concentration k/n. In contrast, our algorithm needs $O(\lg n)$ to mix two fluids.

6   Conclusions

Microfluidic devices are an exciting substrate for biological computing because they allow precise and automatic control of the underlying biological protocols.

However, as the complexity of microfluidic hardware comes to rival that of silicon-based computers, it will be critical to develop effective abstraction layers that decouple application development from low-level hardware details.

This paper presents two new abstraction layers for microfluidic biocomputers: the BioStream protocol language and the Fluidic ISA. Protocols expressed in BioStream are portable across all devices implementing a given Fluidic ISA. We demonstrate this portability by building two fundamentally different microfluidic devices that support execution of the same BioStream code. We also present a new and optimal algorithm for obtaining a given concentration of fluids using a simple on-chip mixing device. This algorithm is essential for efficiently supporting the mix abstraction in the BioStream language.

It remains an interesting area of future work to leverage DNA computing technology to target the BioStream language from a high-level description of the computation. This will create an end-to-end platform for biological computing that is seamlessly portable across future generations of microfluidic chips.

Acknowledgements

We are grateful to David Wentzlaff and Mats Cooper for early contributions to this research. We also thank John Albeck for helpful discussions about experimental protocols. This work was supported by National Science Foundation grant #CCF-0541319. J.P.U. was funded in part by the National Science and Engineering Research Council of Canada (PGSM Scholarship).

References

1. Braich, R.S., Chelyapov, N., Johnson, C., Rothemund, P.W.K., Adleman, L.: Solution of a 20-variable 3-SAT problem on a DNA computer. Science **296** (2002)
2. Farfel, J., Stefanovic, D.: Towards practical biomolecular computers using microfluidic deoxyribozyme logic gate networks. In: DNA 11. (2005)
3. Gehani, A., Reif, J.: Micro flow bio-molecular computation. Biosystems **52** (1999)
4. Grover, W.H., Mathies, R.A.: An integrated microfluidic processor for single nucleotide polymorphism-based DNA computing. Lab on a Chip **5** (2005)
5. Livstone, M.S., Weiss, R., Landweber, L.F.: Automated design and programming of a microfluidic DNA computer. Natural Computing (2006)
6. McCaskill, J.S.: Optically programming DNA computing in microflow reactors. BioSystems **59** (2001)
7. Somei, K., Kaneda, S., Fujii, T., Murata, S.: A microfluidic device for DNA tile self-assembly. In: DNA 11. (2005)
8. van Noort, D.: A programmable molecular computer in microreactors. In: DNA 11. (2005)
9. van Noort, D., Gast, F.U., McCaskill, J.S.: DNA computing in microreactors. In: DNA 8. (2002)
10. van Noort, D., Zhang, B.T.: PDMS valves in DNA computers. In: SPIE International Symposium on Smart Materials, Nano-, and Micro-Smart Systems. (2004)
11. Breslauer, D.N., Lee, P.J., Lee, L.P.: Microfluidics-based systems biology. Molecular BioSystems **2** (2006)

12. Erickson, D., Li, D.: Integrated microfluidic devices. Anal. Chim. Acta **507** (2004)
13. Sia, S.K., Whitesides, G.M.: Microfluidic devices fabricated in poly(dimethylsiloxane) for biological studies. Electrophoresis **24** (2003)
14. Thorsen, T., Maerkl, S., Quake, S.: Microfluidic large scale integration. Science **298** (2002)
15. Hong, J.W., Quake, S.R.: Integrated nanoliter systems. Nature BioTechnology **21**(10) (2003)
16. Allan, L., Morrice, N., Brady, S., Magee, G., Pathak, S., Clarke, P.: Inhibition of caspase-9 through phosphorylation at Thr 125 by ERK MAPK. Nature Cell Biology **5** (2003)
17. Urbanski, J.P., Thies, W., Rhodes, C., Amarasinghe, S., Thorsen, T.: Digital microfluidics using soft lithography. Lab on a Chip **6** (2006)
18. Chou, H., Unger, M., Quake, S.: A microfabricated rotary pump. Biomedical Microdevices **3** (2001)
19. Faulhammer, D., Cukras, A.R., Lipton, R.J., Landweber, L.F.: Molecular computation: RNA solutions to chess problems. PNAS **97**(4) (2000)
20. Ouyang, Q., Kaplan, P.D., Liu, S., Libchaber, A.: DNA solution of the maximal clique problem. Science **278** (1997)
21. Yamamoto, M., Matsuura, N., Shiba, T., Kawazoe, Y., Ohuchi, A.: Solutions of shortest path problems by concentration control. In: DNA 7. (2002)
22. Paik, P., Pamula, V., Fair, R.: Rapid droplet mixers for digitial microfluidic systems. Lab on a Chip **3** (2003)
23. Thies, W., Urbanski, J.P., Thorsen, T., Amarasinghe, S.: Abstraction layers for scalable microfluidic biocomputers (Extended version). Technical Report MIT-CSAIL-TR-2006-034, MIT (2006) http://hdl.handle.net/1721.1/32543.
24. Gascoyne, P.R.C., Vykoukal, J.V., Schwartz, J.A., Anderson, T.J., Vykoukal, D.M., Current, K.W., McConaghy, C., Becker, F.F., Andrews, C.: Dielectrophoresis-based programmable fluidic processors. Lab on a Chip **4** (2004)
25. Su, F., Chakrabarty, K.: Architectural-level synthesis of digital microfluidics-based biochips. In: ICCAD. (2004)
26. Su, F., Chakrabarty, K.: Unified high-level synthesis and module placement for defect-tolerant microfluidic biochips. In: DAC. (2005)
27. King, R.D., Whelan, K.E., Jones, F.M., Reiser, P.G.K., Bryant, C.H., Muggleton, S.H., Kell, D.B., Oliver, S.G.: Functional genomic hypothesis generation and experimentation by a robot scientist. Nature **427** (2004)
28. Gu, W., Zhu, X., Futai, N., Cho, B.S., Takayama, S.: Computerized microfluidic cell culture using elastomeric channels and Braille displays. PNAS **101**(45) (2004)
29. Johnson, C.: Automating the DNA Computer to Solve n-Variable 3-SAT Problems. In: DNA 12. (2006)
30. Dertinger, S.K.W., Chiu, D.T., Jeon, N.L., Whitesides, G.M.: Generation of gradients having complex shapes using microfluidic networks. Anal. Chem. **73** (2001)
31. Neils, C., Tyree, Z., Finlayson, B., Folch, A.: Combinatorial mixing of microfluidic streams. Lab on a Chip **4** (2004)
32. Lin, F., Saadi, W., Rhee, S.W., Wang, S.J., Mittalb, S., Jeon, N.L.: Generation of dynamic temporal and spatial concentration gradients using microfluidic devices. Lab on a Chip **4** (2004)
33. Pollack, M., Fair, R., Shenderov, A.: Electrowetting-based actuation of liquid droplets for microfluidic applications. Applied Physics Letters **77**(11) (2000)
34. Ren, H., Srinivasan, V., Fair, R.: Design and testing of an interpolating mixing architecture for electrowetting-based droplet-on-chip chemical dilution. Transducers (2003)

Fuzzy Forecasting with DNA Computing

Don Jyh-Fu Jeng[1], Junzo Watada[1], Berlin Wu[2], and Jui-Yu Wu[3]

[1] Graduate School of Information, Production and Systems
Waseda University, Fukuoka 808-0135, Japan
jeng@fuji.waseda.jp, junzow@osb.att.ne.jp
[2] Department of Mathematical Sciences
National Chengchi University, Taipei 116, Taiwan
berlin@nccu.edu.tw
[3] Department of Biochemistry, School of Medicine
Taipei Medical University, Taipei 110, Taiwan
jwu@tmu.edu.tw

Abstract. There are many forecasting techniques including: exponential smoothing, ARIMA model, GARCH model, neural networks and genetic algorithm, etc. Since financial time series may be influenced by many factors, conventional model based techniques and hard computing methods seem inadequate in the prediction. Those methods, however, have their drawbacks and advantages. In recent years, the innovation and improvement of forecasting techniques have caught more attention, and also provides indispensable information in decision-making process. In this paper, a new forecasting technique, named *DNA forecasting*, is developed. This may be of use to a nonlinear time series forecasting. The methods combined the mathematical, computational, and biological sciences. In the empirical study, we demonstrated a novel approach to forecast the exchange rates through DNA. The mean absolute forecasting accuracy method is defined and used in evaluating the performance of linguistic forecasting. The comparison with ARIMA model is also illustrated.

1 Introduction

The DNA computing in the forecasting process is applied for several reasons: First, a problem specific computing will be easier to design and implement, with less need for model/functional complexity and flexibility. Secondly, the types of soft computational problems that DNA based computing may be able to effectively solve social competence for economic conditions that a dedicated processor would be naturally reasonable. As well, these problems will be likely to require extensive time that would preclude the need for a more versatile and interactive system that may be able to be implemented with a universal computing machine.

The practical possibility of using DNA molecules as a medium for computation was first demonstrated by Adleman [2]. He demonstrated a proof-of-concept use of DNA as form of computation which was used to solve the seven-point Hamiltonian path problem [1][3]. The primary intention of Adleman's work was to

C. Mao and T. Yokomori (Eds.): DNA12, LNCS 4287, pp. 324–336, 2006.

prove the feasibility of molecular computation, and also gave an indication that the emergence of this new computational paradigm could provide an advantage over conventional electronic computing techniques. Specifically, DNA has shown to have massively parallel processing capabilities which allow a DNA based computer to solve complicated computational problems in a reasonable amount of time.

The main advantage offered by most proposed models of DNA based computation is the ability to handle millions of operations in parallel. The massively parallel processing capabilities of DNA computers may give them the potential to find tractable solutions to otherwise intractable problems, as well as potentially speeding up large, but otherwise solvable, polynomial time problems requiring relatively few operations.

Many different models exist within the paradigm of "classical" DNA computation [4], [8], [10], [15], and [20] each of them with different advantages and degrees of applicability to classes of problems.

On the other hand, in the field of humanity and social science, fuzzy statistics and fuzzy forecasting have attracted many attentions lately. This is a nature result because of the complicated phenomenon of humanity and society which is hard to be fully explained by traditional models. Taking stock market as an example, the essence of closing price is uncertain and indistinctive. Moreover, there are many factors could influence the closing price, such as trading volume and exchange rate, etc. Therefore, if we merely consider the closing price of previous day to construct our forecasting model, we are capable of estimating the future trend, rather than unexpected or unnecessary loss.

Upon applying fuzzy logic in the time series analysis, the first step is to identify how to integrate linguistic variable analysis methods in solving the autoregressive relation problem of the dynamic data. Chiang et. al. [6] presented self-learning methods to modify fuzzy models for dynamic system in linguistic field. Later, Huarng [9] proposed a fuzzy linguistic summary as one of the data mining function to discover useful knowledge from database. In fact, fuzzy relation equations are easier to be understood and applied than decision tables or decision rules.

In view of this, many researchers have adopted fuzzy relation equations for time series analysis and forecasting. For instance, Wu and Hung [26] proposed a fuzzy identification procedure for ARCH and Bilinear models. Kumar and Wu [12] used fuzzy statistical techniques in change period's detection of nonlinear time series. Chen and Hwang [5] proposed the two-factors time-variant fuzzy time series model and developed two algorithms for temperature prediction. Huarng [9] and Tseng et. al. [24] proposed heuristic models by integrating problem-specific heuristic knowledge with Chen's [5] model to improve forecasting.

In this research, we propose a fuzzy time series modeling process with DNA computing. This method is applied to the financial time series data, and then forecast future trend while comparing the forecasting performance. From the empirical studies, it is shown that our proposed method demonstrates an appropriate and efficient performance of prediction for exchange rates.

The paper is organized as follows: In section 2, we briefly review the DNA computing. In section 3, the algorithm of modeling time series and forecasting with DNA computing is proposed. An empirical application is then shown in section 4. Finally, section 5 recaps our conclusions.

2   DNA Computing

2.1   Preliminary

DNA computing is fundamentally comparable to the parallel computing; we take advantage of the many different DNA molecules to try many different possibilities. DNA computing also known as part of molecular computing which is a new approach to massively parallel computation based on a ground-breaking work by Adleman [3]. DNA molecules was first designed to solve a seven-node Hamiltonian path problem, a special case of an NP-complete problem that attempts to visit every node in a graph exactly once.

Since Adleman's pioneering accomplishment in 1994, research on DNA computing is ongoing; Lipton [13] and Adleman [2] have extended on Adleman's original work with more efficient designs to build up possible DNA computers. Adleman observed that a DNA computer sufficient to search for 256 DES keys would occupy only a small set of test tubes [1]. DNA computing has been applied to various fields including combinatorial optimization [17], massive parallel computing [14], Boolean circuit development [18], nanotechnology [25], vary large scale database [22], etc.

DNA computing experiment has been heralded as the "first example of true nanotechnology", and even the "start of a new era," forging an unprecedented link between computational science and life science. Just as the cell pushes the limit of the second law of thermodynamics, which predicts that one joule can fuel a maximum of 34×10^{19} irreversible power operations, the DNA computer's energy consumption from DNA strand synthesis and PCR should also be small compared to that used up by a supercomputer.

2.2   Application in the Statistical Model Construction

In statistics, the patterns of collected data can be numerical, qualitative formats, or linguistic values (such as data derived from testing). These kinds of data are hard to be analyzed by traditional time series models. Nevertheless, with fuzzy sets, the patterns of data will not be restricted and a more suitable model can be established. There is no certain rule for the optimal partition in building fuzzy range sets. Generally, the more partition we do, the more precision we have; however, the more complicated calculation is required. In short, the determination between accuracy and complexity is entirely up to the individual requirements.

Despite being published before the discovery of DNA, Darwin's 1859 classic remains a robust description of evolution by natural selection. Inspired by this, the humanity and social science sustained growth stimulated natural selection that shaped the evolution of the biology. The mechanism of evolutionary change is adopted in this pioneering research, which we applied the DNA computing to forecasting. The process of DNA polymerase chain reaction (PCR) reflects the natural selection in our society.

3 Modeling Time Series and Forecasting with DNA Computing

3.1 Modeling Fuzzy Time Series

In this study, we will make a conventional estimation as well as a fuzzy estimation (with membership function) of the coefficients for the investigated time series. The ARIMA model is a very popular forecasting technique for the time series data. The key process is to find the order for the candidate models ARIMA(p,d,q). After decided the order, the appropriate coefficients are then estimated for getting the lowest value of AIC. Finally, we make a forecasting from the constructed model. For the fuzzy autoregressive process of ordered one, the l-steps prediction becomes $X_n(l) = E(X_{n+l} \mid X_n, \cdots, X_1)$. In order to find the forecasting performance, we will also compare our result with traditional other methods.

How to estimate the appropriate coefficients with the DNA computing is demonstrated as following:

Let $\{X_t; t = 1, 2, \cdots, n\}$ be a time series which has followed by an AR(1) or Markov process; $\{\triangle X_t = X_t - X_{t-1}; t = 2, 3, \cdots, n\}$ be the first order difference of the time series $\{X_t\}$; $T = max|\triangle X_t|$ be the maximum first order difference $\triangle X_t$.

We assign the variation of $\triangle X_t$ into five period with respect to linguistic variables, say *plunge*=-1.0, *drop*=-0.5, *draw*=0, *soar*=0.5, *surge*=1.0, where each linguistic variable denotes an element in fuzzy set and its corresponding membership. The memberships of fuzzy time series transformed by $k_t = \frac{\triangle X_t}{T}$ are based on the Table 1.

Table 1. Memberships of fuzzy time series $k_t = \frac{\triangle X_t}{T}$ w.r.t the linguistic variables

	Membership				
	Plunge=-1	Drop=-0.5	Draw=0	Soar=0.5	Surge=1
$k_t \leq -0.5$	$-2(k_t + 0.5)$	$2(k_t + 1)$	0	0	0
$-0.5 < k_t \leq 0$	0	$-2k_t$	$2(k_t + 0.5)$	0	0
$0 < k_t \leq 0.5$	0	0	$-2(k_t - 0.5)$	$2k_t$	0
$0.5 < k_t$	0	0	0	$-2(k_t - 1)$	$2(k_t - 0.5)$

3.2 Encoding Scheme

Next will be a series of biochemical process. Since we are having a time series data with n records, which means there exist n-1 first order difference, a number of n-1 test tubes are set up in represent each first order difference $\triangle X_t$. Two

DNA sequences are designed in represent the "up" trend and "down" trend, denote as DNA1 and DNA2, respectively. These two DNA sequences have its own special designed PCR primers denote as UF, UR, DF, and DR (UF: DNA1 forward primer; UR: DNA1 reverse primer, DF: DNA2 forward primer, DR: DNA2 reverse primer) as shown in Fig. 1.

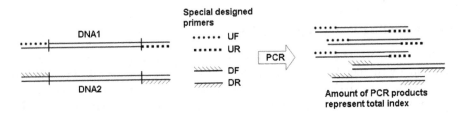

Fig. 1. DNA encoding scheme. DNA1 and DNA2 are designed in represent "up" trend and "down" trend respectively. DNA1 and DNA2 have its own special design primers.

Synthesizing short single-stranded DNA is now a routine process. The molecules can be made by an auto-programming machine called DNA synthesizer. Itineraries can then be produced from the encodings by linking them together in proper order. To accomplish this we can take advantage of the fact that DNA hybridizes with its complementary sequence. Random itineraries can be made by mixing difference encodings.

The ratio of up trend primers (UF, UR) and down trend primers (DF, DR) for each test tube is set up corresponding with $\frac{\triangle X_t}{T}$ according to the Table 2.

Table 2. The ratio of primer for each test tube

$\frac{\triangle X_t}{T}$	[-1,-.9]	(-.9,-.6]	(-.6,-.3]	(-.3,-.1]	(-.1,.1]	(.1,.3]	(.3,.6]	(.6,.9]	(.9,1]
$P_{up} : P_{down}$	9:1	8:2	7:3	6:4	5:5	4:6	3:7	2:8	1:9

3.3   Polymerase Chain Reaction

In each test tube, we now have the ratio of template DNA1 and DNA2 accompanied with their own specific PCR primers that encode UF, UR, and DF, DR. We expect to collect the amplified numbers of DNA1 and DNA2 after DNA amplification as illustrated in Fig. 1. To accomplish this, the technique of PCR is used, which allows us to produce many copies of a specific sequence of DNA. PCR is an iterative process that cycles through a series of copying events using an enzyme called *Taq* DNA polymerase. Polymerase will copy a section of single-stranded DNA starting at the position of a primer, a short piece of DNA complimentary to one end of a section of the DNA that we are interested in. By selecting primers that flank the section of DNA we want to amplify, the polymerase preferentially amplifies the DNA template through specific set of primers, doubling the amount of DNA containing this sequence in each cycle of PCR reaction. After many iterations of PCR (we take 25 cycles in this research), the DNA we are working on is amplified exponentially. The free process of amplification between DNA1 and DNA2 represents the natural selection, which the superior amplifies more, the inferior amplifies less. What we end up with after PCR is a specific sequence of test tube full of double-stranded DNA of two different sequences, DNA1 and DNA2.

3.4 DNA Quantification

The test tube is now filled with DNA encoded itineraries. We now want to calculate the number for the DNA1 and DNA2. Many molecular biology techniques can be utilized to estimate PCR-produced DNA populations, such as real-time PCR (quantitative PCR, or Q-PCR), gel electrophoresis, and fluorescence assay, etc. The detection and quantification of DNA amplification is very important both in research and in a clinical diagnostic setting. Real-time PCR, which is used in this research, has become a well-established procedure for quantifying levels of gene expression. Its power resides in the ability to detect, at every cycle of the PCR, the amount of PCR product (amplicon) using fluorescence.

3.5 Making Prediction

Finally, according to the quantity of DNA1 and DNA2 retrieved from previous step, we define the ratio of up-and-down trend as the auto-correlation coefficient by:

$$U_n = \frac{\sum_{i=2}^{n} DNA1_i}{\sum_{i=2}^{n} DNA1_i + \sum_{i=2}^{n} DNA2_i} \tag{1}$$

$$D_n = \frac{\sum_{i=2}^{n} DNA2_i}{\sum_{i=2}^{n} DNA1_i + \sum_{i=2}^{n} DNA2_i} \tag{2}$$

The first order difference can be represented as:

$$\triangle X_n = (U_n - D_n)T \tag{3}$$

The one step forecasting value can be derived:

$$X_{n+1} = X_n + (U_n - D_n)T \tag{4}$$

The forecasting with l-steps ($l \geq 2$) can be expressed:

$$X_{n+1} = X_n + (U_{n+l-1} - D_{n+l-1})T \tag{5}$$

where

$$U_{n+l-1} = \frac{\sum_{i=2}^{n} DNA1_i + \sum_{i=n+1}^{n+l-1} DNA1_i}{\sum_{i=2}^{n} DNA1_i + \sum_{i=2}^{n} DNA2_i + \sum_{i=n+1}^{n+l-1} DNA1_i + \sum_{i=n+1}^{n+l-1} DNA2_i} \tag{6}$$

$$D_{n+l-1} = \frac{\sum_{i=2}^{n} DNA2_i + \sum_{i=n+1}^{n+l-1} DNA2_i}{\sum_{i=2}^{n} DNA1_i + \sum_{i=2}^{n} DNA2_i + \sum_{i=n+1}^{n+l-1} DNA1_i + \sum_{i=n+1}^{n+l-1} DNA2_i}$$

$$(7)$$

In the fuzzy forecasting for one step, we transfer $(U - D)T$ into fuzzy number according to Table 1. Hence, the one step fuzzy forecasting becomes:

$$FX_{n+1} = X_n \oplus (\frac{m_i}{L_i} + \frac{m_{i+1}}{L_{i+1}})T \qquad (8)$$

where $(\frac{m_i}{L_i} + \frac{m_{i+1}}{L_{i+1}})T$ means the memberships of the variation with respect to the linguistic variables L_i, L_{i+1} based on the total variation T, and $\oplus$ stands for the fuzzy addition.

The proposed DNA computing algorithm for forecasting is summarized as following:

Step 1. For time series $\{X_t\}$, decide the total variation of $\{\triangle X_t\}$ and linguistic variables $\{L_1, L_2, \cdots, L_5\}$ of $\triangle X_t$
Step 2. Design the memberships of fuzzy time series w.r.t the linguistic variables
Step 3. Determine the ratio of primer
Step 4. Design the DNA sequences and primers
Step 5. Molecular reaction [PCR]
Step 6. DNA quantification [Real-time PCR]
Step 7. Calculate the forecasting value

3.6   Forecasting Performance with DNA Computing

This experiment solved a forecasting problem, but there are two major short-comings preventing a large scaling up of this computation. The complexity of the forecasting problem simply does not disappear when applying a different method of solution - it still increases exponentially. For our method, what scales exponentially is not the computing time, but rather the amount of DNA. Unfortunately this places some hard restrictions on the number of variables that can be solved. Another factor that places limits on this method is the nonlinear trend for each operation. Since these operations are not deterministic but stochastically driven (from biochemistry), each step contains statistical errors, limiting the number of iterations you can do successively before the probability of producing an error becomes greater than producing the correct result.

The Mean Absolute Forecasting Accuracy (MAFA) is used in evaluating the forecasting performance in this research. To compare the performance between different forecasting methods, each linguistic variable need to be assigned with an ordered rank. In this study, for instance, a *plunge* as -1.0, *drop* as -0.5, *draw* as 0, *soar* as 0.5 and *surge* as 1.0. By doing so, MAFA can be defined.

Definition 1. *Suppose $\{RL_t, t = 1, \cdots, n\}$ and $\{FL_t, t = 1, \cdots, n\}$ denote the real and outputting linguistic variables respectively. The mean absolute forecasting accuracy can be defined as:*

$$MAFA = 1 - \frac{\sum_{t=1}^{n} \frac{|FL_t - RL_t|}{r-1}}{n} \quad (9)$$

where r denotes the number of linguistic variables.

Example 1. Suppose that real linguistic variables of the time series are {drop, draw, drop, surge, soar, drop, surge, drop, draw, plunge}, then the corresponding values of linguistic variables are {-0.5, 0, -0.5, 1.0, 0.5, -0.5, 1.0, -0.5, 0, -1.0}. The outputting linguistic variables are {drop, draw, plunge, surge, draw, draw, surge, drop, surge, soar}, then the corresponding values of linguistic variables are {-0.5, 0, -1.0, 1.0, 0, 0, 1.0, -0.5, 1.0, 0.5}. By Definition 1, the MAFA can be obtained.

$$MAFA = 1 - \frac{\sum_{t=1}^{10} \frac{|FL_t - RL_t|}{5-1}}{10} = 1 - \frac{1}{10} = 0.90$$

4   An Empirical Application

Following data source comes from the Center Bank of China monthly records of *JPY/USD* and *EUR/USD* exchange rate from 2004/1 to 2005/12. The tendencies of these data are shown in Fig. 2. From these data, $T = max|\triangle X_t| = 5.27$ for *JPY/USD*, and $T = max|\triangle X_t| = 0.034$ for *EUR/USD* can be obtained.

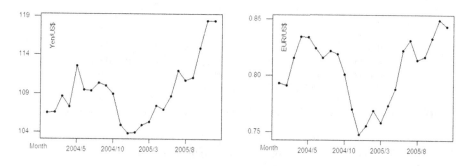

Fig. 2. Trend for Exchange Rate: *JPY/USD* and *EUR/USD*

4.1   Fuzzy Model Construction

After fuzzifying these data of monthly exchange rate, we can apply the method mentioned in Section 3.1 to calculate data's corresponding memberships in linguistic variables L_i as illustrated in Table 3 and 4.

4.2   Forecasting Performance

The memberships of monthly variations of *JPY/USD* and *EUR/USD* between January 2004 and December 2005 can be observed from Table 3 and 4. Finally,

Table 3. The ratios of primer and memberships for *JPY/USD*

Month	$\frac{\triangle X_t}{T}$	$P_{up} : P_{down}$	plunge	drop	draw	soar	surge
2004/1-2	0	5:5	0	0	1	0	0
2004/2-3	0.39	3:7	0	0	0.22	0.78	0
2004/3-4	-0.26	6:4	0	0.5	0.5	0	0
2004/4-5	1.00	1:9	0	0	0	0	1
2004/5-6	-0.58	7:3	0.16	0.84	0	0	0
2004/6-7	-0.02	5:5	0	0.04	0.96	0	0
⋮	⋮	⋮	⋮	⋮	⋮	⋮	⋮
2005/9-10	0.71	2:8	0	0	0	0.59	0.41
2005/10-11	0.68	2:8	0	0	0	0.64	0.36
2005/11-12	0	5:5	0	0	1	0	0

Table 4. The ratios of primer and memberships for *EUR/USD*

Month	$\frac{\triangle X_t}{T}$	$P_{up} : P_{down}$	plunge	drop	draw	soar	surge
2004/1-2	-0.06	5:5	0	0.12	0.88	0	0
2004/2-3	0.73	2:8	0	0	0	0.48	0.52
2004/3-4	0.54	3:7	0	0	0	0.92	0.08
2004/4-5	-0.01	5:5	0	0.02	0.98	0	0
2004/5-6	-0.28	6:4	0	0.56	0.44	0	0
2004/6-7	-0.25	6:4	0	0.50	0.50	0	0
⋮	⋮	⋮	⋮	⋮	⋮	⋮	⋮
2005/9-10	0.48	3:7	0	0	0	0.96	0.04
2005/10-11	0.47	3:7	0	0	0	0.94	0.06
2005/11-12	-0.16	5:5	0	0.32	0.68	0	0

Table 5. The ratios of two DNAs and memberships for *JPY/USD* after reaction

Month	DNA1:DNA2	plunge	drop	draw	soar	surge
2004/1-2	52:48	0	0.08	0.92	0	0
2004/2-3	25:75	0	0	0	1	0
2004/3-4	69:31	0	0.76	0.24	0	0
2004/4-5	6:94	0	0	0	0.24	0.76
2004/5-6	58:42	0	0.32	0.68	0	0
2004/6-7	54:46	0	0.16	0.84	0	0
⋮	⋮	⋮	⋮	⋮	⋮	⋮
2005/9-10	52:48	0	0.08	0.92	0	0
Average	41:59	0	0	0.36	0.64	0
2005/11	38:62	0	0	0.52	0.48	0
2005/12	45:55	0	0	0.20	0.80	0
2006/1	47:53	0	0	0.88	0.12	0
2006/2	42:58	0	0	0.68	0.32	0

the result from DNA forecasting is shown at Table 5 and 6. The memberships are transformed from the concentration of DNA by $C = \frac{50 - DNA1\%}{50}$ as illustrated at Table 7, which is generated from Table 1.

Because this research is to explore the qualitative trend of time series, we computed the transformed memberships through fuzzy rule base in fuzzy systems for getting their corresponding linguistic variables to facilitate analysis. We compared it with best ARIMA model (p,d,q). The results derived from above principles are shown at Table 8 and 9.

Table 6. The ratios of two DNAs and memberships for EUR/USD after reaction

Month	DNA1:DNA2	plunge	drop	draw	soar	surge
2004/1-2	46:54	0	0	0.16	0.94	0
2004/2-3	13:87	0	0	0	0.52	0.48
2004/3-4	23:77	0	0	0.04	0.96	0
2004/4-5	51:49	0	0.04	0.96	0	0
2004/5-6	70:30	0	0.80	0.20	0	0
2004/6-7	68:32	0	0.72	0.28	0	0
⋮	⋮	⋮	⋮	⋮	⋮	⋮
2005/9-10	41:59	0	0	0.36	0.64	0
Average	52:48	0	0.08	0.92	0	0
2005/11	50:50	0	0	1	0	0
2005/12	45:55	0	0	0.20	0.80	0
2006/1	62:38	0	0.48	0.52	0	0
2006/2	58:42	0	0.32	0.68	0	0

Table 7. Memberships of DNA concentration $C = \frac{50 - DNA1\%}{50}$ w.r.t the linguistic variables

	Membership				
	Plunge=-1	Drop=-0.5	Draw=0	Soar=0.5	Surge=1
$C \leq -0.5$	$-2(C + 0.5)$	$2(C + 1)$	0	0	0
$-0.5 < C \leq 0$	0	$-2C$	$2(C + 0.5)$	0	0
$0 < C \leq 0.5$	0	0	$-2(C - 0.5)$	$2C$	0
$0.5 < C$	0	0	0	$-2(C - 1)$	$2(C - 0.5)$

Table 8. Forecasting comparison of JPY/USD

Time	Real value	Best ARIMA 1,0,0[1]	DNA computing[2]	Fuzzy DNA computing[3]
2005/11	118.41	113.67	115.77	$114.82 \oplus (\frac{0.36}{draw} + \frac{0.64}{soar})5.27$
2005/12	118.43	112.75	117.14	$115.77 \oplus (\frac{0.48}{draw} + \frac{0.52}{soar})5.27$
2006/1	115.25	112.01	117.71	$117.14 \oplus (\frac{0.80}{draw} + \frac{0.20}{soar})5.27$
2006/2		111.41	118.05	$117.71 \oplus (\frac{0.88}{draw} + \frac{0.12}{soar})5.27$
2006/3		110.93	118.96	$118.05 \oplus (\frac{0.68}{draw} + \frac{0.32}{soar})5.27$

[1] $X_{t+1} = 21.45 + 0.8X_t + \varepsilon_t$
[2] $X_{t+1} = X_t + (U - D)T$
[3] $FX_{n+1} = X_n \oplus (\frac{m_i}{L_i} + \frac{m_{i+1}}{L_{i+1}})T$

Table 9. Forecasting comparison of *EUR/USD*

Time	Real value	Best ARIMA 1,0,0[4]	DNA computing[5]	Fuzzy DNA computing[6]
2005/11	0.849	0.835	0.831	$0.832 \oplus (\frac{0.08}{draw} + \frac{0.92}{soar})0.034$
2005/12	0.843	0.837	0.831	$0.831 \oplus (\frac{0}{draw} + \frac{1}{soar})0.034$
2006/1	0.825	0.838	0.835	$0.831 \oplus (\frac{0.20}{draw} + \frac{0.80}{soar})0.034$
2006/2		0.839	0.827	$0.835 \oplus (\frac{0.48}{draw} + \frac{0.52}{soar})0.034$
2006/3		0.840	0.822	$0.827 \oplus (\frac{0.32}{draw} + \frac{0.68}{soar})0.034$

[4] $X_{t+1} = 0.203 + 0.759X_t + \varepsilon_t$
[5] $X_{t+1} = X_t + (U - D)T$
[6] $FX_{n+1} = X_n \oplus (\frac{m_i}{L_i} + \frac{m_{i+1}}{L_{i+1}})T$

Table 10. Forecasting accuracy of Best ARIMA and DNA computing from Mean Absolute Forecasting Accuracy

Exchange Currency	Best ARIMA 1,0,1	DNA computing
JPY/USD	81.25%	81.25%
EUR/USD	75%	81.25%

The MAFA is defined in Section 3.6 for measuring the accuracy of forecasting methods. From Table 8 and 9, we can find the *DNA(1)* model has better forecasting performance than ARIMA model. The MAFA result is shown at Table 10.

The major reason why the prediction cannot hit real value is that we only consider the greatest membership and omit others memberships. Therefore, only with reasonable forecasting model can we decide investment strategy from forecasting results. Otherwise, without the direction of clear outlines, investors will face a plight as to which information they should take.

5   Concluding Remarks and Future Works

DNA computing is so exciting because of the collaboration of chemists, biologists, mathematicians, and computer scientists to understand and simulate fundamental biological processes and algorithms taking place within cells. Although DNA computing might not replace conventional computing in the near future, they still have endless potentials for other applications. The DNA computing has clear advantages over conventional computing when applied to problems that can be divided into separate, non-sequential tasks. The reason is that for non-decomposable problems, those that require many sequential operations are much more efficient on a conventional computer due to the length of time required to conduct the biochemical operations.

In this research, we tried to make an appropriate process of constructing fuzzy time series model and use this model to forecast the exchange rate of *JPY/USD* and *EUR/USD*. Compare the DNA forecasting and DNA fuzzy forecasting model with traditional ARIMA model by the performance of forecasting accuracy, we can find that the DNA model has better forecasting performance than that of traditional ARIMA model. We hope this method will provide a new forecasting technique for investors to make optimal decision with fuzzy information.

In spite of the forecasting performance for DNA forecasting model, there are some problems for further studies. For example:

1. To make a general rule for fuzzy order identification instead of the Markov relation. DNA computing can be more accurately described as a collection of new computing paradigms rather than a single focus. Each of these different paradigms within molecular computing can be associated with different potential applications that may prove to place them at an advantage over conventional methods. Many of these models share certain features that lend them to categorization by these potential advantages. However, there exists enough similarities and congruencies that hybrid models will be possible, and that advances made in both "classic" and "natural" areas.

2. To extend our result to the multivariate fuzzy time series case. In fact, how to solve the nonstationary or seasonal factors for the time series are still open questions.

3. In this research, we adopt five-ranking classification and transform the time series data into fuzzy numbers through membership functions. However, seven-ranking classification used in social sciences may be used in future studies for special situation, and it is yet to prove where it will provide significant improvement on forecasting performance?

4. Future applications might make use of the error rates and instability of DNA based computation methods as a means of simulating and predicting the emergent behavior of complex systems. This could pertain to weather forecasting, economics, and lead to more a scientific analysis of social science and the humanities. Such a system might rely on inducing increased error rates and mutation through exposure to radiation and deliberately inefficient encoding schemes. Similarly, methods of DNA computing might serve as the most obvious medium for use of evolutionary programming for applications in design or expert systems. DNA computing might also serve as a medium to implement a true fuzzy logic system.

References

1. Adams, J.: On the Application of DNA Based Computation. (1998) http://publish.uwo.ca/~jadams/dnaapps1.htm
2. Adleman, L.: Computing with DNA. Scientific American **279** (1988) 34–41
3. Adleman, L.: Molecular computation of solutions to combinatorial problems. Science **226** (1994) 1021–1024
4. Boneh, D., Dunworth, C., Lipton, R., Sgall, J.: On the Computational Power of DNA. DAMATH: Discrete Applied Mathematics and Combinatorial Operations Research and Computer Science **71** (1996)
5. Chen, S.M., Hwang, J.R.: Temperature prediction using fuzzy time series. IEEE Transactions on Systems, Man, and Cybernetics **30** (2000) 263–275

6. Chiang, D., Chow, L., Wang, Y.: Mining time series data by a fuzzy linguistic summary system. Fuzzy Sets and Systems **112** (2000) 419–432

7. Cho, A.: DNA Computing: Hairpins Trigger an Automatic Solution. Science **288** (2000) 1152–1153

8. Guarnieri, F., Fliss, M., Bancroft, C.: Making DNA Add. Science **273** (1996) 220–223

9. Huarng, K.: Heuristic models of fuzzy time series for forecasting. Fuzzy Sets and Systems **123** (2001) 369–386

10. Johnson, R.C.: Time to Engineer DNA Computers. EE Times (2001) http://www.eetimes.com/story/OEG20001221s0032

11. Kari, L., Gloor, G., Yu, S.: Using DNA to solve the Bounded Post Correspondence Problem. Theoretical Computer Science **231** (2000) 192–203

12. Kumar, K., Wu, B.: Detection of change points in time series analysis with fuzzy statistics. International Journal of Systems Science **32** (2001) 1185–1192

13. Lipton, R.: DNA Solution of Hard Computational Problems. Science **268** (1995) 542–545

14. Liu, Q., Wang, L., Frutos, A.G., Condon, A.E., Corn, R.M., Smith, L.M.: DNA computing on surface. Nature **403** (2000) 175–179

15. Miller, C.: Using DNA Algorithms to Solve NP-Complete Problems. http://www.csd.uwo.ca/~jamie/.Refs/Courses/CS881/charlotte.html

16. Normile, D.: Molecular Computing: DNA-Based Computer Takes Aim at Genes. Science **295** (2002) 951

17. Ouyang, Q., Kaplan, P.D., Liu, S., Libchaber, A.: DNA Solution of the Maximal Clique Problem. Science **278** (1997) 446–449

18. Owenson, G.G., Amos, M., Hodgson, D.A., Gibbsons, A.: DNA-based logic. Soft Computing **5** (2001) 102–105

19. Parker, J.: Computing with DNA. European Molecular Biology Organization Reports **4** (2003) 7–10

20. Păun, G.: Computing with Bio-Molecules: Theory and Experiments. Springer-Verlag (1998)

21. Păun, G., Rozenberg, G., Salomaa, A.: DNA Computing - New Computing Paradigms. Springer-Verlag (1998)

22. Reif, J.H., LaBean, T.H., Pirrug, M., Rana, V.S., Guo, B., Kingsford, C., Wickham, G.S.: Experimental construction of a very large scale DNA database with associatice search capability. In The 7th International Workshop on DNA-Based Computers (2001) 241–250

23. Tseng, F., Tzeng, G.: A fuzzy SARIMA model for forecasting. Fuzzy Sets and Systems **126** (2002) 367–376

24. Tseng, F., Tzeng, G., Yu, H., Yuan, B.: Fuzzy ARIMA model for forecasting the foreign exchange market. Fuzzy Sets and Systems **118** (2001) 9–19

25. Winfree, E., Lin, F., Wenzler, L.A., Seeman, N.C.: Design and self-assembly of two-dimensional DNA crystals. Nature **394** (1998) 539–545

26. Wu, B., Hung, S.: A fuzzy identification procedure for nonlinear time series: with example on ARCH and bilinear models. Fuzzy Sets and Systems **108** (1999) 275–287

"Reasoning" and "Talking" DNA: Can DNA Understand English?

Kiran C. Bobba, Andrew J. Neel, Vinhthuy Phan, and Max H. Garzon

Computer Science, The University of Memphis, TN 38152-3240, U.S.A.
{kbobba, aneel, vphan, mgarzon}@memphis.edu

Abstract. Memory is a fundamental challenge in computing, particularly if they are to store large amounts of interrelated data based on content and be queried associatively to retrieve information useful to the owners of the storage, such as self-assembled DNA structures, cells, and biological organisms. New methods to encode large data sets compactly on DNA chips have been recently proposed in (Garzon & Deaton, 2004) [6]. The method consists of shredding the data into short oligonucleotides and pouring it over a DNA chip with spots populated by copies of a basis set of noncrosshybridizing strands. In this paper, we probe into the capacity of these memories in terms of their ability to discern semantic relationships and discriminate information in complex contexts in two applications, as opposed to their raw capacity to store volumes of uncorrelated data. First, we show that DNA memories can be designed to store information about English texts so that they can "conduct a conversation" about their *content* with an interlocutor who wants to learn about the subject contained in the memories. In this preliminary approach, the results are competitive, if not better, with state-of-the-art methods in conventional artificial intelligence. In a second application in biology, we show how a biomolecular computing analysis based on similar techniques can be used to re-design DNA microarrays in order to increase their sensitivity to the level required for successful discrimination of conditions that may escape detection by standard methods. Finally, we briefly discuss the scalability of the common technique to large amounts of data given recent advances in the design of noncrosshybridizing DNA oligo sets, as well other applications in bioinformatics and medical diagnosis.

Keywords: Semantic analysis and information retrieval; DNA chips and microarrays; question answering; sensitivy analysis; data classification and discrimination.

1 Introduction

Large associative memories have been envisioned since the early days of DNA-based computing (Baum, 1995)[1]. Recently, much progress has been made in the design and analysis of their performance based on recent advances in word design and encodings for biomolecule-based computing (BMC) (Garzon Deaton, 2004; Bi et al., 2003; Garzon et al., 1997) [6,2,8]. Naive linear encodings of data

C. Mao and T. Yokomori (Eds.): DNA12, LNCS 4287, pp. 337–349, 2006.

and information onto biomolecules have given way to more sophisticated methods based on sound theoretical analyses (Garzon et al., 1995) [8] and practical constraints on implementation on DNA or current biotechnology (Garzon et al., 2005) [10]. The biology of life shows that this is indeed feasible to levels of complexity as clearly evident (*e.g.*, in the cycle transcribe → express → translate) as yet very poorly understood, despite enormous progress, encompassing such wide array of phenomena as genetics, morphogenesis, and even perhaps ontogenetic processes such as learning and adaptation. The problem of handling massive amounts of extremely rich and complex data using DNA still bears an intriguing potential as a major contribution of biomolecular computing in both biological and non-biological applications. The search for the "killer app" is still on.

(Garzon et al., 2005; Garzon & Deaton, 2004) [10,6] showed how abiotic and biological data can both be represented by signatures on a DNA chip, the spots of which are copies of a so-called noncrosshybridizing basis. By placing adequate thresholds on the stringency of reaction condition necessary for acceptable levels of variability of the representation (i.e., the capacity to distinguish inputs through their representations), the signatures of arbitrary inputs are completely determined and require no precomputation or synthesis of any DNA strands, other than the basis strands. In other words, this method provides a *universal* and *scalable* method to represent data. For example, because of the superposition (linearity) property (modulo the variability implicit in the representation), a corpus of English text can be automatically encoded just by finding representations for the words in the basic vocabulary (words) in the corpus. Thereafter, the representation of a previously unknown piece of text can be inferred by superposition of the component words. There is evidence that these representations can be used for semantic processing of text corpora in lieu of the original text (Neel & Garzon, 2006; Garzon et al, 2003)[15,9]. Given the newly available large basis sets (Chen et al., 2005; Chen et al., 2004) [3,5] approaching megasets, devices with the ability to process data for information extraction appear now within reach in a relatively short time.

In this paper, we present two novel applications of BMC based on recent developments in word design and data representation in DNA proposed in (Garzon et al., 2005; Garzon & Deaton, 2004) [10,6]. The first one is an application to information retrieval of natural language (English) texts. The problem is whether we can build a DNA-based memory of a corpus of data good enough for us to be able to query it just by "*having a conversation with it*," i.e., take turns in asking it questions and getting answers, albeit at an extremely slow pace. In others words, a question can be fairly complicated, but we may have to go through a lot of chemistry to get the answer (in the order of hours, or even days perhaps.) The prototype corpus used is qualitative Newtonian physics, as learned by freshman college students. To illustrate the breadth of the technique, in Section 2 we provide an experimental evaluation of the semantic capacity of the technique for information retrieval in the area of natural language processing. We compare it to what is considered the best method in conventional AI, namely Latent Semantic Analysis(LSA), on the same task. The performance of DNA is based on the judgment of physics professors (experts) on a selection of typical questions asked to students when tutoring them on the subject. In Section 3, we present a novel application to DNA chip design for bioinformatic analysis based on the same technique. We show how current DNA chips could be re-designed to increase their sensitivity and make possible discrimination between genomic data currently impossible with current designs because of the noise inherent in crosshybridization with the maximum number of genes normally packed on the chip. Finally, in Section 4 we briefly discuss the their scalability to large amounts of data given current advances in the design of noncrosshybridizing DNA oligo sets on which they are based, as well as the potential of the techniques for similar applications in bioinformatics, microbiology and genetics.

In the following section we summarize the structure of the memory presented in (Garzon & Deaton, 2004) [6]. In the remaining two sections we will indicate the refinements required, as well the results, for the two applications .

1.1   DNA Chips and Memory Devices

The obvious method to encode data on DNA, namely a one-one mapping of alphabet symbols (e.g., bits) or words (e.g., bytes or English words in a dictionary) to DNA fragments could possibly be used to encode symbolic data (strings) in DNA single strands. Longer texts can be mapped homomorphically by ligation of these segments to represent larger concatenations of symbolic text. A fundamental problem with this approach is that abiotic data would appear to require massive synthesis of DNA strands of the order of the amount of data to be encoded. Current lab methods may produce massive amounts of DNA copies of the same species, but not of too many diverse species selected and assembled in very specific structures such as English sentences in a corpus of data (e.g., a textbook), or records in a large data warehouse. Even if the requisite number of species were available, the mapping between the data and the DNA strands is hard to establish and maintain; as the species get transformed by the reactions, they must get involved in and must be translated back to humanly usable expression.

An alternative more effective representation using recently available large sets of noncrohybridizying oligonucleotides obtainable *in vitro* (Chen et. al., 2005; Bi et. al, 2003) [3,2] has been suggested in (Garzon and Deaton, 2004) [6]. We repeat next the basic definitions to make this paper self-contained. This method can be regarded as a new implementation of the idea in (Head et al., 1999; 2001) [13,12] of aqueous computing for writing on DNA molecules, although through a simpler set of operations (only hybridization.) Since binary strings can be easily mapped to a four letter alphabet, we will simply assume that the data are given in DNA form over $\{A, C, G, T\}$. Representations using sets with crosshybridization present are usually ambiguous and cannot be reliably used. More details on this point can be found in (Garzon and Deaton, 2004) [6].

Let B be a set of DNA molecules (the encoding basis, or "stations" in Head's terminology (Head et al., 1999) [12], here not necessarily bi-stable), which is assumed to be finite and noncrosshybridizing according to some model of hybridization, denoted $h(*, *)$ (for example, the Gibbs energy, or the h-distance in (Garzon et al, 1997) [7,8].) We will also assume that we are provided some parameter coding for the stringency of reaction conditions τ (for example, a threshold on the Gibbs energy or the h-distance) under which hybridization will take place. For simplicity, it is further assumed that the length of the strands in B is a fixed integer n, and that B contains no hairpins. For example, if the h-distance is the hybridization criterion and $\tau = 0$, two strands x, y can only hybridize if they are perfectly complementary (i.e., $h(x, y) \leq 0$), so a maximal such set B can be obtained by selecting one strand from every (non-palindromic) pair of Watson-Crick complementary strands; but if, on the other extreme, $\tau = n$, the mildest hybridization condition, any two strands can hybridize, so a maximal set B consists of only one strand of length n, to which every other strand may "hybridize" without further restrictions. Let $m = |B|$ be the cardinality of B. The basis strands will also be referred as *probes*. For easy visualization, we will assume in the illustrating examples below that m is a perfect square ($m = 121$ or 49) and that the base set of probes has been affixed onto a DNA chip.

Given a string x (ordinarily much longer than the probe length n and even perhaps the number of probes m), x is said to be *h-dependent on B* if there is some concatenation c of elements of B that will hybridize to x under stringency τ, i.e., such that $h(x, c) \leq \tau$. Shredding x to the corresponding fragments according to the components of c in B leads to the following slightly weaker but more manageable definition. The *signature* of x with respect to B is a vector X of dimension m that is obtained as follows. Shredding x to $|x|/n$ fragments of size n or less, X_i is the number of fragments f of x that are within threshold τ from a strand i in B, i.e., such that $h(f, i) < \tau$. The value X_i will thus be referred to as a *pixel* at probe spot i. The input strands x will also be referred as *targets*. Signatures can be just easily implemented using standard DNA oligo chip technology.

Several questions about this representation were resolved in (Garzon et al., 2005) [10]. First, the vector X is indeed well-defined if we make the representation be a sphere centered around the ideal representation of radius the standard deviation of the distribution of all possible representations under the uncertainty of the biochemical reactions in a test tube. Second, the sensitivity of this representation may be high enough to make fine discrimination of microarray data. We provide further evidence that this is the case in the applications under consideration in the following sections.

2   Reasoning and Talking DNA

To construct the chip, $10K+$ paragraphs were extracted from a physics text appropriate for a freshman student. The paragraphs were sanitized to remove common words, articles, etc and shredded into words. Each word was transcribed into DNA by random assignment to a unique species from a family of noncrosshybridizing 8−mers. PCR Selection (Deaton et al., 2004) [4] was then performed on this input paragraph set (about 5000 species) to extract a subset of noncrosshybridizing oligos. Finally, two DNA oligo chips were produced, the "large chip" by fixing on the spots all DNA oligos extracted from the physics text (full corpus) and the "small chip" by fixing on the spots only those oligos species that survived the PCR Selection (the processed corpus). Now, the entire corpus of paragraphs in a physics text book is compacted to a large set of signatures found on a DNA chip. Further, two pieces of text can be compared semantically by looking at the similarity of the signatures on the corresponding DNA chips.

In order to compare the capabilities of DNA to organize this type of information in a context-addressable fashion, we made a systematic comparison with the best conventional method to store information in content-addressable form, namely LSA (Latent Semantic Analysis) by (Laundauer et al., 1998)[17]. The effectiveness of LSA has been evaluated by comparison to human evaluation (four experts on the knowledge in the corpus). Each expert human would judge a set of 1187 student answers to a selected set of 57 questions about the material in the corpus. The degree of agreement between the evaluations of LSA and the expert human for each student answer yields a correlation of about 0.40 over the entire set of questions (Graesser et al., 2006)[11].

2.1 Experimental Design

In a similar manner, we evaluated DNA memories using the same set of physics paragraphs as above. Here we substituted DNA memories for LSA and performed an analogous comparison to human judgments.

The paragraphs of ideal and student answers were transcribed into DNA by two protocols. First, the ideal or student answer was transcribed as a unit. Like words were mapped to their bases (e.g. paying was mapped to pay) before encoding into DNA. Second, a copy of the DNA from the first method was shredded. Their signatures were computed by pouring the resulting paragraphs over the Large and Small Chip and allowing hybridizations to take hold. Next, we objectively measured the similarity (or semantic quality) of each student answer as compared to the ideal answer by taking the correlation of the signals from each chip pair. The correlations of student answers (to ideal answers) were then correlated with the expert human evaluations in (Graesser et al., 2006)[11].

Fig. 1 shows the results of 18 rounds of PCR Selection on a shredded physics corpus (about 5000 DNA words). The min, max, average, and standard deviation between every pair of strands after each successive round of PCR Selection were calculated. An ideal result would be a maximally noncrosshybridizing set of DNA where the minimum Gibbs free energy is greater than the threshold for hybridization, which determines whether two oligos hybridize to a spot on the chip (Gibbs energy under -6 Kcal/mole). The minimum free energy is never greater than -9 Kcal/mole even though the energy stabilizes after round 10. Similarly, the standard deviation decreases slightly after each round. Fig. 1 shows that, while the minimum and standard deviation of pairwise energies in the set are improving, the number of species is declining rapidly and eventually stabilizes around $2, 400$ oligos. Thus, a maximal noncrosshybrizing subset has been found after 18 rounds of PCR selection.

2.2   Results

Fig. 2 presents the results of the comparison to human judgments. Low correlations indicated poor semantic quality when transcribing the entire paragraph
for both the large and small chip or rather indicates poor quality regardless of
whether the entire corpus or a subset of the corpus is placed on the chip. After
shredding each paragraph, the semantic quality improves but is still negative
for the full corpus (large chip). However, the the quality of the small chip approaches that of LSA. Therefore, processing the DNA chip with PCR Selection
improved the results by two orders of magnitude. In the end, this application
helps demonstrate the competitiveness of DNA memories for semantic retrieval.

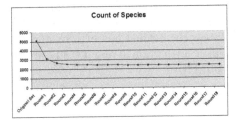 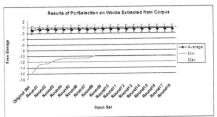

Fig. 1. Left: The simulations converge to a noncrosshybridizing subset of the seed set
in about four (4) rounds. This is in good agreement with the experimental value in
runs of the protocol *in vitro*. Right: A count of the number of species was made after
each successive round of PCR Selection. Here, the number of species remaining after
each round decreases as the minimum free energy (i.e., the noncrosshybridizing quality
of the set) increases.

3   DNA Microarray Design and Analysis

DNA microarrays, or their variant, DNA chips, are powerful tools that can provide snapshots of genetic activity inside a cell. Physically, a microarray is a small
solid slide (typically, glass or mica) on which single DNA strands (the probes)
are immobilized at regularly spaced spots on the slide. Typically, these strands
represent whole structural genes of a species and are either short oligonucleotides
(in DNA chips) or much longer cDNAs (in microarrays). We will focus our discussion to DNA chips in the remainder of the paper. Similar results can be
obtained with microarrays.

Genetic activity is measured by extracting a large number of mRNAs from a
cell under a given condition and pouring them onto a microarray to be studied.
Since these DNA strands immobilized on the array are complementary to these
mRNAs, they are expected to hybridize to the spots on the chip where enough
affinity is found. By adding fluorescent tags to the mRNAs and measuring the
intensity of each spot on the chip, the gene activity can be estimated. By looking
at the activity of a large number on genes on the chip, we can tell which genes
are active at that particular time and which sets are not. Further, by contrasting

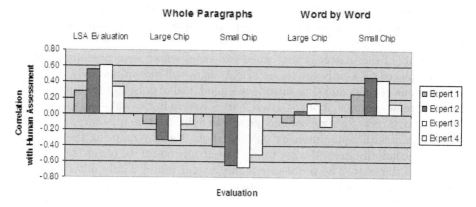

Fig. 2. The quality of DNA semantic memory is improved when shredding paragraphs into word-by-word representations for semantic retrieval. When compared to the best conventional method available (LSA, left), the performance of long DNA strands (full paragraphs from student answers) is dismal, regardless of whether a full corpus of data or a noncrosshybridizing subset of words is placed on the chip (second and third columns.) Shredding probes into words gives a notable improvement, but still yields negative correlations with human standards on a large chip (fourth column). Finally, on the small chip (noncrosshybridizing basis) the performance of the DNA memory comes to levels comparable to that of LSA for the same task (last column).

with the expression of the same cell under different conditions, we can compare the expressions of each group of genes and be able to tell which group up-regulate or down- regulate over time.

Longer cDNAs produced by PCR can improve the specificity of microarray signals, but unfortunately the synthesis of thousands of genes is time consuming and can easily be error-prone. Thus, the usage of oligonucleotides has become popular. Much shorter oligonucleotides are not as specific as cDNAs when it comes to hybridization to targeted mRNAs. And the design of oligonucleotide probes is a challenging task, as it affects the sensitivity and specificity of microarray signals. In an ideal design, a probe should hybridize with and only with one target mRNA. As a probe is usually very short (order of $25-$mers) they may hybridize to untargeted mRNAs that happen to be very similar to the targeted mRNAs at certain local regions, which are almost, but not quite complementary to the probe. There are different techniques to cope with this. Affymetrix, for example, follows PM-MM (perfect match-mismatch) strategy to design their probes in such a way that for each probe, there is one mismatched probe, which contains a single mismatch located directly in the middle of the $25-$mer probe. This mismatch probe serves as a mechanism to detect non-specific sequences and therefore helps to reduce spurious signals. To increase sensitivity, Affymetrix also uses 22 copies for each probe for each expression measurement. Other approaches propose different ways of design oligonucleotide probes. For example, (Rimour et. al., 2005)[16] proposed concatenating disjoint complementary short oligos to improve specificity and retain sensitivity of microarray signals. The probe selection process must be done very carefully in selecting the best probes for each mRNAs as it determines effectively the reliability, sensitivity and specificity of microarrays.

The crux of the matter is that when two genes whose sequences are homologous are placed on a chip, the target cDNA oligos will indiscriminately hybridize to either one of them and thus introduce a good deal of noise in the chip readout. In an extreme example, if oligos u and v are identical in two different spots, complementary target oligos x close enough to u in hybridization (distance) similarity will fluoresce with equal intensity on both spots, thereby reducing the signal in each to one half the strength they would have if only one of the oligos (say u) were present on the spot. The problem remains even if u and v only differ by a few nucleotides (or, technically, are close to complement in hybridization affinity, or distance), and is of course generalized over the entire chip if this kind of "crosstalk" was to occur among a number of different pairs u, v. The consequences on the practical use of the microarrays for diagnostic purposes, for example, is likely to be a decreased sensitivity of the chip and an inability to offer enough of a statistical margin of error to distinguish two different conditions in the target.

In order to address these problems, we propose a new approach to analyze microarray data. This analysis will suggest a new architecture for a microarray (or DNA chip). The method removes the crosstalk by extracting a noncrosshybridizing subset of the genes by an application of the PCR selection protocol in (Bi et al, 2003)[2]. The full set of genes in the original chips is each shredded into small fragments (say 100−mers). After PCR selection, the remaining fragments are re-grouped into the original genes and they are used on a smaller chip. The resulting signatures should produce a much better signal-to-noise ratio (SNR) of the original targets. We put this idea to the test as described next.

3.1 Experimental Design

Seven (7) diseases were selected to gauge the gains in SNR, as shown in Table 1. The data was obtained from the Kyoto Encyclopedia of Genes and Genomes (KEGG) pathway database (Kanehisa, 2002) [14]. This database contains curated metabolic and signaling pathways, which show a network of interacting proteins. These networks depict the interactions involved in processes that carry

Table 1. Test set of seven (7) neurodegenerative diseases given by a number of critically active genes

ReferenceNumber	Disease	Number of genes
1	Alzheimers disease	23
2	Amyotrophic lateral sclerosis (ALS)	17
3	Dentatorubropallidoluysian atrophy	15
4	Huntington's disease	25
5	Neurodegenerative Disorders	16
6	Parkinson's disease	14
7	Prion disease	11

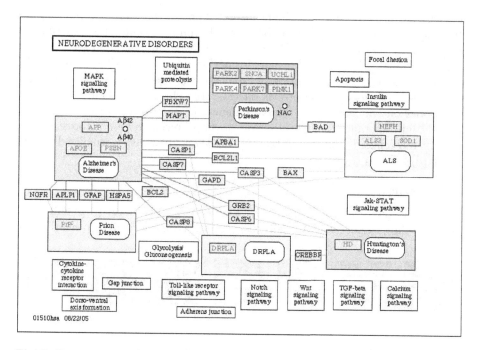

Fig. 3. Common neurodegenerative disorder pathway for the seven diseases in Table 1

out biological functions such as metabolism or signal transduction. Information on enzymatic reactions, enzymes, small molecules and genes are also available from KEGGS. For our test data, we chose the set of genes encoding for the proteins that are involved in the currently known neurodegenerative disorder pathway. This pathway consists of five sub-pathways of five major neurodegenerative diseases. Table 1 shows the number of proteins involved in each disease. Only a fraction of proteins in each of these sub-pathways are present in the common neurodegenerative disorder pathway. In particular, only 3 out of 21 of the Alzheimer proteins are present in the common pathway; and 3 out of 19, 1 out of 7, 1 out of 28, 6 out of 18, and 1 out of 11 of the other respective diseases proteins are present in the common pathway. Fig. 3 below shows the common neurodegenerative disorder pathway.

Based on these data, the signatures were generated *in silico* on two chips. The first chip, referred to as the "large chip", had all 121 genes occurring in any of the seven diseases. The second chip, referred to as the "small chip", has on it only selected fragments of 121 genes by careful selection of noncrosshybridizing fragments from 7 genes with more frequent fragments from each disease over all. So, each row in the small chip represents 7 genes that highly occur in noncrosshybridizing fragments sets. This filtering goes beyond the Affymetrix PM-MM strategy. In PM-MM strategy, mismatch probes are introduced with the assumption that no foreground probes exist matching the mismatch probe. Our strategy performs similar elimination but at a larger scale. According to our strategy, we not only eliminate single mismatch probes but all the probes that are more homologous to a selected probe. Several rounds of tuning the experiments were done to obtain correct results in terms of being unbiased and selecting the right hybridization criterion threshold for performing the experiment. To make a fair comparison, the concentration of input strands was increased for large chip signatures so as to give the probes on the large chip equal chances to match the probes on the small chip. The stringency of the reactions, as given for by a threshold h-distance for hybridization, was optimized to be not too stringent and not too relaxed. Moreover, the resulting signatures were normalized by the standard Z-scores transformation.

3.2 Results

Fig. 4 shows the resulting signatures for the diseases on the large (top) and small (bottom) chip, respectively. Visually, the signatures on the small chip clearly look

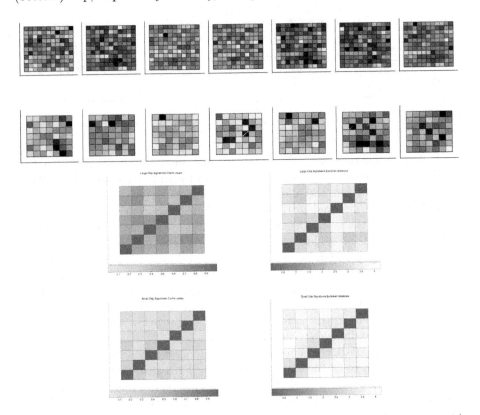

Fig. 4. Top: Signatures of seven diseases (given by a number of critically active genes) on a large chip (top row) and a small chip (second row). Bottom: Comparison of the SNR reduction on the same chips using cosines (left) and Euclidean distance (right). Lighter intensities indicate larger angles or distances. The signatures have been normalized by the standard Z-scores transformation.

more easily distinguishable than the signatures on the large chip in terms of the intensity and distribution of the pixel signals. However, this approach is bound to fail for much larger chips needed in practice. Therefore, a more objective analysis is required to quantify the gains and their significance. To this end, we can regard the signatures as vectors in high dimensional Euclidean spaces (121D and 49D, respectively). The cosine of the angle between these vectors signatures and their Euclidean distances can be used to get the objective analysis of the signatures if they are properly normalized to provide a faircomparison, using standard Z-scores transformation. The left matrix in the bottom center in Fig. 4 shows the cosine values are computed between each pair of disease vectors in graphical form. Euclidean distances are also shown on the bottom right between all possible pair of diseases. The top row shows the cosine matrix and Euclidean distance matrix for large chip, while the bottom row is for small chip. Euclidean distances were also normalized so as to compare the high dimensional (121 for large chip) Euclidean distances with lower dimensional (49 for small chip) Euclidean distances.

The difference in the cosine values matrices for large and small chip signatures is now obvious, as is the improvement in the distribution of vectors in Euclidean space. The brighter signatures shown imply either orthogonal vectors or large spatial distances between signatures. (Dark values would imply that the signature vectors are parallel or close by in Euclidean distance.)

4   Conclusions and Future Work

This paper probes the memory capacity of DNA oligochips beyond the raw capacity to store volumes of independent data. It gives experimental (in simulation) and theoretical analyses of the ability of DNA to encode and discern semantic relationships in contexts as complex and wide ranging as natural language and bioinformatic analysis. First, we show how DNA memories can be designed to store information about English texts so that they might be able to "conduct a conversation" (perhaps extremely slowly, as slow as the chemistry on DNA chips compared to natural language) about their content with an interlocutor who wants to learn about the subject contained in the memories. In this preliminary approach, the results suggest that these memories can capture enough semantics of natural language to be competitive with, if not better than (due to their scalability to very large corpora on large DNA chips), state-of-the-art methods in conventional artificial intelligence.

A second application of the same essential idea and technique in bioinformatics gives a consistent result in a different area. We show how a biomolecular computing analysis can be used to re-design DNA microarrays in order to increase their sensitivity to the level required for successful discrimination of conditions that may escape detection by standard methods, in disease diagnosis for example. A direct application of the same technique used for semantic analysis constitutes a new approach to genomic analysis that increases the signal-to-noise (SNR) ratio in microarrays commonly used in bioinformatics. The method yields higher resolution and accuracy in the analysis of genomic data, and only requires some processing in what can be termed an "orthogonalization" procedure to the given set of targets/genes before placing them on the microarrays. These advantages may be critical for problems such as classification problems (disease/healthy data). More details can be found in (Garzon et al., 2005)[10].

Advantages on the common technique used in this paper for both applications are worth pointing out. First, the selection protocol has been performed *in vitro* (Chen et al., 2004)[5] and hence scales well to massive numbers of biotic data, as well as, although not evidently to abiotic data (Garzon et al., 2005) [10]. Second, further applications of this technique can be easily envisioned to large amounts of data given current advances in the design of noncrosshybridizing DNA oligo sets on which they are based.

Acknowledgements

Partial support from the National Science Foundation grant QuBiC/EIA-0130385.

References

1. E. Baum. Building an associative memory vastly larger than the brain. *Science*, 268(5210):583–585, 1995.
2. H. Bi, J. Chen, R. Deaton, M. Garzon, H. Rubin, and D. Wood. A pcr-based protocol for in vitro selection of non-crosshybridizing oligonucleotides. *J. of Natural Computing*, 2:4:417–426, 2003.
3. J. Chen, R. Deaton, M. Garzon, J.W. Kim, D.H. Wood, H. Bi, D. Carpenter, J.S. Le, and Y.Z. Wang. Sequence complexity of large libraries of dna oligonucleotides. In *11th International Conference on DNA Computing*, page in press, 2005.
4. J. Chen, R. Deaton, Max Garzon, D.H. Wood, H. Bi, D. Carpenter, and Y.Z. Wang. Characterization of non-crosshybridizing dna oligonucleotides manufactured in vitro. Proc. 8th Int Conf on DNA Computing DNA8.
5. J. Chen, R. Deaton, Max Garzon, D.H. Wood, H. Bi, D. Carpenter, and Y.Z. Wang. Characterization of non-crosshybridizing dna oligonucleotides manufactured in vitro. In L. Smith G.C. Mauri, editor, *10th International Workshop on DNA Computing*, pages 50–61, 2004.
6. M. Garzon and R. Deaton. Codeword design and information encoding in dna ensembles. *J. of Natural Computing*, 3:253–292, 2004.
7. M. Garzon, R. Deaton, P. Neathery, D. R. Franceschetti, and R. C. Murphy. A new metric for dna computing. In *Second Annual Genetic Programming Conference*, pages 472–478, 1997.
8. M. Garzon, R. Deaton, P. Neathery, R.C. Murphy, D.R. Franceschetti, and E. Stevens Jr. On the encoding problem for dna computing. In *The Third DIMACS Workshop on DNA-based Computing*, pages 230–237, 1997.
9. M. Garzon, A. Neel, and K. Bobba. Efficiency and reliability of semantic retrieval in dna-based memories. In *9th International Workshop on DNA Based Computers*, pages 157–169, 2003.
10. M. Garzon, V. Phan, K. Bobba, and R. Kontham. Sensitivity analysis of microarray data: A new approach. In *Proc. IBE Conference, Athens GA.*, 2005. Biotechnology Press.

11. A.C. Graesser, P. Penumatsa, M. Ventura, Z. Cai, and X. Hu. Using lsa in autotutor: Learning through mixed initiative dialogue in natural language. *In: T. Landauer, D. McNamara, S. Dennis, and W. Kintsch (Eds.), LSA: A Road to meaning. Mahwah, NJ: Erlbaum*, page in press, 2006.

12. T. Head, M. Yamamura, and S. Gal. Aqueous computing: Writing on molecules. 1999. Proceedings of the Congress on Evolutionary Computing (CEC'99).

13. T. Head, M. Yamamura, and S. Gal. Relativized code concepts and multi-tube dna dictionaries. In *Finite vs Infinite: Contributions to an eternal dilemma (Discrete math and Theoretical Computer SCience)*, pages 175–186, 2001.

14. M. Kanehisa, S. Goto, S. Kawashima, and A. Nakaya. The kegg databases at genome net. *Nucleic Acid Res.*, 30:42–46, 2002.

15. A. Neel and M.H. Garzon. Semantic retrieval in dna-based memories with gibbs energy models. *Biotechnology Progress*, 21:in press, 2006.

16. S. Rimour, D. Hill, C. Militon, and P. Peyret. Goarrays -highly dynamic and efficient microarray probe design. *Bioinformatics*, 21(7):1094–1103, 2005.

17. D. Laham T.K. Landauer, P.W. Foltz. Introduction to latent semantic analysis. *Discourse Processes*, 25:259–284, 1998.

A New Readout Approach in DNA Computing Based on Real-Time PCR with TaqMan Probes

Zuwairie Ibrahim[1], John A. Rose[2], Yusei Tsuboi[3], Osamu Ono[4], and Marzuki Khalid[1]

[1] Center for Artificial Intelligence and Robotics, Department of Mechatronics and Robotics,
Faculty of Electrical Engineering, Universiti Teknologi Malaysia,
81310 UTM Skudai, Johor Darul Takzim, Malaysia
zuwairie@fke.utm.my, marzuki@utmkl.utm.my
http://fke.utm.my/~zuwairie
[2] Institute of Information Communication Technology,
Ritsumeikan Asia Pacific University, 1-1 Jumonjibaru, Beppu-shi, Oita 874-8577 Japan
Japan Science and Technology Agency-CREST
jarose@apu.ac.jp
[3] Bio-Mimetic Control Research Center, RIKEN, 2271-130 Anagahora,
Shimoshidami, Moriyama-ku, Nagoya, 463-0003, Japan
tsuboi@bmc.riken.jp
[4] Institute of Applied DNA Computing, Meiji University, 1-1-1 Higashi-mita, Tama-ku,
Kawasaki-shi, Kanagawa-ken, 214-8571 Japan
ono@isc.meiji.ac.jp

Abstract. A new readout approach for the Hamiltonian Path Problem (HPP) in DNA computing based on the real-time polymerase chain reaction (PCR) is investigated. Several types of fluorescent probes and detection mechanisms are currently employed in real-time PCR, including SYBR Green, molecular beacons, and hybridization probes. In this study, real-time amplification performed using the TaqMan probes is adopted, as the TaqMan detection mechanism can be exploited for the design and development of the proposed readout approach. Double-stranded DNA molecules of length 120 base-pairs are selected as the input molecules, which represent the solving path for an HPP instance. These input molecules are prepared via the self-assembly of 20-mer and 30-mer single-stranded DNAs, by parallel overlap assembly. The proposed readout approach consists of two steps: real-time amplification *in vitro* using TaqMan-based real-time PCR, followed by information processing *in silico* to assess the results of real-time amplification, which in turn, enables extraction of the Hamiltonian path. The performance of the proposed approach is compared with that of conventional graduated PCR. Experimental results establish the superior performance of the proposed approach, relative to graduated PCR, in terms of implementation time.

1 Introduction

Since the discovery of the polymerase chain reaction (PCR) [1], numerous applications have been explored, primarily in the life sciences and medicine, and importantly, in DNA computing as well. The subsequent innovation of real-time PCR has rapidly

C. Mao and T. Yokomori (Eds.): DNA12, LNCS 4287, pp. 350–359, 2006.
© Springer-Verlag Berlin Heidelberg 2006

gained popularity and plays a crucial role in molecular medicine and clinical diagnostics [2]. All real-time amplification instruments require a fluorescence reporter molecule for detection and quantitation, whose signal increase is proportional to the amount of amplified product. Although a number of reporter molecules currently exist, it has been found that the mechanism of the TaqMan hydrolysis probe is very suitable for the design and development of readout method for DNA computing, and is thus selected for the current study.

A TaqMan DNA probe is a modified, non-extendable dual-labeled oligonucleotides. The 5' and 3' ends of the oligonucleotide are terminated with an attached reporter, such as FAM, and quencher fluorophore dyes, such as TAMRA, respectively, as shown in Fig. 1 [3]. Upon laser excitation at 488 nm, the FAM fluorophore, in isolation emits fluorescence at 518 nm. Given proximity of the TAMRA quencher, however, based on the principle of fluorescence resonance energy transfer (FRET), the excitation energy is not emitted by the FAM fluorophore, but rather is transferred along the sugar-phosphate-backbone to TAMRA. As TAMRA emits this absorbed energy at a significantly longer wavelength (580 nm), the resulting fluorescence is not observable in Channel 1 of real-time PCR instruments [4].

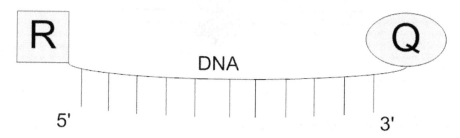

Fig. 1. Illustration of the structure of a TaqMan DNA probe. Here, R and Q denote the reporter and quencher fluorophores, respectively.

The combination of dual-labeled TaqMan DNA probes with forward and reverse primers is a must for a successful real-time PCR. As PCR is a repeated cycle of three steps (denaturation, annealing, and polymerization), a TaqMan DNA probe will anneal to a site within the DNA template in between the forward and reverse primers during the annealing step, if a subsequence of the DNA template is complementary to the sequence of the DNA probe. During polymerization, *Thermus aquaticus* (*Taq*) DNA polymerase will extend the primers in a 5' to 3' direction. At the same time, the *Taq* polymerase also acts as a "scissor" to degrade the probe via cleavage, thus separating the reporter from the quencher, as shown in Fig. 2 [5], where R and Q denote the reporter and quencher dyes, respectively. This separation subsequently allows the reporter to emit its fluorescence [6]. This process occurs in every PCR cycle and does not interfere with the exponential accumulation of PCR product. As a result of PCR, the amount of DNA template increases exponentially, which is accompanied by a proportionate increase in the overall fluorescence intensity emitted by the reporter group of the excised TaqMan probes. Hence, the intensity of the measured fluorescence at the end of each PCR polymerization is correlated to the total amount of PCR product, which can then be detected, using a real-time PCR instrument for visualization.

In this paper, we propose a new readout method tailored specifically to HPP in DNA computing, which employs a hybrid *in vitro-in silico* approach. In the *in vitro* phase, $O(|V|^2)$ TaqMan-based real-time PCR reactions are performed in parallel, to investigate the ordering of pairs of nodes in the Hamiltonian path of a $|V|$-node instance graph, in terms of relative distance from the DNA sequence encoding the known start node. The resulting relative orderings are then processed *in silico*, which efficiently returns the complete Hamiltonian path. To the best of our knowledge, the proposed approach is the first experimentally validated optical method specifically designed for the quick readout of HPP instances, in DNA computing. Previously, graduated PCR, which was originally demonstrated by Adleman [7], was employed to perform such operations. While a DNA chip based methodology, which makes use of biochip hybridization for the same purpose has been proposed [8-10], this method is more costly, and has yet to be experimentally implemented.

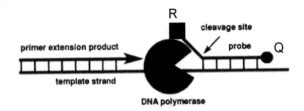

Fig. 2. Degradation of a TaqMan probe, via cleavage by DNA polymerase

2 Notation and Basic Principle

First of all, $v_{1(a)}v_{2(b)}v_{3(c)}v_{4(d)}$ denotes a double-stranded DNA (dsDNA) which contains the base-pairs subsequences, v_1, v_2, v_3, and v_4, respectively. Here, the subscripts in parenthesis (a, b, c, and d) indicate the length of each respective base-pair subsequence. For instance, $v_{1(20)}$ indicates that the length of the double-stranded subsequence, v_1 is 20 base-pairs (bp). When convenient, a dsDNA may also be represented without indicating segment lengths (*e.g.*, $v_1v_2v_3v_4$).

A reaction denoted by TaqMan(v_0,v_k,v_l) indicates that real-time PCR is performed using forward primer v_0, reverse primer $\overline{v_l}$, and TaqMan probe v_k. Based on the proposed approach, there are two possible reaction conditions regarding the relative locations of the TaqMan probe and reverse primer. In particular, the first condition occurs when the TaqMan probe specifically hybridizes to the template, between the forward and reverse primers, while the second occurs when the reverse primer hybridizes between the forward primer and the TaqMan probe. As shown in Fig. 3, these two conditions would result in different amplification patterns during real-time PCR, given the same DNA template (*i.e.*, assuming that they occurred separately, in two different PCR reactions). The higher fluorescent output of the first condition is a typical amplification plot for real-time PCR. In contrast, the relatively lower fluorescent output of the second condition, which reflects the cleavage of a lower number of TaqMan probes via DNA polymerase due to the 'unfavourable' hybridization position of the reverse primer, is due to linear rather than exponential amplification of the template. Thus, TaqMan(v_0,v_k,v_l) = YES if an amplification plot similar to the first condition is observed, while TaqMan(v_0,v_k,v_l) = NO if an amplification plot similar to the second condition is observed.

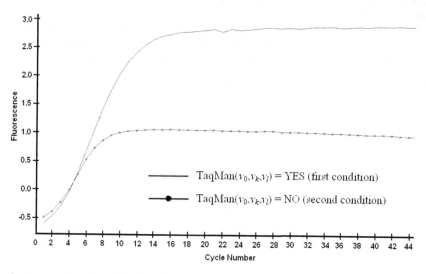

Fig. 3. An example of amplification plots corresponding to TaqMan(v_0,v_k,v_l) = YES (first condition) and TaqMan(v_0,v_k,v_l) = NO (second condition)

3 The Proposed Readout Approach

Let the output of an *in vitro* computation of an HPP instance of the input graph be represented by a 120-bp dsDNA $v_{0(20)}v_{2(20)}v_{4(20)}v_{1(20)}v_{3(20)}v_{5(20)}$, where the Hamiltonian path $V_0{\rightarrow}V_2{\rightarrow}V_4{\rightarrow}V_1{\rightarrow}V_3{\rightarrow}V_5$, begins at node V_0, ends at node V_5, and contains intermediate nodes V_2, V_4, V_1, and V_3, respectively. Note that in practice, only the identities of the starting and ending nodes, and the presence of all intermediate nodes will be known in advance to characterize a solving path. The specific order of the intermediate nodes within such a path is unknown.

The first part of the proposed approach, which is performed *in vitro*, consists of $[(|V|-2)^2-(|V|-2)]/2$ real-time PCR reactions, each denoted by TaqMan(v_0,v_k,v_l) for all k and l, such that $0 < k < |V|-2$, $1 < l < |V|-1$, and $k < l$. For this example instance, so that the DNA template is dsDNA $v_0v_2v_4v_1v_3v_5$, these 6 reactions, along with the expected output in terms of "YES" or "NO" are as follows:

(1) TaqMan(v_0,v_1,v_2) = NO
(2) TaqMan(v_0,v_1,v_3) = YES
(3) TaqMan(v_0,v_1,v_4) = NO
(4) TaqMan(v_0,v_2,v_3) = YES
(5) TaqMan(v_0,v_2,v_4) = YES
(6) TaqMan(v_0,v_3,v_4) = NO

Note that the overall process consists of a set of parallel real-time PCR reactions, and thus requires $O(1)$ laboratory steps for *in vitro* amplification. The accompanying SPACE complexity, in terms of the required number of capillary tubes is $O(|V|^2)$. Clearly, only one forward primer is required for all real-time PCR reactions, while the number of reverse primers and TaqMan probes required with respect to the size of input graph are each $|V|-3$.

After all real-time PCR reactions are completed, the *in vitro* output is subjected to a pseudo-code for *in silico* information processing, producing the satisfying Hamiltonian path of the HPP instance in $O(n^2)$ TIME (here, n denotes vertex number) as follows:

```
Input:  A[0...|V|-1]=2  // A[2, 2, 2, 2, 2, 2]
A[0]=1, A[|V|-1]=|V|  // A[1, 2, 2, 2, 2, 6]
        for k=1 to |V|-3
        for l=2 to |V|-2
            while l>k
                if    TaqMan(v0,vk,vl) = YES
                      A[l] = A[l]+1
                else  A[k] = A[k]+1
                endif
            endwhile
        endfor
        endfor
```

It is assumed that a Hamiltonian path is stored *in silico*, in an array (e.g., A[0...|V|-1]), for storage, information retrieval, and processing, such that $A[i] \in A$ returns the exact location of a node, $V_i \in V$, in the Hamiltonian path. Based on the proposed pseudo-code, and the example instance, the input array A is first initialized to A = {1, 2, 2, 2, 2, 6}. During the loop operations of the pseudo-code, the elements $A[0] \in A$ and $A[|V|-1] \in A$, are not involved, as these two elements may conveniently be initialized to the correct values, as the distinguished starting and ending nodes of the Hamiltonian path are known in advance. The loop operations are thus strictly necessary only for the remaining elements, $A[1,2,3..., |V|-2] \in A$. Again, for the example instance, the output of the *in silico* information processing is A = {1, 4, 2, 5, 3, 6}, which represents the Hamiltonian path, $V_0 \rightarrow V_2 \rightarrow V_4 \rightarrow V_1 \rightarrow V_3 \rightarrow V_5$. For instance, in this case, it is indicated that V_3 is the fifth node in the Hamiltonian path, since $A[3] = 5$, and so on.

4 Experiments

4.1 Preparation of Input Molecules

A pool of 120-bp input molecules $v_{0(20)}v_{2(20)}v_{4(20)}v_{1(20)}v_{3(20)}v_{5(20)}$ is prepared, via standard protocol of parallel overlap assembly (POA) of single-stranded DNA strands (ssDNAs). For this purpose, 11 ssDNAs are required, including additional ssDNAs, which act as link sequences for self-assembly. These strands are listed in Table 1. After completion, amplification via PCR was performed using the same protocol as POA. The forward primers and reverse primers used for the PCR reaction were 5'-CCTTAGTAGTCATCCAGACC-3' and 5'-CCACTGGTTCTGCATGTAAC-3', respectively.

Table 1. The required single-stranded DNAs for the generation of input molecules

Name	DNA Sequences (5'-3')	Length
v_0	CCTTAGTAGTCATCCAGACC	20
v_2	CGCGCACCTTCTTAATCTAC	20
v_4	ATGCGCCAGCTTCTAACTAC	20
v_1	TGGACAACCGCAGTTACTAC	20
v_3	TCCACGCTGCACTGTAATAC	20
v_5	GTTACATGCAGAACCAGTGG	20
v_2v_4	GCAGCGTGGAGTAGTTAGAA	20
v_4v_1	AAGGTGCGCGGTATTACAGT	20
v_1v_3	CGGTTGTCCAGTAGATTAAG	20
v_0v_2	GCTGGCGCATGGTCTGGATGACTACTAAGG	30
v_3v_5	CCACTGGTTCTGCATGTAACGTAGTAACTG	30

The PCR product was subjected to gel electrophoresis and the resultant gel image was captured, as shown in Fig. 4. The 120-bp band in lane 2 shows that the input molecules have been successfully generated. Afterwards, the DNA of interest is extracted. The final solution for real-time PCR was prepared via dilution of the extracted solution, by adding ddH$_2$O (Maxim Biotech, Japan) into 100 μl.

4.2 Real-Time PCR Experiments

The real-time PCR reaction involves primers (Proligo, Japan), TaqMan probes (Proligo, Japan), and LightCycler TaqMan Master (Roche Applied Science, Germany). The sequences for forward primers, reverse primers, and TaqMan probes are listed in Tables 2 and Table 3. In Table 2, the GC contents (GC%) and melting temperature (T_m), are also shown. The LightCycler TaqMan Master essentially contains 1 vial of enzyme, 3 vials of master mix, and 2 vials of PCR grade/water. The master mix contains FastStart Taq DNA polymerase, reaction buffer, MgCl$_2$, and dNTP mix (with dUTP instead of dTTP). PCR grade/water is important to adjust the final reaction volume for real-time PCR. Subsequently, a reaction mix is prepared by pipetting 10 μl of enzyme into the master mix.

For real-time PCR, as recommended by the manufacturer, the final concentration of primers should be between 0.1-1 μM, whereas the final concentration of the DNA probes should be between 0.05-0.1 μM. The final concentration for primers is set to 0.5 μM in this study, and for the TaqMan probes, the maximum final concentration, which is 0.1 μM, is chosen and prepared.

In this study, real-time PCR was performed on a LightCycler 2.0 Instrument (Roche Applied Science, Germany) where amplification is carried out in a 20 μl LightCycler capillary tube (Roche Applied Science, Germany). Two solutions were prepared for each reaction: (1) 3 μl of a 10x primer/probe solution (5 μM of primers and 1 μM of probes), prepared by mixing 0.75 μl of 20 μM forward primer solution, 0.75 μl of a 20 μM reverse primer solution, and 1.5 μl of a 2 μM probe solution; and, (2) 15 μl of PCR mix, containing 4 μl of reaction mix, 9 μl PCR grade/water, and 2 μl of the previous 10x primer/probe solution. Note that even though 3 μl of 10x

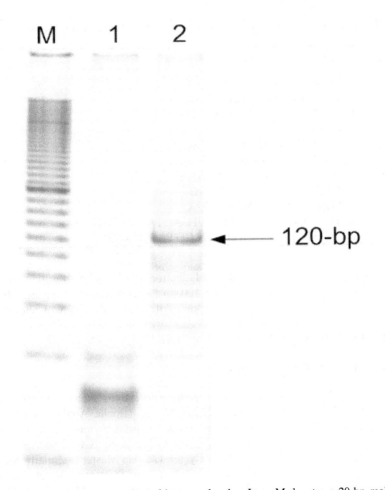

Fig. 4. Gel image for the preparation of input molecules. Lane M denotes a 20-bp molecular marker, lane 1 is the product of initial pool generation based on parallel overlap assembly, and lane 2 is the amplified PCR product.

Table 2. Sequences for forward primer and reverse primers employed for the real-time PCR

Primer	DNA Sequences (5'-3')	GC%	T_m (°C)
Forward primer, v_0	CCTTAGTAGTCATCCAGACC	0.5	53.5
Reverse primer, $\overline{v_2}$	GTAGATTAAGAAGGTGCGCG	0.5	59.8
Reverse primer, $\overline{v_4}$	GTAGTTAGAAGCTGGCGCAT	0.5	59.4
Reverse primer, $\overline{v_1}$	GTAGTAACTGCGGTTGTCCA	0.5	58.6
Reverse primer, $\overline{v_3}$	GTATTACAGTGCAGCGTGGA	0.5	59.7

Table 3. Sequences for TaqMan dual-labeled probes

TaqMan Probes	Sequences
Taqw	R-5'-CGCGCACCTTCTTAATCTAC-3'-Q
Taqx	R-5'-ATGCGCCAGCTTCTAACTAC-3'-Q
Taqy	R-5'-TGGACAACCGCAGTTACTAC-3'-Q
Taqz	R-5'-TCCACGCTGCACTGTAATAC-3'-Q
Taqw	R-5'-CGCGCACCTTCTTAATCTAC-3'-Q

primer/probe solution was prepared, only 2 μl of that solution was used for the preparation of PCR mix. 15 μl of PCR mix was then injected via pipette into a capillary tube. Afterwards, 5 μl of the input molecule solution was injected into the same capillary tube. The capillary tube was then sealed with a stopper, and placed in an adapter. The adapter containing the capillary was placed into a microcentrifuge, and the centrifugation was performed at 3000 rpm.

Seven separate real-time PCR reactions, including a negative control were performed, in order to implement the first stage of the proposed HPP readout. Note that the seventh reaction is for a negative control, which contains PCR grade/water instead of input molecules. The amplification consists of 45 cycles of denaturation, annealing, and extension, performed at 95°C, 48°C, and 72°C, respectively. The annealing temperature is primer-dependent, and should be selected at 5°C below the calculated primer melting temperatures. In the current study, the lowest primer melting temperature was estimated at 53.5°C. Accordingly, an annealing temperature of 48°C was selected. The resulting real-time PCR amplification plots are illustrated in Fig. 5.

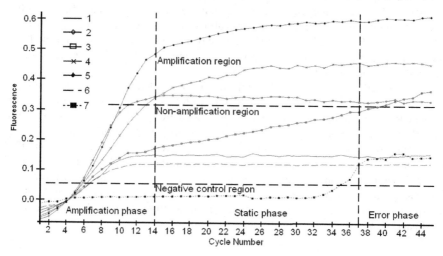

Fig. 5. Output of real-time PCR and grouping of output signals into three regions: amplification region (YES), non-amplification region (NO), and negative control region. The amplification phase, static phase, and error phase are also shown. The numbering 1 to 6 indicate the $[(|V|-2)^2-(|V|-2)]/2$ reactions TaqMan(v_0,v_k,v_l) of input instance, while the seventh reaction is for negative control.

6 Discussion

As discussed in Section 2, in the *in vitro* stage of the proposed approach, each real-time PCR reaction is mapped to a binary output (i.e., either "YES" or "NO"), based on the occurrence or absence of the typical exponential amplification. More specifically, a combination of forward primer, reverse primer, and TaqMan probe, operating on an input template molecule produces an output signal, which may then be recognized as an amplification plot corresponding to either the first or second condition, respectively. Based on the output plot of Fig. 5, the output of the negative control (seventh reaction) can be distinguished easily. As shown in Fig. 5, the real-time PCR output signals can be grouped into three regions: an amplification region, a non-amplification region, and a negative control region. Given the existence of this grouping, the subsequent *in silico* information processing is able to determine the Hamiltonian path of the input instance (e.g., $V_0 \rightarrow V_2 \rightarrow V_4 \rightarrow V_1 \rightarrow V_3 \rightarrow V_5$, for the example instance).

In this study, each real-time PCR reaction consists of 45 cycles, and requires about 1 hour and 10 minutes. The resulting set of real-time PCR output signals may collectively be subdivided into three phases: an amplification phase, a static phase, and an error phase. Here, the amplification phase is defined as the initial time period, during which amplification-like signals are observed. The static phase, on the other hand, is defined as the subsequent time period, in which nearly all of the output signals exhibit only slight increases in fluorescence, with an increasing number of cycles. Lastly, the error phase is defined as the time period in which the negative control signal behaves as an amplification-like signal. Consideration of this effect is important, as shown by the current example instance. In particular, during the error phase (only), the output signal of reaction 3, which is properly clustered into the non-amplification region, enters the amplification region erroneously. In practice, real-time PCR may be halted, if desired, after the 11th cycle (i.e., after 25 min), as the output signals may already be easily distinguished and grouped into the above defined regions, and a set of binary output (i.e., either "YES" or "NO") data may be obtained, as the static and error phases are not important for this purpose. Note that at present, however, this grouping itself is not computerized.

If the ability to extract molecular information is graded based on time factors, the proposed TaqMan-based real-time PCR approach is superior to conventional graduated PCR. In particular, for the case of graduated PCR, following DNA extraction, roughly 30 min is normally required to perform each PCR reaction, 30 min for gel electrophoresis, and another 30 min to stain the gel with SYBR Gold solution. Thus, graduated PCR is proven to be a very time consuming method. As discussed previously, it required about 25 min for the real-time PCR, and the *in silico* information processing can be done in less than 1 min. Hence, it would appear that the proposed approach is superior, in terms of implementation time compared to conventional graduated PCR.

7 Conclusion

This research offers an improved and effective readout approach for DNA computing, which consists of *in vitro* real-time amplification and *in silico* information processing, respectively. It is clear that the use of real-time PCR, as proposed in this paper differs from the conventional application of real-time PCR in the life sciences and medicine. Note that although the current approach has been developed here specifically for HPP readout, variations of this approach could also be applied to other computation models of DNA computing, in which automated and rapid visualization of the computational output are required.

Acknowledgments. Zuwairie Ibrahim is very thankful to Universiti Teknologi Malaysia (UTM) for granting a study leave in Meiji University, Japan, under SLAB-JPA scholarship. This work was supported, in part, by a Grant-in-Aid for Scientific Research B (18300100; J. Rose), from the Japan Society for the Promotion of Science (JSPS).

References

1. Mullis, K. *et al.*: Specific Enzymatic Amplification of DNA *in vitro*: The Polymerase Chain Reaction, Cold Spring Harbor Symposium on Quantitative Biology, Vol. 51 (1986) 263-273
2. Overbergh, L. *et al.*: The Use of Real-Time Reverse Transcriptase PCR for the Quantification of Cytokine Gene Expression, Journal of Biomolecular Techniques, Vol. 14 (2003) pp. 33-43
3. Walker, N.J.: A Technique Whose Time Has Come, Science, Vol. 296 (2002) 557-559
4. Bubner, B. and Baldwin, I.T.: Use of Real-Time PCR for Determining Copy Number and Zygosity in Transgenic Plants, Plant Cell Reports, Vol. 23 (2004) 263-271
5. Heid, C.A. *et al.*: Real-Time Quantitative PCR, Genome Research, Vol. 6 (1996) 986-994
6. Holland, P.M. *et al.*: Detection of Specific Polymerase Chain Reaction Product by Utilizing the 5'→3' Exonuclease Activity of Termus Aquaticus DNA Polymerase, Proceedings of the National Academy of Sciences of the United States of America, Vol. 88 (1991) 7276-7280
7. Adleman, L.: Molecular Computation of Solutions to Combinatorial Problems, Science, Vol. 266 (1994) 1021-1024
8. Rose, J.A. *et al.*: The Effect of Uniform Melting Temperatures on the Efficiency of DNA Computing, DIMACS Workshop on DNA Based Computers III (1997) 35-42
9. Wood, D.H.: A DNA Computing Algorithm for Directed Hamiltonian Paths, Proceedings of the Third Annual Conference on Genetic Programming (1998) 731-734
10. Wood, D.H. *et al.*: Universal Biochip Readout of Directed Hamiltonian Path Problems, Lecture Notes in Computer Science, Vol. 2568 (1999) 168-181

Automating the DNA Computer: Solving n-Variable 3-SAT Problems

Clifford R. Johnson

Dept. of Chemistry
New York University
crjohnso@usc.edu

Abstract. In the decade since the first molecular computation was performed, it has been shown that DNA molecules can perform sophisticated, massively parallel computations avoiding the Von Neumann bottleneck. However, progress in the field has been slow. The largest problem solved to date is an instance of the 20-variable 3-CNF SAT problem. Performing the computation took more than two man-weeks to complete because every aspect of the computation was performed by hand. Molecular computations are extremely labor intensive and error prone--automation is necessary for further progress.

The next step, (the second generation DNA computer – that of taking the laborious, laboratory bench protocols performed by hand, and automating them), has been achieved with the construction of an automated DNA computer dubbed EDNAC. It employs the same paradigm that was used to solve the labor-intensive instance of the 20-variable 3-CNF SAT problem. Using a combinatorial DNA library and complementary probes, EDNAC solves instances of the n-variable 3-CNF SAT problem. A 10 variable instance of the 3-CNF SAT problem was essayed. The computation took 28 hours to perform. EDNAC correctly computed nine of the ten variables, with a tenth variable remaining ambiguous. This result is comparable to current results in the molecular computation community. This research tested the critical properties, such as complexity, robustness, reliability, and repeatability necessary for the successful automation of a molecular computer.

1 Introduction

In the decade since the first molecular computation [1], the largest problem solved to date is a 20 variable instance of the 3-CNF SAT problem performed by the Adleman Laboratory for Molecular Science [2]. Using the theoretical model and techniques discussed in the two papers, [2], and [3], first an instance of the 6-variable 3-CNF SAT problem was solved, then an instance of the 20 variable 3-CNF SAT. Both computations were performed by hand. The 20 variable computation took more than two man-weeks to perform, not including time for set up and sequencing. It almost goes without saying that this is laborious and very error prone. Molecular computations require "many laborious separation and detection steps, which will only increase as the scale increases." [4] Automation is essential for progress in the field.

C. Mao and T. Yokomori (Eds.): DNA12, LNCS 4287, pp. 360–373, 2006.

An automated DNA computer, dubbed EDNAC has been designed, built, and tested. An instance of the 10 variable 3-CNF SAT problem was essayed. EDNAC correctly computed 9 out of 10 variables, with the tenth variable remaining ambiguous. This is comparable to current results in the molecular computation community. The automated computation took 28 hours to perform. By extrapolation, a 20 variable problem would take 2 days on EDNAC as compared to more than two weeks by hand. Furthermore, EDNAC is amenable to miniaturization with a concomitant reduction in computation time.

2 The Computational Paradigm

The computational paradigm used by EDNAC is the same one used for previous 6 variable and 20 variable 3 SAT computations, based on Lipton encoding of DNA [5], and, in fact uses exactly the same chemistry as in references [2] and [3]. The basic unit of computation is separation, performed by using a combinatorial library and Watson-Crick complementary probes. Probes, immobilized in plugs of acrylamide gel, perform the separation operation by annealing to complementary subsequences on strands of the combinatorial library.

The library sequences and details on synthesis, are found in references [2] and [3].

Recalling the terminology used in [2], variables are represented as X_i , $1 \leq i \leq n$. Each variable has 2 distinct 15-mer sequences one for X_i = True (represented as X_i^T) and another sequence for X_i = False (represented as X_i^F). The 15-mer Watson-Crick complements of these sequences (i.e., the probes) are represented by $\mathbf{X}_i^T$ and $\mathbf{X}_i^F$ respectively.

A typical instance of the 3-CNF SAT problem presented to EDNAC is the following 10 variable 14-clause problem:

$$
\begin{aligned}
\phi = \ & (X_2 \vee X_4 \vee X_9) \wedge (X_8 \vee \neg X_{10} \vee X_5) \wedge (\neg X_6 \vee \neg X_8 \vee \neg X_{10}) \wedge \\
& (X_2 \vee \neg X_4 \vee \neg X_9) \wedge (\neg X_9 \vee \neg X_3 \vee X_6) \wedge (X_{10} \vee X_5 \vee X_7) \wedge \\
& (\neg X_7 \vee X_1 \vee \neg X_2) \wedge (X_2 \vee \neg X_4 \vee X_9) \wedge (X_3 \vee X_6 \vee \neg X_8) \wedge \\
& (\neg X_5 \vee X_7 \vee X_1) \wedge (\neg X_2 \vee \neg X_4 \vee \neg X_9) \wedge (X_2 \vee X_4 \vee \neg X_9) \wedge \\
& (\neg X_1 \vee \neg X_2 \vee X_4) \wedge (X_2 \vee \neg X_4 \vee X_9)
\end{aligned}
\qquad \text{Eq. (1)}
$$

Here ϕ has the unique solution:

$$X_1 = F, X_2 = T, X_3 = T, X_4 = F, X_5 = F, X_6 = F, X_7 = F, X_8 = T, X_9 = F, X_{10} = T.$$

In the above equation, each expression enclosed by a parenthesis is called a clause. In the molecular world, these clauses are formed by covalently bonding (and thus immobilizing) Acrydited® oligonucleotide probes (represented by $\mathbf{X}_i^T$ and $\mathbf{X}_i^F$) to the acrylamide gel in a computation module. The encoded, combinatorial ssDNA library (representing every possible solution) is moved via electrophoresis through the module. The module is maintained at 15° C ± 1° C under hybridizing conditions. DNA strands that are complementary to the probe strands anneal and remain in the module while strands that are not complementary pass through. The annealed strands are strands that satisfy the clause. The non-annealing strands do not satisfy the clause, and pass on through to waste.

For example, in Figure 1, probes X_2^T, X_4^F, X_9^F (which are the probes complementary to the subsequences representing X_2^T, X_4^F, X_9^F) are covalently bonded to the acrylamide gel. Thus, library strands that have the X_2 bit True, OR the X_4 bit False, OR the X_9 bit False, will satisfy the clause and will be captured by hybridizing to these probes. Strands with X_2^F AND X_4^T AND X_9^T will not satisfy the clause and will pass through the computational module to the waste well.

I The combinatorial library enters the computation module, which acts as a test for the clause.

II Strands that satisfy the clause, hybridize to the covalently bound probes and remain in the module. Strands that do not satisfy the clause pass through to waste.

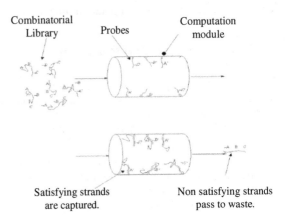

Combinatorial Library   Probes   Computation module

Satisfying strands are captured.   Non satisfying strands pass to waste.

Fig. 1. The molecular implementation of a 3-CNF SAT computation

To continue the computation, the captured strands are released by heating to 65° C, and then pass via electrophoresis to the next (cooled) computation model. Once again, those strands that satisfy the clause hybridize and are captured; those that fail the clause pass through the module and go on to waste. The computation concatenates in this fashion to the final answer module. The final module will contain the answer strands, i.e., those strands that have successfully satisfied all clauses.

3 Implementation of Automation

EDNAC has three major subsystems: I. The Computation Core, II. The Motion Control System, and III. The Control System Shell.

I. The Computation Core

Figure 2 shows the basic architecture used for EDNAC. Module 1 is heated (65° C ± 5° C). The strands pass to module 2 (15° C ± 1° C) via the electrophoretic current going from buffer well A to buffer well C. Those strands not satisfying the module-2 clause, pass through the module, to end up in buffer waste well C, where they are destroyed. As the computation progresses, each module in the system is first cooled to capture satisfying strands, then heated to

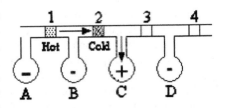

Fig. 2. Basic Architecture

release them for the next computation. To perform this function, a heater and a Peltier junction slide underneath the modules on a sled. This sled also carries a set of electrical commutators that progresses along with heating/cooling unit to provide the correct voltages at the buffer wells A, B, C, and D in Figure 2.

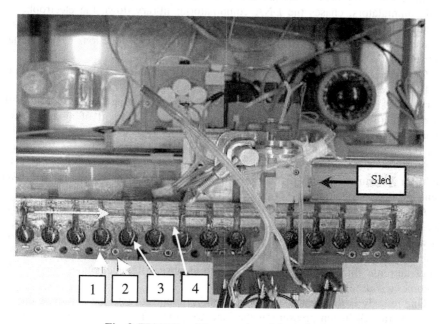

Fig. 3. EDNAC architecture viewed from above

(1) Stator (2) Computation Core (3) Buffer well (4) Computation Module White arrow (lower left) shows direction of computation.

Computation Core construction - In Figure 3 above, we see EDNAC's computation core, viewed from above, in the process of a computation. The buffer wells are visible, as are the copper heating/cooling components of the computation modules. The computation core comprises a 20" long, 4" wide, 1" thick, slab of High Density Polyethylene (HDP). The HDP was milled so that 1" x 1" x ¾" slugs of copper could be inserted at 1" intervals. These were secured with threaded bolts and epoxied in place. The ¼" electrophoresis computation trough is milled through the center of the copper slugs. Electrophoresis channels ¼" wide were milled between the copper slugs to the buffer wells. The electrophoresis channels were then coated with a thin layer of vacuumed epoxy to prevent current bypass.

The underside of the computation core electrophoresis unit, along the copper slug edge, was milled and polished to allow the aluminum sled, securing the heating and cooling units, to slide easily under the copper slugs. Platinum electrode wire was tacked into each buffer well with aquarium silicon cement and soldered to 1" long gold plated stator rails. Recessed head bolts holes were drilled every inch along the edge for securing the core unit to a surfaced U-beam aluminum section.

In Figure 3, the "sled" arrow points to the heater/cooler sled that slides under the copper computational modules. The eighth module from the right is being heated to 65° C, while the seventh module from the right is being cooled to 15° C. Library strands released from the eighth module pass via electrophoresis to the seventh module, where the next computation is performed. The sled, with the heater/cooler and commutators, chases the DNA combinatorial library down the electrophoresis channel, left to right, through the clauses. The white arrow (lower left in photo) indicates the channel center and shows the computation direction. The channels between the copper slugs lead to the buffer wells.

Construction of a computation module - A plastic gel comb is placed in the trough, and 0.8% agarose is poured into the troughs and channels. No gel is used in the buffer wells. Once the agarose is set, the comb is lifted leaving a ¼ inch^3 well for the placement of the computation module. Approximately 150 µl of ag-ac gel with probes is pipetted into the well, then photo-catalyzed. Ag-ac is a mixture of the two gel types, agarose (0.8%) and acrylamide (8.0%), and three probes at 100 pmoles each. Once the agarose component has set, the acrylamide is catalyzed under UV light for 3 minutes. In this manner, a ¼ inch^3 computation gel module is formed and centered in the trough of each of the copper slugs.

Preventing laminar flow disruption at channel junction points - Experiments showed that when an opening occurs in the main electrophoresis channel, laminar flow is disrupted. In particular, at the points where a buffer well channel joins the main electrophoresis channel, eddy currents form, disrupting the laminar flow. The eddy currents diffuse into the channel and deposit oligonucleotides there. This becomes a source of contamination.

One remedy is to inject a small electrophoretic current into the adjoining channel to counteract diffusion pressure at the juncture point. Experiments performed with dyes and radioactively tagged oligonucleotides have shown that an injected counter current equal to 1/10 of the main electrophoretic current is sufficient to maintain laminar flow. In Figure 2, the small negative sign in buffer well B denotes this injected counter current.

Similarly, to prevent non-satisfying strands from by-passing the channel leading to buffer waste well C (where they are to be destroyed) and continuing on towards module 3, a small electrophoretic counter current is applied from buffer well D to buffer waste well C (denoted by a small negative sign in buffer well D in Figure 2).

The Sled – The frame of the automated computer is a modified HP plotter frame. The sled is attached to the HP plotter pen bracket, which gives horizontal motion. The sled contains a heater, Peltier junctions for cooling, a custom-machined commutator system, and brackets for holding delivery lines from peristaltic pumps. The heater and cooler units are separated by ¼" of insulation. This allows efficient heating and cooling of juxtaposed computational modules. Because of the proximity of the two units though, a single Peltier junction was insufficient for reliable cooling. Two Peltier junctions, in a front to back configuration, were used to maintain the necessary module temperature at 15° C.

- *Heating unit* – the heater is made from a milled block of copper with fiberglass coated heating wires, running internally through the copper block. The temperature is adjusted with a Variac controlling an ac voltage ranging from 0 to 140 volts AC. The heater temperature can be adjusted from room temperature to 250°C.

- *Cooling unit* – the cooling unit is formed from doubled Peltier junctions, hermetically sealed to prevent condensation from shorting the junctions. A forced air heat sink is used to improve heat convection off the hot side of the Peltier. The Peltier temperature is adjusted with a DC power supple capable of supplying 13 volts DC at 6 amps. At running temperature, with the heater maintaining 65° C at the adjacent module, the Peltier junction set can maintain the capture module at 15° C.

- *Commutators* – Four flexible strips of copper move along with the sled (in Figure 2, they would supply the current for wells A, B, C, and D.) They are held over the computation core by an L-shaped brace, making contact from above with the gold plated stators. (In Figure 3, the stators are visible as silvery bars tangential to the bottom of the buffer wells. Though they look silver, they are gold plated.) Each commutator strip has a 1kΩ variable resistor placed in series to allow individual regulation of current flow through the well electrode.

- *Buffer delivery and evacuation* – Individual peristaltic pumps can fill and evacuate buffer fluid from each of the wells. This keeps fresh buffer in the wells to avoid ion depletion, and allows the removal of unsuccessful DNA library strands from the target waste well.

- *Thermal transfer* - Even though the computer was machined to precise tolerances (1/10,000"), it is subjected to temperature gradients that cause minute physical deformations preventing effective thermal conduction from the heater to the release module. Even though the Computation Core is tightly bolted in place, there is still enough of a geometrical deformation when heated so that a space is formed between the heater sled and the release module. The heat transfer from the heater to the release module is affected. Though the gap formed is only 1/5,000 of an inch wide, disruption of heat flow occurs. A water bearing was incorporated into the machine to insure good thermal conduction between the heating unit and the release module. This consists of a pumping system delivering 100 µl of distilled water between the two surfaces at 2-minute intervals. The Peltier-junction cooling unit did not require any heat transfer assistance. When properly set up, it takes on the order of 75 seconds for the temperature of the release module to reach 60° C, well above the Tm (denaturing temperature) for the hybridized library strands, and for the capture module to reach 16° C, well within the capture temperature range. It takes about 30 minutes for the released DNA library strands to reach the capture module and another hour for non-satisfying strands to pass through the module and to pass into the waste well.

Visualizing a computation - To substantiate the oligonucleotide advance as the computation is being performed, an aliquot of the 10-variable library was radioactively labeled with ^{32}P ddATP, and a series of images was obtained using a phosphor imaging system (Molecular Dynamics Storm 860). Figure 4 is a composite of multiple phosphor screen images, showing the tagged library progressing through a computation. Shown are the first seven computations of a 10-variable 3-SAT problem. The three images of Figure 5 are: 1) An image taken of the library prior to the computation, 2) An image taken at the end of the first computation, and 3) The image of the partial solution after 7 computations. The reason for the extreme radioactivity in the first well (Figures 4 and 5) is that it contains labeled free mononucleotides as well as labeled library. Only failing strands are seen in the waste wells thereafter.

Fig. 4. Composite images showing the progression of a computation

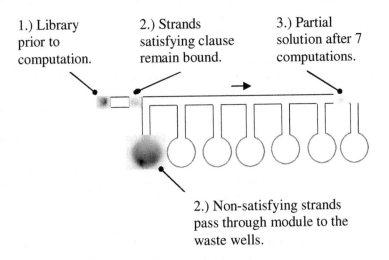

1.) Library prior to computation.

2.) Strands satisfying clause remain bound.

3.) Partial solution after 7 computations.

2.) Non-satisfying strands pass through module to the waste wells.

Fig. 5. Diagram of the composite image

II. The Motion Control System – The motion control system manages the movement of the heating/cooling unit. This unit slides under the computation gels supplying the correct temperatures and voltages for a computation. As a computation ends (for a 20 variable problem, there are 24 computations performed), the heating cooling unit will move down to the next computation unit to perform a new computation. The sled carries with it temperature sensors, dispensers for buffer replenishment, electrodes to provide the correct voltages to each well, as well as the heating and cooling devices. Nominally, the sled moves once every two hours - however through the operator control panel this time can be adjusted from 500 milliseconds to 999.99 hours.

An Astrosyn MiniAngle stepper motor is attached, via a custom-machined bracket and a dual cog synchronous belt system, to the HP plotter's horizontal plotting-pen mechanism. The sled is attached to the plotting-pen bracket and thus its position can be controlled backward and forward by the stepper motor. Every 100 steps on the stepper motor equals 1" of horizontal motion of the sled.

The stepper motor is coupled with a Pontech STP100 stepper motor controller board, with an on-board e-prom to decode serial commands from the Labview Management Program. An XT PC power supply powers the board and stepper motor.

III. The Control System Shell – To provide an easy, intuitive operator interface and to handle the various processes for the molecular computation, a controller shell was developed using LabView 6.0. The program was developed on a generic Pentium 4, 1 GHz computer. Digital and analog I/O is handled by a National Instruments PCI-6025E digital I/O board (a 200 kS/s, 12-Bit, 16 Analog Input, Multifunction DAQ board). With the Shell Program and the PCI-6025E DAQ board, all important aspects of the computation process can be monitored and controlled. The Shell Program provides a GUI (graphical user interface) for data acquisition, data analysis, and process control. Sled position, ion depletion control, process timing, pump timing, and voltage adjustment are all monitored and controlled. (Temperature control was not completely implemented for the first experiments.) It also provides an intuitive operator interface for easy manipulation of the control parameters.

Figures 6a and 6b show the Control Panel (6a) and the EDNAC frame (6b) during a computation.

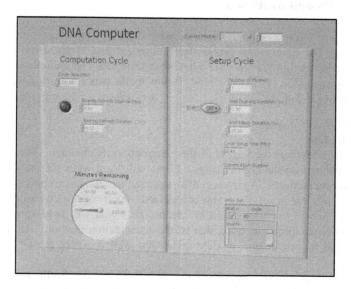

Fig. 6a. Control Panel displaying computation parameters

Computation   Control Panel   Sled
Core

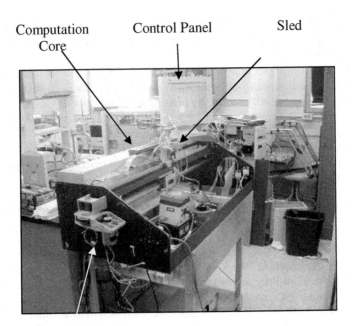

Stepper motor
assembly

Fig. 6b. EDNAC performing a computation

4   Materials and Methods

The Library and Probes - The library used was one half of the twenty variable library used in [2]. The library and probes were designed and synthesized according to procedures described in [2] and [3]. They were tested for capture/release and for the equi-molar distribution of answer strands following the procedures as described in the two references [2] and [3].

Running the DNA Computer
Set-Up Procedure
A gel comb is inserted the length of the electrophoresis channel. The teeth of the comb are centered on the copper slugs. These will form wells for the placement of the computation modules. Agarose, 0.8%, is poured into the electrophoresis channel.

When the agarose is set, the comb is removed. Each copper slug will have a well centered in it about ¼" square and ¼" deep. This well will hold about 150 μl of agarose/acrylamide/probe mixture, that is, the computation module. All but the first well will hold a computation module. The first well is where the DNA library is introduced.

When the computation modules have gelled, the buffer wells are filled with 1xTBE. The DNA library is pipetted into the first well.

The commutator heater/cooler sled is positioned under the first module.

Data is entered into the Management Shell Program for the number of variables, release time, and buffer refresh time (and eventually, for the temperature and voltage parameters). The computation is ready to be performed.

Selecting the start button starts the computation.

The release module is heated to 65° C ± 5° C while the capture module is cooled to 15° C ± 1° C. Because the length of the electrophoresis distance remains constant, the voltage is maintained at about 125 V ± 15 V, to give a constant current of 3 mA. As already mentioned, because of the reduced size of the unit, buffer in the wells is replenished automatically, under control of the Shell Program, at regular intervals to prevent buffer ion depletion.

5 Results

To obtain a read out of the answer strand, after the computation has finished, the last module is excised from the electrophoresis gel. The DNA is leached out of the final module by crushing the module in 1 ml of DD H_2O and then incubating the material at 65° C overnight. The DNA solution is decanted from the gel and desalted, then assayed for the answer by graduated PCR or DNA sequencing.

Three different instances of the 10 variable 3-CNF SAT were run on EDNAC. All three instances gave approximately the same result: EDNAC resolved nine of the ten variables, with the 10th variable remaining ambiguous. Equation 1 was one of the instances of the 10 variable 3-CNF SAT problem essayed on the DNA computer. An analysis of the result of that computation follows.

Determination of the Answer
Once the computation is finished, a read-out of the answer strands, that is, those strands found in the final module, is performed by graduated PCR (detailed in references [2] and [3]). In graduated PCR, the answer strand is PCR amplified using primer sets. The first primer set is for the 5' most subsequence, i.e., for bit X_1 and for the 3' most subsequence i.e., X_{10}. That is, aliquots are PCR amplified using as the primer sets $<X_1^F, X_{10}^F>, <X_1^T, X_{10}^F>, <X_1^F, X_{10}^T>, <X_1^T, X_{10}^T>$. Theoretically, only one set of primers should show any amplification. In reality, more than one set shows a PCR signal. The procedure for answer readout is as follows: Take the decanted solution and serially dilute it in steps of 1:10. PCR amplify each of the serial dilutions with the four primer sets. The primer set that amplifies the most dilute aliquot is deemed to represent the truth-value assignments of the variables X_1 and X_{10}. The number of dilution steps between the answer signal and the next strongest signal is a measure of computation error.

The Full length PCR amplification - Assigning truth values to X_1 and X_{10}
Figure 7a and 7b show the results of the PCR amplification using the primers $<X_1^F, X_{10}^F>, <X_1^F, X_{10}^T>, <X_1^T, X_{10}^F>, <X_1^T, X_{10}^T>$. These are full-length strand amplifications. The aliquot stock was 10-fold serially diluted from 10^{-2} to 10^{-4}, annotated on the images as -2, -3, and -4, and then PCR amplified. No PCR amplification appeared after the 10^{-5} dilution for any primer pair. All primer pairs had a signal appear for at least 10^{-2} dilutions. The primer pair $<X_1^F, X_{10}^T>$ shows the strongest signal-a signal can be observed at 10^{-4}. Based on this, X_1 is assigned the value False; and X_{10} is assigned the value True.

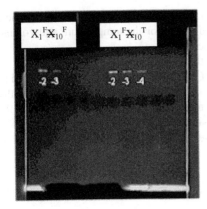

Fig. 7a.

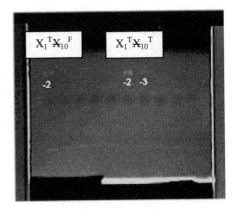

Fig. 7b.

Ideally, the primer pair $<X_1^F, X_{10}^T>$ should have been the only primer pair to exhibit PCR amplification. Here the $<X_1^F, X_{10}^T>$ primer pair simply appears to be the dominant species (right hand side, Figure 7a). The numbers $-2, -3, -4$ represent 1:10 dilutions. Thus, it takes a dilution of 10^{-5} of the final answer solution before the $<X_1^F, X_{10}^T>$ primer pair PCR no longer exhibits a signal. This is about ten-fold better than the $<X_1^T, X_{10}^T>$ and the $<X_1^F, X_{10}^F>$ primer pair. This ten-fold difference between correct answer strands and wrong answer strands is poor, and is probably the reason for ambiguous resolution later on. Ideally, we would like to see absolutely no PCR amplification for any primer pairs, outside of those for the answer strands.

Graduated PCR of inner subsequences - Assigning truth values to X_2, X_3, ... , X_9

Once we have determined the truth assignments for the outer variables X_1 and X_{10}, we continue with graduated PCR on the answer strand to identify the remaining variable values.

Aliquots of this answer stock are then amplified via PCR with the appropriate primers to assign truth-values to the variables X_2 through X_9. Keeping the 5' primer as X_1^F, we test the closest 3' subsequence with the primer sets (X_1^F, X_2^F) and (X_1^F, X_2^T) to assign the X_2 variable. Then we test the subsequent 3' subsequence with the primer set (X_1^F, X_3^F), (X_1^F, X_3^T), to assign the X_3 variable, and so on, finally testing the X_9 variable with the primer sets (X_1^F, X_9^F), (X_1^F, X_9^T). This gives each bit has a signature oligonucleotide length; thus, we can readily identify the appropriate bits by the length of their PCR product.

Again, looking for the dominant species (the PCR lane that gives the highest PCR amplification) we can assign the variables their values.

The labels on the gels shown in 8a and 8b are in-house identifiers for the variables. They represent:

E-->X_2, F-->X_3, G-->X_4, H-->X_5, I-->X_6, J-->X_7, K-->X_8, L-->X_9, and M-->X_{10}.

Thus, for example, E0 represents X_2^F, and E1 represents X_2^T.

Using the in-house terms for the moment, by inspection, E1, F1, G0, H0, I0 are dominant, (Figure8a), and in fact give the correct variable assignment.

Fig. 8a.

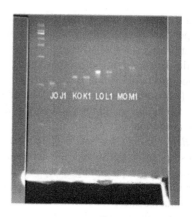

Fig. 8b.

However, in the inspection of the second gel (Figure 8b) obtaining the K value poses a problem. The brightness of K0 and K1 are very close together, and an objective evaluation of the correct value is impossible. In fact, K0 (the wrong value) although smaller, appears brighter, and K1 (the correct value) appears dimmer, but has a larger area. Inspection of the other variables gives the values of J0, L0, and M1, which are the correct variable assignments. Of the 10 variables, 9 can be discerned objectively. K0 and K1 cannot be convincingly distinguished, and the variable K is ambiguous.

The unique truth assignment satisfying Φ as determined by EDNAC is,

$$X_1 = F, X_2 = T, X_3 = T, X_4 = F, X_5 = F, X_6 = F, X_7 = F, X_8 = ?, X_9 = F, X_{10} = T.$$

For this computation, we were unable to unambiguously assign a value to X_8.

6 Discussion

Three different instances of the 10 variable 3-CNF SAT were run on EDNAC. (Equation 1 was one of the three.) All three instances gave approximately the same result: EDNAC resolved nine of the ten variables, with the 10th variable remaining ambiguous.

These results leave room for improvement. However, with each computational run of EDNAC, the variable differences became easier to identify. Since exactly the same chemistry was used for EDNAC as was used in the successful 20 variable problem [2], the ambiguity that EDNAC has in solving a 10 variable SAT problem is probably not due to a fundamental flaw in the paradigm; but rather, is due to difficulties in translating the bench protocols into automation. The project was terminated before these difficulties could be elucidated and corrected.

The following are considered good candidates for improving performance:

Changing the electrophoresis trough from epoxy to glass. Preliminary experiments using glass show promise. Epoxy was first used to coat the electrophoresis trough because it was easy to apply; it acted as an electrical insulator; and the surface was thin enough to not inhibit thermal transfer from the sled. However, microscopic inspection of the epoxy surface uncovered the formation of pinholes. Pinholes disappeared somewhat when the epoxy was catalyzed under a vacuum and applying it to the surface just before hardening occurred. This required following a precisely timed protocol. With glass, thermal transfer is not as good; however, we can control the hydro-phobicity of the surface better; the trough is easier to clean between runs; and we would avoid the problems of the epoxy peeling, and of micro-fissures forming due to thermal stress.

Better characterization of the gel pre-run and gel density conditions. If the gel density can be optimized, the computation will run faster. That is, DNA travels faster through lower percentage acrylamide gels, but the gel is more fragile. The problem becomes balancing the fragility of the gel against the rate of movement of the oligonucleotides.

Better characterization of the agarose-acrylamide polymerization. The gel used in this paper, a photo-catalyzed agarose-acrylamide gel, is a non-standard gel, and is not well characterized in the literature. Experiments need to be undertaken to optimize the gel for DNA computation.

Characterize the capture temperature. 1) The capture temperature is currently set at 15° C. This is a temperature that is heuristically chosen and that seems to work. However, optimizing the capture temperature may mean a faster computation and better capture efficiency. 2) Joule heating of the gel needs to be better characterized for the same reasons.

7   Conclusion

The 3 CNF SAT computation provides a model system that allows the development and study of protocols and methods for large molecular computations. It is the "hydrogen atom" for molecular computations in that it provides a simple model for experimentation in biology's inherently noisy environment.

However, simply optimizing FSAT computations by increasing problem size and diminishing computation time is not a goal by itself - for one thing, solving FSAT problems with a fixed memory, as performed in this paper, is computationally limited. Yet, this research is significant in that

1.   The ability to automate DNA computation is demonstrated.
2.   EDNAC demonstrates that DNA computation is scalable. (In fact, EDNAC was designed physically to perform 3-CNF SAT problems with up to 50 variables.)
3.   EDNAC is easy to use. The control system was designed to be user friendly and flexible, giving the operator straightforward access to all of the reaction parameters: temperatures, voltages, timing, buffer fills, etc. After 15 minutes of instruction, a complete novice was able to perform a 10 variable molecular computation.
4.   Finally, there are easy paths to miniaturization for this architecture. For example, a nano-technology group at NIST is currently working with micro-fluidic chips very similar to EDNAC's architecture [6]. One can almost cut and paste their micro-channel chip implementation into EDNAC while maintaining the same

control shell system. Shrinking the travel lengths from centimeters to millimeters at the micro-chip level will decrease computation time. A 20 variable 3-CNF SAT problem using microfluidics is estimated to run in about 2 hours time.

Acknowledgements

I would like to thank friends, and colleagues at the USC Laboratory for Molecular Science for their support, in particular Len Adleman, Rebecca Anderson, Ravi Braich, Nickolas Chelyapov, Ramon Del Gadillo, Victor Jordan, Howard Lukefahr, Jim Merritt, Dustin Reishus, Paul Rothemund, Bilal Shaw, Areio Soltani, and Don Wiggins.

References

1. Adleman, L.; Molecular computation of solutions to combinatorial problems. Science:266 1021-1024 (1994)
2. Braich, R., Chelyapov, N., Johnson, C., Rothemund, P., Adleman, L.; Solution of a 20-Variable 3-SAT Problem on a DNA Computer. Science:296 499-502 (2002)
3. Braich R., Johnson C., Rothemund P.W.K., Hwang D., Chelyapov N., Adleman, L.; Satisfiability Problem on a Gel Based DNA Computer. DNA Computing – DNA 6 2000, Springer Verlag Volume 2054, New York (2000)
4. Reif, J.H.; Computing. Success and challenges. Science:268 478-479 (2002)
5. Lipton, R.J.; DNA solution of hard computational problems. Science:268 542-545 (1995)
6. Olsen, K., Ross, D., Tarlov, M.; Immobilization of DNA Hydrogel Plugs in Microfluidic Channels. Anal. Chem. 74: 1436-1441 (2002)

Local Area Manipulation of DNA Molecules for Photonic DNA Memory

Rui Shogenji[1,*], Naoya Tate[1], Taro Beppu[2], Yusuke Ogura[2], and Jun Tanida[2]

[1] Japan Science and Technology Agency (JST-CREST),
[2] Department of Information and Physical Sciences, Graduate School of Information
Science and Technology, Osaka University
{tate, ogura, tanida}@ist.osaka-u.ac.jp

Abstract. The address space in DNA memory can be extended by combining information of spatial position and base sequences. Controlling the states of DNA in a local area is an essential technique to use positional information. In this paper, we focus on a photonic DNA memory, which uses optical techniques for addressing on the basis of positional information. We present the concept of photonic DNA memory and describe the read out method using local area manipulation of DNA molecules.

1   Introduction

Generally, when information is stored in memory, the information is assigned to an individual address. The stored information can be read out by reference to this address. A huge address space is needed to store large amounts of information.

DNA memory, embodied by DNA molecules and DNA reactions, has been proposed as a specific example of DNA computing[1,2]. If the base sequence of a DNA molecule in the DNA memory is different, it behaves as a different molecule. Therefore, on individual molecule can be identified with a unique address. That is to say, the sequence of DNA molecules determines the address in the DNA memory. In this paper, this type of address is referred to as a molecular address. Addressing with the molecular address is based on Watson-Crick base pairing. Many requirements on the design of the sequences should be considered for highly accurate addressing, so that extending the address space is difficult when only using the molecular address. On the other hand, conventional memory media, such as an optical disk, utilizes positional information.

Positional information can be applied to the DNA memory when DNA molecules are fixed to a solid, such as a glass substrate. The address space can be extended as compared to the single use of molecular addressing. For utilizing positional information, the states of DNA must be controlled in the local area. Especially, the transfer of DNA from one substrate to another in a local area is an important technique.

We have proposed the concept of photonic DNA computing, which is a parallel computing scheme that uses DNA cooperatively with light[3,4]. To achieve

---

* Now at Shizuoka University, 3-5-1 Johoku, Hamamatsu, Shizuoka 432-8011.

C. Mao and T. Yokomori (Eds.): DNA12, LNCS 4287, pp. 374–380, 2006.

photonic DNA computing, photonic techniques for DNA manipulation are studied and useful data accumulated.

In this paper, we present an implementation of the photonic DNA memory and describe the read-out method using photonic techniques. Implementation of the idea might realize compact and high-capacity memory media.

2 Photonic DNA Memory

Figure 1 shows a schematic diagram of photonic DNA memory. DNA molecules are used as a data storage medium in photonic DNA memory as well as in conventional DNA memory. When DNA molecules are fixed to a solid, such as a glass substrate, the positional information can be used as an address. The storing space of DNA is hypothetically divided into multiple small spaces. We call this sort of address a spatial address.

The address space can easily be extended by combining positional information and molecular base sequences. The reactions of DNA molecules are controlled independently in the individual spatial addresses. The address space can be divided into any shape. For example, the address spaces can be scrambled by division into a random shape. This is useful for achieving secure DNA memory.

Base sequence, structure, and fluorescent label of a DNA molecule in the DNA memory can be treated as information. Large amounts of information can be stored by joining a data sequence to an end of the sequence of a molecular address. Also, the existence of a DNA molecule or the structure of a DNA molecule, such as a hairpin-like structure, can be treated as information.

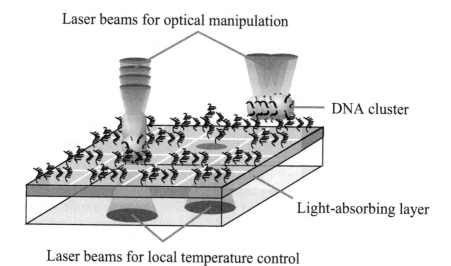

Fig. 1. Schematic diagram of photonic DNA memory

As mentioned above, DNA molecules in the DNA memory are used as a datum in contrast to the regular memory. This means that the DNA memory can be used not only for storage of information but also for storage of macro molecules, such as DNAs, RNAs, and proteins. Thus, applying to DNA microreactor and DNA computer can be expected.

In this paper, a data sequence joined with a molecular address sequence, which is called a tag DNA, was used and is called a data DNA. That is to say, a data DNA consisted of tag and data parts. A data DNA was hybridized with an anti-tag DNA, which had the complimentary sequence of the tag part.

Writing to memory corresponds to attaching data DNA to a substrate. Erasing a datum corresponds to detaching data DNA from a substrate. Reading a datum corresponds to storing data DNA, which is detached from a substrate, in another place as reusable memory.

To use positional information, control of the states of DNA in a local area is essential. We have proposed photonic DNA computing, which is a parallel computing scheme that uses DNA cooperatively with light. Optical techniques used for photonic DNA computing are applicable to photonic DNA memory. Optical manipulation and local temperature control via light irradiation are useful optical DNA controlling techniques to implement photonic DNA memory.

Optical manipulation is method to manipulate nano or microscopic object that uses radiation pressure force induced by the interaction between light and the object. Optical manipulation has the following useful characteristics.

- Optical trap is a non-contact manipulation method.
- Surface tension is not generated.
- Precise operation of an object can be achieved because of its pico-newton order forces.
- Strength and direction can be changed by changing the irradiating pattern of the optical field.

An optical trap has many advantages as a method to manipulate microscopic objects. However, the length of DNA molecules used for photonic DNA memory is very short, typically a few nanometers long. Direct control of DNA molecules is difficult by optical trap. Therefore, a DNA cluster, which combines data DNAs with the surface of a microscopic bead, is used for optical trap. A schematic diagram of a DNA cluster is shown in Fig. 2. The DNA cluster makes it easy to control the position of data DNA. Also, a large amount of DNA molecules can be transfered at a time.

A local temperature control method by light irradiation is one of the most important techniques to achieve photonic DNA memory. In reactions of DNA molecules, such as hybridization and denaturation, the temperature setting is extremely important. These reactions in the photonic DNA memory correspond to writing, reading, and erasing data to the memory. To control reactions in the local area, a substrate is coated with a material to absorb light energy. The temperature of the solution can be controlled by changing the power of a laser beam.

We have previously demonstrated experiments for attaching/detaching data DNAs to/from a substrate or a bead in a solution with optical techniques[3,5].

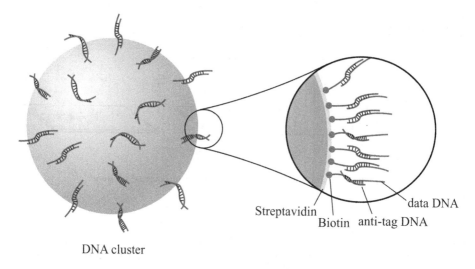

DNA cluster

Fig. 2. Schematic diagram of DNA cluster

Transfer of data DNA from a substrate to a bead is necessary to read data DNA in the photonic DNA memory.

The flow of reading data DNA by using optical DNA controlling techniques is as follows: A bead with no data DNA is moved to an area where the data DNA is combined to a substrate by optical manipulation. This operation corresponds to addressing the spatial address. The substrate is then irradiated with a laser to control the temperature over the bead. The substrate absorbs the laser energy and converts it into heat, then the data DNA is transfered from the substrate to the bead by an autonomous response of DNA molecules in the local area. Reading out the data DNA from the memory can be performed by these operations.

3   Experiments

In the photonic DNA memory, reading data corresponds to transferring a data DNA from a substrate to a bead. To demonstrate validity of the reading techniques for photonic DNA memory, we performed experiments on transferring a data DNA from a substrate to a bead.

The procedure for transferring data DNA from a substrate to a bead is shown in Fig. 3. We irradiated the bead from above with the lasers to control the position of the bead and the temperature of the solution.

A cross sectional view of the substrate for controlling is shown in Fig. 4. The substrate consisted of a gold layer, a titanyl phthalocyanine (TiOPc) layer, and a glass plate. Gold was deposited on the TiOPc layer to fix anti-tag DNA. As an anti-tag DNA, which is DNA thiolated at the 5' end, a substrate was used to fix the combined molecules by gold-thiol bonds. The gold-thiol bond is extremely stable and hardly desorbs[6]. The TiOPc layer absorbs the light energy and converts it into heat.

Translate bead

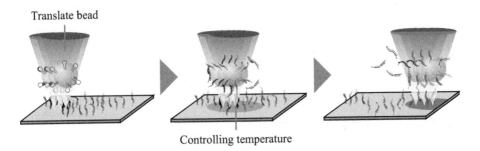

Controlling temperature

Fig. 3. Transferring of data DNAs from substrate to bead

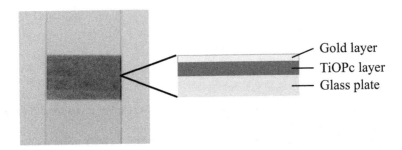

Gold layer
TiOPc layer
Glass plate

Fig. 4. Appearance and cross sectional view of substrate

A 6 μm diameter streptavidin-coated microscopic polystyrene bead was used as a DNA cluster. The anti-tag DNA, which is biotinated at the 5' end, for the DNA cluster can be attached to a bead via a streptavidin-biotin bond. The streptavidin-biotin interaction is the strongest noncovalent biological interaction. The bond formation between streptavidin and biotin is unaffected by most extremes of pH, organic solvents, and denaturing reagents[7].

The data DNA were fluorescent-labeled with Alexa Fluor 546 (Molecular-Probes, U-21652). The absorption and fluorescence emission maxima of Alexa Fluor 546 were 555 nm and 570 nm.

A bead was first moved to the area to be transfered onto DNA molecules on the substrate by optical manipulation. The wavelength of laser for optical manipulation was 800 nm. The laser for controlling the temperature irradiated the substrate over the bead and detached the data DNAs. The laser power was set to 2 mW. By translating the bead, data DNAs detached from a substrate could be attached with the whole bead surface of the bead. The translating speed was 1 μm/sec. Therefore, irradiated energy on each local area was 2mJ.

Figure 5 shows a fluoresce image before and after laser irradiation. The bead to be operated is arrowed. The results show that the fluorescence intensity of the bead increased. In contrast, the fluorescence intensity of the bead's route on the substrate decreased. The results show that data DNAs were transfered from the substrate to the bead by laser irradiation.

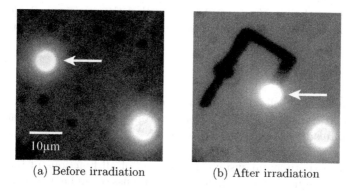

(a) Before irradiation   (b) After irradiation

Fig. 5. Fluorescent images of transferring data DNA from substrate to bead

(a) 1 mW   (b) 2 mW   (c) 3 mW   (d) 4 mW

Fig. 6. Experimental results on data DNA detaching by changing laser power

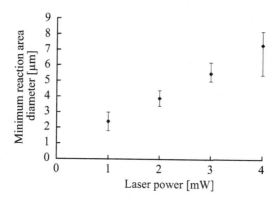

Fig. 7. Relationship between laser power and reaction area diameter

To measure the minimum diameter of the reaction area, the laser power was changed, and denaturation was performed. The power of the irradiation beam used was 1, 2, 3, or 4 mW. The fluorescent images are shown in Figs. 6 (a) - (d). The results show that the reaction area extended as the laser power increased. Also, the results show that the reaction area could be controlled by changing the laser power. Figure 7 shows the relationship between the laser power and the diameter of the reaction area. We found that the minimum diameter of the reading data DNA was approximately 2 μm in this experimental system.

4   Conclusion

As a fundamental technology for implementing photonic DNA memory, we demonstrated transferring DNA molecules from a substrate to a bead by laser irradiation. Reading data of the photonic DNA memory was performed by this technique. The minimum diameter of the reading data DNA was approximately 2 μm in this experimental system. Realization of all functions of the photonic DNA memory requires, for example, further discussion and experiment on accurate and prompt control of DNA molecules from a bead to a substrate.

Acknowledgments

This work was supported by JST CREST and the Ministry of Education, Science, Sports, and Culture, Grant-in-Aid for Scientific Research (A), 15200023, 2003 - 2005. The authors would thank Akira Suyama, Masami Hagiya, and Masahiro Takinoue (The University of Tokyo) for valuable discussion.

References

1. J. Chen, R. Deaton, and Y. Wang, "A DNA-based memory with *in vitro* learning and associative recall," in Proc. of *Ninth Annual Meeting on DNA-Based Computers*, pp. 127–136 (2003).
2. M. Takinoue and A. Suyama, "Molecular reactions for a molecular memory based on hairpin DNA," Chem-Bio Informatics Journal, **4** (3), pp. 93–100 (2004).
3. Y. Ogura, T. Beppu, F. Sumiyama, and J. Tanida, "Translation of DNA molecules based on optical control of DNA reactions for photonic DNA computing," in Proc. SPIE **5931**, pp. 110–120 (2005).
4. Y. Ogura, R. Shogenji, S. Saito, and J. Tanida, "Evaluation of Fundamental Characteristics of Information Systems Based on Photonic DNA Computing," in Proc. of *Second International Workshop, Bioadit 2006*, pp. 192–205 (2006).
5. Y. Ogura, T. Beppu, F. Sumiyama, and J. Tanida, "Toward photonic DNA computing: developing optical techniques for parallel manipulation of DNA," in Proc. SPIE **5897**, pp. 34–43 (2005).
6. S. O. Kelley, J. K. Barton, N. M. Jackson, L. D. McPherson, A. B. Potter, E. M. Spain, M. J. Allen, and M. G. Hill, "Orienting DNA helices on gold using applied electric fields," Langmuir, **14**, pp. 6781–6784 (1998).
7. A. Jenne and M. Famulok, "Disruption of the Streptavidin Interaction with Biotinylated Nucleic Acid Probes by 2-Mercaptoethanol," BioTechniques **26** (2), pp. 249–254 (1999).

Unravel Four Hairpins!

Atsushi Kameda[1], Masahito Yamamoto[1,2], Azuma Ohuchi[1,2],
Satsuki Yaegashi[1], and Masami Hagiya[1,3]

[1] Japan Science and Technology Corporation
[2] Graduate School of Information Science and Technology, Hokkaido University
[3] Graduate School of Information Science and Technology, University of Tokyo

Abstract. DNA machines consisting of consecutive hairpins, which we have previously described, have various potential applications in DNA computation. In the present study, a 288-base DNA machine containing four consecutive hairpins was successfully constructed by ligation and PCR. PAGE and fluorescence spectroscopy experiments verified that all four hairpins were successfully opened by four opener oligomers, and that hairpin opening was dependent on the proper openers added in the correct order. Quantitative analysis of the final results by fluorescence spectroscopy indicated that all four hairpins were open in about 1/4 to 1/3 of the DNA machines.

1 Introduction

DNA machines consisting of consecutive hairpins, which have been introduced previously by the authors, have various potential applications in DNA computation [3]. As depicted in Fig. 1, the inputs for this type of machine are single-stranded oligomers called *openers,* each of which consists of leading and invading sections. The *leading* section of an opener hybridizes with the single-stranded region of the DNA machine, and the *invading* section disrupts the stem of the adjacent hairpin by branch migration, thus opening the hairpin. Consequently, the region flanking the succeeding hairpin becomes single-stranded, and the opener for the adjacent hairpin can hybridize with the machine (Fig. 1b). The single-stranded region flanking a hairpin is also called the *leading* section of the hairpin.

As described previously [3], this type of DNA machine can serve as a memory unit with a hierarchical address wherein each opener corresponds to one address digit. The memory unit is considered accessible only when all of the address digits match the machine address and all of the hairpins are opened. Since the second hairpin can be opened only after the first hairpin has opened, a machine consisting of two hairpins can serve as an AND gate that detects the existence of both openers. When the loop of the second hairpin becomes exposed as single-stranded DNA, it can hybridize with another structure, such as the hairpin shown in Fig. 2, to yield the output of the AND gate [5].

DNA machines that make state transitions by conformational changes have been the subject of active research in the field of DNA computing [9,4,6,5]. The

C. Mao and T. Yokomori (Eds.): DNA12, LNCS 4287, pp. 381–392, 2006.
© Springer-Verlag Berlin Heidelberg 2006

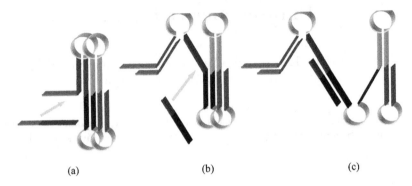

Fig. 1. Opening of a DNA machine consisting of four hairpins. (a) The leading section of the first opener hybridizes with the leading section of the first hairpin. (b) The first hairpin has been opened, and the second opener is ready to hybridize. (c) The second hairpin has been opened.

Fig. 2. An AND gate consisting of two hairpins. The second hairpin is opened only when both openers are present. Since the loop of the second hairpin is exposed as single-stranded DNA, it can hybridize with another structure, such as a hairpin. This reaction is the output of the AND gate.

simple machine made of consecutive hairpins presented in the present report could easily be combined with other machines proposed thus far. The use of hairpins has also been advocated in the hybridization chain reaction [1]; this application is of interest in light of the present study because consecutive hairpins can also trigger the chain reaction.

In the present study, we successfully constructed a DNA machine consisting of four consecutive hairpins with a total length of 288 bases by ligation and PCR. The PCR conditions were carefully selected so that oligomers containing hairpin structures would be properly amplified. PAGE and fluorescence experiments then confirmed that the four openers successfully opened all four hairpins, and that the hairpins could be opened only when the openers were added in the correct order. From spectrophotometric measurements, we estimate that all four hairpins were opened in a ratio of about 1/4 to 1/3.

2 Materials and Methods

2.1 Sequence Design

The DNA sequences of the four hairpins were designed by a method based on selectivity and ordinality, as reported in our previous papers [3,7]. "Selectivity" means that each hairpin can be opened only by the particular opener that has leading and invading sections complementary to the leading section and stem of the hairpin, respectively. "Ordinality" means that a hairpin can be opened only when the adjacent hairpin has already been opened, and its leading section is exposed as single-stranded DNA.

The entire 288-mer sequence of the DNA hairpin machine, designated the "Hairpin Template," is listed in Table 1. In addition to the four hairpins (designated Hairpins A, B, C, and D, starting from the 5'-end), the Hairpin Template has a 20-base, single-stranded region on each side. Two sets of four openers were prepared: Openers A, B, C, and D were designed to open the hairpins from the 5'-side of the Hairpin Template, whereas Openers RD, RC, RB, and RA were designed to open them from the 3'-side. Each 30-mer opener is composed of a 10-base leading section that hybridizes with the corresponding hairpin and a 20-base invading section that is complementary to the hairpin stem sequence.

2.2 Preparation of Hairpin Template

Ligation. We tried to synthesize the Hairpin Template using two different strategies. In one strategy, the Hairpin Template was divided into four fragments (shown in Fig. 3a), each of which was chemically synthesized. The four fragments were then concatenated by a ligase reaction. In the other strategy, the

Table 1. Sequences of the Hairpin Template and the openers. One set of four openers was prepared for opening hairpins from the 5'-side of the Hairpin Template, and another set was prepared for opening from the 3'-side.

Name	Sequence
Hairpin Template	5'- TACGAATATCGCTCCCAACCTAGCAAAAGCGGCCTCACTCTAACTATGAGTGAGGCCGCTTTTGCTACCAG TATATACCAGCACCTGTTAGCCCCAGGTGCTGGTATATACTGGGCAGCAAGAATAGTTCACCTTCGCGGTA GGTGAACTATTCTTGCTGCCAGAACGACATTACGAGAGACGCTGTCTCTCTCGTAATGTCGTTCTGATGTGG TTAAGCAACTGCCG -3'

Name	Sequence	Name	Sequence
Opener A	GAGTGAGGCCGCTTTTGCTA GGTTGGGAGC	Opener RA	TATATACTGG TAGCAAAAGCGGCCTCACTC
Opener B	CAGGTGCTGGTATATACTGG TAGCAAAAGC	Opener RB	TTCTTGCTGC CCAGTATATACCAGCACCTG
Opener C	AGGTGAACTATTCTTGCTGC CCAGTATATA	Opener RC	TGTCGTTCTG GCAGCAAGAATAGTTCACCT
Opener D	TCTCTCGTAATGTCGTTCTG GCAGCAAGAA	Opener RD	TTAACCACAT CAGAACGACATTACGAGAGA

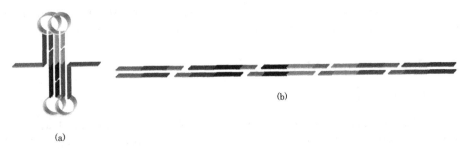

(b)

(a)

Fig. 3. Ligation of the Hairpin Template. We employed two strategies to synthesize the Hairpin Template. (a) The Hairpin Template was divided into four fragments. (b) The double-stranded Hairpin Template was divided into five fragments with loop sequences as sticky ends.

double-stranded Hairpin Template was divided into five fragments in which the loop sequences were sticky ends.

PCR amplification. The Hairpin Template thus obtained was amplified by PCR with 20-mer primers corresponding to the single-stranded regions of the Hairpin Template. The reverse primer was 5'-biotinylated. Reaction conditions were carefully selected with the consideration that hairpin-forming DNA molecules might not be correctly amplified by PCR. PCR reactions were carried out in a final volume of $100\mu l$ with 0.5 U of KOD DASH DNA Polymerase (Toyobo) and 0.5 pmol/μl each primer. Thirty cycles of denaturation (94°C for 15 s), annealing (55°C for 15 s), and extension (74°C for 15 s) were carried out in a thermal cycler. The PCR product was then verified by the standard sequencing protocol and it was confirmed that its sequence was correct.

Alkali treatment with magnetic beads. The completed PCR reaction solution was incubated with BioMag Streptavidin beads (Qiagen) at room temperature according to the manufacturer's instructions. Beads bound to double-stranded Hairpin Template were then re-suspended in purified water. To obtain non-biotinylated, single-stranded Hairpin Template, double-stranded DNA molecules were denatured with sodium hydroxide, which was added until the mixture turned alkaline. Since the reverse primer was 5'-biotinylated in this case, the strands complementary to the Hairpin Template were adsorbed to the BioMag beads. Single strands of the Hairpin Template were then obtained from the supernatant fraction and transferred to a new tube. Hydrochloric acid was added to neutralize the resulting sample.

2.3   Gel Electrophoresis

Solutions containing the Hairpin Template and various combinations of openers (as described in Table 2) were incubated at various temperatures (20, 40, 45, 50, or 55°C). The total volume of the mixture solution was 10 μl in all cases. Each mixing step was followed by a 15-min waiting period to allow time for the opener

Table 2. Opener combinations used for three PAGE experiments. In the first experiment (A), the reaction temperature was varied, whereas in the other two experiments (B and C), the reaction temperature was fixed at 45°C. Concentration values shown represent the final concentration of the added opener in pmol/μl. The volume of the final mixture solution was 10 μl in all cases. A 15-min waiting period followed each mixing step.

| | No. | Mixed opener's order | | | | | | | |
		First	conc.	Second	conc.	Third	conc.	Fourth	conc.
(A)	1	Opener A	0.04						
	2	Opener A	0.04	Opener B	0.08				
	3	Opener A	0.04	Opener B	0.08	Opener C	0.12		
	4	Opener A	0.04	Opener B	0.08	Opener C	0.12	Opener D	0.16
	5	Openre RD	0.04						
	6	Openre RD	0.04	Opener RC	0.08				
	7	Openre RD	0.04	Opener RC	0.08	Opener RB	0.12		
	8	Openre RD	0.04	Opener RC	0.08	Opener RB	0.12	Opener RA	0.16
(B)	1	Opener A	0.04						
	2	Opener A	0.04						
	3	Opener A	0.04						
	4	Opener A	0.04						
	5	Openre RD	0.04						
	6	Openre RD	0.04						
	7	Openre RD	0.04						
	8	Openre RD	0.04						
	9	Opener A	0.04	Opener B	0.08				
	10	Opener A	0.04	Opener C	0.08				
	11	Opener A	0.04	Opener D	0.08				
	12	Openre RD	0.04	Opener RC	0.08				
	13	Openre RD	0.04	Opener RB	0.08				
	14	Openre RD	0.04	Opener RA	0.08				
	15	Opener A	0.04	Opener B	0.08	Opener C	0.12		
	16	Opener A	0.04	Opener B	0.08	Opener D	0.12		
	17	Openre RD	0.04	Opener RC	0.08	Opener RB	0.12		
	18	Openre RD	0.04	Opener RC	0.08	Opener RA	0.12		
	19	Opener A	0.04	Opener B	0.08	Opener C	0.12	Opener RB	0.16
	20	Openre RD	0.04	Opener RC	0.08	Opener RB	0.12	Opener RA	0.16
(C)	1, 11	Opener A	0.04						
	2, 12	Opener A	0.04	Opener B	0.08				
	3, 13	Opener A	0.04	Opener B	0.08	Opener C	0.12		
	4, 14	Opener A	0.04	Opener B	0.08	Opener C	0.12	Opener D	0.16
	5, 15	Opener A	0.04	Opener B	0.08	Opener C	0.12	5'-FITC Opener D	0.16
	6, 16	Openre RD	0.04						
	7, 17	Openre RD	0.04	Opener RC	0.08				
	8, 18	Openre RD	0.04	Opener RC	0.08	Opener RB	0.12		
	9, 19	Openre RD	0.04	Opener RC	0.08	Opener RB	0.12	Opener RA	0.16
	10, 20	Openre RD	0.04	Opener RC	0.08	Opener RB	0.12	5'-FITC Opener RA	0.16

to hybridize with the Hairpin Template and to ensure that the target hairpin structure was opened. The resulting DNA duplexes were separated according to their length and structure by PAGE. Five microliters of each mixture were subjected to 10% PAGE for 40 min at a constant voltage of 200 V. After electrophoresis, DNA in the gel was visualized by SYBR Gold staining (Molecular Probes, Inc.) or FITC fluorescence detection (for FITC-modified openers).

2.4   Spectrophotometric Detection

Fluorescence spectroscopy is preferable to PAGE for detecting hairpin opening for three reasons: First, fluorescence detection is generally more precise and simpler to use in verifying conformational changes of DNA molecules; second, fluorescence detectors can be inexpensively produced [8]; and third, fluorescence detection yields data that are suitable for quantitative analysis.

To determine whether the final hairpin was opened, two types of fluorescence-modified oligomers were used. One type hybridized with the 20-base, singled-stranded region on one side of the Hairpin Template, and the other type hybridized with the stem sequence of the final hairpin. Fig. 4 illustrates the case in which the four hairpins are opened from the 3′-side of the Hairpin Template. The 5′-FAM-modified 20-mer is expected to hybridize with the single-stranded region on the 5′-side of the Hairpin Template, whereas the 3′-BHQ-modified 20-mer is expected to hybridize with the stem sequence flanking the single-stranded 20-base region. Therefore, if the final hairpin is open, the BHQ-modified oligomer should hybridize with the Hairpin Template, and the fluorescence from FAM should be quenched by BHQ.

Fig. 4. Opening of the final hairpin was verified using oligomers modified with FAM (yellow circle) or with BHQ (black circle). If the final hairpin is open, both oligomers should hybridize with the Hairpin Template, and the FAM fluorescence should be quenched by BHQ.

An F-2500 spectrophotometer (Hitachi) was set to time-scale fluorescence measurement mode, with excitation and emission wavelengths of 494 and 518 nm, respectively. The excitation and emission slit widths were each 5 nm. The spectrophotometer was connected to an isothermal bath (Lauda) to maintain the quartz cell at 45°C.

The Hairpin Template solution was first mixed with openers RD, RC, RB, and RA (in that order) at 45°C, with a 30-min waiting period after each mixing step. The concentrations of the Hairpin Template and the openers were adjusted to 0.1 μM. The total volume of the solution was 400 μl to accommodate the quartz spectrophotometer cell. The solution was then put into a cell containing 4 μl of a 10-μM solution of the FAM-modified oligomer. After a 15-min incubation, measurement of FAM fluorescence was initiated. The FAM baseline was measured for a few minutes, and then 4 μl of a 10-μM solution of the BHQ-modified oligomer was added to the cell.

3 Results

The Hairpin Template and the openers were mixed in various combinations (as described in Table 2) to investigate the secondary structures of the four template hairpins. The various structures assumed by the Hairpin Template were separated by 10% PAGE (Fig. 5), and the bands were visualized by SYBR Gold staining and FITC detection. Analysis of the gel bands confirmed that opening of each hairpin required a specific combination of openers. Quantitative analysis of the resulting structures by fluorescence spectroscopy led us to conclude that only about 1/4 to 1/3 of the Hairpin Template molecules reached the final state in which all four hairpins were open.

In the PAGE analysis, the initial state of the Hairpin Template with four closed hairpin structures generated the single band shown in Lane HT of Fig. 5. As more hairpins were opened, the mobility of this band was shifted.

3.1 Influence of Reaction Temperature on Hairpin Opening

We examined the influence of reaction temperature on hairpin opening by performing experiments at five different temperatures (20, 40, 45, 50, and 55°C). The results are shown in Fig. 5A. At 20°C, little change in the mobility of the Hairpin Template band was observed when openers were added. At 40°C, the HT band shifted slightly upward when Opener A was added (Lane 1), indicating that Opener A opened Hairpin A. When Opener B was subsequently added, a new band appeared (Lane 2), indicating that Opener B opened Hairpin B. In some of the Hairpin Template molecules, however, Hairpin B remained closed, as shown by the fact that the band observed in Lane 1 was still present in Lane 2. Similarly, addition of Opener C caused another band to appear (Lane 3), corresponding to the structure in which Hairpins A, B, and C were all open. Unexpectedly, when Opener D was then added, a new band corresponding to opening of the final hairpin was not observed (Lane 4), possibly due to an overlap with the band observed in Lane 3. This issue was further examined by detection with FITC, as described in the next section. Opening of hairpins from the 3′-side of the Hairpin Template was observed to occur in a similar fashion (Lanes 5–8).

At 40°C, some bands that appeared in one step persisted in succeeding steps, suggesting that not all of the target hairpins opened when the corresponding

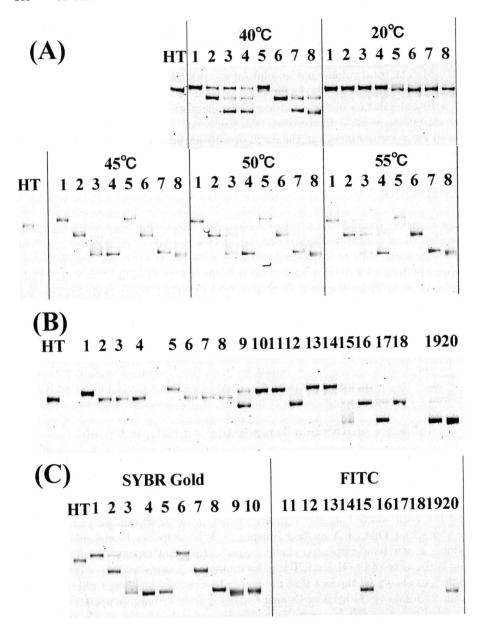

Fig. 5. PAGE analysis of the experiments summarized in Table 2. The gel was stained with SYBR Gold, with the exception of Lanes 11–20 in (C). Lanes labeled HT indicate the mobility of the Hairpin Template in the absence of openers, so that all hairpins are closed. (A) Five reaction temperatures (20, 40, 45, 50, and 55°C) were investigated. (B) Various combinations of openers and the effect of order of addition were investigated. (C) Fluorescence detection was used to determine whether the fourth opener hybridized with the Hairpin Template. In Lanes 11–20, FITC fluorescence was detected in the same sample as that of Lanes 1–20.

opener was added. At 45°C or higher temperatures, persistent bands either did not appear or were smeared.

3.2   Influence of Openers and Order of Addition

We next examined the influence of the added openers and their order of addition. Except for the particular openers used and their orders of addition, experimental conditions were the same as those described above (Section 3.1) at 45°C. The openers and their orders of addition are described in Table 2, and the results are shown in Fig. 5B.

As shown in Lanes 1–4 of Fig. 5B, only Opener A was able to open Hairpin A. Since the position of the HT band (which represents four closed hairpins) did not shift in Lanes 2–4, we can conclude that other openers could not open Hairpin A. Similarly, only Opener RD opened Hairpin D from the 3′-side of Hairpin Template (Fig. 5B, Lanes 5–8). When openers B, C, or D were added to the state in which Hairpin A was already opened, only Opener B was observed to open Hairpin B, as shown in Lanes 9–11 of Fig. 5B. In Lanes 12–14, on the other hand, only the combination of openers RD and RC could open the two hairpins, resulting in the new band in Lane 12. In Lanes 15 and 16, Openers C and D were added to the Hairpin Template in which Hairpins A and B had already been opened by Openers A and B, respectively. As a result, the next hairpin was opened only by Opener C. Similarly, Opener RB could open Hairpin B (Lane 17), but Opener RA could not open the next hairpin (Lane 18).

Using fluorescence detection, we attempted to address the overlap problem for the bands corresponding to the states with three and four open hairpins. We modified the final opener with a 5′-FITC tag so that hybridization of the FITC-tagged opener with the Hairpin Template could be detected in the gel image using fluorescence detection, as shown in Fig. 5C. In Lanes 15 and 20, fluorescent bands of the same mobility as those observed in Lanes 5 and 10, respectively, were clearly detected, demonstrating that the FITC-modified opener successfully hybridized with the Hairpin Template. These results still do not unequivocally demonstrate whether the final hairpin (Hairpin D or A) was opened, however. Hybridization of the final opener with the Hairpin Template is more definitively addressed below.

3.3   Quantitative Spectrophotometric Analysis

As shown in Fig. 6, addition of the BHQ-modified oligomer to the FAM fluorescence experiment reduced the FAM fluorescence intensity by about 1/4 to 1/3, compared with the FAM fluorescence baseline, due to quenching by BHQ. In the complementary experiment in which openers except RA were added and hairpins D, C, and B were considered opened, almost no reduction of the FAM fluorescence intensity was observed upon addition of the BHQ-modified oligomer, as shown in Fig. 6.

In additional experiments, the FAM- and BHQ-modified oligomers were replaced with the unmodified oligomers FAMseq and BHQseq, which have the

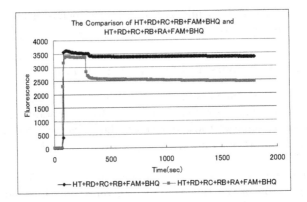

Fig. 6. Fluorescence spectroscopy measurements of hairpin opening. The upper trace represents the case in which Hairpin A is not open, and the lower trace represents the case in which all four hairpins are open.

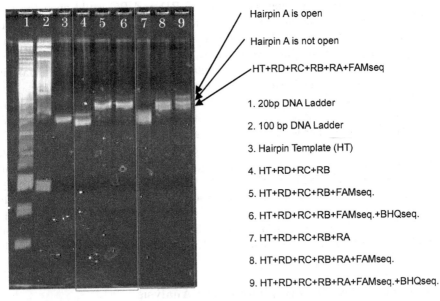

Fig. 7. PAGE analysis of the experiment shown in Fig. 6. Each lane is characterized by a combination of oligomers. FAMseq and BHQseq denote unmodified oligomers that have the same sequences as those of the FAM- and BHQ-modified oligomers, respectively.

same sequences as the FAM- and BHQ-modified oligomers, respectively. The resulting species were examined by 8% (w/v) PAGE (Fig. 7), where the concentration of each species is set to 0.05 μM. When the Hairpin Template, openers RD, RC, RB, and RA, FAMseq, and BHQseq, were mixed, three bands were observed (Fig. 7, Lane 9). Comparison with Lanes 8 and 6 suggests that the lowest band in Lane 9 represents the hybrid species with four open hairpins and containing FAMseq. The middle band in Lane 9 appears to represent the hybrid species in which Hairpin A is not open.

These results appear to explain our fluorescence spectroscopy data indicating that 3/4 to 2/3 of the FAM fluorescence was not quenched by BHQ. The uppermost band in Lane 9 of Fig. 7 represents the hybrid species with four open hairpins and containing both FAMseq and BHQ. The intensity of this upper band is much less than that of the middle band in the same lane. Therefore, this additional gel electrophoresis experiment confirms our previous results, obtained by fluorescence spectroscopy, in which only 1/3 to 1/4 of FAM fluorescence was quenched by BHQ.

In the PAGE experiments described in the previous section, the opener concentrations in later steps were increased because qualitative analysis was the major concern. In this quantitative analysis, we adjusted the concentrations of the openers to be equal to that of the Hairpin Template. With the same concentrations as those in the PAGE experiments, however, we obtained almost the same result (data not shown), indicating that increasing the opener concentrations did not significantly affect the ratio of hairpin opening.

4 Discussion

Opening of two consecutive hairpins was demonstrated in our previous study [3]. In the present study, we used PAGE and fluorescence spectroscopy to demonstrate opening of four consecutive hairpins. We also overcame the initial experimental challenge of obtaining sufficient Hairpin Template to conduct consecutive hairpin-opening experiments. This problem was solved by careful adjustment of PCR and alkaline treatment conditions, as described in the Materials and Methods.

In the study described here, we devised two major improvements in the experimental conditions for hairpin opening, compared to our previous study. First, the leading section of the openers was shortened (from 20 bases to 10 bases) to inhibit formation of undesirable secondary structures that could hinder hairpin opening. Shortening of the leading section also minimized interactions between the openers, as explained below. Second, we raised the reaction temperature to overcome the energy barrier to opening of the target hairpin while avoiding disruption of the intact hairpins.

An important experimental challenge in increasing the number of consecutive hairpins in the model system was avoiding interactions between the openers. Openers for consecutive hairpins should have a complementary subsequence. In our previous study, successive openers had a complementary, 20-base subsequence that led to hybridization of the openers. In the present study, the combination of the shortened leading section and the higher reaction temperature seems to have eliminated this problem. In a PAGE experiment, we confirmed that consecutive openers did not hybridize with one another (data not shown).

We also confirmed that sequential addition of successive openers caused hairpin opening as expected.

The interesting question of how many hairpins can successfully be concatenated remains unanswered. A template molecule designed to form four consecutive hairpins might be expected to form an undesirably complex tertiary structure, but the Hairpin Template used in the present experiments did not appear to form such a structure. Hairpin templates with additional hairpins might form even more complex structures that would hinder proper opening in response to openers.

Acknowledgements

The work presented in this paper was partially supported by Grand-in-Aid for Scientific Research on Priority Area No.14085202, Ministry of Education, Culture, Sports, Science and Technology, Japan.

References

1. Dirks, R.M. and Pierce, N.A.: Triggered amplification by hybridization chain reaction, *Proc. Natl. Acad. Sci.,* Vol.101, No.43, 2004, pp.15275–15278.
2. Hagiya, M., Yaegashi, S. and Takahashi, K.: Computing with Hairpins and Secondary Structures of DNA, *Nanotechnology: Science and Computation,* Natural Computing Series, Springer, 2005, pp.293–308.
3. Kameda, A., Yamamoto, M., Uejima, H., Hagiya, M., Sakamoto, K. and Ohuchi, A.: Hairpin-based state machine and conformational addressing: Design and experiment, *Natural Computing,* Vol.4, No.2, 2005, pp.103–126
4. Seelig, G., Yurke, B., and Winfree, E.: DNA Hybridization Catalysts and Catalyst Circuits, DNA Computing, 10th International Workshop on DNA Computing, Lecture Notes in Computer Science, Vol.3384, 2005, pp.329–343.
5. Takahashi, K., Yaegashi, S., Kameda, A. and Hagiya, M.: Chain Reaction Systems Based on Loop Dissociation of DNA, *DNA Computing: 11th International Workshop on DNA Computing, DNA11,* Lecture Notes in Computer Science, Vol.3892, 2006, pp.347–358.
6. Takahashi N., Kameda A., Yamamoto M. and Ohuchi A.: Aqueous Computing with DNA Hairpin-based RAM, DNA Computing, *10th International Workshop on DNA Computing,* Lecture Notes in Computer Science, Vol.3384, 2005, pp. 355–364.
7. Uejima, H. and Hagiya, M.: Secondary Structure Design of Multi-state DNA Machines Based on Sequential Structure Transitions, *DNA Computing, 9th International Workshop on DNA-Based Computers,* Lecture Notes in Computer Science, Springer, Vol.2943, 2004, pp.74–85.
8. Yan, H.: An Inexpensive LED-Based Fluorometer Used to Study a Hairpin-Based DNA Nanomachine, *DNA Computing, 10th International Workshop on DNA Computing,* Lecture Notes in Computer Science, Vol.3384, 2005, pp.399–409.
9. Yurke, B., Turberfield, A. J., Mills, Jr., A. P., Simmel, F. C., and Neumann, J. L.: A DNA-fuelled molecular machine made of DNA, *Nature,* Vol.406, 2000, pp.605–608.

Displacement Whiplash PCR: Optimized Architecture and Experimental Validation

John A. Rose[1,2,3,*], Ken Komiya[3,4], Satsuki Yaegashi[3], and Masami Hagiya[2,3]

[1] Institute of Information Communication Technology, Ritsumeikan Asia Pacific University
jarose@apu.ac.jp
[2] Department of Computer Science and UPBSB, The University of Tokyo
hagiya@is.s.u-tokyo.ac.jp
[3] Japan Science and Technology Agency-CREST
yaegashi@lyon.is.s.u-tokyo.ac.jp
[4] Dept. of Computational Intelligence and Systems Science, Tokyo Institute of Technology
komiya@dis.titech.ac.jp

Abstract. Whiplash PCR-based methods of biomolecular computation (BMC), while highly-versatile in principle, are well-known to suffer from a simple but serious form of self-poisoning known as back-hybridization. In this work, an optimally re-engineered WPCR-based architecture, *Displacement Whiplash PCR* (DWPCR) is proposed and experimentally validated. DWPCR's new *rule protect* biostep, which is based on the primer-targeted strand-displacement of back-hybridized hairpins, renders the most recently implemented rule-block of each strand unavailable, abolishing back-hybridization after each round of extension. In addition to attaining a near-ideal efficiency, DWPCR's ability to support isothermal operation at physiological temperatures eliminates the need for thermal cycling, and opens the door for potential biological applications. DWPCR should also be capable of supporting programmable exon shuffling, allowing XW-PCR, a proposed method for programmable protein evolution, to more closely imitate natural evolving systems. DWPCR is expected to realize a highly-efficient, versatile platform for routine and efficient massively parallel BMC.

1 Introduction

In Whiplash PCR (WPCR), autonomous molecular computation is implemented by the recursive, self-directed polymerase extension of a DNA hairpin mixture [1]. When combined with a method for generating a combinatorial library of encoded strands, WPCR is theoretically capable of solving instances of a variety of **NP-complete** problems, including *Inductive Inference* [1] and *Hamiltonian Path (HPP)* [2]. WPCR has also been shown to be capable of supporting evolutionary computation, including: solution of *HPP* instances, via restriction-based crossover (*Evolutionary WPCR* (EWPCR)) [3]; co-evolution of poker strategies, via Parallel Overlap Assembly-based crossover [4]; and, programmable protein evolution, via EWPCR-based pseudo-module shuffling and RNA-protein fusion (XWPCR) [5].

---

* To whom correspondence should be addressed.

C. Mao and T. Yokomori (Eds.): DNA12, LNCS 4287, pp. 393–403, 2006.
© Springer-Verlag Berlin Heidelberg 2006

(a) Whiplash PCR

(b) Displacement Whiplash PCR

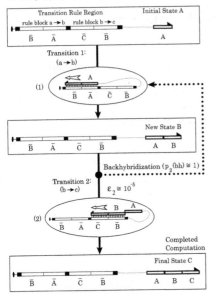

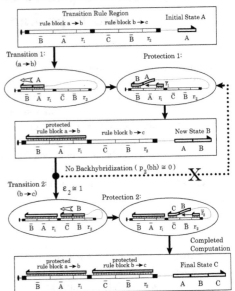

Fig. 1. (a) **Whiplash PCR** Top panel depicts a DNA strand encoding a WPCR program for executing the short computational path, $a \rightarrow b \rightarrow c$. Strand 3' and biotinylated 5' ends are indicated by black circles and arrowheads, respectively. Spacer and stop sequences are indicated by dark squares and light gray lines, respectively. In practice, efficient implementation is prevented by self-poisoning due to competitive hairpin formation, or *back-hybridization* (dotted line). (b) **Displacement Whiplash PCR** In DWPCR, encoded strands are re-engineered to support the addition of a back-hybridization eliminating *rule protect* step, which proceeds via primer annealing, extension, and strand displacement at target sequence, r_i, following transition, i.

Although the fundamental feasibility of using WPCR to implement multiple state transitions and test-scale instances has been demonstrated experimentally [2,6,7], a serious barrier confronting scaling for realistic application is a systematic tendency for WPCR strands to participate in a powerful form of self-inhibition called *back-hybridization* [8,9]. Attempts to improve WPCR efficiency have involved both tuning of reaction conditions [6], and redesign of the basic WPCR architecture [8]. However, because the generation of highly-stable back-hybridized hairpins is favored energetically, elimination via either reaction condition optimization or strand-encoding strategies is impossible. Furthermore, the previous attempt to redesign the WPCR architecture to support targeted PNA_2/DNA triplex formation (PWPCR [8]), thereby reducing the occupancy of back-hybridized structures was not fully optimal.

In this work, an optimally re-engineered WPCR-based architecture, *Displacement WPCR* (DWPCR) is proposed and experimentally validated. DWPCR provides a critical improvement for WPCR, by adding a new *rule protect* operation after each extension. This operation, which replaces thermal cycling, renders the most recently implemented rule-block of each computing strand unavailable via primer-targeted strand-displacement, virtually eliminating back-hybridization. In addition to achieving a near-ideal efficiency, DWPCR presents several other advantages, including: isothermal operation at physiological temperatures, opening the door for biological applications; and, the ability to support programmable exon shuffling, allowing XW-PCR [5] to more closely imitate natural evolving systems. DWPCR is thus expected to realize a highly-efficient, versatile platform for routine and efficient massively parallel BMC.

Paper organization is as follows. Sec. 1.1 presents the original WPCR architecture, followed in Sec. 1.2 with a discussion of back-hybridization and proposal of the DW-PCR architecture. Sec. 2 then presents a set of experiments designed to establish the fundamental feasibility of DWPCR, which rests on the efficiency of the key operation, *rule protect*. Sec. 3 presents experimental results, followed in Sec. 4 with a discussion of experiment results, combinatorial DWPCR implementation, and potential applications.

1.1 The Whiplash PCR Architecture

The original WPCR architecture [1] is illustrated in Fig 1(a). As shown, state transitions are implemented via the recursive template-directed polymerase extension of a DNA hairpin mixture. For clarity, this process is described for a 3-state, 2 step implementation, $a \rightarrow b \rightarrow c$. Prior to computation, a single-stranded (ss) DNA molecule is encoded with three regions (top panel). The *transition rule region* contains a set of state transition *rule-blocks*, each of which encodes a transition from computational state x to state y. If X and $\overline{X}$ respectively denote the DNA word for state x, and its Watson-Crick reverse complement, then the rule-block for executing transition $x \rightarrow y$ is formed by catenated words, $5' - \overline{Y}\overline{X} - 3'$. The completed transition rule region is formed by the concatenation of all rule-blocks, separated by short DNA *stop sequences*. In Fig. 1, the transition rule region contains rule-blocks for transitions $a \rightarrow b$ and $b \rightarrow c$. The 3'-most codeword (word A (top panel)) is the *head*, which encodes the initial state of the WPCR strand. A short DNA *spacer sequence* is placed between these two regions.

Each WPCR state transition is implemented via a three-step process of hybridization, polymerase extension, and denaturation, shown in Fig. 1(a). Each transition is initiated via hybridization of the 3' head with a complementary DNA word in the transition rule region. The transition encoded by the hybridized rule-block is then executed via polymerase extension of the head (horizontal arrow in each transition), which appends the DNA word for the transition target state to the strand's 3' end. Polymerase extension is terminated automatically when the DNA polymerase encounters the transition block's 5' *stop sequence*, usually implemented using a short stop sequence (AAA), combined with the absence of free dTTP in the buffer. Following thermal denaturation of the newly-extended hairpin, the strand is ready for the next round of hybridization and extension (process $b \rightarrow c$). Upon completion of all encoded transitions, a record of the computed path is encoded at the strand's 3' end, as an ordered string of DNA words ($5' - ABC - 3'$ in Fig. 1 (bottom). As each strand is encoded to execute a distinct set of transition rules autonomously, massively parallel computation may be achieved by generating a combinatorial mixture of WPCR strands, and iteratively applying a thermal cycle appropriate for the parallel hybridization, extension, and denaturation of all strands.

1.2 Back-Hybridization and Displacement WPCR

Unfortunately, WPCR state transition efficiency is compromised by a serious form of self-poisoning, known as *back-hybridization*. This effect is illustrated in Fig. 1(a; dotted line) via formation of the extended hairpin structure (hairpin (1)), generated via transition, $a \rightarrow b$, which competes with formation of the extendable structure, (unextended hairpin (2)) required for transition $b \rightarrow c$. The effect of back-hybridization has been modeled theoretically [8], and shown to be sufficiently serious to render massively-parallel WPCR infeasible. Experimental verification of basic model predictions for competitive hairpin formation has also been undertaken [9]. Back-hybridization is fundamentally less tractable than mis-hybridization (*i.e.*, error duplex formation), in that the availability of back-hybridized structures in all rounds after the first is not an error. As a result, while mis-hybridization may be effectively minimized via strand-encoding strategies, difficulties due to back-hybridization will remain, regardless of the encoding method employed. For this reason, the fundamental re-engineering of the WPCR architecture is considered to be required.

The re-engineering of the WPCR architecture to support the primer-directed conversion of each implemented rule-block to double-stranded (ds) DNA, via targeted primer-directed strand displacement after each round of extension, is here proposed. The essential details of the resulting WPCR-based architecture, here referred to as *Displacement WPCR* (DWPCR), are illustrated in Fig. 2. DWPCR implements the same series of basic biotechnological operations applied in standard Whiplash PCR, adding only a protection step, which proceeds via primer annealing, extension, and strand displacement at target sequence, r_i, following each successful transition, i. Given use of an excess of DNA primer, and a DNA polymerase with high strand-displacement activity, such as the Klenow fragment of DNA polymerase I (see [10]), this protection operation is expected to result in the high efficiency conversion of the targeted rule-block to dsDNA, virtually eliminating the potential for back-hybridization in each successive round.

2 Materials and Methods

2.1 Strand Design and Characterization

The feasibility of high-efficiency DWPCR rests upon that of the *rule protect* biostep, which employs primer-directed opening of a DNA hairpin via strand displacement by DNA polymerase. Although the ability of a number of polymerases lacking 5' exonuclease ability (*e.g.* the Klenow fragment of DNA polymerase I) to displace an encountered strand during extension is well-known, the efficiency of using this process for opening a hairpin loop, representing a computational step should be explicitly established. Characterization of this process was accomplished via observation of the FRET system shown in Fig. 1, which consists of two DNA strands. Strand (a) is an 78 nucleotide (nt) template strand, which was encoded so that the 5' and 3'-most 15 bases (codewords 1 and 1', respectively) were Watson-Crick complementary, to facilitate formation of a 15 base-pair (pb) hairpin (panel (c)). To support observation of the binary folding state of this hairpin, the fluorophores FAM and tetramethylrhodamine (TAMRA) were covalently attached to the strand's 5' and 3' ends, respectively. Codewords 1-4 were

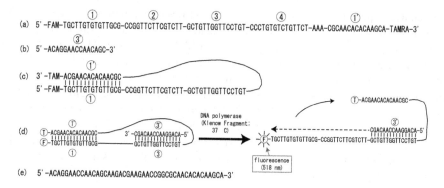

Fig. 2. Model Hairpin System for validating the *Rule Protect* operation (a) The hairpin-forming template strand, composed of catenated DNA words 1, 2, 3, 4, and 1', shown separated with intervening dashes for clarity. This strand is 5' and 3'-substituted with FAM and TAMRA fluorescent groups, respectively; (b) primer strand, complementary to template codeword 3; (c) hairpin formed by normal closure of the template strand. Here, blunt-end duplex formation places the FAM/TAMRA pair in close contact, resulting in loss of FAM fluorescence at 518 nm by FRET or contact quenching; (d) the hairpin displacement process under study, directed by hybridization of primer strand at template codeword 3, followed by primer extension. Successful primer extension, with full penetration into the template hairpin stem is accompanied by hairpin opening, with an observable increase in FAM fluorescence at 518 nm; (e) the fully-extended primer.

derived from those employed in [7], designed to avoid mis-hybridization according to the strategy in [11]. Strand (b) is a 15 nt primer strand (sequence 3') Watson-Crick reverse complementary to bases 46-60 (codeword 3) of the template strand. DNA strands (a,b) and strand (c) were purchased from Nippon EGT and Sigma-Genosys, respectively (strand (c) was kindly gifted from M. Yamamura, Tokyo Institute of Technology).

2.2 System Operation

The essential idea of the hairpin test system is shown in Fig. 2. In the absence of primer and DNA polymerase, hairpin closure is assured by a reaction temperature beneath the characteristic T_m value of template strand (a). Upon excitation of the FAM fluorophore at 494 nm, the typical FAM fluorescence at 518 nm will be inhibited by FRET or contact quenching [12], due to the close proximity of TAMRA, provided by the blunt-end duplex design (d; left). Followed by the addition of the primer strand, and DNA polymerase (Klenow fragment of DNA polymerase I), at 37 °C and under appropriate buffer conditions, successful primer extension (d; right) resulting in the hairpin displacement characteristic of DWPCR will be directly observable by the fluorescence of the excited fluorophore (FAM). On the other hand, failure of the hairpin displacement operation will be observable by the lack of a change in fluorescence, due to the maintenance of FRET and contact quenching. Panel (e) illustrates the fully-extended primer, which was also commercially synthesized separately, for comparison purposes.

A preliminary analysis was performed to ensure that the melting temperatures (T'_ms) of the template hairpin (c) and primer-template pair (d; right) were substantially above the planned experimental temperature of 37°C, at reaction conditions of interest. The thermal stability of the template hairpin was characterized via a standard DNA melting curve obtained by observing the fluorescence of SYBR Green® I (Molecular Probes) using an Real-time PCR machine OPTICON2 (Bio-rad), yielding a hairpin T_m of 78.2°C, at an ionic strength of 0.165 M (data not shown). To avoid difficulties with multiple curve resolution, the thermal stability of the primer-target pair was investigated via simulation using the software package, NucleicPark [13] which employs the parameters in [14], with an enhanced statistical zipper model. Simulations yielded a primer-template T_m of 52.3°C, and a transition width of $\Delta T_m = 6.8$°C at an ionic strength of 0.165 M and [template] = [primer] = 0.1 μM. These results were taken as sufficient to indicate proper formation of the expected DNA helices, at reaction conditions of interest.

2.3   Hairpin Displacement Experiments

Hairpin-displacement via primer extension by DNA polymerase was performed and ob- served, using a fluorescence spectrophotometer F-2500 (Hitachi) with a compact low- temperature thermostats RC6CP (Lauda), upon excitation at the characteristic FAM absorbance frequency (494 nm). A polyacrylamide gel electrophoresis (PAGE) exper- iment was also performed to visualize the variation in the size of the primer-mediated strand-displaced products. A set of wavelength fluorescence scans on polymerization was first performed to investigate the basic feasibility of primer-directed strand dis- placement and hairpin opening. A 0.1 μM final concentration of the template hairpin was prepared in a 300 μl reaction buffer consisting of 1X Klenow buffer (50mM NaCl, 10mM Tris-HCl, 10mM MgCl$_2$, 1mM DTT), 2 mM dNTPs and 0.1 μM strand displace- ment primer. This mixture was incubated at 37°C, and subsequently, 5 units of DNA Polymerase I Klenow Fragment (New England BioLabs) was added, as appropriate. The fluorescence spectrophotometer excitation and emission slit widths were each set to 5 nm. The fluorescence of the following samples were tested: blank (B; buffer only); a template-only sample (T; strand (a) and buffer (no primer)); a template-primer hy- bridized sample (T+P; strands (a), (b), and buffer); and a template/primer/polymerase sample (T+P+K; strands (a), (b), Klenow Fragment, and buffer). Samples other than strand displaced template-primer samples were used as controls. The applied scan range was from 450 nm to 650 nm, and the scan speed was set to 3000 nm/min. For the T+P+K run, the spectral scan measurement was repeated at various time points, beginning from 5 sec after adding Klenow Fragment, to 10 min thereafter.

Secondly, a time-scale fluorescence scan on polymerization was performed to inves- tigate the rapidity of the formation of extended product. The change in the fluorescence of the excited fluorophore (FAM) accompanying hairpin opening was measured at the emission wavelength of 518 nm, in real time. The T+P mixture without Klenow Frag- ment was first monitored at 37 °C to measure the baseline of the fluorescence for the first 90 sec. The time scan was then continued for around 1 hour, after which 5 units of Klenow DNA polymerase was added. To investigate the impact of target saturation with primer strand, this experiment was repeated with a ten-fold saturation of primer strand (*i.e.*, a total primer concentration of 1.0 μM).

A third polymerization experiment was performed to independently confirm the pro- duction of extended DNA product via PAGE, as follows. Polymerization was performed

in 50 μl T+P+K mixture under incubation at 37 ^{o}C, and then halted after various incubation time periods (5, 10, 20, 30, 60, and 600 sec), by adding the equal volume of
Phenol:Chloroform:Isoamyl alcohol (25:24:1) (Nacalai Tesque) to the sample. Samples
were then run on a 12% (w/v) polyacrylamide gel.

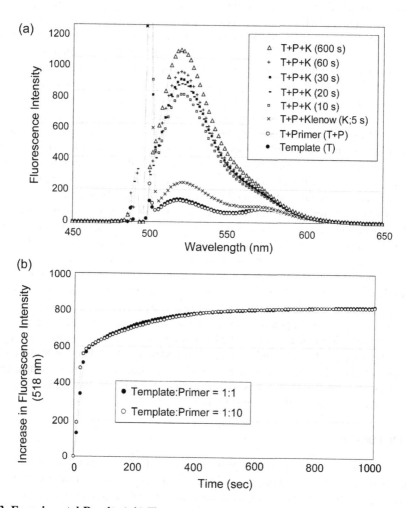

Fig. 3. Experimental Results (a,b) Fluorescence wavelength and time scans from the first and
second polymerization experiments, respectively. See text for discussion.

3   Results

Fig. 3(a) illustrates the results of the first polymerization experiment, and the resulting
set of fluorescence wavelength scans. The fluorescence peak at 518 nm, obtained in all
curves is characteristic of FAM emission, stimulated via excitation at 494 nm, with the
accompanying spike at 494 nm indicative of scattering of the incident excitation beam.

The smaller peak at 580 nm is due to TAMRA (via FRET transfer from FAM, and re-emission). For runs without DNA polymerase (runs T and T+P), equally-low fluorescence values were obtained at 518 nm, indicating FRET or contact quenching, resulting from the closure of hairpin (c) in Fig 2. However, following addition of Klenow fragment, even minimal polymerization time is accompanied by a sizable increase in the fluorescence signal at 518 nm (5 sec; yellow curve). Additional polymerization time is accompanied by successively higher peaks, which rapidly reach saturation, by 600 sec. This result is consistent with the rapid abolition of FRET/TAMRA contact quenching, indicative of successful primer extension, strand displacement, and hairpin opening.

Fig. 3(b) illustrates the results of the second polymerization experiment, in terms of the fluorescence increase of the excited fluorophore (FAM; 518 nm) vs. time. As in the above experiment, the accompanying fluorescence time scan indicates that addition of Klenow fragment is accompanied by a rapid increase in fluorescence at 518 nm, reaching saturation within 600 seconds, for both equimolar and a ten-fold excess of primer:template. This result is again consistent with the rapid abolition of FRET/TAMRA contact quenching, indicative of successful primer extension, strand displacement, and hairpin opening, with essential completion within 600 seconds.

Fig. 4 illustrates the PAGE gel resulting from the third polymerization experiment. The band with high mobility in the control run (lane 7; Template+Primer only; bottom arrow), which runs just under 80 bps, as judged via comparison with the 20 bp ladder (lane 8) is indicative of unpolymerized DNA product (i.e., T+P). On the other hand, lanes 1 (immediate inactivation of Klenow primer) through 6 (600 sec of polymerization by Klenow primer) indicate the essentially immediate disappearance of the band with high mobility beneath 80 bps (difficult to see after only 5 sec of polymerization (lane 1)), and the rapid appearance of a band with lower mobility, running at 80 bps (top arrow), which is consistent with the increasing accumulation of polymerized DNA product (lanes 1-6). To independently verify the identity of the band at 80 bps as the fully-extended product of polymerase extension, the mobility of the hybridized template and fully-extended primer was also investigated separately, and found to run at 80 bps (top arrow), as required (data not shown).

4 Discussion

The presented experimental results provide an unambiguous validation of the fundamental feasibility of targeted primer-directed hairpin displacement. Both the FRET and PAGE experiments indicate the rapid attainment of a high degree of completion of ds-DNA product, corresponding to the hybridized tempate/extended-primer pair. In DW-PCR, this basic operation, employed as the *Rule Protect* biostep, not only achieves hairpin displacement, but also abolishes back-hybridization by converting the most recently executed rule-block of all strands in an encoded mixture to double-stranded form, forcibly preventing hairpin formation within this region. As the template/extended-primer duplex which forms the protected rule-block will clearly remain stable during the isothermal operation characteristic of DWPCR (thermal cycling has been eliminated) the experimental validity of DWPCR as a high-efficiency method for implementing massively parallel autonomous computation is considered to be firmly established, as well.

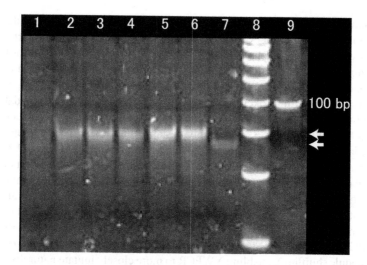

Fig. 4. Experimental Results PAGE gel from the third polymerization experiment. Lane 1-6: T+P+K, with polymerization inactivated just after (5 sec) and following addition of Klenow fragment at 10, 20, 30, 60, and 600 sec, respectively. Lane 7: Control (T+P only; no Klenow). Lane 8: 20 bp ladder. Lane 9: 100 bp DNA ladder. See text for discussion.

Given that the per-transition efficiencies estimated for WPCR and PWPCR are very low (roughly 10^{-5} and 10^{-2} successful transitions per polymerase-hairpin encounter), the per-transition efficiency approaching unity achieved by DWPCR for successfully protected rule-blocks represents a fundamental advance. Completion appears to be rapid and substantial. In contrast with WPCR, which exhibits low reaction efficiency, and requires a high temperature ofr operation (*e.g.*, 80 °C), DWPCR could proceed with an efficiency of nearly unity under physiological conditions (37 °C), and thus open the door for practical biological applications. Further investigation regarding a determination of the absolute concentration of fully-extended product is our concern. Issues related to a quantitative determination of efficiency are deferred to later work.

Implementation of the *rule protect* biostep to achieve high-efficiency DWPCR-based massive parallelism requires a method for preparing a combinatorial mixture of encoded DWPCR strands. Although in principle, various methods could be employed, the 5' to 3' ordering of transition rules provided by the combinatorial assembly method presented for PWPCR in [5] provides a number of clear advantages. First of all, this ordering ensures the round-by-round shortening of the hairpin loops required for successive state transitions on all strands. In this case, computational efficiency for each strand should increase as the computation progresses. Secondly, the accompanying segregation of successively-generated, protected rule-blocks towards each strand's 5' end avoids potential difficulties due to loop stiffness due to duplex islands within the loops used for hairpin computation. Finally, as noted in [5], this ordering supports implementation of a properly coordinated cross-over operation, to generate valid daughter strands in WPCR-based processes evolutionary computation (*i.e.,*, EWPCR and XWPCR. Accordingly, a 5' to 3' transition rule ordering is adopted by the DWPCR architecture.

A striking secondary advantage of DWPCR over PWPCR regards its fundamentally superior ability to support WPCR-based evolutionary protein design. In particular, the polymerization of target sequences for PNA binding between state codeword pairs in the growing 3' tail compromises PWPCR's utility for evolutionary protein design by pseudo-module shuffling (XWPCR) [5]. As RNA-protein fusion-based expression of the 3' tail as a single, unbroken gene is required by XWPCR, the unavoidable presence of glycine loops resulting from translation of transcribed target sequences, left over from the PWPCR process is strongly destabilizing to shuffled polypeptides (see [5] for a discussion). In contrast, target sequences for primer extension in DWPCR are within the rule-region rather than the growing 3' tail, so that destabilization due to the translation of targeting artifacts from the computational process will be completely absent, which itself constitutes a major advance. Furthermore, the architectural freedom gained by the absence of these transcribed glycine coils, which necessitates systematic placement at pseudo-module boundries in PWPCR-based XWPCR, will also allow DWPCR-based XWPCR to support exon shuffling (i.e., protein module rather than pseudo-module shuffling), enabling XWPCR to more closely imitate naturally-evolving systems. Thus, the ability to employ DWPCR, rather than PWPCR as the underlying state transition method represents a significant advance for XWPCR.

The fundamental primer-directed hairpin displacement operation should also be useful for non-WPCR DNA computing applications. For instance, this operation might be employed, rather than simple primer displacement, to direct state transitions in alternative hairpin-based architectures [15]. Beyond directing state transitions, this operation is also useful for parallel-processing of single-stranded DNA molecules during and after computation, such as in directing the conversion to dsDNA form following completed computation, required for gene expression in XWPCR [5]. The current fundamental operation is thus taken to have significance beyond simple WPCR-based applications.

Acknowledgements

Financial support generously provided by Grants-in-Aid for Scientific Research (Kiban B, 15300100 and 18300100), from the Japan Society for the Promotion of Science (JSPS) and by JST-CREST. The authors are also grateful to M. Yamamura of the Tokyo Institute of Technology for the kind gift of primer strand (c).

References

1. Hagiya, M., Arita, M., Kiga, D., Sakamoto, K., Yokoyama, S.: Towards parallel evaluation and learning of boolean μ-formulas with molecules. In Rubin, H., Wood, D., eds.: DNA Based Computers III. (2000) 57–72
2. Sakamoto, K., Kiga, D., Komiya, K., Gouzu, H., Yokoyama, S., Ikeda, S., Sugiyama, H., Hagiya, M.: State transitions by molecules. Biosystems **52** (1999) 81–91
3. Rose, J.A., Hagiya, M., Deaton, R.J., , Suyama, A.: A DNA-based in vitro genetic program. J. Biol. Phys. **28** (2002) 493–498
4. Wood, D., Bi, H., Kimbrough, S., Wu, D.J., Chen, J.: DNA starts to learn poker. In: DNA Computing. 7th Int'l Workshop on DNA-Based Computers. (2002) 22–32

5. Rose, J.A., Takano, M., Hagiya, M., Suyama, A.: A DNA computing-based genetic program for in vitro protein evolution via constrained pseudomodule shuffling. Journal of Genetic Programming and Evolvable Machines **4** (2003)

6. Komiya, K., Sakamoto, K., Gouzu, H., Yokohama, S., Arita, M., Nishikawa, A., Hagiya, M.: Successive state transitions with I/O interface by molecules. In Condon, A., Rozenberg, G., eds.: DNA Computing. 6th Int'l Workshop on DNA-Based Computers. Volume 2054., Springer-Verlag, Berlin (2001) 17–26

7. Komiya, K., Sakamoto, K., Kameda, A., Yamamoto, M., Ohuchi, A., Kiga, D., Yokoyama, S., Hagiya, M.: DNA polymerase programmed with a hairpin DNA incorporates a multiple-instruction architecture into molecular computing. Biosystems **83** (2006) 18–25

8. Rose, J.A., Deaton, R.J., Hagiya, M., Suyama, A.: Equilibrium analysis of the efficiency of an autonomous molecular computer. Phys. Rev. E **65** (2002) 1–13 Article 021910.

9. Komiya, K., Yaegashi, S., Suyama, A., Hagiya, M., Rose, J.A.: Experimental validation of the statistical thermodynamic model for prediction of the behavior of autonomous molecular computers based on DNA hairpin formation. In: DNA Computing. 12th Int'l Workshop on DNA-Based Computers. (2006) in press

10. New England Biolabs: Klenow Fragment ($3L \rightarrow 5L$ exo-). (2004) Technical Bulletin M0212.

11. Arita, M., Nishikawa, A., Hagiya, M., Komiya, K., Sakamoto, K., Gouzu, H., Yokoyama, S.: Improving sequence design for DNA computing. In: Proc. $5^{t}h$ Genetic and Evolutionary Computation Conference (GECCO99), Las Vegas (2000) 875–82

12. Marras, S., Kramer, F., Tyagi, S.: Efficiencies of FRET and and contact-mediated quenching in oligonucleotide probes. Nucl. Acids. Res. **30** (2002) e122

13. Rose, J.A., Deaton, R.J., Suyama, A.: Statistical thermodynamic analysis and design of oligonucleotide based computers. Natural Computing **3** (2004)

14. SantaLucia, Jr., J., Hicks, D.: The thermodynamics of DNA structural motifs. Annu. Rev. Biophy. Biomolec. Struct. **33** (2004) 415–40

15. Kubota, M., Ohtake, K., Komiya, K., Sakamoto, K., Hagiya, M.: Branching DNA machines based on transitions of hairpin structures. In: Proc. Congr. Evol. Comp. (CEC'03). (2003) 2542–2548

MethyLogic: Implementation of Boolean Logic Using DNA Methylation

Nevenka Dimitrova[1] and Susannah Gal[1,2]

[1] Philips Research, 345 Scarborough Rd.
Briarcliff Manor, NY 10510, USA
nevenka.dimitrova@philips.com
[2] Department of Biological Sciences
Binghamton University
Binghamton, NY 13902, USA
sgal@binghamton.edu

Abstract. The MethyLogic method performs flexible and reversible modification of DNA in order to establish the logical value of true or false for a set of clauses. It combines both the biological meaning and experimental procedure with the logical implementation of the basic Boolean operators: OR, AND, and NOT. The original feature of methylation logic, MethyLogic, is the use of the reversibility of DNA methylation of cytosine and adenine. Logic variables can be negated by reversing the DNA methylation status. We introduce four implementation scenarios: three of them use methyl-sensitive restriction enzymes and the fourth uses methyl-binding proteins. Encoding can use either single or double-stranded DNA. In addition, we show how to solve a three variable SAT problem and how to implement a logic circuit.

1 Introduction

DNA Methylation refers to adding a methyl group, CH_3 to the fifth carbon on cytosine or to the sixth carbon of adenine. DNA Methylation is a mechanism known both in animals and plants as an important means for gene expression regulation [1] [2] . In bacteria, it acts as a protection mechanism from attack by foreign DNA [3]. As a biological process, DNA methylation is reversible. DNA methyltransferases catalyze the transfer of a methyl group from S-adenosyl-L-methionine to cytosine or adenine bases in DNA whereas native DNA polymerases do not copy the methylation status when they replicate DNA.

Assays that are experimental tools for the analysis of developmental biology and cancer are mainly used for finding the epigenetic or methylation state of candidate genes and their involvement in a certain biological process [4]. In our method, we propose a conceptual framework of mathematical logic that will allow for more complex operations on the DNA in the process of making diagnostic decisions. As opposed to other methods that are used in DNA computing, our method allows writing and re-writing for DNA computations. Currently aqueous computing suffers from the limitation due to the inability to rewrite or overwrite on existing DNA. Once a piece of DNA has been "processed" it is no longer reusable. The existing methods are not

C. Mao and T. Yokomori (Eds.): DNA12, LNCS 4287, pp. 404 – 417, 2006.

reversible. This new approach would allow reversible methylation of the DNA sequences to change the truth value of the encoded variables.

Molecular computing models have approached different NP-complete problems since Adleman's historic experiment in 1994 [5-7]. Lederman et al. developed an array of seven deoxyribozyme-based molecular logic gates that behaves as a full adder in a single solution [8, 9]. Liu et al. used a surface chemistry based DNA computer to solve a four-variable four-clause SAT problem [10]. Recently Su et al. have implemented a DNA computer capable of simulating Boolean logic circuits. They constructed NOR and OR gates and combined them into a simple logic circuit [11]. Head et al. proposed a novel way for recording information on DNA molecules while they are dissolved in water [12]. The resulting solution of information containing molecules is considered to constitute a "fluid memory." This group also introduced schemes for reading information from these molecules. A simple instance of the Satisfiability Problem of a set of Boolean Clauses was proposed [13]. Given a set of Boolean clauses the problem is to find truth values for which all of the clauses are satisfied, "true". A procedure for solving this SAT problem illustrates the DNA computing method called 'Aqueous Computing' [14]. Benenson et al. implemented an automaton where the computation is performed by a reversible software molecule with input molecule hybridization followed by an irreversible software-directed cleavage of the input molecule [15].

The original feature of methylation logic, MethyLogic, is the use of the reversibility of DNA methylation of cytosine and/or adenine. Here we use DNA methylation of cytosine as an example. The advantage of this approach is that it can be implemented with the use of methylation sensitive restriction enzymes or with methyl-binding proteins. The DNA sequences that encode the "true" and "false" values of a particular logic variable do not have to be encoded with different sequences. Instead, the negation of a variable is encoded with the opposite state: if a variable has value of "true" and is encoded with a methylated sequence, then the negation of this variable is encoded with the same DNA sequence but unmethylated.

There are two types of nucleotides that have the methylation mark: adenine and cytosine, and the methylation logic implementations that are based on either adenine or cytosine are called A- implementation and C- implementation, respectively. In this paper we will focus on C-implementation. This is distinct from the approach taken previously by Gal et al. where unidirectional methylation of specific restriction enzyme sites was used to implement a specific SAT problem [16]. Methylation as a computational strategy for aqueous computing was previously introduced by Head [17].

The rest of this paper is organized as follows. Section 2 gives an overview of the biochemical tools available for implementing DNA methylation logic. Section 3 gives an introduction into various implementation scenarios, encoding schemes and implementation of basic logical operators: AND, OR and NOT. Section 4 gives two example experimental plans. Section 5 concludes and gives directions for future work.

2 Biochemical Tools for MethyLogic

There are a large number of commercially available tools that we can use in our MethyLogic approach to computation. One can prepare methylated DNAs of specific sequence simply by ordering oligonucleotides and requesting specific nucleotides as methyl-cytosine (Integrated DNA Technologies, http://www.idtdna.com). There are well studied methyl transferases, methyl binding proteins and methyl-specific restriction enzymes that can be used in the implementation of MethyLogic.

There are a variety of enzymes that can methylate DNA at specific 4-6 base pair recognition sites [18] [19]. Over 13 different DNA methyl transferases are commercially available at this time (from New England Biolabs [http://www.neb.com] or from Takara Mirus Bio [http://www.takaramirusbio.com/]). The human Dnmt1 enzyme methylates the cytosine in the C-G context, but only if one strand is already methylated, so called hemi-methylated DNA, to make it fully methylated on both strands [20] .

To physically separate methylated from unmethylated DNA, one can use methylated DNA binding proteins. Several different ones are known: Kaiso, MBD1, MBD2, MBD3, MBD4, and MeCP, some of which are sequence specific [21] [22]. There are also antibodies to methylated cytosine residues that are available (http://www.avivasysbio.com/).

One can also carry out specific biochemical reactions to distinguish methylated from unmethylated DNA. Bisulfite treatment modifies unmethylated cytosines and converts them to uridine residues [23] [24]. Methylated cytosines are unmodified. This conversion of a cytosine to a uridine creates a single base mismatch between a uridine on one strand and a guanine on the other. A few endonucleases are available that can cleave this structure specifically [25, 26]. Alternatively one can use DNA sequencing, oligonucleotide hybridization or PCR to distinguish the different levels of the two types of sequences as has previously been done [27, 28]. Recently a specific DNA endonuclease, McrBC has been isolated that only cuts hemi-methylated or methylated DNA [29]. This enzyme has been used to screen for methylated DNA sequences in human DNA   [30]. There are also sequence specific DNA cleavage enzymes, restriction endonucleases that either can cleave methylated DNA or only unmethylated DNA (for example MspI and HpaII). When a pair of these enzymes are used with the same sequence specificity, one that can cut methylated DNA and the other that can not, comparison of the cleavage status in each reaction can indicate whether a specific DNA is methylated or not even in a complicated mixture such as the human genome [13].

All of these well-studied tools are available to use for the analysis of logic algorithms using DNA methylation as described next.

3   Boolean Logic Implemented with DNA Methylation

We introduce Boolean logic using DNA methylation. Since DNA methylation is a reversible process, it allows for an abstract framework and several physical implementations. The goal is to have an implementation procedure that gives more freedom on the choice of the DNA sequences. DNA methylation is important because the write-erase steps can be implemented as methylate-unmethylate in solution. We define a "methylation logic" that allows for using differently encoded strings. The requirement is that encoded logical variables contain at least one cytosine – for the C-implementation or at least one adenine for the A- implementation.

One of the DNA methylation states is taken as true while the other methylation state is taken as false. For example, we can consider methylation of cytosine to be equivalent to "True".

Encoding: Logic variables can be encoded using single or double stranded DNA.

Write: Writing corresponds to applying DNA methylation in vitro or in vivo. In vitro this corresponds to applying one of the methyl-transferase enzymes listed above. In vivo, there is a maintenance methyltransferase DNMT1 which methylates C within a CpG dinucleotide only if one of the strands is already methylated and de novo methyltransferases DNMT3a and DNMT3b that methylate all the CpG dinucleotides.

Erase: Erasing corresponds to any procedure listed in section 2 that removes DNA methylation mark in vitro or in vivo. Copying of DNA using a DNA polymerase like in PCR is one way of losing the methylation modification on a residue.

Destroy: This is any procedure that destroys either unmethylated or methylated DNA. This refers to applying enzymes that digest specifically methylated or unmethylated DNA.

Separate: Methylated DNA binding proteins can separate strands of DNA that have methylated nucleotides from those without any methyl groups attached

Read: The readout procedure can distinguish if a piece of DNA is fully, hemi, partially methylated or completely unmethylated. Methylation sensitive restriction enzymes, bisulfite treatment and PCR can be used for this purpose.

In any computation process, there is always duality between encoding/reading procedures and computation procedures. A computation procedure can be implemented using various physical and chemical processes, and the reading procedure can be used to interpret the result – again acting as a computation step. Here we propose four different implementation scenarios, the first three can be implemented using methyl-sensitive restriction enzymes, whereas the fourth case uses methyl-binding proteins. We describe implementation of AND and OR logical operators. Implementation of NOT is by reversing the methylation status of the input sequence (variable). This could be done with the "write" and "erase" processes mentioned above.

Implementation case 1:

- Encoding: Sequences are encoded with single-stranded DNA, the "logical operators" are evaluated after allowing sequences to hybridize
- AND: both strands and all Cs should be methylated to have a truth value of "True" else the truth value is "False";
- OR: hemi-methylated or fully methylated DNA are treated as "True" whereas unmethylated DNA is treated as "False."

Implementation case 2:

- Encoding: Sequences are encoded as double-stranded DNA, the operation is the same for AND and OR, but the readout is interpreted differently based on the intended operator:
- AND: requires the entire length of the sequence to be methylated to have "True", else it is "False". Of course both strands have to continue to be methylated.
- OR: requires any region to be methylated to have "True" else it is "False."

Implementation case 3: Combined implementation of 1 and 2 where single stranded DNA represents logical variables, and ligating double-stranded DNA is used to implement complex logical expressions.

Implementation case 4:

- *Encoding*: Logic variables are encoded as single or double-stranded DNA. Using methyl binding proteins including methyl specific antibodies, we can separate double-stranded DNA into a "bound" fraction (having methylated DNA) and an "unbound" fraction (having only unmethylated DNA). Encoded sequences are allowed to hybridize and then methyl-binding proteins are used to fish out any DNA sequence that is methylated. Using PCR, we can distinguish in a sensitive and sequence specific manner whether sequences are in the bound or unbound fraction or both. With not very complicated mixtures, we might be able to just see the DNA on a gel, but if we want to implement logical variables that involve representations from the human genome, we need to use PCR to see in which fraction a given sequence is present.
- AND: if the DNA sequences are both in the "bound" fraction, the truth value is "True". Otherwise, it evaluates to "False".
- OR: if either DNA sequence is in the "bound" fraction, means that at least some of the DNA sequence is methylated, so it evaluates to "True". If both DNA sequences are in the unbound fraction, means that the DNA sequence is unmethylated and it evaluates to False.

3.1 Methylation Logic Using Single-Stranded DNA (ssDNA)

Logical Operator AND using ssDNA

Table 1 shows the Boolean logic and methylation logic equivalent for the logical operator AND. The logical variables are encoded as single-stranded DNAs that are converted to double-stranded DNA by hybridizing the strands. In this case, A and B are two single-stranded DNAs that are hybridized. The truth value of the hybridized product is "True" if and only if the double-stranded DNA is methylated on both strands.

There are various implementation considerations to be mentioned here. Implementation of AND will need an experimental procedure that checks for full methylation. One way to do that would be to apply digestion by the enzyme HpaII to maintain intact only completely methylated DNA. This restriction enzyme is sensitive to methylation and so can not cut methylated DNA. It remains to be tested whether it can cut hemi-methylated DNA. If that is the case, to distinguish hemi-methylated from methylated DNA, we would apply the bisulfite treatment first, then use enzymes that cut at a mismatch. The bisulfite treatment would convert an unmethylated-C to a U and make a mis-paired base with the G on the opposite strand. Those mis-paired bases can then be cut with the specific enzymes recognizing the mismatch. This protocol should give us only intact fully methylated DNA.

Table 1. Methylation logic table for AND operator using ssDNA

Boolean Logic			MethyLogic			Realization
A	B	A AND B	A	B	A AND B	
T	T	T	M	M	M	fully methylated on both strands
T	⊥	⊥	M	U	U	unmethylated at least on one strand (hemi-methylated DNA)
⊥	T	⊥	U	M	U	unmethylated at least on one strand (hemi-methylated DNA)
⊥	⊥	⊥	U	U	U	fully unmethylated on both strands

Logical Operator OR using ssDNA

Table 2. shows the Boolean logic and methylation logic equivalent for the logical operator OR. The logical variables are encoded as single-stranded DNA then converted to double-stranded DNA using hybridization. The truth value of the hybridized product is equal to "True" if the double stranded DNA is methylated on at least one strand. Implementation of OR uses the experimental procedure to check if a sequence is hemi- or fully methylated. We can apply the McrBC enzyme to cut all methylated or hemi-methylated sequences. This will keep intact only the unmethylated sequences. Alternatively, we can use the methyl binding proteins, as mentioned in implementation scenario 4, to fish out anything that has methylation. The unmethylated DNA would be in the unbound fraction.

Table 2. Methylation logic table for OR operator using ssDNA

Boolean Logic			MethyLogic			Realization
A	B	A OR B	A	B	A OR B	
T	T	T	M	M	M	fully methylated on both strands
T	⊥	T	M	U	M	unmethylated at least on one strand (hemi-methylated)
⊥	T	T	U	M	M	unmethylated at least on one strand (hemi-methylated)
⊥	⊥	⊥	U	U	U	fully unmethylated on both strands

As in the case of AND, A and B are two single-stranded DNAs that are hybridized.

Logical Operator NOT using ssDNA

Table 3. shows the Boolean logic and methylation logic equivalent for the logical operator NOT. A logical variable is encoded as single-stranded DNA. The truth value

Table 3. Methylation logic table for the NOT operator using ssDNA

Boolean Logic		MethyLogic		Realization
A	NOT A	A	NOT A	
T	⊥	M	U	unmethylated ssDNA
⊥	T	U	M	methylated ssDNA

is reversed by using PCR if the sequence is methylated because during PCR the methylation mark gets lost. Changing the truth value from false to true is equivalent to applying a DNA methyltransferase that sets the methylation mark.

3.2 Methylation Logic Using Double-Stranded DNA (dsDNA)

Logical Operator AND using dsDNA

The logical variables are encoded as double-stranded DNAs that are ligated. Table 4. shows the Boolean logic and methylation logic equivalent for the logical operator AND. The truth value of the ligated product is "True" if and only if the whole DNA sequence is methylated.

Table 4. Methylation logic table for AND operator using dsDNA

Boolean Logic			MethyLogic			Realization
A	B	A AND B	A	B	A AND B	
T	T	T	M	M	M	fully methylated sequence
T	⊥	⊥	M	U	U	at least one C or A unmethylated
⊥	T	⊥	U	M	U	at least one C or A unmethylated
⊥	⊥	⊥	U	U	U	fully unmethylated sequence

Implementation of AND will need an experimental procedure that checks for full methylation and detects if even a single C within a sequence is unmethylated. The bisulfite treatment converts an unmethylated-C to a U and makes a mis-paired base with the G on the opposite strand. Those mis-paired bases can then be cut with the specific enzymes recognizing the mismatch. This protocol should give us only intact fully methylated DNA.

Logical Operator OR using dsDNA

Table 5. shows the Boolean logic and methylation logic equivalent for the logical operator OR using dsDNA. As in the case of AND, A and B are double-stranded DNAs that are ligated or where A and B are two different subsequences on a longer double-stranded DNA sequence. The truth value of the ligated product is equal to "True" if the double stranded DNA is methylated at least partially.

Table 5. Methylation logic table for OR operator using dsDNA

Boolean Logic			MethyLogic			Realization
A	B	A OR B	A	B	A OR B	
T	T	T	M	M	M	fully methylated
T	⊥	T	M	U	M	partially methylated
⊥	T	T	U	M	M	partially methylated
⊥	⊥	⊥	U	U	U	fully unmethylated

Implementation of OR uses the experimental procedure to check if a sequence is partially methylated. Bisulfite treatment followed by DNA sequencing is a method that can check for methylation of single sites and therefore could be used for this purpose. Alternatively, we can use the methyl binding proteins including methyl C-specific antibodies, as mentioned in implementation scenario 4, to fish out anything that is methylated. The unmethylated DNA would be in the unbound fraction.

Logical Operator NOT using dsDNA

A logical variable is encoded as double-stranded DNA. Table 6. shows the Boolean logic and methylation logic equivalent for the logical operator NOT. The truth value is reversed by using PCR if the sequence is methylated because during PCR the methylation mark gets lost. Changing unmethylated to methylated DNA can be done by applying a DNA methyltransferase.

Table 6. Methylation logic table for the NOT operator using dsDNA

Boolean Logic		MethyLogic		Realization
A	NOT A	A	NOT A	
T	⊥	M	U	unmethylated dsDNA or unmethylated at a particular position
⊥	T	U	M	methylated dsDNA

4 Two Examples of an Experimental Plan for MethyLogic

Example 1: Let p, q, r be Boolean variables and let p', q', r', be their respective negations. Does there exist an assignment of truth values (T / F) to the variables p, q and r for which each of the four clauses p OR q, p' OR q OR r', q' OR r', p' OR r has the value true? The same example was used by Head et al. and Gal and Head to demonstrate aqueous computing [12, 16]. In this example, we use the variables encoded as three distinct double-stranded DNAs and ligate them together as necessary. Here we equate the methylated p (Mp) site with p, the unmethylated p site (Up) with p', Mq with q, Uq with q' and Mr with r and Ur with r'. We would create all 8 possible combinations of these variables using ligation of the methylated and unmethylated double-stranded elements (MpMqMr, MpUqMr, MpMqUr, MpUqUr, etc.). There would be a vast amount of these molecules available. The elements will be defined so that there exists a binding protein that can specifically bind to the methylated form of each element independently. For these clauses, to save the p' form, we will apply the DNA to the methyl-binding protein specific for the p site and save the DNA that does not bind (saves only the unmethylated or Up form). To save the p or methylated form, we will apply the DNA to the same protein, but save the bound DNA. Details for each clause are given below. We visualize the first two computational steps in Figures 1 and 2. The rest of the steps are carried out in a similar fashion.

Step 1: Compute p OR q. Figure 1 describes this process.

    1.1 Separate the vast mixture of the starting DNAs into two pots.

    1.2 In one, apply the mixture to the methyl-binding protein specific for the p site and in the other pot, apply the mixture to the methyl-binding protein specific for the q site.

    1.3 In both cases, save the bound material, those that contain either Mp or Mq (p OR q). This sample would contain p OR q OR r and p OR q OR r'.

    1.4 Recombine these two bound samples. This mixture now contains six different double-stranded DNAs: MpUqMr, MpUqUr, UpMqMr, UpMqUr, MpMqMr, and MpMqUr.

Step 2: Compute p' OR q OR r'. Fig. 2. describes this computational step.

    2.1 Separate the mixture of the DNA from the last step into three pots.

    2.2 In one, apply the mixture to the methyl-binding protein specific for the p site and save the unbound material. In another pot, apply the mixture to the methyl-binding protein specific for the q site and save the bound material. In the third pot, apply the mixture to the methyl-binding protein specific for the r site and save the unbound material.

    2.3 Recombine the three saved samples. This sample now contains MpUqUr, UpMqMr, UpMqUr, MpMqMr and MpMqUr.

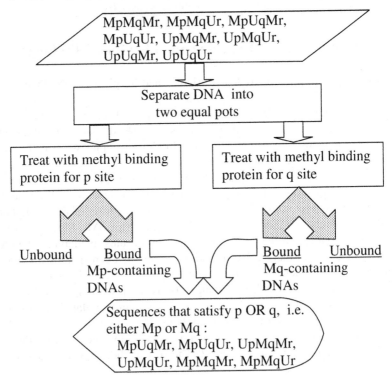

Fig. 1. Steps in computing p OR q

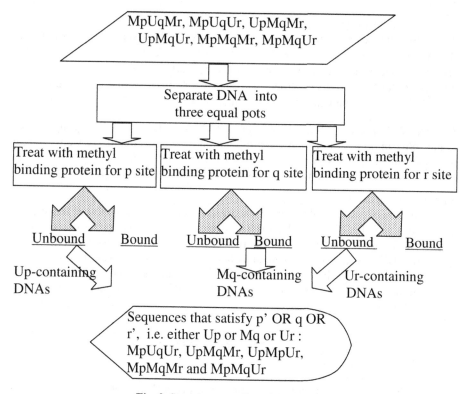

Fig. 2. Steps in computing p' OR q OR r'

Step 3: Compute q' OR r'.
  3.1 Separate the mixture of DNA from the last step into two pots.
  3.2 In one, apply the mixture to the methyl-binding protein specific for the q site and save the unbound material. In the other pot, apply the mixture to the methyl-binding protein specific for the r site and save the unbound material.
  3.3 Recombine the two saved samples. This sample now contains MpUqUr, UpMqUr and MpMqUr.

Step 4: Compute p' OR r
  4.1 Separate the mixture of DNA from the last step into two pots.
  4.2 In one, apply the mixture to the methyl-binding protein specific for the p site and save the unbound material. In the other pot, apply the mixture to the methyl-binding protein specific for the r site and save the bound material.
  5.1 Recombine the two saved samples. This sample should only contain UpMqUr from the bound material from the methyl-p site binding protein. The bound material from the methyl-r binding protein will yield no DNA as all the molecules from the previous step contain Ur.

Step 5: Read the answer

Apply bisulfite treatment to the material and sequence the resulting DNA fragments. Bisulfite treatment converts unmethylated Cs to Us while it has no effect on methylated Cs. Where the sequence is the same as the starting material, that site was methylated in the final product. Where the sequence is different and a U is substituted for a C, that site was unmethylated in the final answer.

Example 2: Represent a logical formula: (a OR b) AND (c') AND d using MethyLogic. We can think of it also as a representation of a logic circuit. We want to know for which inputs (values of a, b, c and d) the logic circuit produces a "true" value. Here we use a representation of logical variables with single-stranded DNA.

Step1: Compute a OR b

We encode the variable a with a sequence and then b with another sequence in such way that they would hybridize. For example, a would be encoded with 5'-ACGCGA-3' then b encoded with 5'-AAATCG-3'. The hybridized form of this DNA would be represented as below: (in reality we want more than a 3-base overlap for better hybridization). We should also note that all sequences need to contain at least one C so it can be methylated. One can also work with methylated As if necessary.

```
a    5'-ACGCGA-3'
        ||| (hydrogen bonds between strands)
b    3'-GCTAAA-5'
```

1.1 Combine four pots containing unmethylated a (Ua), methylated a (Ma), Ub, and Mb and create four different kinds of double stranded DNA (Ua/Ub, Ua/Mb, Ma/Ub, and Ma/Mb).

1.2 Use methyl- binding proteins (e.g. MeCP or MBD1 or antibodies to methyl-C) to fish out the sequences that have hybridized and at the same time contain methylated Cs by saving the bound fraction. This corresponds to performing a OR b.

Step 2: Compute c' AND d

2.1 We encode c and d with different sequences in such a way that they would hybridize together, and as a hybrid ligate with the overhang of the a OR b hybrid (see below). For example, c could be encoded with 5'-TTTCGC-3' then d would be encoded with 5'-ATAGCG-3' such that when hybridized they form a structure as below: (in reality we want more than a 3-base overlap for better hybridization).

```
c    5'-TTTCGC-3'
        ||| (hydrogen bonds between strands)
d    3'-GCGATA-5'
```

As above, create 4 types of single-stranded DNAs, Uc, Ud, Mc and Md.

2.2 Apply NOT to c by applying PCR to the methylated c pot and a methyltransferase to the unmethylated c pot. For simple variables we can just exchange pots.

2.3 Hybridize the two results with unmethylated d and methylated d.

2.4 Apply AND operator to the four pots: bisulfite treatment followed by incubating with an enzyme that destroys mismatched DNA. The bisulfite treatment converts unmethylated Cs to U and does not affect the methylated Cs. Thus in a hybrid where an unmethylated C is hybridized with a G, following bisulfite treatment, you will have a U-G mismatch instead of the normal C:G basepair. The mismatched DNA can be destroyed using specific enzymes that cleave the mismatched double stranded DNA (see section 2).

Step 3: Compute the AND of the product from the previous two computations by combining the pots resulting from step 1 and 2 and ligate. The product of this reaction would have the DNA sequence structure as below:

```
a-c    5'-ACGCGATTTCGC-3'
                ||||||||||| (hydrogen bonds between strands)
b-d       3'-GCTAAAGCGATA-5'
```

Apply bisulfite treatment followed by the enzyme that destroys mismatched DNA. As mentioned above, the bisulfite treatment should convert unmethylated Cs to Us and therefore generate a mismatched DNA sequence where U is opposite a G. Methylated Cs will not be modified by this treatment and therefore should remain correctly basepaired with Gs on the other strand. Mismatched DNA can then be destroyed using specific enzymes. The resulting DNA should satisfy the complex clause: a OR b AND c' AND d.

Step 4: Read the answer. For this, we would divide the mixture into two pots and treat one of them with bisulfite. As mentioned above, this treatment converts unmethylated C's to U's. We would then sequence the DNA strands in each pot. We need to sequence both strands in order to find the truth values of the logical variables in the circuit. Any difference will be because of unmethylated C at that position. In our case, the state of site c should be negated when we read the answer.

5   Conclusions and Future Work

We have introduced the concept of implementing Boolean logic with DNA methylation using both single and double-stranded DNA. We proposed two experimental plans for implementation of a universal DNA computer based on "methylation logic". This approach is potentially generalizable more easily than our past approaches. The past approaches always required the presence of appropriate restriction enzymes and, recently methylases. That would naturally limit the types and numbers of sequences with which one can work. Although some of our new approaches require sequence specific methyl-binding proteins, other implementations use only the presence of methylated residues and as such this expands the kinds of sequences we could use for these approaches greatly. A natural follow-up is to test this procedure in laboratory experiments. All the materials are commercially available so we do not foresee any obstacles in carrying out the experimental procedures.

References

1. Bird, A., *DNA methylation patterns and epigenetic memory.* Genes Dev, 2002. **16**(1): p. 6-21.
2. Rollins, R.A., et al., *Large-scale structure of genomic methylation patterns.* Genome Research, 2006 16(2): p. 157-63.
3. Jeltsch, A., *Beyond Watson and Crick: DNA methylation and molecular enzymology of DNA methyltransferases.* ChemBioChem, 2002. **3**(4): p. 274-93.
4. Esteller, M., et al., *Cancer epigenetics and methylation.* Science, 2002. **297**(5588): p. 1807-8.
5. Conrad, M., *Information processing in molecular systems.* Currents in Modern Biology, 1972. **5**: p. 1-14.
6. Livstone, M.S., D.van Noort, and L.F. Landweber, *Molecular computing revisited: a Moore's Law?* Trends Biotechnol., 2003. **21**(3): p. 98-101.
7. Head, T., *One Mathematician's Tour from Biology into Computing and Back to Life.* submitted, 2006.
8. Lederman, H., et al., *Deoxyribozyme-Based Three-Input Logic Gates and Construction of a Molecular Full Adder.* Biochemistry, 2006. **45**(4): p. 1194-1199.
9. Margolin AA, S.M., *Boolean calculations made easy (for ribozymes).* Nat Biotechnol., 2005. **23**(11): p. 1374-6.
10. Liu, Q., et al., *DNA computing on surfaces.* Nature, 2000. **413**: p. 175-9.
11. Su, X. and L.M. Smith, *Demonstration of a universal surface DNA computer.* Nucleic Acids Res., 2004. **32**(10): p. 3115-23.
12. Head, T., et al. *Aqueous solutions of algorithmic problems: emphasizing knights on a 3X3.* in *DNA Computing - 7th International Workshop on DNA-Based Computers.* Jonoska, Natasa; Seeman, Nadrian C., Editors, 2002: Springer. p. 191-202.
13. Hatada, I., et al., *Genome-wide profiling of promoter methylation in human.* Oncogene, 2006, e-published Jan 9.
14. Head, T., et al., *Aqueous computing: a survey with an invitation to participate.* J. Computer Science & Technology, 2002. **17**: p. 672-681.
15. Benenson, Y., et al., *DNA molecule provides a computing machine with both data and fuel.* Proc Natl Acad Sci U S A, 2003. **100**(5): p. 2191-6.
16. Gal, S. and T. Head. *Exploring Methylation as a Tool for DNA Computing.* in *DNA11: Conference on DNA based computers.* 2005. London, Ontario.
17. Head, T., *Writing by Methylation Proposed for Aqueous Computing,* in *Where Mathematics, Computer Science, Linguistics and Biology Meet,* Carlos Martín-Vide; Victor Mitrana, Editors. 2001, Kluwer Academic Publishers. p. 353-360.
18. Marinus, M.G. and N.R. Morris, *Isolation of deoxyribonucleic acid methylase mutants of Escherichia coli K-12.* J Bacteriol, 1973. **114**(3): p. 1143-50.
19. Geier, G.E. and P. Modrich, *Recognition sequence of the dam methylase of Escherichia coli K12 and mode of cleavage of Dpn I endonuclease.* J Biol Chem., 1979. **254**(4): p. 1408-13.
20. Pradhan, S., et al., *Recombinant human DNA (cytosine-5) methyltransferase. I. Expression, purification, and comparison of de novo and maintenance methylation.* J Biol Chem, 1999. **274**(46): p. 33002-10.

21. Bowen, N.J., M.B. Palmer, and P.A. Wade, *Chromosomal regulation by MeCP2: structural and enzymatic considerations.* Cell Mol Life Sci, 2004. **61**(17): p. 2163-7.
22. Khan, R., et al., *Human methylated DNA-binding protein. Determinants of a pBR322 recognition site.* J Biol Chem, 1988. **263**(28): p. 14374-83.
23. Shapiro, R., R.E. Servis, and M. Welcher, *Reactions of uracil and cytosine derivatives with sodium bisulfite:a specific deamination method.* J. Am. Chem. Soc., 1970. **92**: p. 422–424.
24. Hayatsu, H., Y. Wataya, and K. Kai, *The addition of sodium bisulfite to uracil and to cytosine.* J. Am. Chem. Soc., 1970. **92**: p. 724–726.
25. Fuhrmann, M., et al., *Removal of mismatched bases from synthetic genes by enzymatic mismatch cleavage.* Nucleic Acids Res., 2005. **33**(6): p. e58.
26. Till, B.J., et al., *Mismatch cleavage by single-strand specific nucleases.* Nucleic Acids Res., 2004. **32**(8): p. 2632-41.
27. Kimura, N., et al., *Methylation profiles of genes utilizing newly developed CpG island methylation microarray on colorectal cancer patients.* Nucleic Acids Research, 2005. **33**(5): p. e46.
28. Fatemi, M., et al., *Footprinting of mammalian promoters: use of a CpG DNA methyltransferase revealing nucleosome positions at a single molecule level.* Nucleic Acids Research, 2005. **33**(20): p. e176.
29. Sutherland, E., L. Coe, and E.A. Raleigh, *McrBC: a multisubunit GTP-dependent restriction endonuclease.* J Mol Biol., 1992. **225**(2): p. 327-48.
30. Strichman-Almashanu, L.Z., et al., *A Genome-Wide Screen for Normally Methylated Human CpG Islands That Can Identify Novel Imprinted Genes.* Genome Research, 2002. **12**: p. 543-554.

Development of DNA Relational Database and Data Manipulation Experiments

Masahito Yamamoto[1,2], Yutaka Kita[3], Satoshi Kashiwamura[1],
Atsushi Kameda[2], and Azuma Ohuchi[1,2]

[1] Graduate School of Information Science and Technology, Hokkaido University
North 14, West 9, Kita-ku, Sapporo, Hokkaido 060-0814, Japan
{ki-yu, kashiwa, masahito, ohuchi}@complex.eng.hokudai.ac.jp
http://harmo.complex.eng.hokudai.ac.jp/
[2] Suyama Lab., Department of Life Sciences, The University of Tokyo
Komaba 3-8-1, Meguro-ku, Tokyo 153-8902
kameda@genta.c.u-tokyo.ac.jp

Abstract. An enormous amount of data such as genomic data can be
stored into DNA molecules as base sequences. DNA database is important for organizing and maintaining these data, because extracted data
from DNA database can be directly manipulated by chemical reactions.
In this paper, we develop a DNA relational database with a simple data
model and realize a computational model (relational algebra) of data
manipulation as a sequence of chemical experiments. By using the developed database, it is shown that we can execute query operations based on
the contents of data (the values of attributes). Furthermore, we propose
a conversion scheme of query input to a series of experiment operations.

1   Introduction

Genomes represent genetic information and are just arrangement of DNA base
sequences. In addtion, DNA molecules are stable and nanoscale materials. Recently, using these characteristics, ideas that treat DNA molecules as functional
materials have been executed. DNA memory is a recording medium that stores
data in DNA molecules. Baum proposed associative memory in which an associative search is realized by the parallel hybridization of DNA molecules [1].
Kashiwamura et al. described the use of nested PCR for hierarchical memory
operations to construct large scale DNA memory [2].

If a genome database is only constructed by DNA molecules that appended
information tags (synthesis DNA) to DNA the fragments of the genome DNA,
they can be directly manipulated by chemical reactions. In addition, storing and
processing individual genetic information in electronic data have many problems
from the viewpoint of protecting individual information. Reif et al. invented
a method that makes a database of DNA molecules without changing them
into digital media and retrieved the data from DNA molecules [3]. Even in a
large scale DNA database, processing time is fairly constant because chemical

C. Mao and T. Yokomori (Eds.): DNA12, LNCS 4287, pp. 418–427, 2006.

reactions are also fairly constant due to the massive parallel reactions of DNA molecules.

In studies of DNA relational databases, Arita et al. showed the feasibility of relational algebra (RA) through *in vitro* experiments [4], and Katsányi showed the feasibility of relational algebra with theoretical models that consist of basic biomolecular reactions [5]. However, in their models, one DNA molecule is represented as a tuple of databases. Therefore, such operations that only extract the information of attributes is troublesome. To do this, cleavage or permutation reactions are required, and such recombinant processes cause error accumulations. In addition, they did not address the feasibility of query processing.

In this paper, we construct a DNA relational database with a simple data model in which one DNA molecule stores one piece of data. We confirmed the success of five unique kinds of relational algebra, and executed query processing that refers the contents of the data from a DNA relational database. Supposing the actually utilization of a DNA relational database, we propose a SQL conversion scheme to a series of experiment operations and discuss the estimation of execution time and error based on a series of experiment operations.

2   Model

2.1   Data Model

A relation is denoted by R, and a test tube that consists of DNA molecules representing R is denoted by U. Given two variables $i = 1, 2, ..., n$ and $j = 1, 2, ..., m$,

$$R(A_1, A_2, ..., A_n) = \{(v_1^{(1)}, v_2^{(1)}, ..., v_n^{(1)}),$$
$$(v_1^{(2)}, v_2^{(2)}, ..., v_n^{(2)}), ...,$$
$$(v_1^{(m)}, v_2^{(m)}, ..., v_n^{(m)})\},$$

where A_i is an attribute and $v_i^{(j)}$ is an attribute value of the relation. A tuple of each instance is denoted by $t_j = (v_1^{(j)}, v_2^{(j)}, ..., v_n^{(j)})$. Given t_j, we denote ID_j that stores the value of j. One single-stranded DNA (ssDNA) consists of attribute A, attribute value v, and a tuple information ID and corresponds to a cell of the relational database table shown in Fig. 1. Compared to conventional models, our model freely extracts target molecules from attribute information.

2.2   DNA Computational Model

We define a DNA computational model (DCM) of the basic experimental operations for database processing and use a mathematical scheme based on Reif's RDNA model [6], which has slightly different notation for operations than the models of Reif's and Katsányi's. We describe each operation that includes not only the test tube but also DNA molecules that used in the experiments as variables to represent the continuous operation of query processing to a DNA relational database.

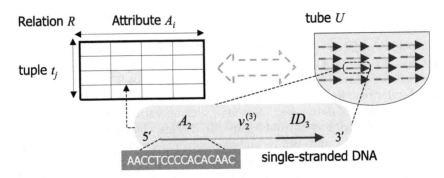

Fig. 1. Representation of data with ssDNA

- $Merge(U_1, \ldots, U_k)$: Mix the DNA molecules in test tubes $U_1, \ldots, U_k$, where k is an arbitrary integer.
- $Divide(U, U_1, \ldots, U_k)$: Extract a portion of the solution of test tube U, and divide it into test tubes $U_1, \ldots, U_k$.
- $Amplify(U, FW, RE)$: Execute amplification reactions for DNA molecules in test tube U with sequence sets of forward primers FW and reverse primers RE.
- $Append(U, S, E))$: Append corresponding sequence set S to DNA molecules in test tube U and contain any subsequence E at the 3' end.
- $CollectID(U)$: Collect all ID of tuple information in test tube U.

For the double-stranded DNA (dsDNA) in which one of the ssDNA's ends are biotinylated in test tube U, operations are the following:

- $Separate_+(U)$: Extract ssDNA whose ends are not biotinylated.
- $Separate_-(U)$: Extract ssDNA whose ends are biotinylated.

For the DNA molecules in test tube U, operations are the following:

- $Get(U, +S)$: Extract ssDNA that contains any subsequence S.
- $Get(U, -S)$: Extract ssDNA that does not contain any subsequence S.

Using the basic operations, we can describe data manipulation and RA operations. We describe the packaging methods of five unique RA in Table 1. As well as related works, we confirmed the success of the in vitro operation of relational algebra with our model. These results were reported in our previous study [7].

3   Query Processing Experiments

To accomplish query processing to the DNA relational database, we assumed the relation of patients $R(3 \times 3)$ and conducted the database search shown in Fig. 2. As conventional relational databases, we used SQL in which we referred to the information of male patients whose diagnosis was negative.

This experimental significance is to verify whether successive operations, AND retrieval, and the extraction of a number of tuple that have equivalent attribute values are feasible.

Table 1. Packaging method for five unique RA operations. Other operations can be expressed by utilizing these ones.

Relational algebra	Chemical experiments
Union	Mix solutions
Difference	Affinity separation
Projection & Selection	PCR
Cartesian product	ST-PCR [8]

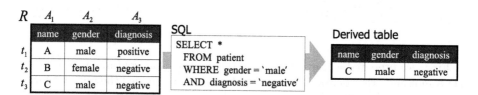

Fig. 2. Query processing for relation of patients $R(3 \times 3)$

3.1 Sequence Design

In this paper, we consider 16-mer ssDNA as a unit and assign them to attribute A_i, tuple information ID_j, and attribute value $v_i^{(j)}$. Here, we denote these 48-mer ssDNA as data strands. We need to design 13 units in total to execute query processing experiments.

We employed a sequence set of GC-Templates in the Template Method [9] for sequence units. Sequences of GC-Templates not only have the same *GCcontent* (the content of bases G and C in a sequence) but the same *GCposition* (the position of bases G or C in a sequence). Especially due to the *GCposition*, these units are not expected to cause much kinetic difference.

However, since the sequences of GC-Templates aren't designed to consider a thermodynamic viewpoint, some sequences may form secondary structure in which a DNA molecule anneals with itself, possibly preventing the intended experimental reactions. To solve this problem, we predict free energy (ΔG) and secondary structure by Mfold [10]. We produced 20,000 sets of data strands by combining three randomized units from all sequence units. Next, we calculated the predicted value of ΔG and selected high scored data strands that consist of unique sequence units. The sequences of each unit of relation R are shown in Table 2. We also describe ssDNA that contains A_i and ID_j as $data(i, j)$, a set of primers that correspond to A_i and ID_j as $p(i, j)$.

3.2 Experimental Procedure

Our chemical experimental procedure of query processing is shown in Fig. 3.

Table 2. Sequences used in chemical experiments. Each data strand $data(i, j)$ is represented $\text{Tag}(A_i) + \text{Data} + \text{Tag}(ID_j)$.

Tag	Sequence (5' to 3')
A_1 (name)	AACCTCCCCACACAAC
A_2 (gender)	AAGCTCGCCACTGTTC
A_3 (diagnosis)	ATGGTCCGCTGTGTTC

Tag	Sequence (5' to 3')
ID_1	CAACACACCCCTCCAA
ID_2	GTAGTCACGCCTCCTA
ID_3	GATCTCTCCGCACGAT

Data	Sequence (5' to 3')
A	CTACACTCGCGACCTA
B	CATCACTCGGGTCCAT
C	CTAGTCAGCCGACCAA
male	ATCGACCGGTCAGTTG
female	CATGTCTCGCCACCAA
positive	CATGTGACGGCTCCTT
negative	TTCCAGGGGAGACTTC

First, we prepared a test tube of initial condition that contains nine types of DNA molecules representing relation R. Here, once we divided the initial solution to collect IDs. In collecting IDs, we mixed complementary strands corresponding to the data of male and negative, respectively. After affinity separation, we collected IDs and select reverse perimer from them. Next, we divided the initial solution and executed continuous PCR, that is, PCR with male IDs and PCR with negative IDs. After continuous PCR, we obtained derived solution.

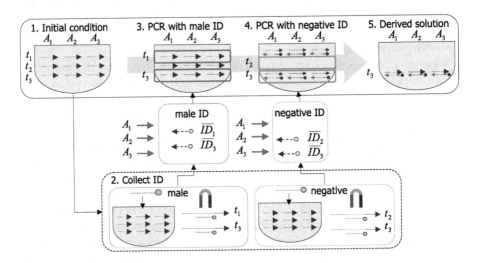

Fig. 3. Experimental procedure of query processing from DNA relational database

3.3 Experimental Results

To collect IDs, we propose utilizing DNA chips. By utilizing DNA chips, we can collect IDs of one solution at a time, however we collected IDs with PCR in this experiment. To detect the experimental results, we used PolyAcrylamide Gel Electrophoresis (PAGE), and the results of collect ID is shown in Fig. 4. Affinity separations operated correctly because data strands $d(2, 1)$ and $d(2, 3)$, representing "male", and $d(3, 2)$ and $d(3, 3)$ representing "negative", were amplified. From these results, we collected male IDs (ID_1, ID_3) and negative IDs (ID_2, ID_3).

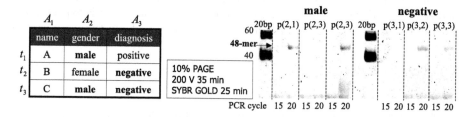

Fig. 4. Results of collected IDs experiment. In PAGE figure, lanes of male and negative representing sample PCR with $p(2, *)$ and $p(3, *)$ for each solution of after affinity separation. Bands appear in using reverse primer of target ID and not in others.

Using IDs collected in the previous step, we executed continuous PCR. Forward primers of continuous PCR are A_1, A_2, and A_3, and reverse primers are selected from IDs. First, we performed PCR for the initial solution with male IDs. Next, we performed PCR for the post-PCR solution with negative IDs. In the derived solution, most DNA molecules in the tube were expected to be tuple $t3$ of $data(*, 3)$. To verify this, we individually performed PCR for the derived solution with all primer pairs. Amplification is observed when DNA molecules in the solution correspond to the primer set, as shown in Fig. 5. Target data strands $data(*, 3)$ were amplified, and others are not. Therefore, most DNA molecules in the derived solution are $data(*, 3)$, and we accomplished chemical experiment of query processing to the DNA database.

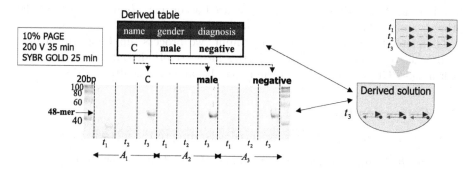

Fig. 5. Verifying PCR results of data reference. From initial solution, we derived solution by collecting IDs and continuous PCR. In PAGE figure, each lane represents sample of PCR with individual primer pairs $p(*, *)$ (PCR cycles are 15 and 20). Bands only appear in using $p(*, 3)$. Moreover, proportion of data strands remained fairly constant through successive operations.

4   SQL Converter

To execute query processing to the DNA relational database, we need to convert query input into a series of experimental operations. As in the conventional relational database, we use basic SQL as input and the series of experimental operation expressed by DCM defined in Section 2. This conversion is expected to indicate to automation of the experiment and estimation of run-time and error.

4.1   SQL Mapping

Basic SQL is described as follows.

SELECT	column 1, column 2, ..., column N
FROM	table 1, table 2, ..., table M
WHERE	[search condition]

SQL means that it selects rows that satisfy search conditions from tables $1 \sim M$ and derives a table that consists of columns $1 \sim M$. In the SELECT phrase, we specify forward primers that utilize the PCR of the Projection operation, in the FROM phrase, we specify test tubes that represent corresponding relations, and in the WHERE phrase, we input search conditions by a logical formula. By search conditions, reverse primers that utilize the PCR of the Selection operation are specified. In this logical formula, literal p is described as "$A_i =' v_i^{(j)'}$". In the case of negative literal $\sim p$, the reverse primers of $\sim p$ are the subtracted reverse primers of p from the reverse primer set.

To process logical queries, Reif et al. proposed processing them in a conjunctive normal form (CNF) [11], that is, using repeated PCR for each clause with primers that satisfy a given clause. However, in our model, it is difficult to process logical formulas in clauses that contain NOT operator of one PCR. In addition, duplication of the literals caused a concentration difference of the primers and could not amplify target molecules.

Therefore, to respond to arbitrary input, we process input logical formula as disjunctive normal form (DNF). In this form, we obtain output by executing continuous PCR for conjunctive term in parallel and mixing after PCR solutions.

4.2   Conversion Procedure

The conversion SQL to DCM is executed according to the following procedure.

1. Convert the input of a logical formula into DNF P, and check the duplication of the literals in P.
2. Divide the initial solution by the number of unique literals and collect ID in parallel.
3. If some column name is specified in the SELECT phrase, then do step 4, otherwise step 5.
4. Execute PCR for the Projection operation, and replace the forward primer to the specified subset.

5. Divide the initial solution by the number of conjunctive terms, and execute continuous PCR in parallel.
6. If an OR operator exists in P, then do step 7, otherwise step 8.
7. Mix all solutions of post continuous PCR.
8. Execute denaturation to put DNA molecules into ssDNA.

In case of SQL used in the query processing experiments, we obtained a series of experimental operations by DCM as shown in Fig. 6. We omit the Cartesian product and Join operation conversion because they can be converted by the uniform procedure.

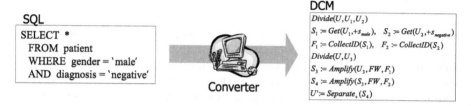

Fig. 6. Conversion from SQL used in Section 3 to DCM defined in Section 2

4.3   Analysis from a Series of Experimental Operations

By a given series of experimental operations, we can estimate the execution time of the query processing experiments. Total execution time T_{total} is estimated as follows:

$$T_{total} \simeq T_{affinity}(1 + \frac{L_{unique}}{C_{affinity}}) + T_{chip}(\frac{L_{unique}}{C_{chip}}) + T_{PCR}(\frac{L_{total}}{C_{PCR}}),$$

where $T_{affinity}$, T_{chip}, and T_{PCR} are the execution times of each experiment, and $C_{affinity}$, C_{chip}, and C_{PCR} are the capacity of each experiment, and L_{unique} and L_{total} are the number of unique and total literals in P. This time is estimated in view of using DNA chips to collect ID. Moreover, when the capacity of experiments is unlimited, a series of experiments is represented as pattern diagrams, as shown in Fig. 7 and T_{total}:

$$T_{total} \simeq 2T_{affinity} + T_{chip} + kT_{PCR},$$

where k is a max size of a term in P of k-DNF. Therefore, execution time is independent of the number of the attributes and tuples and only increases linearly by the number of literals. Compared to conventional database, our database clearly takes a long time. However, in conventional genome databases, require chemical experiments. In addition, we must produce necessary materials to execute experiments with analysis results. Therefore, the characteristics of a DNA relational database that can directly manipulate biological materials positively affects total analysis time.

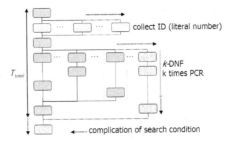

Fig. 7. Pattern diagrams of series of experimental operation. Dotted line arrows represent complication of search condition.

In addition, chemical experiments always include error, and successive operations can cause error accumulations. By using the output of a series of experiment operations from a converter, we can estimate the error of successive operations. In our model, we require k times PCR to process k-DNF derived from the WHERE phrase. On this point, we modeled the proportion of target and non-target on continuous PCR. A characteristic of PCR that exponentially amplifies target molecules eliminates previous errors. We confirmed that the continuous PCR in our model was robust with general PCR parameters. Moreover, in mixing solutions, the proportion of target and non-target is invariant before and after operations.

5   Discussion

For AND\OR retrieval quality in our model, we execute the number of AND operator times PCR in AND retrieval and simply mixing solutions in OR retrieval by using DNF. The number of PCR is greater than when using CNF. However, in DNF, we can execute PCR for each term in parallel.

The maximum size of our DNA database depends on the specificity of hybridization. If using sequences with low specificity, we can not correctly extract target molecules. To use many sequences, we must consider sequence length and constraints. For example, we used 64 16-mer sequences generated by GC-Templates in this paper. However, by using 23-mer orthogonal DNA sequences [12], we can use 300 sequences. Even smaller than a conventional genome database, our DNA relational database is useful as a secure database that can store the critical part of human genome information that are closely linked to some diseases.

6   Conclusion

We developed a DNA relational database model in which one DNA molecule stored one piece of data. We confirmed successive operations of query processing to the DNA relational database *in vitro*. Additionally, we proposed a conversion scheme of SQL to DCM. From a derived series of experimental operations, we discussed execution time and error.

However, to realize a DNA relational database, many problems need to be solved. First, we would like to collect ID experimetns with DNA chips to process at one time. The size of our database is still too small to use bioinformatics. On this point, we must cooperate with large scale DNA memory. In addition, in view of the accuracy of reactions, we would like to replace affinity separation to PCR in collecting ID. As future work, we will solve these problems and construct a DNA relational database system with real genome data.

References

1. E. B. Baum, "Building an Associative Memory Vastly Larger Than the Brain", Science, Vol. 268, pp. 583-585, 1995.
2. S. Kashiwamura, M. Yamamoto, A. Kameda, T. Shiba, and A. Ohuchi, "Hierarchical DNA memory based on nested PCR", 8th International Workshop on DNA-based Computers (DNA8), in Lecture Notes In Computer Science, pp. 112-123, 2003.
3. J. H. Reif, T. H. LaBean, M. Pirrung, V. S. Rana, B. Guo, C. Kingsford, G. S. Wickham, "Experimental Construction of Very Large Scale DNA Databases with Associative Search Capability", 7th International Workshop on DNA-Based Computers (DNA7), in Lecture Notes in Computer Science 2340, pp. 231-247, 2002.
4. M. Arita, M. Hagiya, and A. Suyama, "Joining and Rotating Data with Molecules", Proc. of IEEE 4th International Conference on Evolutional Computaion (ICEC'97), pp. 243-248, in IEEE press, 1997.
5. I. Katsányi, "On implementating Relational Database on DNA Strands", Acta Cybernetica, vol. 16(2), pp. 259-270, 2003.
6. J. H. Reif, "Parallel biomolecular computation:Models and simulations", Algorithmica, Vol. 25, pp.142-175, 1999.
7. Y. Kita, S. Kashiwamura, A. Kameda, M. Yamamoto, A. Ohuchi, "Data Manipulation of DNA Relational Database", Proceedings of The Ninth International Symposium on Artificial Life and Robotics (AROB 11th '06), pp. 629-632, 2006.
8. K. Hasimoto, A. Kameda, M. Yamamoto, and A. Ohuchi, "State Transition Model Based on DNA Polymerization", *Proc. of the International Technical Conference on Circuit/Systems, Computers and Communications (ITC-CSCC'03)*, pp. 1889-1892, 2003.
9. M. Arita and S. Kobayashi, "DNA Sequence Design Using Templates", New Generation Computing, vol. 20, pp. 263-277, 2002.
10. M. Zuker, "Mfold web server for nucleic acid folding and hybridization prediction", Nucleic Acids Research, vol. 31, pp. 3406-3415, 2003.
11. John H. Reif, Michael Hauser, Michael Pirrung, and Thomas LaBean, "Application of Biomolecular Computing to Medical Science: A Biomedical Database System for Storage and Retrieval of Genetic Information and Material, Chapter 3 of Complex Systems Science in Biomedicine (Edited by Tom Deisboeck)", Kluwer-Springer Publishers, Chapter 3, pages 701-735, 2005
12. H. Yoshida and A. Suyama, "Solution to 3-SAT by breadth first search" Proc. 4th International Meeting on DNA Based Computers, pp. 9-20, 1999.

Experimental Validation of the Statistical Thermodynamic Model for Prediction of the Behavior of Autonomous Molecular Computers Based on DNA Hairpin Formation

Ken Komiya[1,2,*], Satsuki Yaegashi[3], Masami Hagiya[2,3], Akira Suyama[3,4], and John A. Rose[2,3,5,*,**]

[1] Dept. of Computational Intelligence and Systems Science, Tokyo Institute of Technology
komiya@dis.titech.ac.jp
[2] Department of Computer Science, The University of Tokyo
hagiya@is.s.u-tokyo.ac.jp
[3] Japan Science and Technology Agency-CREST
yaegashi@lyon.is.s.u-tokyo.ac.jp
[4] Department of Life Sciences and Institute of Physics, The University of Tokyo
suyama@dna.c.u-tokyo.ac.jp
[5] Institute of Information Communication Technology, Ritsumeikan Asia Pacific University
jarose@apu.ac.jp

Abstract. Due to the multi-state nature of autonomous computing systems, it is important to develop a simulation model which accounts for process coupling, and allows the precise prediction of the behavior of a composite system formed by a series of competing reactions, in which each intermediate step is difficult to probe. In this work, the statistical thermodynamic apparatus for predicting the efficiency of DNA hairpin-based computers is validated experimentally. The model system employed is a simple competitive folding system, formed by two competing hairpin structures (sub-optimal vs. optimal), with the intent of testing the ability to predict the efficiency of target structure formation in the presence of a non-target structure. System behavior was characterized via a set of fluorescence measurement experiments, to directly determine the fractional occupancy of target structures versus temperature. Predicted and experimental behaviors are compared for both the melting of each of the two isolated hairpin structures (control), and the efficiency of the competitive composite system. Results indicate that the applied equilibrium model provides predictions which consistently agree with experimental results, supporting design for the control and programming of DNA-based systems.

1 Introduction

During formation of hairpin structures, an ensemble of single-stranded (ss) DNA molecules independently execute an autonomous, intramolecular search of the set of accessible folds formed by hybridization between complementary pairs of subsequences.

---

* These authors contributed equally to the current work.
** Corresponding author.

C. Mao and T. Yokomori (Eds.): DNA12, LNCS 4287, pp. 428–438, 2006.
© Springer-Verlag Berlin Heidelberg 2006

In the absence of significant strand association, the formation of a hairpin structure has an efficiency which is independent of strand concentration [1], and can largely be controlled via careful sequence design [2,3]. When properly controlled, this process may be harnessed for information processing, providing a sophisticated form of parallel computation, in which each single DNA molecule acts as an independent computing unit [4], endowed with robustness in computing speed which is robust to changes in strand concentration. Note that the transformations between distinct hairpin forms could be regarded as the state transitions of a molecular machine. In an actual application [5], a DNA-based state machine, the 'Whiplash' Machine was used to autonomously implement multiple state transitions via recursive polymerase extension, triggered by transformations between repeatedly-extended hairpin structures.

In order to achieve practical DNA-based computing, bio-molecular reactions implementing the underlying model of computation are expected to be appropriately efficient and of high fidelity. Although the issue of sequence design to ensure high fidelity has been well-studied for both folding and strand association processes [1,2], the development of quantitative methods allowing a estimation of computational efficiency, along with experimental validation remains rare.

In the present study, the statistical thermodynamic apparatus for the precise prediction of the behavior of DNA hairpin-based computers is validated experimentally. The theoretical model employed is an improved version, in terms of parameter selection, of the coupled equilibrium model previously reported for the Whiplash PCR (WPCR) system [6,7], and is applied to a competitive hairpin system, allowing a direct comparison between predicted and experimental efficiency. The model system employed is a simple competitive folding system, formed by two competing hairpin structures (sub-optimal vs. optimal hairpin formation), each of which may potentially be formed by each strand. Note that the specific intent of this design was to test the ability to predict the efficiency of target structure formation in the presence of a non-target structure, which could be applied to such unfavorable competitive effects during hairpin-based computation, due to competition between a sub-optimal target and more stable non-targets [6]. An autonomous molecular computation is, in general a composite system, formed by a series of (often competing) reactions, in which each intermediate step is difficult to probe. In a practical sense, it is important to establish the validity, applicability, and practical limitations of the physical model to allow precise prediction of system behavior. Simulation of the folding system enables us to estimate the efficiency of each computational step performed via hairpin formation, to determine the optimal reaction conditions, and to evaluate strategies for improving the efficiency and architecture [8].

Competitive hairpin formation is an intramolecular folding process, which occurs with no changes with respect to DNA chemical structure or the total concentration of strands. We therefore took advantage of the fluorescent resonance energy transfer (FRET) or contact quenching technique [9], by which the change in the distance between the donor and acceptor fluorescent dyes can be monitored via the change in the fluorescence intensity (FI), to determine the fractional occupancy versus temperature. Results indicate that the applied equilibrium model provides predictions which consistently agree with experimental results. Predicted and experimental behaviors are compared for both: (1) the folding/melting of each of the two isolated hairpin structures, in

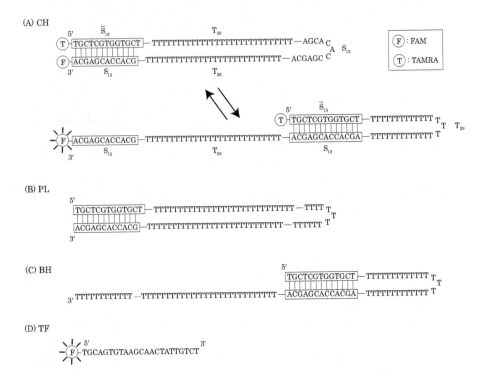

Fig. 1. Sequences used in the experiment (A) Strand **CH** may form two different hairpin structures (*pl* and *bh*). Vertical lines represent the formed base pairs. Upon formation of the upper hairpin (*pl*), emission from FAM is quenched by contact quenching. (B)(C) Strands **PL** and **BH** each form a unique hairpin structure. The impact on thermal stability due to the difference only in the central loop-base sequence is negligible. Therefore, equivalent structures in **CH** and **PL**, and in **CH** and **BH** are referred to collectively as *pl* and *bh*, respectively. (D) **TF** does not form any specific structure.

terms of the melting temperatures (T_m's) predicted for each in the absence of competition (control); and (2) the efficiency behavior of the competitive system formed by the combination of the two, in terms of: (i) the temperature, $T^{\maltese}$ at which the efficiency is maximized (*i.e.*, maximum occupancy of the target structure); (ii) the shape of the distribution vs. temperature, via the width (*i.e.*, the full-width at half-maximum FWHM; denoted here by $\Delta T^{\maltese}$); and (3) the magnitude of the reduced efficiency at $T^{\maltese}$, due to the occupancy of non-target hairpins.

2   Materials and Methods

2.1   Oligonucleotides

The sequences used in the experiment are: TGCAGTGTAAGCAACTATTGTCT (S_{23}), AGCACCACGAGCA (S_{13}), GCACCACGAGCA (S_{12}). S_{23} is a sequence selected from a 23-mer code word set [1,10] designed to meet the following three requirements. First of all, codewords should exhibit similar T_m values, to ensure 'normality' in the stability of binding. Secondly, the sequence homology among the codewords should be minimized to ensure 'orthogonality' (*i.e.*, encoding to avoid unexpected interactions). Thirdly, each codeword should not form secondary structure, by itself. S_{13} is a 13-mer subsequence generated by truncation of a 23-mer codeword from the 'orthonormal' set, while S_{12} is identical to S_{13}, except that it lacks the 5'-most base. These sequences were carefully selected to avoid accidental generation of complementary subsequences upon codeword concatenation. The oligonucleotide, **CH** which was employed for competitive hairpin formation, was encoded to have a sequence of the form: $\overline{S_{13}}$ - T_{29} - S_{13} - T_{28} - S_{12} (5' to 3'). Note that $\overline{S_{13}}$ indicates a sequence complementary to S_{13}. T_{29} and T_{28} indicate poly-T sequences of length 29 and 28 bases, respectively. The fluorescent dyes, FAM and TAMRA were attached to the strand's 5' and 3' strand ends, respectively (Fig. 1(A)). The emission from FAM is quenched via FRET in close proximity to TAMRA, since the absorption spectrum of TAMRA overlaps the fluorescence emission spectrum of FAM, or by contact quenching, when placed in very close contact (*e.g.*, via blunt-ended hairpin formation [9]). The poly-T sequence, T_{28} was inserted to avoid FRET upon formation of the more stable hairpin structure, which forms a duplex region of 13 base pairs (bp), causing the distance between FAM and TAMRA flourophores to become sufficiently longer than the Förster radius of the pair ($R_o = 55\,\text{Å}$), defined as the distance at which the FRET efficiency is 50% [11]. On the other hand, the FAM emission is efficiently quenched by contact quenching upon formation of the less stable target hairpin structure, having a duplex region of 12 bp. Note here that distribution among the less stable (target) and more stable (non-target) hairpin structures corresponds to the transforming efficiency of the Whiplash machine between the extendable hairpin structure, 'planned (*pl*) hairpin', and the unextendable hairpin structure, 'back-hybridized (*bh*) hairpin' [5]. The oligonucleotides **PL** and **BH** for the (uncoupled control) T_m analysis have sequences of a form similar to **CH**, but with S_{13} or S_{12} replaced with a poly T sequence for **PL** or **BH**, respectively, to allow formation of only one hairpin structure (Fig. 1(B)(C)). The oligonucleotide, **TF** used for the thermal profiling of FAM has the sequence, S_{23}, with FAM attached to the 5' end (Fig. 1(D)). All nucleotides were commercially synthesized and purified by Nippon EGT.

The design considerations involved in the specific selection of the 28 base spacer sequence, to effect a 40 base spacing between the FAM and TAMRA fluorophores upon formation of the more stable hairpin structure, *bh* are critical, and deserve further comment. In [11], the persistence length of poly-T ssDNA was measured via direct observation of the decrease in FRET efficiency accompanying separation of a fluorophore donor-acceptor pair, Cy3/Cy5 (Förster radius, $R_o = 59 - 61\,\text{Å}$) via poly-T dangling ends of lengths from 0 (blunt-ended) to 70 bases, yielding persistence length values ranging from $a = 30\,\text{Å}$ (at 2.0 M [NaCl]) to $a = 15\,\text{Å}$ (at 25 mM [NaCl]). Combining the linearly interpolated value at 0.165 M [NaCl] of $a = 16\,\text{Å}$ with the measured value for the ssDNA inter-chain phosphate spacing also reported in [11] ($l = 6.3\,\text{Å}$), and the well-known limiting expression for the mean square distance of a freely rotating chain in the infinite-chain limit derived in [12],

$$\langle r^2 \rangle_o = nl^2 + 2nl(a - l), \tag{1}$$

yielded an approximate FAM/TAMRA rms spacing of $\langle r^2 \rangle_o^{1/2} \approx 80.5\,\text{Å}$ for our design's effected insertion of a ssDNA spacer of length $n = 40$ bases, upon formation of the bh hairpin. Note that due to the very short persistence length of ssDNA, this expression was expected to provide an adequate approximation of $\langle r^2 \rangle_o$ for ssDNA chains longer than about 30 bases. At this spacing, the FRET efficiency of the FAM/TAMRA pair, estimated via the expression,

$$E = 1/[1 + (R/R_o)^6], \tag{2}$$

where the separation distance is approximated by $R = \langle r^2 \rangle_o^{1/2}$, is roughly $E = 0.092$. This predicted 91% decrease in FRET efficiency, accompanying formation of the more stable hairpin structure, bh rather than the less stable (target) hairpin structure, pl was taken as very adequate for the current purpose of distinguishing between formation of the two species based on the accompanying FAM emission, and strongly motivated our selection of the 28 nucleotide spacer. In addition, the prediction of excellent functionality for our system was equally well supported by the roughly $80 - 85\%$ drop in FRET efficiency reported for use of a 40 base poly-T spacer, between the Cy3/Cy5 fluorophore-quencher pair employed in [11] under similar salt conditions, which exhibits a greater Förster radius of $59 - 61\,\text{Å}$.

2.2   Statistical Thermodynamic Modeling

The equilibrium behaviors of the competitive hairpin system shown in Fig. 1(A), and the two simple hairpins in Fig. 1(B)(C), were each modeled using a standard statistical thermodynamic approach. First, a statistical weight of folding for each hairpin structure, which is independent of folding context, was separately estimated. Due to the short lengths of the accessible duplex structures involved (12-13 bps), an all-or-none model was taken to be adequate for each duplex, so that contributions to hairpin stability made by partially-melted intermediates were neglected. For each hairpin stem, a sequence-dependent Gibbs free energy, ΔG_{stem}^o of duplex stacking was first estimated via the standard Watson-Crick nearest-neighbor model, using the doublet parameters reported in [13], which estimates ΔG_{stem}^o as a sum of the temperature and ionic strength-dependent free energy contributions of each doublet, with contributions from dangling ends included as an energetic perturbation. The accompanying statistical weight of stacking for each hairpin stem was then estimated by the Gibbs factor, $\omega_{st} = \exp(\Delta G_{stem}^o/RT)$, where R is the molar gas constant and T is the absolute temperature. The statistical weight of loop formation was modeled via the explicit composition of the statistical weights expected due to helix cooperativity/unwinding and loop closure. The statistical penalty of unraveling at both duplex ends was modeled via the cooperativity parameter, $\sigma \approx \exp(\Delta G_{init}^o/RT)$, where ΔG_{init}^o was estimated as the sum of two sequence-dependent helix initiation parameters (i.e., one for each duplex end) listed in [13]. The statistical weight of hairpin loop closure was estimated via a Jacobson-Stockmeyer inverse-1.5 power law, $\omega_{loop} = (1 + n)^{-1.5}$, where n is the number of bases in the loop, and the impacts of volume exclusion and chain stiffness on loop closure

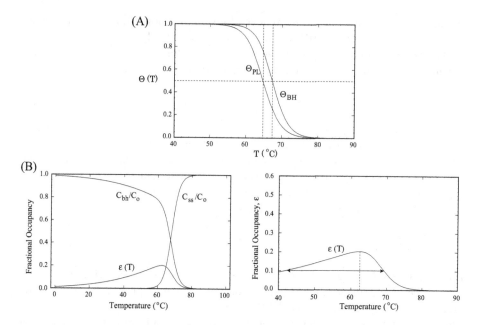

Fig. 2. Results of Simulation (A) Predicted thermal profile of the fraction of stacked base pairs. The left and right curves represent the melting curves of hairpin structures formed by **PL** and **BH** strands, respectively. (B) Predicted thermal profiles of the fractional population occupying the pl hairpin structure (*i.e.*, the efficiency, $\varepsilon(T)$), the bh hairpin structure (C_{bh}/C_o), and the unfolded form (C_{ss}/C_o). Here, C_o denotes the total strand concentration. The right figure shows a magnification of the pl hairpin occupancy curve, $\varepsilon(T)$ shown in the left figure. A dashed vertical line and a horizontal arrow indicate the peak and FWHM of the curve, respectively.

was neglected, due to the use of large hairpin loops. The net equilibrium constant for formation of each hairpin structure then takes the standard form, given by:

$$K_{eq} = \sigma\omega_{st}\omega_{loop} = \frac{\sigma\exp\left(\Delta G^o_{stem}/RT\right)}{(1+n)^{1.5}}. \tag{3}$$

The estimated equilibrium constants for the two hairpin species K_{pl} and K_{bh} were then employed to model the melting behavior for the two isolated, control hairpins shown in Fig. 1(B)(C), as well as the efficiency behavior of the competitive strand in Fig. 1(A).

For each of the melting analyses of the simple control hairpins in Fig. 1(B)(C), the fraction Θ of folded DNA strands was estimated by an expression of the form (all-or-none model):

$$\Theta = \frac{C_{hp}}{C_o} = \left(1 + \frac{1}{K_{hp}}\right)^{-1}. \tag{4}$$

Here, the total strand concentration for each simple equilibrium is given by $C_o = C_{ss} + C_{hp}$, where C_{ss} and C_{hp} denote the equilibrium concentrations of fully-opened and fully-folded hairpin structures, respectively. Note that a law of mass action for simple hairpin formation, $C_{hp} = C_{ss}K_{hp}$ has been employed for simplification. Here, K_{hp} corresponds to K_{pl} and K_{bh} in the **PL** and **BH** melting analyses, respectively.

The efficiency of formation of the target hairpin structure for the competitive hairpin system, **CH** was estimated via the coupled equilibrium expression,

$$\varepsilon = \frac{C_{pl}}{C_o} = \left[1 + \frac{(1 + K_{bh})}{K_{pl}}\right]^{-1}. \tag{5}$$

Here, the total strand concentration for the coupled equilibrium is given by $C_o = C_{ss} + C_{pl} + C_{bh}$, where C_{ss}, C_{pl}, and C_{bh} denote the equilibrium concentrations of opened, fully-formed pl hairpins, and fully-formed bh hairpin structures, respectively. Once again, a mass action expression for each component equilibrium (*i.e.*, $C_{pl} = C_{ss}K_{pl}$ and $C_{bh} = C_{ss}K_{bh}$) has been employed for simplification.

2.3 T_m Analysis

The melting temperature, T_m of a DNA duplex is defined as the temperature at which 50% of the base pairs are melted, and is specific for both duplex sequence and loop structure. SYBR® GreenI (Molecular Probes) intercalates to DNA molecules, and exhibits a large fluorescence enhancement upon formation of a DNA duplex. For each of the strands forming single hairpin structures, **PL** and **BH**, we therefore measured the variation in the fluorescence intensity (FI) of SYBR® GreenI with change in temperature to obtain the thermal profile that represents the melting of the hairpin structure, using a real-time PCR machine, OPTICON2 (Bio-Rad). Measurement was performed in a 20 μl solution of 1X SSC buffer (150 mM NaCl, 15 mM Sodium citrate, pH 7.0 (23°C), Invitrogen), containing 5 μM **PL** or **BH** and 1X SYBR® GreenI. The solution was incubated at 94°C for 1 min and gradually cooled to 10°C at a rate of -0.1°C / sec, to ensure formation of the expected hairpin structure. The FI variation was then monitored by increasing the temperature slowly to 99°C, at a rate of 0.2°C / sec.

2.4 Competitive Hairpin Formation

The intensity of the fluorescence emitted from fluorescent dyes varies with temperature. We therefore measured the FI of the oligonucleotide, **TF** to obtain the thermal profile of FAM, to be used for revision of the FI profile monitored during competitive hairpin formation. Competitive hairpin formation of **CH** was then investigated by monitoring the FI of the emission from FAM versus temperature. At higher temperatures, as the hairpin structures of all DNA molecules were expected to be melted, the FAM emission was expected to be detected without reduction by FRET or contact quenching. At lower temperatures, as most of the DNA molecules are expected to form the bh hairpin, the FAM emission was also expected to be detected with little reduction. However, at moderate temperatures, if a significant fraction of the DNA molecules formed the pl hairpin which causes contact quenching as predicted via preliminary simulations, the accompanying reduction of the FAM emission by contact quenching could be detected. Measurement was performed in a 20 μl solution of 1X SSC buffer, containing 1 μM **CH**. The solution was incubated at 94°C for 1 min, and gradually cooled to 10°C at a rate of -0.1°C / sec. FI was then monitored by increasing the temperature slowly to 99°C, at a rate of 0.2°C / sec.

3   Results

3.1   Simulations

Fig. 2(A) illustrates the melting behavior of each of the simple hairpins formed by the **PL** and **BH** strands, respectively encoded to form hairpins *pl*-only and *bh*-only, as simulated by the model in Sec. 2.2. Simulated ionic strength was at 0.165 M [Na$^+$]. The predicted melting behavior for each strand is illustrated in terms of the fraction of folded strands, denoted by Θ. Note that under an all-or-none model, this quantity is equal to the normalized fraction of stacked base pairs. For each curve, the simulated sigmoidal structure is characteristic of the cooperative melting of a single DNA duplex region. The simulated T_m values of the *pl* and *bh* hairpins were 64.8°C and 67.3°C, respectively.

Fig. 2(B) illustrates the simulated efficiency behavior for the competitive strand **CH**, in terms of the predicted fractional occupancy, ε of the *pl* hairpin structure, as a function of reaction temperature. For clarity, the predicted fractions of *bh* hairpin, and unfolded ssDNA are also shown. All predictions were estimated via the equilibrium thermodynamic model described in Sec. 2.2, assuming an ionic strength of 0.165 M [Na$^+$]. As illustrated, model predictions indicate a non-symmetric, singly-peaked efficiency behavior, reaching a maximum value of 0.2 at optimal temperature, $T^{\maltese} = 62.3$°C, corresponds to a 20% occupancy of the *pl* hairpin. The accompanying curve width, estimated via the full width at half-maximum (FWHM), is predicted to be $\Delta T^{\maltese} = 27.3$°C.

3.2   Experimental T_m Analysis

The T_m of each of hairpin structures *pl* and *bh* was determined by the conventional method, as the temperature at which the negative first derivative $(-dFI/dT)$ on the plotted melting curve reaches its maximum value. The T_m of the *pl* and *bh* hairpins were 66.8°C and 70.8°C, respectively. For comparison to simulation, the thermal profiles of the hairpin structures obtained by monitoring the FI of SYBR® GreenI were each transformed into a plot of the fraction of stacked base pairs (Fig. 3(A)), according to the normalizing method in [14]. A sudden increase or decrease to beneath zero (not shown) observed near 90°C was due to the normalizing treatment.

3.3   Competitive Hairpin Formation

The thermal profile of FAM emission is shown in Fig. 3(B). For normalization, a series of FI values of **TF** were divided by the value at 80°C (*i.e.*, values were normalized by setting the FI value at 80°C to unity). Note that here, normalization was performed exclusively for the purpose of clarifying relative FI values. Thus, an arbitrary temperature can be set as the standard. As shown, the FI of FAM increased monotonically with increasing temperature, up to 70°C, and remained nearly constant at values above 70°C. Accordingly, the thermal profile of **CH**, obtained by monitoring the FI of FAM, and

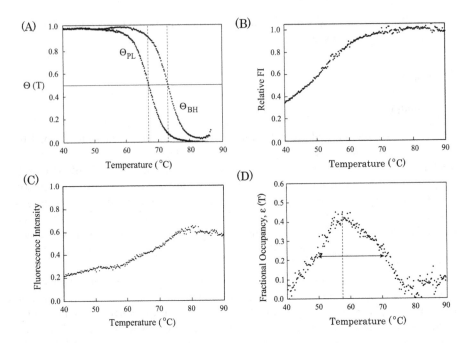

Fig. 3. Experimental Results (A) Experimentally determined thermal profiles of the fraction of stacked base-pairs, Θ. Left and right curves represent the melting curves of the hairpin structures formed by the **PL** and **BH** strands, respectively. (B) Thermal profile of FAM emission. (C) Experimentally determined thermal profile of **CH**. (D) Revised thermal profile of **CH** representing the fractional population of the *pl* hairpin structure, $\varepsilon(T)$ vs. T. The dashed vertical line and a horizontal arrow indicate the peak and FWHM of the curve, respectively.

shown in Fig. 3(C), was revised. For this purpose, a series of FI values were divided by the relative FI values of **TF**. Finally, the revised values of **CH** were again normalized. The series of revised FI values were divided by the value at 80°C, since the value at 80°C was the maximum value within the temperature range from 50 to 90°C. Note that for temperatures beneath roughly 40°C, the revised FI values were considerably dispersed (data not shown). Each resulting value represents an estimate of the fractional occupancy of DNA molecules in structural forms that hardly cause FRET or contact quenching, which is equal to the sum of the populations occupying the *bh* hairpin and the single-stranded form. By subtracting each value from 1, the thermal profile of the fractional occupancy of DNA molecules in the *pl* hairpin was obtained (Fig. 3(D)). The resulting plot exhibits a hill-shaped curve, characterized by a peak located at 57.6°C, having a maximum value of 0.45, and a peak width, estimated by the FWHM, of 21.4°C.

4   Discussion

An explicit experimental validation of the effectiveness of the underlying thermodynamic theory for modeling each of the specific component equilibria is an essential benchmark towards an advantageous treatment of the more complex, coupled hairpin equilibrium. In addition, the hairpin structure (bh) of one component equilibrium was designed to form a loop substantially larger than those studied in [13], which motivated independent characterization. The experimentally-determined T_m values for the simple hairpin structures, pl and bh were slightly higher (by 2.0 to 3.5°C) than those predicted via simulation. This result is consistent with the presence of modest stabilization by partially-melted intermediates, which is neglected by an all-or-none model. When each T_m was determined as the temperature at which Θ, the normalized fraction of stacks is 0.5, the experimentally determined values increased slightly (to 67.2°C and 73.1°C for pl and bh, respectively). For comparison, simulations using the loop parameterization method in [13] were also undertaken, yielding T_m values comparable to the experimental values determined based on Θ (69.5°C and 72.4°C for pl and bh, respectively). Given the performance generally expected to accompany use of the employed nearest-neighbor parameter set, with an all-or-none model of duplex formation (*e.g.*, agreement to within 3.9°C, for the hairpin T_m prediction method in [13]), it was concluded that differences between experimental and simulated T_m values were well within the range of uncertainty of the current models.

Overall predictions of the efficiency behavior vs. temperature for the competitive hairpin system, as provided by fluorescence measurements, exhibited substantial quantitative similarity to model predictions. The most notable point of agreement was in the predicted and experimental peak temperatures, which differed by only 4.7°C ($T^{\maltese}$ = 62.3°C (pred) vs. 57.6°C (exp)), achieving roughly the same level of accuracy expected when predicting the melting of single hairpin species. The predicted peak width (FWHM) also demonstrated reasonable agreement with the experimental value, differing by less than 6°C ($\Delta T^{\maltese}$ = 27.3°C (pred) vs. 21.4°C(exp)). On the other hand, the absolute value of the peak efficiency, $\varepsilon(T^{\maltese})$ provided only good order-of-magnitude agreement, with the peak magnitude of the predicted efficiency attaining only half of that determined in the FRET experiment. This difference might be attributed to multiplication of uncertainty, intrinsic to each measurement in revision.

As a concluding remark, the most significant point of comparison lies in the clear experimental observation of the characteristic hill-shaped behavior predicted both here and for WPCR [6] in the presence of competition with stable alternative hairpins. The monotonically decreasing efficiency predicted to occur beneath $T^{\maltese}$ is contradictory to intuitive expectations based on considerations of the uncoupled, component equilibria, as a naive consideration of the pl and bh hairpin T_m values, estimated separately instead suggests a tendency towards an equal occupancy of the two species at temperatures well beneath the T_m values of both component equilibria (since both hairpins will be regarded as separately 'stable', in a qualitative sense). The experimental validation of this fundamental prediction regarding competitive hairpin systems is a striking caveat against the unqualified use of uncoupled T_m values to predict the overall behavior of 'coupled DNA systems', as is usual in the case in biomolecular reactions, and highlights the need for further work on the efficient and accurate modeling of DNA-based systems. Notably, semi-quantitative agreement between the predicted and experimental gross thermal behaviors, which is obtained here, is generally taken to be indicative of the feasibility of controlling molecular system behavior, which is required for molecular programming beyond molecular computation (*i.e.*, the programming of biomolecular reactions, beyond simple usage for elementary computation).

Acknowledgements

Financial support generously provided by Grants-in-Aid for Scientific Research (Kiban B, 15300100 and 18300100), from the Japan Society for the Promotion of Science (JSPS) and by JST-CREST.

References

1. Rose, J.A., Deaton, R.J., Suyama, A.: Statistical thermodynamic analysis and design of oligonucleotide based computers. Natural Computing **3** (2004)
2. Rose, J.A., Suyama, A.: Physical modeling of biomolecular computers: Models, limitations, and experimental validation. Natural Computing **4** (2004) 411–26
3. Kubota, M., Ohtake, K., Komiya, K., Sakamoto, K., Hagiya, M.: Branching DNA machines based on transitions of hairpin structures. In: Proc. Congr. Evol. Comp. (CEC'03). (2003) 2542–2548
4. Sakamoto, K., Kiga, D., Komiya, K., Gouzu, H., Yokoyama, S., Ikeda, S., Sugiyama, H., Hagiya, M.: State transitions by molecules. Biosystems **52** (1999) 81–91
5. Komiya, K., Sakamoto, K., Kameda, A., Yamamoto, M., Ohuchi, A., Kiga, D., Yokoyama, S., Hagiya, M.: DNA polymerase programmed with a hairpin DNA incorporates a multiple-instruction architecture into molecular computing. Biosystems **83** (2006) 18–25
6. Rose, J.A., Deaton, R.J., Hagiya, M., Suyama, A.: Equilibrium analysis of the efficiency of an autonomous molecular computer. Phys. Rev. E **65** (2002) 1–13 Article 021910.
7. Rose, J.A., Takano, M., Hagiya, M., Suyama, A.: A DNA computing-based genetic program for in vitro protein evolution via constrained pseudomodule shuffling. Journal of Genetic Programming and Evolvable Machines **4** (2003)
8. Rose, J.A., Komiya, K., Yaegashi, S., Hagiya, M.: Displacement Whiplash PCR: optimized architecture and experimental validation. In: DNA Computing. 12th Int'l Workshop on DNA-Based Computers. (2006) in press
9. Marras, S., Kramer, F., Tyagi, S.: Efficiencies of FRET and and contact-mediated quenching in oligonucleotide probes. Nucl. Acids. Res. **30** (2002) e122
10. Yoshida, H., Suyama, A.: Solution to 3-sat by breadth first search. In: DNA Based Computers V. (2000) 9–22
11. Murphy, M.C., Rasnik, I., Cheng, W., Lohman, T.M., Ha, T.: Probing single-stranded DNA conformational flexibility using fluorescence spectroscopy. Biophysical Journal **86** (2004) 2530–2537
12. Cantor, C., Schimmel, P.: Biophysical Chemistry, Part III: The Behavior of Biological Macromolecules. W. H. Freeman, New York (1983)
13. SantaLucia, Jr., J., Hicks, D.: The thermodynamics of DNA structural motifs. Annu. Rev. Biophy. Biomolec. Struct. **33** (2004) 415–40
14. Wartell, R., Benight, A.: Thermal denaturation of DNA molecules: A comparison of theory with experiment. Phys. Rep. **126** (1985) 67–107

Author Index

Lecture Notes in Computer Science

For information about Vols. 1–4247

please contact your bookseller or Springer